Online and Distance Learning:
Concepts, Methodologies, Tools, and Applications

Lawrence Tomei
Robert Morris University, USA

Volume VI

INFORMATION SCIENCE REFERENCE

Hershey · New York

KH

Assistant Executive Editor:	Meg Stocking
Acquisitions Editor:	Kristin Klinger
Development Editor:	Kristin Roth
Senior Managing Editor:	Jennifer Neidig
Managing Editor:	Sara Reed
Typesetter:	Sharon Berger, Jennifer Neidig, Sara Reed, Laurie Ridge, Jamie Snavely, Michael Brehm, Elizabeth Duke, and Diane Huskinson
Cover Design:	Lisa Tosheff
Printed at:	Yurchak Printing Inc.

Published in the United States of America by
Information Science Reference (an imprint of IGI Global)
701 E. Chocolate Avenue, Suite 200
Hershey PA 17033
Tel: 717-533-8845
Fax: 717-533-8661
E-mail: cust@igi-pub.com
Web site: http://www.igi-pub.com/reference

and in the United Kingdom by
Information Science Reference (an imprint of IGI Global)
3 Henrietta Street
Covent Garden
London WC2E 8LU
Tel: 44 20 7240 0856
Fax: 44 20 7379 0609
Web site: http://www.eurospanonline.com

Library of Congress Cataloging-in-Publication Data

Online and distance learning : concepts, methodologies, tools, and applications / Lawrence Tomei, editor.
      p. cm.
   Summary: "This comprehensive, six-volume collection addresses all aspects of online and distance learning, including information communication technologies applied to education, virtual classrooms, pedagogical systems, Web-based learning, library information systems, virtual universities, and more. It enables libraries to provide a foundational reference to meet the information needs of researchers, educators, practitioners, administrators, and other stakeholders in online and distance learning"--Provided by publisher.
   Includes bibliographical references and index.
   ISBN 978-1-59904-935-9 (hardcover) -- ISBN 978-1-59904-936-6 (ebook)
   1. Distance education--United States. 1. Tomei, Lawrence A.
   LC5805.O55 2007
   371.350973--dc22
                            2007023793

British Cataloguing in Publication Data
A Cataloguing in Publication record for this book is available from the British Library.

8/25/08

Associate Editors

Editorial Advisory Board

List of Contributors

Contents
by Volume

Section 1: Fundamental Concepts and Theories in Online and Distance Learning

This section serves as a foundation for this exhaustive reference tool by addressing crucial theories essential to the understanding of online and distance learning. Chapters within this segment provide an excellent framework in which to position distance learning within the field of information science and technology. With 60 chapters comprising this foundational base of knowledge, the reader can learn and chose from a compendium of expert research on the elemental theories underscoring the online and distance learning discipline.

VOLUME II

Section 2: Online and Distance Learning Development and Design Methodologies

This section offers in-depth coverage of conceptual architectures and distance learning frameworks to provide the reader with a comprehensive understanding of the emerging technological developments within the field of online distance learning. From basic designs to abstract developments, chapters found in this section serve to expand the reaches of development and design technologies within the online and distance learning community. Included in this section are over 40 contributions from researchers throughout the world on the topic of online development and methodologies in distance learning.

VOLUME III

Section 3: Online and Distance Learning Tools and Technologies

This section presents an extensive coverage of various tools and technologies available in the field of distance learning that practicing educators can utilize to develop different techniques in support of the development of distance learning educational programs. Research within this section enlightens readers about fundamental research on some of the many tools used to facilitate and enhance the distance learning experience. Also explored are some of the recent technologies that have been deployed in support of distance learning course offerings. With more than 35 chapters, this section offers a broad treatment of some of the many tools and technologies within the online and distance learning community.

VOLUME IV

Section 4: Utilization and Application of Online and Distance Learning

This section discusses a variety of applications and opportunities available that can be considered by practicing educators in developing viable and effective distance learning educational programs. This section includes more than 40 chapters which incorporate applications of distance learning into institutions within the educational system as well as application of online learning in the corporate realm. Contributions included in this section also provide excellent coverage of today's global community and how distance learning technologies and education are impacting the social fabric of our present-day global village.

Section 5: Organizational and Social Implications of Online and Distance Learning

This section offers a wide range of research regarding the social and organizational impact of online and distance learning technologies around the world. Chapters included in this section epitomize some of the most contested issues within society where access to technology is concerned. One of the most prominent issues discussed in this section is the integration of technology to allow access for all classrooms regardless of socioeconomic status; arguably the most important social and organizational barrier that this field of study has yet to overcome. With over 30 chapters, this discussion offers new insights into the incorporation of distance education within organizations and its impact on the social scheme within our global community.

VOLUME V

Section 6: Managerial Impact of Online and Distance Learning

This section presents contemporary coverage of the social implications of online and distance learning, more specifically related to the corporate and managerial utilization of online learning technologies and applications, and how online learning can be facilitated within these organizations. Core ideas such as training and continuing education of human resources in modern organizations are discussed through these more than 25 chapters. Discussions of strategic planning related to the organizational elements as well as the e-learning program requirements that are necessary to build a framework for the institutionalization and sustainment of e-learning as a core business process are discussed within the chapters found in this section. Equally as crucial, within the educational management system, there is discussion of the virtual classroom which addresses the latest research concerning the management of digital networking via the internet to schools, particularly those in rural communities.

VOLUME VI

Section 7: Critical Issues in Online and Distance Learning

This section contains 30 chapters addressing issues such as gender barriers, web accessibility, quality assurance and development of e-learning in under-developed countries presenting readers with an in-depth analysis of the most current and relevant issues within this growing field of study. Models for researchers and practitioners are offered as attempts are made to expand the reaches of online and distance learning within the higher education community. The Formation of Frameworks in which to position the issues faced in this growing field is provided by research found in this section while the core psychological paradigms of education are translated into applicable ideas within the exploding realm of online and distance education. Crucial examinations of the cultural biases innate in online and distance learning are presented in this section while simultaneously enticing and inspiring the reader to research further and participate in this increasingly pertinent debate.

Section 8: Emerging Trends in Online and Distance Learning

This concluding section highlights research potential within the field of online and distance learning while exploring uncharted areas of study for the advancement of the discipline. The introductory chapters set the stage for future research directions and topical suggestions for continued debate. Providing a fresh, alternative view of distance education, colleagues from universities all over the world explore the adaptive traits necessary as disseminators of knowledge within this evolving platform of education; a reminder that not only is the role of the learner rapidly evolving, but so too is the role of the facilitator. Educational programs throughout the world have witnessed fundamental changes during the past two decades—changes that are emphasized in the 25 rigorously researched chapters included in this section. With continued technological innovations in information and communication technology and with on-going discovery and research into newer and more innovative techniques and applications, the online and distance learning discipline will continue to witness an explosion of knowledge within this rapidly evolving field.

Preface

Technological advancements of the past two decades have allowed educators to deliver various effective academic programs to knowledge seekers and students around the world free of the traditional dependency on campus based programs. These advancements are formidable technological innovations that have profoundly impacted all realms of society including business, education, health care, and interpersonal and intercultural interfacing.

During this period of time numerous researchers and educators have developed a variety of techniques, methodologies, and measurement tools that have allowed them to develop, deliver and at the same time evaluate the effectiveness of several online and distance learning programs. The explosion of these technologies and methodologies in this new field of web-based education and online learning have created an abundance of new, state-of-art literature related to all aspects of this expanding discipline, allowing researchers and practicing educators to learn about the latest discoveries in the filed of distance learning and online teaching.

Due to rapid technological changes that are continually taking place, it is a constant challenge for researchers and educators in distance learning to stay abreast of the far-reaching effects of this change, and to be able to develop and deliver more innovative methodologies and techniques utilizing new technological innovation. In order to provide the most comprehensive, in-depth, and recent coverage of all issues related to web-based education and online distance learning, as well as to offer a single reference source on all conceptual, methodological, technical and managerial issues, as well as the opportunities, future challenges and emerging trends related to distance learning, *Information Science Reference* is pleased to offer a six-volume reference collection on this rapidly growing discipline, in order to empower students, researchers, academicians, and practitioners with a comprehensive understanding of the most critical areas within this field of study.

This collection entitled, *Online and Distance Learning: Concepts, Methodologies, Tools, and Applications*, is organized in eight (8) distinct sections, providing the most wide-ranging coverage of topics such as: (1) Fundamental Concepts and Theories; (2) Development and Design Methodologies; (3) Tools and Technologies; (4) Utilization and Application; (5) Organizational and Social Implications; (6) Managerial Impact; (7) Critical Issues; and (8) Emerging Trends. The following provides a summary of what is covered in each section of this multi volume reference collection:

Section 1, *Fundamental Concepts and Theories*, serves as a foundation for this exhaustive reference tool by addressing crucial theories essential to the understanding of online and distance learning. Chapters such as, *Technology's Role in Distance Education* by Caroline Howard, Murray Turoff and Richard Discneza, as well as *Distance Education in the Era of the Internet* by Giorgio Agosti provide an excellent framework in which to position distance learning within the field of information science and technology. Sara Dexter's, *Principles to Guide the Integration and Implementation of Educational Technology* offers excellent insight into the critical incorporation of technology into the classroom for educators and administers alike, while chapters such as, *Learning IT: Where Do Lecturers Fit?* by Tanya McGill and Samantha Bax address some of the basic, yet crucial stumbling blocks of distance learning. With 60 chapters comprising this foundational section, the reader can learn and chose from a compendium of expert research on the elemental theories underscoring online and distance learning discipline.

Section 2, *Development and Design Methodologies*, provides in-depth coverage of conceptual architectures and distance learning frameworks to provide the reader with a comprehensive understanding of the emerging technological developments within the field of online and distance learning. *A Conceptual Architecture for the Development of Interactive Educational Media* by Claus Pahl offers research fundamentals imperative to the understanding of the design of educational tools. Conversely, Gregg Asher's, *Inadequate Infrastructure and the Infusion of Technology into K-12 Education* explores the shortcomings of those schools under-prepared for emersion into the world of online and distance learning. From basic designs to abstract development, chapters such as *Systems Model of Educational Processes* by Charles E. Beck and *Reliving History with 'Reliving the Revolution': Designing Augmented Reality Games to Teach the Critical Thinking of History* by Karen Schrier serve to expand the reaches of development and design technologies within the online and distance learning community. This section includes over 40 contributions from researchers throughout the world on the topic of online development and methodologies in distance learning.

Section 3, *Tools and Technologies*, presents an extensive coverage of various tools and technologies available in the field of distance learning that practicing educators and researchers alike can utilize to develop different techniques in support of offering distance learning educational programs. Chapters such as *Core Principles of Educational Multimedia* by Geraldine Torrisi-Steele enlightens readers about the fundamental research on one of the many tools used to facilitate and enhance the distance learning experience whereas chapters like, *Hyper Video for Distance Learning* by Mario Bochicchio and Nicola Fiore explore the latest technological offerings. It is through these rigorously researched chapters that the reader is provided with countless examples of the up-and-coming tools and technologies emerging from the field of online and distance learning. Another contribution entitled *Vega Grid Technology, Hyper Video, Learning Portals, Wireless Technologies, Simulation and Gaming and Videoconferencing* explores some of the recent technologies that can have been deployed in support of distance learning course offerings. With more than 35 chapters, this section offers a broad treatment of some of the many tools and technologies within the online and distance learning community.

Section 4, *Utilization and Application*, discusses a variety of applications and opportunities available that can be considered by practicing educators in developing viable and effective distance learning educational programs. This section includes more than 40 chapters such as *Overcoming Organizational Barriers to Web Accessibility in Higher Education: A Case Study* by Amy Scott Metcalfe which incorporates applications of distance learning into the higher education society, while chapters such as Dorris Lee's, *Implementing*, discusses the utilization of online learning within the corporate realm. Also considered in this section are the challenges faced when utilizing distance learning as outlined by Martha Cleveland-Innes, Randy Garrison, and Ellen Kinsel's, *Role Adjustment for Learners in an Online Community of Inquiry: Identifying the Challenges of Incoming Online Learners*. The adaptability of developing countries is also given consideration in chapters like, *Issues of E-Learning in Third World Countries* by Shantha Fernando which investigates the major hurdles faced by the socio-economic under-privileged within our global community. Contributions included in this section provide excellent coverage of today's global community and how distance learning technologies and education are impacting the social fabric of our present-day global village.

Section 5, *Organizational and Social Implications*, includes a wide range of research pertaining to the social and organizational impact of online and distance technologies around the world. Introducing this section is Pier Cesare Rivoltella's chapter entitled, *Education and Organization: ICT, Assets, and Values* providing a comprehensive introduction of education and its technological role within organizations as a social construct. Additional chapters included in this section such as *Narrowing the Digital Divide: Technology Integration in a High-Poverty School* by June K. Hilton epitomize one of the most contested issues within society concerning access to technology—the digital divide. Also introducing a rising concern within the education organization is Bryan D. Bradley's, *Legal Implications of Online Assessment: Issues for Educators* which provides an alternative approach to research regarding the legality of online assessment. The discussions presented in this section offer research into the integration of technology to allow access for all classrooms regardless of socioeconomic status; arguably the most important social and organizational barrier that this field of study has yet to overcome.

Section 6, *Managerial Impact*, presents contemporary coverage of the social implications of online and distance learning, more specifically related to the corporate and managerial utilization of online learning technologies and applications, and how online learning can be facilitated within these organizations. Core ideas such as training and continuing education of human resources in modern organizations are discussed through these more than 25 chapters. *Implementing and Sustaining E-Learning in the Workplace* by Zane Berge discusses strategic planning related to the organizational elements and the e-learning program requirements that are necessary to build a framework in order to institutionalize and sustain e-learning as a core business process. Equally as crucial, within the educational management system, chapters such as *The Management of Virtual Classes in School District Digital Intranets* by Ken Stevens address the latest research concerning the management of digital networking via the internet to schools, particularly those in rural communities. Concluding this section is a chapter by Catherine C. Schifter of Temple University, *Faculty Participation in Distance Education Programs.* This chapter refocuses the issue of the managerial impact of distance learning to the facilitators by examining the crucial role that faculty members play in the success of a virtual classroom.

Section 7, *Critical Issues*, containing 30 chapters addressing issues such as gender barriers, Web accessibility, quality assurance and development of e-learning in under-developed countries presents readers with an in-depth analysis of the most current and relevant issues within this growing field of study. Barbara A Frey, Ashli Molinero and Ellen Cohn's, *Increasing Web Accessibility and Usability in Higher Education*, develops an excellent model for researchers and practitioners as attempts are made to expand the reaches of online and distance learning within the higher education community. Forming a frameworks in which to position the issues faced in this growing field are provided by research found in chapters such as, *Inquisitivism: The Evolution of a Constructivist Approach for Web-Based Instruction*, by Dwayne Harapnuik—a chapter that takes the core psychological paradigms of education and translates them into applicable ideas within the exploding realm of online and distance education. Crucial examinations such as that presented in David Catterick's chapter, *Do the Philosophical Foundations of Online Learning Disadvantage Non-Western Students?* serves to reinforce the ideas presented in this section while simultaneously enticing and inspiring the reader to research further and participate in this increasingly pertinent debate.

The concluding section of this authoritative reference tool, *Emerging Trends*, highlights research potential within the field of online and distance learning while exploring uncharted areas of study for the advancement of the discipline. Introducing this section is Bruce Rollier's, *Trends and Perspectives in Online Education*, which sets the stage for future research directions and topical suggestions for continued debate. Providing a fresh, alternative view of distance education is the chapter, *Faculty Perceptions and Participation in Distance Education*, by James R. Lindner, Kim E. Dooley, Chanda Elbert, Timothy H. Murphy and Theresa P. Murphrey of Texas A&M University. These colleagues explore the adaptive traits necessary as disseminators of knowledge within this evolving platform of education; a reminder that not only is the role of the learner rapidly evolving, but so too is the role of the facilitator. Another a debate which currently finds itself at the forefront of research within this field is presented by David Gibson's research which centers on simSchool and gaming within online and distance learning as a discipline, whereas Bernhard Ertl, Katrin Winkler, and Heinz Mandl's, *E-Learning: Trends and Future Development*, summarizes contemporary trends while projecting future developments.

Although the primary organization of the contents in this multi-volume is based on its eight sections, offering a progression of coverage of the important concepts, methodologies, technologies, applications, social issues, and emerging trends, the reader can also identify specific contents by utilizing the extensive indexing system listed at the end of each volume. Furthermore to ensure that the scholar, researcher and educator have access to the entire contents of this multi volume set as well as additional coverage that could not be include in the print version of this publication, the publisher will provide unlimited multi-user electronic access to the online aggregated database of this collection for the life of edition, free of charge when a library purchases a print copy. This aggregated database provides far more contents than what can be included in the print version in addition to continual updates. This unlimited access, coupled with the continuous updates to the database ensures that the most current research is accessible knowledge seekers.

Educational programs at the college level have witnessed fundamental changes during the past two decades, moving more toward campus-free approaches and allowing millions of non-traditional as well traditional students around the globe to have access to educational programs which two decades ago, were inaccessible. In addition to this transformation, many traditional educational programs have taken advantage of the technologies offered by distance and online learning in order to expand and augment their existing programs. This has allowed educators to serve their student base more effectively and efficiently in the modern virtual world. With continued technological innovations in information and communication technology and with on-going discovery and research into newer and more innovative techniques and applications, the online and distance learning discipline will continue to witness an explosion of knowledge within this rapidly evolving field. This continued trend will also lead to expansion of the literature related to all areas of this discipline. Access to more up-to-date research findings and knowledge of established techniques and lessons to be learned from other researchers and practicing educators will facilitate the discovery and invention of more effective methodologies and applications of online and distance learning technologies.

The diverse and comprehensive coverage of online and distance learning in this six-volume authoritative publication will contribute to a better understanding of all topics, research, and discoveries in this developing, significant field of study. Furthermore, the contributions included in this multi-volume collection series will be instrumental in the expansion of the body of knowledge in this enormous field, resulting in a greater understanding of the fundamentals while fueling the research initiatives in emerging fields. We at Information Science Reference, along with the editor of this collection, and the publisher hope that this multi-volume collection will become instrumental in the expansion of the discipline and will promote the continued growth of online and distance learning.

Section 7
Critical Issues in Online and Distance Learning

This section contains 30 chapters addressing issues such as gender barriers, web accessibility, quality assurance and development of e-learning in under-developed countries presenting readers with an in-depth analysis of the most current and relevant issues within this growing field of study. Models for researchers and practitioners are offered as attempts are made to expand the reaches of online and distance learning within the higher education community. The Formation of Frameworks in which to position the issues faced in this growing field is provided by research found in this section while the core psychological paradigms of education are translated into applicable ideas within the exploding realm of online and distance education. Crucial examinations of the cultural biases innate in online and distance learning are presented in this section while simultaneously enticing and inspiring the reader to research further and participate in this increasingly pertinent debate.

Chapter 7.1
Critical Barriers to Technology in K–12 Education

Christine Sweeney
NCS Pearson, USA

INTRODUCTION

Those who are fortunate enough to be associated with K-12 education during this first decade of the 21st century will witness tremendous evolutionary—even revolutionary—changes throughout those institutions. The interrelated dynamics of public education, the IT industry, and the evolving "digital society" are already combining to produce a variety of entirely new models for K-12. Although those models are indeed emerging, significant change will come at a pace that is perhaps somewhat slower initially than some would prefer. K-12 education is, after all, an institution rich in tradition and culture, and often slow to change. Nonetheless, as the presence and reach of new technologies—the Internet in particular—reach critical mass, that pace will quicken, and by the year 2010, school age children will enjoy an educational experience profoundly different from anything previously known. Profound change usually occurs when not one, but several change agents come together, either deliberately or coincidentally, and interact—often sparked by some sort of catalyst. This type of interaction is occurring throughout public education today. In this case, the change agents at work include K-12 institutions, the evolving IT industry, and the rapidly emerging digital society.

K-12 INSTITUTIONS

Public education leaders are facing tremendous challenges and unclear demands as we begin the new millennium. The call for improved student performance—education's "bottom line"—is pervasive and louder than ever. At the same time, state and federal departments of education have, or are creating, high stakes examinations around tough new curriculum standards designed to determine "how well" our students are learning—as well as which schools are not performing as well as they should. *Accountability* is a word that is part of virtually every current discussion about education, yet there is little consensus around its precise meaning or how to measure it. The need for productive school-community collaboration—the so-called school-to-home connection—is greater than ever. More parents are becoming engaged in their children's education and expect to have ready access to information about grades, attendance, discipline, content mastery, test scores, and so forth. Privacy and security concerns, however, are prevalent, with some parents adamantly opposed to making that information available via the Internet (even when appropriately secured). Local and state leaders now see the value in *data-driven decision making*. This is creating an insatiable need for program-level information and seamlessly integrated information systems that produce it. Simultaneously, costs for IT support continue to rise, making the challenge of providing and supporting a technology environment rich in educational content, valuable information, and easy-to-use tools quite daunting.

THE IT INDUSTRY

This is an industry that essentially reinvents itself every 12-18 months. And while technology is without doubt an enabler of change, the Internet is truly the catalyst that has sparked (and is fueling) the emergence of new models for K-12 education. Still, back in our schools and offices, the need for interoperability among disparate technology-based systems is increasing. The Internet, and specifically the Web, is greatly reducing the effects of this issue. Most of what is sold as "integration" on the Web today is nothing more than Web pages with multiple URL links to other independent sites. As educational leaders recognize the value of seamlessly integrated systems for managing curriculum, instruction, and assessment, the demand for appropriate and powerful integration technologies will follow. In short, as the same people who are engaged in public education today—teachers, parents, students, and administrators—realize that they are conducting much of their lives online, they

will want to know why public education cannot be the same.

THE FUTURE—YEARS 2003 AND 2010

While technology will affect virtually every aspect of K-12 education during this timeframe, three particular areas will feel the effects most profoundly. First, educators will realize—indeed, are already realizing—that managing curriculum, instruction, assessment, and associated individual and aggregated student information as a seamlessly integrated thread, extending from the classroom through the school, district, and state, will have dramatic and positive effects on student performance. Correspondingly, the traditional divide between instructional and administrative technology will be seen as nonproductive as educators learn that the benefits of delivering instruction with technology cannot be measured or realized without integration with systems that manage curriculum, assessment, and student information.

Second, educational leaders will discover the availability and benefits of high-value, strategic information in support of decision making. The same integration activities described above will result in a base of information that, when coupled with powerful decision support and knowledge management tools, will enable leaders to gain new and profound insights into the educational process. In addition to the ability to make—often for the first time—truly informed decisions, these capabilities will help educational leaders understand *what makes education work*. Interestingly enough, development of these capabilities will occur not only because educators realize the value, but because community members who want to hold their school leaders accountable will demand it.

Third, powerful *electronic collaboration* technologies will enable the most vital stakeholder groups—teachers, parents, administrators, and students—to become productively engaged. They will rapidly form local communities of people who can effectively collaborate on everything from the mundane to the critical and controversial: What is on the calendar this week for my children? What are the best techniques for teaching 9th grade science to ESL students? How will changes in the district budget affect next year's attendance boundaries? Again, rapid changes in how people in each of those stakeholder groups use digital technology everyday to conduct their

lives—specially the Internet—will drive the demand for these collaborative capabilities in the educational experience.

While numerous other technological advances will have found their way into public education by 2003, their uses will still largely be constrained by the strong cultures that have evolved over the centuries in education. The reality in 2003 will be that while the early adopters and many mainstream participants will be using these technologies everyday to help students achieve and to improve the quality of their institutions, many will continue with a "same script, new props" mentality.

This work was previously published in the Encyclopedia of Distance Learning, Vol. 1, edited by C. Howard, J. Boettcher, L. Justice, K. Schenk, P. L. Rogers, and G. A. Berg, pp. 481-482, copyright 2005 by Idea Group Reference (an imprint of IGI Global).

Chapter 7.2
Computer Skills, Technostress, and Gender in Higher Education

Sonya S. Gaither Shepherd
Georgia Southern University, USA

INTRODUCTION

The creation of computer software and hardware, telecommunications, databases, and the Internet has affected society as a whole, and particularly higher education by giving people new productivity options and changing the way they work (Hulbert, 1998). In the so-called "information age" the increasing use of technology has become the driving force in the way people work, learn, and play (Drake, 2000). As this force evolves, the people using technology change also (Nelson, 1990).

Adapting to technology is not simple. Some people tend to embrace change while others resist change (Wolski & Jackson, 1999). Before making a decision on whether to embrace technology or not, people may look at the practical and social consequences of accepting change. Therefore, the technology acceptance model, the accepting or resisting of technology is considered to be a form of reasoned behavior (Wolski & Jackson, 1999).

BACKGROUND

Technology changes the way people work and learn. As the role of technology is being defined and technology is constantly being improved, change is inevitable (Brand, 2000; Davis-Millis, 1998). As a result, those involved in higher education have to find ways to adapt to technological change. Administrators, faculty, academic librarians, and students should define the role of technology for the purposes of (1) sharing new ideas and techniques for teaching and learning; (2) encouraging enthusiasm and innovativeness; and (3) learning about opportunities and challenges, and how to deal with them (Landsberger, 2001).

In fact, college faculty are spending more time with those from the business sector to ensure what is taught in the classroom is applicable in the workforce (Gavert, 1983; Katz, 1999; Lynton, 1984). This partnership is providing opportunities for faculty to remain current in rapidly changing technical disciplines because both are collaborating on curriculum that meet education standards and job related skills required in industry (Gavert, 1983; Katz, 1999). And the researcher presumes that professors in Colleges of Business Administration are more adept and comfortable using technology than those in other colleges within universities. On the other hand, other disciplines such as liberal arts have had less need to adapt as quickly, and perhaps have been more reluctant to change (Miller & Rojewski, 1992).

Likewise, education faculty are preparing future teachers, counselors, and administrators to go into elementary, middle, and secondary schools. These teacher programs may or may not require their students to obtain and use technological skills. Similarly, there may or may not be an expectation among the education faculty to obtain or utilize these same skills. Some education faculty and students may only learn and use technology because they wanted to and not because there was an expectation (Miller & Rojewski, 1992).

University library staff also has had to adapt to a wide variety of technological demands unimaginable just a few years ago (e.g., processing library materials and teaching research skills online). Other disciplines such as liberal arts have had less need to adapt as quickly, and perhaps have been more reluctant to change. All, however, are faced with the necessity to change. Therefore, in all likelihood, all faculty and librarians are experiencing some level of technological stress.

Furthermore, the rapid growth in technology over the last three decades has been well documented. Accompanying that growth has been an equally rapid increase in the struggle to keep up with technology. The way services are provided by society and to society (e.g., fast, instantly, remotely) is changing. While virtually all facets of society are affected by technology, its impact can be clearly seen in the way higher education clientele have been served. Colleges and universities are being changed in multiple and profound ways, ways almost unrecognizable to students, faculty, academic librarians, administrators, and alumni.

The move to the Information Age, with its changes and need for adaptation to technology, has been rapid and stressful for many people. While many people have increased their technology use and are comfortable with it, many others still do not use much technology and are not comfortable using it when they must do so. For those who are not amenable to change, who find it difficult to adapt, there are often a variety of responses or results. One such response is called technostress. Technostress is the inability to adapt to or cope with new computer technologies which reveals itself in one of two ways: (1) computer users struggle to accept the technologies or (2) computer users over-identify with the technology (Brod, 1984).

Studies relating to technostress have been fairly limited. Those conducted have sought to determine correlations between such variables as personality type, academic performance, self-concept, and why certain faculty decided to use technology while others do not. Study participants have included people from the business industry, students majoring in business and education, and a limited number of faculty members and librarians. However, there are few studies that look at the severity of stress for various types of computer users (e.g., faculty, academic librarians) in postsecondary settings based on the computer users' gender.

For example, differences between females and males regardless of discipline were identified in the way they accepted or resisted technology. Even though Sievert, Albritton, Roper, and Clayton (1988) and Ballance and Ballance (1992, 1993) found computer-related stress was not related to computer experience and sex, other researchers found a relationship. For instance, females experienced technostress or resisted information technology (IT) more than males (Fine, 1979; Elder, Gardner, & Ruth, 1987; Hudiburg, Brown, &

Jones, 1993; Ogan & Chung, 2003). Additionally, Heinssen, Glass, & Knight (1987) believed the less computer experience a female had the more computer anxiety she experienced. Murphy, Coover, and Owen (1989) revealed men were better able to perform certain computer skills more successfully than females. Similarly, Reed and Overbaugh (1993) found men to have less computer anxiety as their computer experience increased. According to Baroudi and Levine (1997), "women were generally more scared of computers ..." (p. 178). Finally, females rated information technology as the fourth cause of stress while men rated IT as the fifth cause of stress (Sax, Alexander, Korn, & Gilmartin, 1999).

Male and female faculty members and librarians also identified the IT they used as well as certain coping skills to help them handle the increased stress. The information technology identified included e-mail, spreadsheets, the Internet, statistical software, presentation software, and multimedia software (Groves & Zemel, 2000) where e-mail, spreadsheets, and the Internet were highly used. They suggested eating, relaxing, staying healthy, having a positive attitude, managing time, setting realistic goals, and seeking additional training as ways to cope with their stress as they continued to use information technology (Hickey, 1992; Kupersmith, 1992; McKenzie, Davidson, Bennett, & Clay, 1997).

With this, the intent of the current study was to explore the relationship between technology skills and the possible causes of technostress among academic librarians, and education and business faculty. The exploration looked at the role, if any, of how gender may have also made a difference in this relationship between technology skills and technostress.

FUTURE TRENDS

Research Method

This study was originally designed to answer the following question based on several demographics: Do computer skills relate to the levels of technostress among faculty in the Colleges of Business and Education, and academic librarians? However, the main focus of this article is on the gender of faculty in the Colleges of Business and Education, and academic librarians and the relationship between their computer skills and levels of technostress. As a result, 994 eligible participants were identified and 316 usable surveys were returned (32.8% return rate).

Participants were given the option of completing a survey instrument electronically and having the responses e-mailed to the researcher (n=234), or receiving numbered, color-coded paper copies, and mailing the results back to the researcher in a self-addressed stamped envelope (n=93). The numbered, color-coded paper copies were used to keep track of participants who responded so the researcher could do follow-up requests for survey participation. The survey was a new instrument containing four sections: (1) Computer Hassle Scale-revised (CHS-R); (Hudiburg, 1999) (2) Computer Skills Survey (May, 1998); (3) two open-ended questions; and (4) demographic items. When completing the CHS-R section of the instrument, respondents were asked to circle the number corresponding to the severity of the computer hassle they have experienced. Choice of numbers were 0=not at all, 1=rarely severe, 2=moderately severely, and 3=extremely severe. They were asked to complete the Computer Skills section by rating his/her skill level. Answer choices were 1=low, 2, 3=medium, 4, and 5=high. The faculty and academic librarians were then asked to answer two open-ended questions about what they perceived to be possible causes of

Table 1. Computer skills and technostress levels by gender of COBA, COE faculty, and academic librarians

		Computer Skills					
		Males			Females		
	M	SD	n	M	S	D	n
COBA	3.88	73	5 1	3.97	76	3	0
COE	3.58	92	6 9	3.74	78	6	6
LIB	3	.46	.84 11	3	.57	.70	29
		Technostress Levels					
		Males			Females		
	M	SD	n	M	S	D	n
COBA	41.43	19.73	5 1	43.27	19.95	5	3 0
COE	45.17	20.76	6 9	41.52	20.04	6	6
LIB	45.91	15.80	1 1	47.10	21.29	2	9

technostress and possible solutions for relieving technostress. Lastly, faculty and academic librarians were asked to provide certain demographics including gender.

Alternatively, participants completed the instrument electronically by filling out a Web-based form posted on the Internet. Using the same numeric code found on the paper copy of the survey, each faculty member or academic librarian wishing to complete the instrument online was able to enter that code on the Web form for tracking purposes. The code was used to keep track of those who responded to the survey so the researcher could request participation from non-respondents after follow-up contact had been made with those not responding initially. Each participant completed the CHS-R section by clicking the radio button corresponding to the appropriate severity level of each of the computer hassles they have experienced. The choices were the same as the ones on the paper copy. Similarly, the Computer Skills section had clickable radio buttons corresponding to the skill level for each computer skill. The choices were the same as those on the paper copy. Two open text boxes were provided for respondents to type in their responses to the open-ended questions. Lastly, clickable radio buttons were provided for responding to the demographics section. All responses from the survey were snail mailed or e-mailed to the researcher for data analysis.

Research Findings

The major findings of the current study were:

1. Male and female business faculty reported their computer skills as the highest over education faculty and academic librarians (see Table 1).
2. Male and female academic librarians perceived themselves to experience more severe levels of technostress than male and female business and education faculty (see Table 1).
3. Male academic librarians reported the lowest computer skills level, while female academic librarians perceived to experience higher levels of technostress (see Table 1).
4. All males reported lower computer skill levels and they perceived to experience more severe levels of technostress (see Table 2).

Table 2. Correlations between computer skills and technostress levels of males and females

	r	p* n		Computer Skills		Technostress
Males	-.29	.00	155	3.67	4	3.34
Females	-.32	.00	161	3.74	4	2.67

*Note: *Significance is p≤0.05*

Table 3. Ordered rank of applications used by COBA, COE faculty, and academic librarians

1. E-mail
2. Word Processing
3. Internet
4. Presentation
5. Library Databases
6. Spreadsheets
7. Library Catalog
8. Databases

5. The levels of technostress among males and females regardless of discipline decreased as their levels of computer skills increased (see Table 2).

6. Academic librarians, education and business faculty regardless of gender used a wide variety of software applications or other computer technology but they mainly used e-mail, word processing, and the Internet (see Table 3).

7. Participants identified computer information and computer runtime problems more than any other problem as causes of their technostress (see Table 4).

8. Solutions for reducing technostress as reported by the participants included calling for help, screaming or yelling, walking away, leisurely talking to someone, and doing something non-technical or non-computer related (see Table 5).

Findings-Past and Present

According to Fine (1979), Elder, Gardner, and Ruth (1987), Hudiburg and Jones (1993), Heinssen, Glass, and Knight (1987), and Ogan and Chung (2000), females experienced more technostress and resisted technology more than males which contradicts the findings of this study. However, the current study's results supported those revealed in Reed and Overbaugh's (1993) study that men had less computer anxiety as their computer experience increased. Additionally, faculty and librarians identified word processing, spreadsheets, and e-mail as highly used information technology (Groves & Zemel, 2000) while e-mail, word processing, and the Internet were identified in the current study. In summary, further research may be needed to understand how gender may relate to computer users' technostress levels and their computer skills. Furthermore, researchers could investigate the roles males and females

Table 4. Technostress causes as perceived by COBA, COE faculty, and academic librarians

Cause	F requency
computer information problems	178
• difficulty keeping up, too many passwords	
computer runtime problems	119
• hardware failure, computer crashes	
computers' impact on society	70
• increase in expectation to use computers, increase in demand or time to use computers	
Internet/E-mail problems	48
• too much e-mail, spam	
everyday computer technology	42
• confusing, threatening computer terminology, answer cannot be found	
computer processing speed	41
• slow CPU/Internet connection	
computer as person	8
• lack of human interaction	
computer costs	2
• software costs	

Note: Hudiburg (1997) identified eight categories for measuring causes of technostress

Table 5. Coping solutions as perceived by COBA, COE faculty, and academic librarians

Solution	Frequency
increase knowledge and skills	114
• ask for help, attend training workshops	
relax or socialize	77
• take nap, talk to people	
manage time or projects/tasks	77
• multi-task, back up data	
complain	54
• threaten computer, yell and curse	
try to fix the problem	29
• reboot computer, start project over	
exercise	24
• yoga, play basketball	
change attitude/expectations	24
• find humor in situation, control anger	
eat	12
• drink tea, eat popcorn/candy	
perform non-technology related tasks	1
• clean office	

play in the decision making process of technology implementation and training, the increased expectation to use technology in classrooms, and the increased expectation to use technology for tenure and promotion.

CONCLUSION

This study attempted to investigate whether computer skills relate to the levels of technostress among male and female faculty members in the

Colleges of Business and Education, and academic librarians. The analysis of the data revealed a negative relationship that as computer skills increased, technostress levels decreased among males and females. However, further study may be necessary to determine and understand more the relationship of computer skills and technostress among males and females. Moreover, in order for computer users to experience less stress, they will have to keep up with the rapid change of technology and take part in some form of training on a regular basis. "Changes break patterns that we are comfortable to, and that can be rather threatening. The key is to make sure that we are the masters, and that computer and other formats of technology are tools we manipulate. IN SHORT, WE ARE THE ONES WHO ARE IN CHARGE!" (Rocha, 2001).

REFERENCES

Ballance, C. T., & Balance, V. V. (1993). Psychology of computer use: XXVII. Relating self-rated computer experience to computer stress. *Psychological Reports, 72*(2), 680-682.

Ballance, C. T., & Ballance, V. V. (1992). Psychology of computer use: XXVI. Computer related stress and in-class computer usage. *Psychological Reports, 71*(1), 172-174.

Baroudi, C., & Levine, J. (1997). Technophobia. In L. K. Enghagen (Ed.), *Technology and higher education* (pp. 177-184). Washington, DC: NEA Professional Library.

Brand, S. (2000 July 3). Is technology moving too fast? *Time Atlantic, 156*(1), 66-67.

Brod, C. (1984). *Technostress: The human cost of the computer revolution*. Reading, MA: Addison Wesley.

Davis-Millis, N. (1998). *Technostress and the organization: A manager's guide to survival in the information age*. Retrieved from http://web.mit.edu/ninadm/www/mla.htm

Drake, M. (2000). Technological innovations and organizational change revisited. *Journal of Academic Librarianship, 26*(1), 53-59.

Elder, V. B., Gardner, E. P., & Ruth, S. R. (1987). Sex and age in technostress: Effects on white collar productivity. *Government Finance Review, 3*(6), 17-21.

Fine, S. (1979). *Resistance to technological innovations in libraries: Part III reviews of the study* (ERIC Document Reproduction Service No. ED 310 776).

Gavert, R. V. (1983). Business and academe: An emerging partnership. *Change, 15*(3), 23-28.

Groves, M. M., & Zemel, P. Z. (2000). Instructional technology adoption in higher education: An action research case study. *International Journal of Instructional Media, 27*(1), 57.

Heinssen, R. K., Glass, C. R., & Knight, L. A. (1987). Assessing computer anxiety: Development and validation of the computer anxiety scale. *Computers in Human Behavior, 3*(1), 49-59.

Hickey, K. D. (1992). Technostress in libraries and media centers: Case studies and coping strategies. *Tech Trends, 37*(2), 17-20.

Hudiburg, R. A. (1997). Computer hassles scale. In C. P. Zalaquett & R. J. Woods (Eds.), *Evaluating stress: A book of resources*. Lanham, MD: Scarecrow Press.

Hudiburg, R. A. (1999). *Revision of the computer hassles scale. Arts and Sciences Grant*. Florence, AL.

Hudiburg, R. A., Brown, S. R., & Jones, T. M. (1993). Psychology of computer use: XXIX. Measuring computer users' stress: The computer hassles scale. *Psychological Reports, 73*(3), 923-929.

Hulbert, D. J. (1998). *Libraries and librarianship in the information age* (ERIC Document Reproduction Service No. ED 420 019).

Kupersmith, J. (1992). *Technostress and the reference librarian*. Retrieved from http://home.pacbell.net/jjkup/tstr_ref.html

Landsberger, J. (2001). The producer, the critics, and the glitch factor. *Tech Trends, 45*(5), 18-20.

Lynton, E. A. (1984). *The missing connection between business and the universities*. New York: Macmillan Publishing Co.

Katz, R. N. (1999). *Dancing with the devil: Information technology and the new competition in higher education*. San Francisco: Jossey-Bass, Inc.

May, S. A. (1998). *Evaluation of faculty competencies in the delivery of contracted workforce training with recommendations for faculty development at Fox Valley Technical College*. (Doctoral dissertation, Nova Southeastern University, 1998). Dissertation Abstracts International, 60, 9919788.

McKenzie, B., Davidson, T., Bennett, P., & Clay, M. (1997). Trying to reduce your technostress? *School Library Media Activities, 13*(9), 24-26.

Miller, M. T., & Rojewski, J. W. (1992). Integrating technology into the liberal arts: The perspective of liberal arts administrators. *Journal of Studies in Technical Careers, 14*(2), 115-126.

Murphy, C. A., Coover, D., & Owen, S. V. (1989). Development and validation of the computer self-efficacy scale. *Educational and Psychological Reports, 59*(3), 1199-1204.

Nelson, D. L. (1990). Individual adjustment to information-driven technologies: A critical review. *MIS Quarterly, 14*(1), 78-108.

Ogan, C., & Chung, D. (2003). Stressed out! A national study of women and men journalism and mass communication faculty, their uses of technology, and levels of professional and personal stress. *Journalism and Mass Communication Education, 57*(4), 352-368.

Reed, W. M., & Overbaugh, R. C. (1993). The effects of prior experience and instructional format on teacher education students' computer anxiety and performance. *Computers in the Schools, 9*(3), 75-89.

Rocha, T. (2001). *Technostress in libraries*. Retrieved from http://www.slais.ubc.ca/courses/libr500/01_02_wt1/www/F_Rocha/whozncharge.htm

Sax, L. J., Alexander, W., Korn, W. S., & Gilmartin, S. K. (1999). *The American college teacher: National norms for the 1998-1999 HERI faculty survey* (ERIC Document Reproduction Service ED 435 272).

Sievert, M. C., Albritton, R. L., Roper, P., & Clayton, N. (1988). Investigating computer anxiety in an academic library. *Information Technology and Libraries, 7*(3), 243-252.

Wolski, S., & Jackson, S. (1999). *Technological diffusion within educational institutions: Applying the technology acceptance model* (ERIC Document Reproduction Service ED 432 301).

KEY TERMS

Academic Librarian (LIB): Person who has completed an outlined course of study from an accredited library school and performs one or more of the following: (1) purchase and catalog resources for public use; (2) help find information for search or study; (3) plan, operate, and maintain computer systems; and (4) manage and plan library operations; and may or may not hold faculty status.

Business Faculty (COBA): Professor/instructor who teaches business courses at a college or university.

Computer Technology (aka Technology or Information Technology): Machines with cd-rom, DVD, zip, and/or floppy disk drives with software applications which are used to enhance human efficiency and workflow.

Education Faculty (COE): Professor/instructor who teach education courses at a college or university.

Information Age: Creation of computer software and hardware, telecommunications, databases, and the Internet.

Information Problems (aka Computer Information Problems): Having little or no information or sometimes having too much information when trying to utilize computer technology.

Over-Identification with Technology: Constant use of technology or over reliance on a computer to complete a task especially if the task is simple and may be performed by an individual much faster than performing the task by computer.

Post-Secondary Settings: Any college or university that prepares a person for a career beyond high school at the undergraduate and/or graduate level.

Runtime Problems: Difficulties occurring while software applications are being used.

Technostress: Modern disease of adaptation caused by an inability to cope with computer technologies in a healthy manner and manifests itself in one of two ways: (1) struggle to accept computer technology and (2) over identification with computer technology.

This work was previously published in the Encyclopedia of Gender and Information Technology, edited by E. Trauth, pp. 122-128, copyright 2006 by Idea Group Reference (an imprint of IGI Global).

Chapter 7.3
Increasing Web Accessibility in Higher Education

Barbara A. Frey
University of Pittsburgh, USA

Ashli Molinero
University of Pittsburgh, USA

Ellen Cohn
University of Pittsburgh, USA

INTRODUCTION

Just as wheelchair ramps and elevators provide access to wheelchair users, good Web design provides "electronic curb ramps" to the Internet for individuals with visual or other disabilities (Waddell, 1997). Research shows it is easier and less expensive to initially construct accessible Web pages rather than to retrofit the pages with corrections. Most of the technical requirements for accessible Web design can be met if Web designers adhere to the straightforward principles suggested by the World Wide Web Consortium's Web Accessibility Initiative.

Accessible Web site design benefits all users, not just persons with disabilities. This is because users with slow Internet connections, users who access the Internet via personal Web devices and users who are speakers of foreign languages may also experience accessibility challenges (Rose & Meyer, 1996). In short, accessible Web sites increase usability. Accessibility, a component of usability, suggests "information systems flexible enough to accommodate the needs of the broadest range of users ... regardless of age or disability" (Waddell, 1997). Usability is achieved by designing with the end user in mind, to ensure that a user has access to any Web site, no matter when or how the access is sought (Pearrow, 2000).

This chapter addresses the current status of Web accessibility and usability in higher education. Specifically, it includes (1) why accessibility and usability concepts are important; (2) who is affected; and (3) some basic strategies to design accessible Web sites.

CURRENT STATUS

Most universities offer application-to-graduation services via the World Wide Web. Students access the Internet to read course descriptions, register for classes, pay tuition, purchase books, submit assignments, take quizzes and check grades. Students appreciate this ability to perform such functions at any time and from any place.

Faculty members seek to enhance student learning via online PowerPoint lecture notes, graphics and Web site links; and both faculty and students routinely access printed, audio and visual resources from around the world. Many faculty members develop their own course Web sites or use course management software packages such as Blackboard or WebCT to supplement their resident courses. Just-in-time (JIT) classroom-based learning now coexists with anytime, anywhere Web-supported learning.

Many persons with disabilities (i.e., visual, auditory, physical and/or cognitive) have limited or no access to the Web. Though approximately 29% of Internet users with disabilities take courses over the Internet or use online resources for their schoolwork (Kaye, 2000), Web-based "schools" are not open to all. Cohn, Molinero and Stoehr (1999) analyzed the Web sites of 25 United States (US) major universities and 76 US pharmacy schools using the CAST Bobby 3.1.1 validation tool. Results showed that 76% of Web sites were *not* accessible. Two years later, Cohn and Wang (2001) examined 114 US sites of top universities with doctoral programs identified via *U.S. News and World Reports*; 39% Web sites were not accessible.

Given the globalization of education, international Web site accessibility is also of interest. Cohn and Wang found that university Web sites are even less accessible in China. Only 8% of 62 Chinese university sites identified via the yahoo.com search engine were accessible.

The accessibility of "virtual campuses" is not just an issue of fairness or good business, but is addressed by legislation Section 508 of the Rehabilitation Act. Federally supported institutions now must comply with accessibility guidelines, and government Web sites must be accessible. The US Architectural and Transportation Barriers Compliance Board set forth requirements for federal Web sites under Section 508 of the Rehabilitation Act. Furthermore, Section 255 of the Telecommunications Act mandates universal access to computer networks (www.w3.org/wai/policy/#USA).

RATIONALE

The major categories of disability that can impede access to Internet-based information include vision, hearing, motor and cognitive impairments. Persons with disabilities sometimes use various assistive technologies to access computer information. Unfortunately, recent Internet trends and developments have surpassed the capabilities of the assistive technologies (e.g., audible screen readers), often leaving these individuals with insufficient access.

Common problems that create difficulties for people who rely upon assistive devices to obtain Web-based information include the improper use of image-based navigation, frames and multimedia. However, when properly structured, each of these elements can be employed in an accessible manner. It is a common misconception that an accessible page needs to be "text-only." In fact, even a "text-only" page can be inaccessible if it includes ASCII art. Conversely, a well-designed page displaying different multimedia and graphics can be completely accessible.

EFFECTIVE WEB DESIGN STRATEGIES

The World Wide Web Consortium's Web Accessibility Initiative (www.w3c.org/wai/) established guidelines for images, frames, tables, multimedia components, hypertext links, page orientation, Java scripts and applets, graphs and charts. This section on Web design describes some simple yet effective design practices.

Page Organization

A well-organized page promotes Web accessibility. Text should be clear and simple, and headings and lists should follow a consistent structure throughout the Web site. Navigation buttons should be placed in the same page location so the user can anticipate their position. Also, large buttons are easier for users to see and use.

Many Web designers use cascading style sheets (CSS) to control how elements are displayed within a Web page. The style sheets include style specifications for fonts, colors and spacing to Web documents. However, older browsers or assistive technologies cannot read all style sheet presentation features. Alternative Web pages should therefore be presented without CSS. This requires adding text equivalents for any image or text generated via style sheets.

The presence of style sheets can enable the user to suspend the movement of flashing text or graphics by simply turning off the style sheets. This is an important accessibility feature, because displays that flash or blink can cause epileptic seizures in susceptible individuals if the flash has a high intensity and is within the frequency range of 2 Hz to 55 Hz. Instructions on how to "turn off" the style sheets are usually present in the "Help Section" of the user's browser (www. webaim.org).

Images

Images can easily be defined or described to persons with visual disabilities by using the HTML-based ALT IMAGE attribute, commonly referred to as ALT tags. ALT tags are small blocks of text that describe the graphics on a screen. The tags work in most browser environments, including nonstandard browsers such as pwWebSpeak or Lynx. The text that follows shows how an ALT attribute appears in HTML:

<IMG SRC="…/yourcollegelogo. gif" ALIGN=MIDDLE ALT="[yourcollegelogo]">

Figure 1.

Figure 2.

Figure 3. Accessible table format (top) and the HTML code (bottom)

	Column 1 header	Column 2 header
Row 1 header	Column 1 Row 1 C	olumn 2 Row 1
Row 2 header	Column 1 Row 2 C	olumn 2 Row 2

```
<TABLE border=1>
<        CAPTION>Simple table created using HTML markup</CAPTION>
        <TR><TD></TD>
        <TH>Column1header</TH>
        <TH>Column2header</TH>
        </TR>
<       TR><TH>Row1 header</TH>
                <TD>Column1 Row1</TD>
                <TD>Column2 Row1</TD>
        </TR>
<       TR><TH>Row2 header</TH>
                <TD>Column1 Row2</TD>
                <TDColumn2 Row2</TD>
        </TR>
</TABLE>
```

The ALT tag should convey the content and/or purpose of an image via short text description. When the graphic is required to understand the page content, the designer can use the ALT tag with a description placed between two quotation marks. Conversely, visitors to the site should also be informed when a graphic conveys no content-related information. For example, in the case of a blue divider bar, sighted visitors will see the bar, while those visiting via text browser will see nothing. The HTML would be written as:

<IMG SRC=".../bluebar.gif" ALIGN=MIDDLE ALT=" ">

Figures 1 and 2 show the same Web page, viewed in Explorer with the images "on" and "off." Notice how the ALT tags used in the second frame still provide information to allow a user to navigate the page.

Image Maps

In an image map, the user clicks the part of a picture referred to as a hotspot. This activates a link to another part of the page or Web site or triggers a connection to a new Web site. For example, a graphic including all the counties in Pennsylvania could be constructed as an image map in which each county serves as an active link to the official county Web site.

Image maps can present accessibility challenges. While sighted students using graphical browsers such as Netscape or Internet Explorer would see a county map of Pennsylvania, students using text-based browsers would be provided the visual image. A better Web design strategy is to use the client-side MAP and text for hotspots. Web designers should provide equivalent text links for server-side maps. Text alternatives to image-based links will ensure that the image maps are available to all users.

Graphs and Charts

Graphs and charts can present even greater accessibility challenges. It is important that ALT tags used for graphics and charts are not lengthy, since some browsers have difficulty displaying long stings of text. When inserting text to describe more complex graphs or charts (or to transcribe sound files), the Web designer can employ the LONGDESC attribute, which provides for more detailed text than the ALT attribute. Descriptions can be linked to an external document or can immediately follow the graphic or sound element.

The following text describes a pie chart:

This pie chart shows the percentage of ducks and geese flying across the Allegheny River in 1 hour. Seventy-five percent of the birds were ducks and 25% of the birds were geese. The total number of ducks and geese is not displayed on the pie chart.

Tables

Most screen reader programs scroll down columns and read from the top of the table to the bottom. They then progress to the top of the next column. This vertical reading pattern can obscure the meaning of information tables for which the person is required to read in a horizontal manner. Inaccessible tables can be avoided by authoring the Web pages using HTML 4. Figure 3 shows an example of a well-constructed table, followed by the HTML that identifies row and column headers. The "TD" designates the data cells, "TH" the headers and "TR" the table rows.

It is also important to keep tables as simple as possible, because students with screen enlargers will view only part of the table at a time. Finally, the Web designer can provide a textual summary of the table content using HTML text or the LONGDESC attribute.

Following is an example of the latter:

<TABLE border="1"summary="This table compares the retention rates of students in resident and distance education courses taken in the Fall of 2004. Students in the distance education courses ...">

Multimedia Components

Multimedia components are frequently used in distance education, adding new dimensions to the learning experience. However, just as images can create problems for persons with *visual* impairment, sound must also be rendered accessible for those with *hearing* impairment. Even though some persons who are deaf can read lips, this is not generally a practical solution for watching video over the Internet, as it requires the presenter to maintain a position where his or her face is in front of the camera at all times. Therefore, both audio and video clips should be closed captioned and/or accompanied by a text transcript. Captioning also aids persons whose native language differs from the presenters, as they can both "read" and listen to the content.

The Media Access Generator (MAGpie) is an authoring tool for making multimedia materials accessible to persons with disabilities. Using MAGpie, authors can add captions to three multimedia formats: Apple's QuickTime, the World Wide Web Consortium's Synchronized Multimedia Integration Language (SMIL) and Microsoft's Synchronized Accessible Media Interchange (SAMI) format (http://ncam.wgbh.org/accessncam.html).

Scripts, Applets and Plug-Ins

Inaccessibility can also occur when the user is prompted to complete Web-based forms to provide information online. CGI (common gateway interface) scripts are mini-programs used to handle information submitted in Web-based forms, such as surveys and registration forms.

Figure 4. Depicts a Web page with and without frames

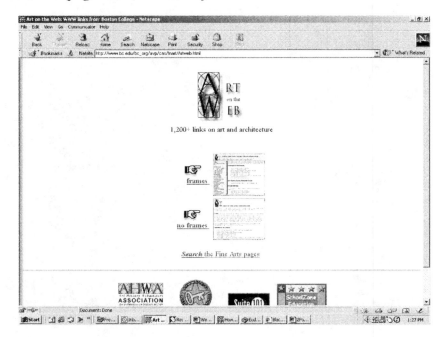

Scripts are written in programming languages such as Visual Basic, C++, Perl or Apple-Script. Since not all screen readers can properly interpret online forms, it is important that Web designers alternatively include an e-mail link to enable the reader to submit information.

Further barriers to accessibility arise because all browsers do not support applets and plug-ins. Applets are programs executed or launched from within a Web page, and plug-in technologies (e.g., QuickTime or Flash movies, Shockwave files or Adobe Acrobat Reader) are mini-programs that play within the browser to support visual and audio effects. Whenever possible, the Web designer should select plug-in formats (such as .wav or .mov) that have browser support. It is helpful to include links to the appropriate source so the user can download the appropriate plug-in. The Web designer should also present the information in an alternative format. When applet programming is not accessible via the software, inform the user of the applet and provide an equivalent alternative in HTML.

Frames

Frames are used to simultaneously display multiple documents. This technique is useful only when the browser is running on a graphical display. Other platforms, such as a speech browser, either do not recognize the concept of "multiple displays" or have only limited screen space.

Frames are often difficult for text-based screen reading software to interpret, and older browsers do not support frames. Furthermore, browsers and screen readers cannot follow multiple events in multiple frames; they can follow only one frame at a time. The Web designer should therefore add the HTML designation <NOFRAMES> and meaningful titles to pages that feature frames. Figure 4 shows a Web site that presents content both with and without frames.

Figure 5. Viewers may employ the left-sided pull-down menu to increase (as shown) or decrease font size

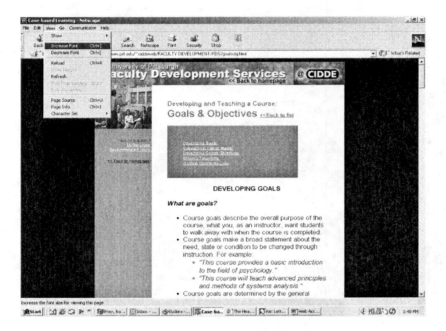

When no alternative to the use of frames is available, ensure that frames are labeled with the TITLE attribute. Provide a text alternative with NOFRAMES; and use the TARGET="_top" attribute to ensure useful Uniform Resource Location (URL) addresses.

Use of Color

When color is used for navigational purposes, persons who are blind, color blind or simply have a monochrome computer screen will not receive the information. Therefore, it is important to not rely on color alone to convey Web-based information.

Pages should be designed using high-contrast colors and color palettes should be minimized to decrease a page's download time.

Default Font Settings

Web documents should always incorporate default font settings. This allows users with impaired vision to increase the font size. Figure 5 demonstrates use of a pull-down menu to increase or decrease font size.

CONCLUSION

The use of Web-based educational resources is on the increase, as are the numbers of persons with disabilities who will require accessible Web sites. It is imperative that faculty render their Web-based materials accessible to all learners. In doing so, they will comply with federal accessibility regulations for electronic and information technology, which, in turn, will benefit all users.

REFERENCES

Castro, E. (2000). *HTML for the World Wide Web* (4th ed.). Berkley, CA: Peachpitt Press.

Cohn, E., Molinero, A., & Stoehr, G. (1999). *Is School Really Open? A Report Card of Web Based Accessibility in Higher Education.* International Conference on Technology and Education.

Cohn, E., & Wang, J. (2001). *Web-based accessibility in 2001: Representative rehabilitation, education and health related sites.* Joint Conference of the Chinese Rehabilitation Medical Society, The Health and Rehabilitation Medicine Engineering Branch 4th Conference and China Disabled Person Rehabilitation Association, Health and Rehabilitation Engineering Branch 5th Conference.

Jolliffe, A., Ritter, J., & Stevens, D. (2001). *The online learning handbook: Developing and using Web-based learning.* Sterling: Kogan Page Publishers.

Kaye, H.S. (2000). *Computer and Internet use among people with disabilities.* Disabilities Statistics Center, Institute for Health and Aging (National Institute on Disability and Rehabilitation Research, US Department of Education). San Francisco: University of California.

National Center for Accessible Media (NCAM). Retrieved August 10, 2004, from http://ncam.wgbh.org/

Neilsen, J. (2000). *Designing Web usability.* Indianapolis: New Riders Publishing.

Paciello, M.G. (2000). *Web accessibility for people with disabilities.* Lawrence: CMP Books.

Pearrow, M. (2000). *Web usability handbook.* Rockland: Charles River Media.

Rose, D., & Meyer, A. (1996). *The future is in the margins: The role of technology and disability in educational reform.* Retrieved August 10, 2004, from www.cast.org/udl/index.cfm?i=542

Waddell, C.D. (1998). *Applying the ADA to the Internet: A Web Accessibility Standard.* American Bar Association National Conference.

Web Accessibility Initiative. Retrieved August 10, 2004, from www.w3c.org/wai

Webaim. *Webaim: Web Accessibility in Mind.* Retrieved August 10, 2004, from www.Webaim.org

This work was previously published in the Encyclopedia of Distance Learning, Vol. 3, edited by C. Howard, J. Boettcher, L. Justice, K. Schenk, P. L. Rogers, and G. A. Berg, pp. 1069-1075, copyright 2005 by Idea Group Reference (an imprint of IGI Global).

Chapter 7.4
Evaluating Computer–Supported Learning Initiatives

John B. Nash
Stanford University, USA

Christoph Richter
University of Hannover, Germany

Heidrun Allert
University of Hannover, Germany

INTRODUCTION

The call for the integration of program evaluation into the development of computer-supported learning environments is ever increasing. Pushed not only by demands from policy makers and grant givers for more accountability within lean times, this trend is due also to the fact that outcomes of computer-supported learning environment projects often fall short of the expectations held by the project teams. The discrepancy between the targets set by the project staff and the outcomes achieved suggests there is a need for formative evaluation approaches (versus summative approaches) that facilitate the elicitation of information that can be used to improve a program while it is in its development stage (c.p., Worthen, Sanders & Fitzpatrick, 1997). While the call for formative evaluation as an integral part of projects that aim to develop complex socio-technical systems is widely accepted, we note a lack of theoretical frameworks that reflect the particularities of these kind of systems and the ways they evolve (c.p., Keil-Slawik, 1999). This is of crucial importance, as formative evaluation will only be an accepted and effective part of a project if it provides information useful for the project staff. Below we outline the obstacles evaluation faces with regard to projects that design computer-supported learning environments, and discuss two promising approaches that can be used in complimentary fashion.

BACKGROUND

According to Worthen et al. (1997), evaluation is "the identification, clarification, and application of defensible criteria to determine an evaluation object's value (worth or merit), quality, utility, effectiveness, or significance in relation to those criteria." In this regard evaluation can serve different purposes. Patton (1997) distinguishes between judgment-, knowledge- and improvement-oriented evaluations. We focus on improvement-oriented evaluation approaches. We stress that evaluation can facilitate decision making and reveal information that can be used to improve not only the project itself, but also outcomes within the project's target population. The conceptualization of evaluation as an improvement-oriented and formative activity reveals its proximity to design activities. In fact this kind of evaluative activity is an integral part of any design process, whether it is explicitly mentioned or not. Accordingly it is not the question if one should evaluate, but which evaluation methods generate the most useful information in order to improve the program. This question can only be answered by facing the characteristics and obstacles of designing computer-supported learning environments.

Keil-Slawik (1999) points out that one of the main challenges in evaluating computer-supported learning environments is that some goals and opportunities can spontaneously arise in the course of the development process and are thus not specified in advance. We believe that this is due to the fact that design, in this context, addresses ill-structured and situated problems. The design and implementation of computer-supported learning environments, which can be viewed as a response to a perceived problem, also generates new problems as it is designed. Furthermore every computer-supported learning experience takes place in a unique social context that contributes to the success of an intervention or prevents it. Therefore evaluation requires that designers pay attention to evolutionary and cyclic processes and situational factors. As Weiss notes, "Much evaluation is done by investigating outcomes without much attention to the paths by which they were produced" (1998, p. 55).

For developers designing projects at the intersection of information and communication technology (ICT) and the learning sciences, evaluation is difficult. Evaluation efforts are often subverted by a myriad of confounding variables, leading to a "garbage in, garbage out" effect; the evaluation cannot be better than the parameters that were built in the project from the start (Nash, Plugge & Eurlings, 2001). Leaving key parameters of evaluative thinking out of computer-supported learning projects is exacerbated by the fact that many investigators lack the tools and expertise necessary to cope with the complexity they face in addressing the field of learning.

We strongly advocate leveraging the innate ability of members of the computer science and engineering communities to engage in "design thinking" and turn this ability into a set of practices that naturally becomes program evaluation, thereby making an assessment of the usefulness of ICT tools for learning a natural occurrence (and a manifest activity) in any computer-supported learning project.

Design-Oriented Evaluation for Computer-Supported Learning Environments

There are two approaches that inherently relate themselves to design as well as to evaluation. Therefore they are useful tools for designers of computer-supported learning initiatives. These two perspectives, discussed below, are scenario-based design and program theory evaluation. Both approaches assume that the ultimate goal of a project should be at the center of the design and evaluation discussion, ensuring a project is not about only developing a usable tool or system, but is about developing a useful tool or system that improves outcomes for the user. Beyond this common ground, these approaches are rather complementary to each other and it is reasonable to use them in conjunction with one another.

Scenario-Based Approaches

Scenario-based approaches are widely used in the fields of software engineering, requirements engineering, human computer interaction, and information systems (Rolland et al., 1996). Scenarios are a method to model the universe of discourse of an application, that is, the environment in which a system, technical or non-technical, will be deployed. A scenario is a concrete story about use of an innovative tool and/or social interactions (Carroll, 2000). Scenarios include protagonists with individual goals or objectives and reflect exemplary sequences of actions and events. They refer to observable behavior as well as mental processes, and also cover situational details assumed to affect the course of actions (Rosson & Carroll, 2002). Additionally it might explicitly refer to the underlying culture, norms, and values (see Bødker & Christiansen, 1997). That said, scenarios usually focus on specific situations, only enlighten some important aspects, and generally do not include every eventuality (e.g., Benner, Feather, Johnson & Zorman, 1993).

Beside their use in the design process, scenarios can also be used for purposes of formative evaluation. First of all, as a means of communication, they are a valuable resource for identifying underlying assumptions regarding the program under development. Stakeholder assumptions might include those related to instructional theories, the learner, the environmental context, and its impact on learning or technical requirements. Underlying assumptions such as these are typically hidden from view of others, but easily developed and strongly held within individuals developing computer-supported learning environments. Scenarios help to reveal the thinking of designers so that others can participate in the design process and questionable assumptions can

come under scrutiny. The use of scenarios also allows identification of pros and cons of a certain decision within the design process. In this vein Carroll (2000) suggests employing "claim analysis." Claims are the positive or negative, desirable and undesirable consequences related to a certain characteristic of a scenario. Assuming that every feature of a proposed solution usually will entail both positive and negative effects helps to reflect on the current solution and might provoke alternative proposals. The analysis of claims is thereby not limited to an intuitive ad hoc evaluation, but also can bring forth an explicit hypothesis to be addressed in a subsequent survey.

Program Theory Evaluation

Program theory evaluation, also known as theory-based evaluation, assumes that underlying any initiative or project is an explicit or latent "theory" (or "theories") about how the initiative or project is meant to change outcomes. An evaluator should surface those theories and lay them out in as fine detail as possible, identifying all the assumptions and sub-assumptions built into the program (Weiss, 1995). This approach has been promoted as useful in evaluating computer-supported learning projects (Strömdahl & Langerth-Zetterman, 2000; Nash, Plugge & Eurlings, 2001) where investigators across disciplines find it appealing. For instance, for designers (in mechanical engineering or computer science), program theory evaluation reminds them of their own use of the "design rationale." And among economists, program theory evaluation reminds them of total quality management (TQM). In the program theory approach (Weiss, 1995, 1998; Chen, 1989; Chen & Rossi, 1987), one constructs a project's "theory of change" or "program logic" by asking the various stakeholders, "What is the project designed to accomplish, and how are its components intended to get it there?" The process helps the project stakeholders and the evaluation team to identify and come to consensus on the project's theory of change. By identifying and describing the activities, outcomes, and goals of the program, along with their interrelationships, the stakeholders are then in position to identify quantifiable measures to portray the veracity of the model.

Theory-based evaluation identifies and tests the relationships among a project's inputs or activities and its outcomes via intermediate outcomes. The key advantages to using theory-based evaluation are (Connel & Kubisch, 1995; Weiss, 1995):

- It asks project practitioners to make their assumptions explicit and to reach consensus with their colleagues about what they are trying to do and why.
- It articulates a theory of change at the outset and gains agreement on it, by all stakeholders reducing problems associated with causal attribution of impact.
- It concentrates evaluation attention and resources on key aspects of the project.
- It facilitates aggregation of evaluation results into a broader context based on theoretical program knowledge.
- The theory of change model identified will facilitate the research design, measurement, data collection, and analysis elements of the evaluation.

Both scenario-based design and program theory stress the importance of the social context while planning computer-supported environments. They also represent means to facilitate the communication among the stakeholders and urge the project team to reflect their underlying assumptions in order to discuss and test them. Furthermore, both approaches are particularly suitable for multidisciplinary project teams. Scenarios and program logic maps are not static artifacts; they are a starting point for discussion and have to be changed when necessary. With these similarities

there are also differences in both approaches. The major difference between them is that program theory offers a goal-oriented way to structure a project, while scenario-based design proffers an explorative approach that opens the mind to the complexity of the problem, alternatives, and the diversity of theories that try to explain social and socio-technical process. That is, scenario-based design highlights the divergent aspects of project planning, and evaluation program theory stresses the convergent aspects. Program theory evaluation helps to integrate each scenario, decision, and predefinition into the whole process. Scenarios force users not just to use terms, but to give meaningful descriptions. They force users to state how they actually want to instantiate an abstract theory of learning and teaching. This helps to implement the project within real situations of use, which are complex and ill structured. Program theory helps to focus on core aspects of design and prevent getting 'lost in scenarios'. Scenarios and program theory evaluation can be used in an alternating way. Thereby it is possible to use both approaches and improve the overall development process.

The program theory of an initiative can be a starting point for writing scenarios. Especially the interrelations between the goals and interrelation between ultimate goal and inputs can be described with a scenario. The scenario can help to understand how this interrelation is meant to work and how it will look in a concrete situation. Scenarios on the other hand can be used to create program theory by pointing out main elements of the intended program. They can also be used to complete already existing program theory by presenting alternative situations of use. For developers of computer-supported learning environments, scenario-based design and program theory represent complementary approaches, which when used together or separately, can add strength to the implementation and success of such projects.

FUTURE TRENDS

It is clear that formative evaluation will become more important in the future, and it will be especially crucial to think about how to integrate evaluation into the design process. Essentially, designers will need to answer the question "Why does the program work?" and not just "Did it work?" It becomes obvious that the design of a computer-supported learning environment, like the development of any other complex socio-technical system, is a difficult process. In fact the necessity for changes in the original plan is practically preordained due to the ill-structured and situated nature of the domain. The mere act of engaging in a design process suggests that designers will engage in planned as well as evolutionary, unplanned activities. Therefore it is important that the project designers use methods that support divergent thinking and methods that support convergent processes. While scenario-based design and program-theory evaluation represent complementary views on the design and evaluation of computer-supported learning environments that can facilitate these processes, there is still room for improvement.

CONCLUSION

Formative evaluation is an important means to ensure the quality of an initiative's outcomes. Formative evaluation directed towards improvement of an initiative can be understood as a natural part of any design activity. While this is widely recognized, there is still a lack of program evaluation frameworks that reflect the uniqueness of the design process, the most crucial of which is the inherent ambiguity of design. In spite of great inspiration portrayed by project teams, usually manifested by visions of a certain and sure outcome, no project can be pre-planned completely, and midcourse corrections are a

certainty. Scenario-based design and program theory evaluation provide a theoretical foothold for projects in need of collecting and analyzing data for program improvement and judging program success.

In sum, scenario-based design and program theory hold many similarities. The major difference between them is that program theory offers a goal-oriented way to structure a project, while scenario-based design provides an explorative approach that opens the mind to the complexity of the problem, alternatives, and the diversity of theories that try to explain social and socio-technical process.

Scenario-based design highlights the divergent aspects of project planning, and evaluation program theory stresses the convergent aspects. For developers of computer-supported learning experiences, scenario-based design and program theory represent complementary approaches, which when used together or separately can add strength to the implementation and success of ICT learning projects.

REFERENCES

Benner, K.M., Feather, M.S., Johnson, W.L. & Zorman, L.A. (1993). Utilizing scenarios in the software development process. In N. Prakash, C. Rolland & B. Pernici (Eds.), *Information system development process* (pp. 117-134). Elsevier Science Publishers.

Bødker, S. & Christiansen, E. (1997). Scenarios as springboards in design. In G. Bowker, L. Gaser, S.L. Star & W. Turner (Eds.), *Social science research, technical systems and cooperative work* (pp. 217-234). Lawrence Erlbaum.

Carroll, J.M. (2000). *Making use: Scenario-based design of human-computer interactions.* Cambridge: MIT Press.

Chen, H.T. (1989). Issues in the theory-driven perspective. *Evaluation and Program Planning, 12,* 299-306.

Chen, H.T. & Rossi, P. (1987). The theory-driven approach to validity. *Evaluation Review, 7,* 95-103.

Connell, J.P. & Kubisch, A. (1995). Applying a theory of change approach to the evaluation of comprehensive community initiatives: Progress, prospects, and problems. In K. Fulbright-Anderson et al. (Eds.), *New approaches to evaluating community initiatives. Volume 2: Theory, measurement, and analysis.* Washington, DC: Aspen Institute.

Keil-Slawik, R. (1999). Evaluation als evolutionäre systemgestaltung. aufbau und weiterentwicklung der paderborner DISCO (Digitale Infrastruktur für computerunterstütztes kooperatives Lernen). In M. Kindt (Ed.), *Projektevaluation in der lehre—multimedia an hochschulen zeigt profil(e)* (pp. 11-36). Münster, Germany: Waxmann.

Nash, J.B., Plugge, L. & Eurlings, A. (2001). Defining and evaluating CSCL evaluations. In A. Eurlings & P. Dillenbourg (Eds.), *Proceedings of the European Conference on Computer-Supported Collaborative Learning* (pp. 120-128). Maastricht, The Netherlands: Universiteit Maastricht.

Patton, M.Q. (1997). *Utilization-focused evaluation* (3rd Edition). Thousand Oaks, CA: Sage Publications.

Rolland, C., Achour, C.B., Cauvet, C., Ralyté, J., Sutcliffe, A., Maiden, N.A.M., Jarke, M., Haumer, P., Pohl, K., Dubois, E. & Heymans, P. (1996). *A proposal for a scenario classification framework.* CREWS Report 96-01.

Rosson, M.B. & Carroll, J.M. (2002). *Usability engineering: Scenario-based development of human-computer interaction.* San Francisco: Morgan Kaufmann.

Strömdahl, H. & Langerth-Zetterman, M. (2000). *On theory-anchored evaluation research of educational settings, especially those supported by information and communication technologies (ICTs).* Uppsala, Sweden: Swedish Learning Lab.

Weiss, C. (1995). Nothing as practical as good theory: Exploring theory-based evaluation for comprehensive community initiatives for children and families. In J. Connell et al. (Eds.), *New approaches to evaluating community initiatives: Concepts, methods, and contexts.* Washington, DC: Aspen Institute.

Weiss, C. (1998). *Evaluation research: Methods for studying programs and policies.* Englewood Cliffs, NJ: Prentice-Hall.

Worthen, B.R., Sanders, J.R. & Fitzpatrick, J.L. (1997). *Program evaluation—alternative approaches and practical guidelines* (2nd Edition). New York: Addison Wesley Longman.

KEY TERMS

Computer-Supported Learning: Learning processes that take place in an environment that includes computer-based tools and/or electronically stored resources. CSCL is one part of this type of learning.

Evaluation: The systematic determination of the merit or worth of an object.

Formative Evaluation: The elicitation of information that can be used to improve a program while it is in the development stage.

Program: A social endeavor to reach some predefined goals and objectives. A program draws on personal, social, and material resources to alter or preserve the context in which it takes place.

Program Theory: A set of assumptions underlying a program that explains why the planned activities should lead to the predefined goals and objectives. The program theory includes activities directly implemented by the program, as well as the activities that are generated as an response to the program by the context in which it takes place.

Scenarios: A narrative description of a sequence of (inter-)actions performed by one or more persons in a particular context. Scenarios include information about goals, plans, interpretations, values, and contextual conditions and events.

Summative Evaluation: The elicitation of information that can be used to determine if a program should be continued or terminated.

This work was previously published in the Encyclopedia of Information Science and Technology, Vol. 2, edited by M. Khosrow-Pour, pp. 1125-1129, copyright 2005 by Idea Group Reference (an imprint of IGI Global).

Chapter 7.5
Do the Philosophical Foundations of Online Learning Disadvantage Non-Western Students?

David Catterick
University of Dundee, Scotland

ABSTRACT

A product of its historical origins, online learning is firmly rooted in the educational values that dominate post-secondary education in Britain, Australasia, and North America. With the increasing numbers of international students studying degree programs online, this chapter asks whether students from diverse educational cultures are disadvantaged in their learning by the teaching approaches implemented within online teaching environments. Active learning, reflective practice, and collaborative learning are all based on a cognitive, constructivist tradition (Fox, 2001), one which is evidently not shared by much of the rest of the world (Kim & Bonk, 2002; Wright & Lander, 2003). Employing evidence from the field of cross-cultural psychology (Allik & McCrae, 2004) and taking Chinese students as an example (Cheung, Leung, Zhang, Sun, Gan, Song, & Xie, 2001; Lin, 2004; Matthews, 2001), the author suggests that there may be some cause for concern within online instructional practices. The chapter concludes with three possible responses to the issue, two of which might go some way towards ensuring that international students find themselves on a more even playing field in their online degree program of study.

INTRODUCTION

The online delivery of degree programmes is a relatively new phenomenon in the field of higher education. Like so many technologies applied to the field of education, the arrival of the virtual learning environment (VLE) or learning management systems (LMS) seemed to precede the underlying educational philosophy needed to give it both support and credibility within the academic community. This meant there was something of a need to play catch-up, and the result was what the author terms the "Magpie Effect," a process of appropriating a variety of educational philosophies in order to justify the pedagogical value of the emerging technology. This is evidenced by the number of educational technologists who still post to discussion lists like the *Distance Education Online Symposium Listserv* (DEOS-L) asking the list for help in justifying the rationale of VLE use to university policy makers. This may seem like an overly cynical opening for a chapter on distance learning, but even if the reader views things in a slightly different way, there are two notions which would not seem so contentious:

1. It is higher education institutions in Britain, Australasia and North America (BANA) which have been at the forefront of online degree programme development and delivery.

2. The philosophical foundations of online distance learning have arisen out of Western (particularly BANA) educational paradigms.

In this chapter, the author argues (from a theoretical rather than an empirical perspective) that in spite of the supposed global reach of online distance learning, the Western philosophical "software" that runs on the technology might disadvantage students who do not share the same constituent values. In exploring this hypothesis, the example of ethnic Chinese international students studying within an online degree program will be used. This choice stems from a combination of the author's six years' experience of teaching in universities in the People's Republic of China, his eight years' experience of teaching Chinese students in UK higher education, and the fact that many Chinese students joining online distance learning programs in BANA countries are of ethnic Chinese origin.

EXPLORING THE HISTORICAL CONTEXT OF THE PREVAILING EDUCATIONAL PHILOSOPHY

By the time online learning became a feature of higher education in the mid-1990s, the landscape of educational philosophy had already long since been radically transformed as a consequence of the shift from behaviourist beliefs about the nature of learning to cognitivist ones (Mayer, 1996). Essentially, the change was the result of a growing conviction among educational philosophers and theorists that the mental processes which constitute learning and development are the product of symbolic activity within individual minds (Atkinson & Shiffrin, 1968). Supported by breakthroughs in the study of memory and problem-solving, cognitivism quickly became the touchstone for educational research eventually giving rise to a theory of information processing (ibid, 1968). Over the intervening decades, different branches of cognitivism have developed, but the most influential, and arguably the most controversial (Fox, 2001; Liu & Matthews, 2005), is constructivism. Grounded in the philosophical writings of Lev Vygotsky (1896-1934) and, later, Jean Piaget (1896-1980), constructivism portrays learning as a process in which the individual translates information from the people and world around them into a form that is intelligible to them at the personal level. With Vygotsky (1978) himself emphasising the role of social context on cognitive development, positing that "all the higher functions originate as actual relationships between individuals" (1978, p. 57), and Piaget emphasising the role of environmental stimuli (Piaget, 1953), constructivism fundamentally rejected the knowledge transmission model of the behaviourist era. In constructivism, knowledge was portrayed not as an objective reality transmitted from teacher to student, but rather co-constructed in the interactions between teacher and learners. This key philosophical shift placed far more emphasis on the role of the learner in the learning process, as it focused upon providing learners with the opportunity to contextualise knowledge within the framework of their own experience and schema. Words like interaction, collaboration, and facilitation seemed to take on a growing importance in the educational literature which, in turn, engendered a shift from teacher-led instructional strategies to more learner-centred ones.

But while this shift in educational philosophy was already well-established in the BANA nations by the mid-1990s, the same shift had not affected mainstream education in whole swathes of the rest of the world. As a so-called "foreign expert" working in Chinese higher education institutions during the 1990s, the author personally saw the philosophical conflicts which resulted when the usually young, well-meaning foreign teacher adopted cognitivist and constructivist approaches in the classroom. As the barriers to these approaches materialised, some foreign teachers took upon themselves a personal agenda of regime change at the philosophical level. Unfortunately, this frequently damaged the social fabric of the classroom relationship, which is a vital aspect of constructivist approaches. One of the most enlightening studies addressing this fallout is by Li (1999) who details the findings from a three-year study of student responses to and opinions of Western teaching staff in his institution in China. What Li's study points to is a world of conflict in which the students slowly withdraw from the classroom and thereby the learning process. Students cite the fact that the classroom experience did not live up to their expectations of what constitutes quality learning (Li, 1999). In this chapter, it will be argued that the cultural imperialism at the heart of the process which Li describes is also evident in online distance learning, meaning that there is a similar risk of disenfranchising the online learner.

TEACHING APPROACHES RESULTING FROM THE EDUCATIONAL PHILOSOPHY

Before investigating the implications of conflicting educational philosophies, it is important to consider in more detail the connection between educational philosophy and teaching approach, and understand the key teaching approaches commonly associated with a cognitivist educational philosophy. While it would be easy simply to see the adoption of a teaching approach as an entirely bottom-up process which teachers themselves are fully able to rationalise, the reality is that the well-developed quality assurance mechanisms in BANA countries (which are themselves the product of the same philosophical foundations) add a significant top-down influence. In addition, it is important to note that decisions about teaching approach should not necessarily be seen as conscious ones. In fact, a survey of online teaching staff would likely reveal that teaching approaches can be subscribed to without the teacher being fully aware of the philosophy which underpins them. Evidence for this can be found in in-service staff development programmes (at least in UK higher education) which introduce staff to various techniques such as fostering active learning, but present them at face value rather than in their philosophical context. In his own teaching in China, the author found himself in the rather absurd situation of trying to defend his use of Western teaching strategies without having properly considered and debated the values from which they stemmed. Now, years later, it is time to consider whether something similar might be happening in online teaching contexts.

The following (non-hierarchical) list is far from comprehensive, but would probably be recognisable to many BANA teaching staff as the teaching approaches which characterise their own online teaching. In summarising each teaching approach, there has been a conscious attempt not to oversimplify the issues and to avoid critiquing what the author personally believes are (in some cases, at least) approaches that are not systematically supported by thorough, well-grounded research.

Teaching Approach #1: Promoting Active Learning

Active learning is a rather nebulous term used to describe a range of instructional approaches which seek to promote active involvement of the learner in learning process (De Vita, 2000). Even the term "active learning" invites a contrast with the notion of passivity in learning, and is used as one of the main arguments against the transmissional mode of teaching common in behaviourist approaches. Active learning acknowledges the cognitivist view that the learner is not simply an empty vessel, but someone who plays an integral role in the learning process (Vygotsky, 1978). In online learning, the active learning typically finds its expression in contributions to online discussion fora and synchronous discussion (Dalgarno, 2001).

Teaching Approach #2: Promoting Reflective Practice

Reflective practice (Kolb, 1984) has become a key teaching approach, particularly in contexts where the learners are post-experience professionals. At its simplest, reflective practice involves "thinking about one's work" (Parker, 1997, p. 2) and is the product of the positivist notion that mental processes can be observable. Reflective teaching involves a "willingness to engage in constant self-appraisal and development [which] implies flexibility, rigorous analysis and social awareness" (Parker, 1997, p. 2, quoting Pollard & Tann, 1994, p. 9). As knowledge is seen to be constructed

rather than received, there is clearly a need to have a mechanism by which to monitor and evaluate one's own practice. One such tool, the learning cycle, was developed by David Kolb (1984) and is now in widespread use (see Figure 1).

The idea of the learning cycle is that the learner can begin at any point on the cycle and use the framework as an aid to reflection. In online degree programmes, this typically involves learners from a professional background reflecting on a critical incident or case in their professional life and demonstrating reflective practice before integrating it into an academic written report.

Teaching Approach #3: Promoting Collaborative Learning and Group Interaction

Collaborative learning is defined by UNESCO (United Nations Educational, Scientific, and Cultural Organization) as learners working together "in groups on the same task simultaneously, thinking together over demands and tackling complexities" (*Definitions of Open Learning*, n.d., p. 1). Collaborative learning is very much seen as "the act of shared creation and/or discovery" (*Definitions of Open Learning*, n.d., p. 2) with the interaction in small groups being seen as a trigger for metacognitive activity. In online distance

Figure 1. Kolb's learning cycle as quoted in Smith (2001)

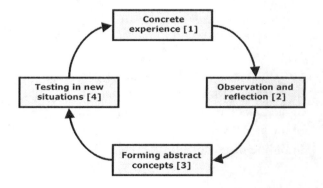

learning, collaborative learning can be facilitated through the use of "break out" groups who discuss an issue without significant input from the tutor before feeding back to the larger forum.

Teaching Approach #4: Promoting Autonomous Learning

Littlewood (1999, p. 74) citing Ryan (1991, p. 210) defines autonomy as a process of "self-determination" or "self-regulation," a process in which "one experiences the self to be an agent, the 'locus of causality' of one's behaviour." Autonomy sees learning as going beyond a partnership between teacher and learner with the locus of control shifting far more to the learner than in previous, more teacher-centred approaches. In online learning, learners are encouraged to use online research to promote their learning, sometimes with the intention that what the learner finds is incorporated into the content of the course. On a wider scale, autonomy is considered a key transferable skill in societies which value independent thought and also fits in with the relativistic worldview of post-modernity.

To the younger, Western learner, these teaching approaches are simply accepted as part of the educational process. To learners from educational cultures still influenced by more traditional practices, these values and teaching strategies may produce a certain cultural dissonance which frequently has a negative impact on the learning experience, at least in the short term.

CULTURE, PERSONALITY, AND THE CHINESE LEARNER

Over the past five years, BANA countries have seen an unprecedented growth in the number of mainland Chinese international students studying degree programs abroad. This growth has been matched by the number of ethnic Chinese students studying for degrees online (International Stu-

dents in Australian Universities, IDP Education, 2004). These learners are likely to encounter some, if not all, of the teaching approaches previously described. But how are these approaches different from the ones previously encountered by the students? Though learning objectives are the key determinant of student success, it is conceivable that one consequence of not conforming to the teaching approaches could possibly be the student failing the program. Space does not allow me to explore the integration of assessment into teaching approaches. Suffice it to say that if as part of an assessment a student is asked to provide evidence of reflective practice (*Teaching Approach #2*) and fails to do so, they will likely fail.

Exploring the Notion of Culture

One of the most common ways of investigating culture is by profiling national cultural characteristics. This has become increasingly common predominantly in the context of global business interaction with key work by Hofstede (1991) and more recently by Inglehart and Baker (2000). While high-context/low-context and male/female cultural dichotomies do go some way towards creating a framework for understanding cultural differences, it must be remembered that national boundaries are sometimes rather arbitrarily drawn and may well be based more upon geography and political expediency rather than shared cultural history or ethnicity. Even ethnicity may prove problematic, as it may indicate a shared heritage rather than cultural homogeneity. Holliday (1999) calls for researchers to recognise the importance of what he terms "small cultures," a paradigm which "attaches culture to small social groupings or activities wherever there is cohesive behaviour, and thus avoids culturist ethnic, national, or international difference" (1999, p. 237). It might be useful, however, to take Holliday's notion of "small cultures" one step further by suggesting that it may be more beneficial to think in terms of the "culture of the individual," a realisation that cultural features need to be examined at the lowest common denominator, which is, of course, the individual rather than the national or the ethnic.

What allows us to make this transition is the growing body of research into personality. The personality inventory is a tool widely used in this type of research and is based on the principles of psychometric testing. A number of these inventories report results in the form of a personal profile, an extended prose text which summarises individuals' main strengths and weaknesses, their communication preferences and how they best function in a team. Personality inventories have until relatively recently been designed around a universalist (etic) view of personality; that is to say, based on the theory that personality will broadly exhibit the same characteristics whatever the cultural background of the individual. Cheung et al. (2001) and her colleagues at the University of Hong Kong have, over the past decade and a half, challenged this view of personality arguing, that while many aspects of personality measured by a standardised, globally-administered instrument are indeed universal, there are a number of key elements which are missed by the etic tools. Using the *Five Factor Personality Inventory* as a starting point, Cheung et al. (2001) created a culture-specific (emic) personality inventory for ethnic Chinese learners and workers called the Chinese Personality Assessment Inventory (CPAI). Cheung's tool, though starting from the culture of the individual, focuses on certain unique elements which she claims points to the existence of what she terms a "Chinese tradition." Cheung is careful not to claim that these components are unique to Chinese culture, but she does portray them as key elements of the psyche of ethnic Chinese individuals.

The Chinese Tradition

Chinese tradition, as it is defined by Cheung et al. (2001), contains six key characteristics: *Ren Qing*; *Ah-Q mentality*; *harmony*; *face*; *thrift vs.*

extravagance; and *modernization*. Of these, only the first four can be considered to be relevant to educational contexts and thus relevant to the content of this chapter. In this next part of the chapter, these four characteristics will be used as predictors of response to the cognitivist teaching approaches referred to earlier.

1. ***Ren Qing* (translated as relationship orientation):** *Ren Qing* literally translates into the English language as "human emotion." *Ren Qing* "covers adherence to cultural norms of interaction based on reciprocity, exchange of social favours, and exchange of affection according to implicit rules (Cheung et al., 2001, p. 408). This complex interplay of interactions focusing on social favours and obligations is difficult (for the outsider) to grasp except perhaps at the surface level, when someone receives a favour they are expected to pay it back at some point in the future. Favours might take the form of gifts carefully chosen to match the givers' own perception of the depth of the relationship (Chan, 2002).

2. ***Ah-Q mentality* (translated as defence mechanism):** Ah-Q is a well-known, satirical Chinese character from the Chinese literary canon. His name is used to indicate the "defense mechanisms of the Chinese people, including self-protective rationalization, externalization of blame, and belittling others' achievements" (Cheung et al., 2001, p. 408).

3. ***Harmony*:** Harmony is defined by Cheung as referring to one's inner peace of mind, contentment, interpersonal harmony, avoidance of conflict, and maintenance of equilibrium (2001, p. 408). Harmony, both personal and interpersonal, is seen to need protecting at all costs. On the surface, we might equate harmony with the Western concept of not rocking the boat, but this under-represents the role of maintaining equilibrium among

all parties rather than simply monitoring one's own input in a situation.

4. *Face*: Although the concept of face exists in a variety of cultures, in the Chinese tradition, face refers to something more complex involving the interplay of relationships. In the Chinese tradition, face is "the pattern of orientations in an interpersonal and hierarchical connection and social behaviours to enhance one's face and to avoid losing one's face" (Cheung et al., 2001, p. 408). One of the key differences between face in the Western sense is that in ethnic Chinese circles, face is not only lost, it is also protected (avoiding your own loss of face) and given (praising another individual in front of their superiors). The rest of the chapter will examine how these characteristics might impact on the response of ethnic Chinese students to cognitivist teaching approaches.

Comparing Teaching Approaches and Chinese Culture

While there are a number of things that ethnic Chinese learners might recognise within the value system of online education, there are certain values inherent within the teaching approaches that might seem as puzzling and as alien to them as the four Chinese tradition characteristics seem to the Westerner. It is possible to see how these may lead to feelings of cultural dissonance for some Chinese learners, not least because they may perceive there to be an essential conflict between their own educational heritage and the values inherent in online learning.

Teaching Approach #1: Promoting Active Learning

Evidence of active learning in higher education is often seen in terms of involvement in discussion or contribution to debate. In promoting active learning, the teacher would, for example, expect students to contribute to an online discussion forum, behaviour often reinforced by rewards or penalties. Despite this, some students remain "silent" or uninvolved, a phenomenon known in face-to-face contexts as communication reticence or communication apprehension and in online learning as lurking. Communication reticence, as the name suggests, results from students remaining uninvolved in what are considered normal classroom processes such as asking or answering questions. Cheng (2000) and Jones (1999) both examine communication reticence in the context of non-native speakers communicating in an English language environment. While Cheng (2000) suggests that linguistic factors play a key role, Jones (1999) cites the influence of cultural background. Whatever the cause, and personality issues aside, the anecdotal and perhaps stereotypical view of many teaching staff in higher education is that communication reticence is more typical of Chinese students than other non-native or native speakers of English. The notion implicit in many such observations is that it is only through the use of productive language and contribution to debate that learning can occur. In Confucius' writings (Book XVI, *Record on the Subject of Education*), the system of college education is described as one in which the students "listened, but they did not ask questions; and they could not transgress the order of study imposed on them" (Ulrich, 1947, p. 20). With communications reticence woven into the cultural heritage of a society, we might expect to see the behaviour justified. One cultural justification for communication reticence was heard by the author. A student said that, in their approach to learning, Western learners were like hot water bottles, hot on the outside and hot on the inside. Chinese learners, in contrast, were like thermos flasks, equally hot on the inside but cool on the outside. In the traditional Chinese classroom, the link between active participation in the class and the process of learning is one which may seem far from obvious.

Teaching Approach #2: Promoting Reflective Practice

Reflection, as we have seen, is an activity focusing on the individual's experience and their own independent evaluation of that experience. One interesting study on cross-cultural aspects of reflective practice is Stockhausen and Kawashima's (2002) study of post-experience Japanese nurses studying at an Australian university. The chapter focuses on the cultural barriers experienced by these nurses in grasping reflective practice as a key concept that would enhance their professional standing. In the opening section of the article, by way of explaining the barriers, the authors point out that,

Japanese culture is principled on interdependence, interconnectedness and interrelations ... [and] the behaviour of Japanese people is governed by social sensitivity and their extreme concern regarding social interactions and relationships and the avoidance of conflict. (Stockhausen & Kawashima, 2002, p. 119)

The authors go on to describe how in Japan, nurses are "devoted to patients while being required to function subserviently to predominantly male physicians" (2002, p. 119) which means that any reflective practice would impact upon these "interconnected" relationships and social roles in the hospital, a case perhaps of the educational value being in contradiction with social values. Though the students in this study are Japanese, and we have no data available to compare the professional working cultures of Japan and China, it would seem that Japanese learners are as concerned about harmony as Chinese learners. This can be explained by the fact that other Confucian Heritage countries share some of the characteristics which constitute Chinese tradition. It would seem, based on this evidence, therefore, that at least as far as harmony is concerned, BANA

teaching approaches may conflict with deeply-held socio-cultural values. Though the credibility gap barrier may not be insurmountable (as we see in Stockhausen & Kawashima's 2002 study), it is important for Western teachers to reflect upon the impact that their teaching approaches might have in other contexts.

Teaching Approach #3: Promoting Collaborative Learning and Group Interaction

In the Confucian tradition, the teacher is portrayed as the "fount of all wisdom," or in Western parlance "the sage on the stage." This seems to contrast sharply with constructivist notions of co-construction of meaning. In the traditional Chinese classroom, teacher-centeredness has a profound influence on the role of the learner. In Book XVI, *Record on the Subject of Education*, the subservience of the learner to the teacher is emphasised by the following analogy "Those who first yoke a young horse place it behind, with the carriage going on in front of it" (Ulrich, 1947, p. 23), that is, that the learner is subservient to the teacher and needs to be carefully developed. As the source of all wisdom, the teacher is the most important person in the classroom, which influences the way the relationship with other learners is viewed. Learner input via collaborative and group interaction might therefore be seen as immediately less expert and the time spent in collaboration could well be considered as detracting from teacher input. This is, again, not to suggest that these are insurmountable barriers, but where collaborative and group interaction does occur, there could be a further problem in the nature of the group dynamic. Ren Qing (relationship orientation) suggests that the complex web of favours might have an influence on the negotiation of team roles as favours are called in. The assumption that the division of labour in a group task is an equal one may be faulty, but whether or not it is equal, cultural norms would suggest it is likely to be fair according to a non-Western code of social ethics.

Teaching Approach #4: Promoting Autonomous Learning

Autonomous learning, as evidenced by the prefix "auto," emphasises the individual's role in learning. In any collectivist society, the concept of self is primarily understood in one's relationship to the wider group, and we have already seen the complex web of interactions and obligations that make up Ren Qing. Autonomous learning in Ryan's (1991, p. 210) sense of "self-regulation" and "self-determination" may be seen as desirable traits to be inculcated in BANA countries, but elsewhere in the world these may be seen as destabilising forces. In Chinese, the word that is used to translate the English word "individual" is one which bears the connotation of selfishness or self-centredness, concepts which seem to run counter to the sense of self as informed by the wider social context. For Chinese learners, autonomous learning may be seen as an abdication of responsibility by the teacher. Though they may agree to go along with this approach, if a problem is encountered such as failing an assessment, then there is a real possibility that the Ah-Q (defensiveness) characteristic may become evident. In this scenario, if the failed assessment leads to the possibility of the student not receiving their award, the teacher might be blamed for not having taken a more active role in assisting the student. Tied in with this is the notion of face. While the author was teaching in China 10 years ago, students would express surprise that he would be prepared to fail a student. If a student were to receive a fail grade, it might suggest that he would lose face because it would reflect badly on his teaching record. The concept of students taking responsibility for and managing their own learning should probably not be assumed given the wide diversity of students currently studying on online programs.

RESPONSES TO CULTURAL DISSONANCE

If one attributes even partial validity to the issues presented thus far in this chapter, then a response strategy needs to be considered. The fact that these teaching approaches are at the very heart of the higher education system in BANA countries and are reinforced by assessment and quality assurance mechanisms means that options for change are relatively limited. In this final section of the chapter, the author proposes three possible responses available to staff.

1. **Non-accommodation response:** The first response is already very common in face-to-face higher education. The response is based on the notion that the teaching approaches and the educational philosophies which underpin them are part of our own higher education heritage and that these approaches have developed for sound reasons. The argument runs that BANA countries offer world-class higher education and state-of-the-art delivery of online learning and this is proving to be a great attraction to international students from across the world. In this view of Western higher education as a model of success, the learner must simply adapt to the learning context. In the strongest non-accommodation response model, no support is offered to help underpin the acculturation process. As evidence, proponents would point to the large number of students successfully completing degree programs online each year. Critics of this response, on the other hand, would point to the methodological problems associated with tracking studies (e.g., How do we measure "success?") and the fact that higher education is becoming more client-driven with a greater emphasis on the overall student experience. Despite these criticisms, in many institutions in

BANA countries this seems to be the default response.

2. **Intervention response:** The intervention response shares many of the same characteristics as the non-accommodation response, promoting the notion that teaching approaches should remain unchanged. Where it differs is in the view that differences in the educational culture need to be acknowledged and addressed by some form of intervention strategy. The intervention typically forms part of the induction program, either prior to or as the learners start the degree study online. While this response is more accommodating of learner diversity, critics might point to the resource implications in the design and delivery of an online support module. Others suggest that drawing a distinction between home and international students in the delivery of support is unnecessary.

3. **Modification response:** The modification response is without a doubt the most far-reaching of the three responses. The modification response is based on the notion that the educational philosophies which inform teaching approaches in BANA countries need to be re-evaluated and possibly modified. The argument is that given the fact that a significant and growing proportion of the online student body have been educated in educational cultures which do not share the same values, we need to consider issues of accessibility. One practical measure would be to reevaluate the emphasis placed on certain teaching approaches such as reflective practice. To what extent are these approaches simply the ones in vogue? Critics of this response point to the importance of stakeholders in the educational process, in particular the role played by employers in determining the type and range of skills they desire in graduates. A further criticism

focuses on the issue of change management, the argument being that change may not be practicable via a unilateral approach given the increasing role played by quality assurance agencies in higher education, who increasingly assess programs according to what is seen as best practice in teaching approach.

CONCLUSION

Cross-cultural research is notoriously prone to oversimplification of issues, ethnocentrism, and stereotyping, and the author is conscious of the fact that in this chapter he has been at risk as far as these issues are concerned. This chapter has been very much a call to thought rather than a call to action, though in this chapter it is hoped that the Western reader will see their own teaching approaches reflected. If they are convinced of the possibility of cultural dissonance, then invitation is to consider which response is most appropriate in their particular context. Future research needs to test the theoretical hypotheses presented in this chapter through a large-scale empirical study of an online degree context. With the expansion in international student numbers in online degree programs set to continue in the near future at least, this is an area of research whose time very much has come.

REFERENCES

Allik, J. M., & McCrae, R.R. (2004). Toward a geography of personality traits: Patterns of profiles across 36 cultures. *Journal of Cross-Cultural Psychology, 35*(1), 13-28.

Atkinson, R. C., & Shiffrin, R. M. (1968). Human memory: A proposed system and its control processes. In K. Spence & J. Spence (Eds.), *The psychology of learning and motivation: Advances in research and theory (Vol. 2).* New York: Academic Press.

Chan, B. (2002, October). A study of the relationship between tutor's personality and teaching effectiveness: Does culture make a difference? *International Review of Research in Open and Distance Learning, 3*(2).

Cheng, X. T. (2000). Asian students' reticence revisited. *System, 28*, 435-446.

Cheung, F. M., Leung, K., Zhang, J. X., Sun, H. F., Gan, Y. Q., Song, W. Z., & Xie, D. (2001). Indigenous Chinese personality constructs: Is the five-factor model complete? *Journal of Cross-Cultural Psychology, 32*(4), 407-433.

Dalgarno, B. (2001). Interpretations of constructivism and consequences for computer assisted learning. *British Journal of Educational Technology, 32*(2), 183-194.

Definitions of Open Learning (n.d.). Retrieved June 20, 2005, from http://www.unesco.org/education/educprog/lwf/doc/portfolio/definitions.htm

De Vita, G. (2000). Inclusive approaches to effective communication and active participation in the multicultural classroom. *Active Learning in Higher Education, 1*(2), 168-180.

Distance Education Online Symposium Listserv (n.d.). Retrieved September 15, 2005, from http://www.ed.psu.edu/acsde/deos/deos-l/deosl.asp

Fox, R. (2001). Constructivism examined. *Oxford Review of Education, 27*(1), 23-35.

Hofstede, G. (1991). *Cultures and organizations: Software of the mind.* London: McGraw Hill

Holliday, A. (1999). Small cultures. *Applied Linguistics, 20*(2), 237-264.

IDP Education. (2004). *International students in Australian universities.* Canberra: IDP Education Australia.

Inglehart, R., & Baker, W. E. (2000). Modernization, cultural change, and the persistence of traditional values. *American Sociological Review, 65*(February), 19-51.

Jones, J. F. (1999). From silence to talk: Cross-cultural ideas on students' participation in academic group discussion. *English for Specific Purposes, 18*(3), 243-259.

Kim, K. J., & Bonk, C. J. (2002). Cross-cultural comparisons of online collaboration. *Journal of Computer-Mediated Communication, 8*(1).

Kolb, D. (1984). *Experiential learning: Experience as the source of learning and development.* NJ: Prentice-Hall

Li, M. S. (1999). Discourse and culture of learning—Communication challenges. Unpublished academic presentation. *AARE-NZARE.* Melbourne.

Lin, E. J. -L., & Church, A. T. (2004). Are indigenous Chinese personality dimensions culture-specific?: An investigation of the Chinese personality assessment inventory in Chinese American and European American samples. *Journal of Cross-Cultural Psychology, 35*(5), 586-605.

Littlewood, W. (1999). Defining and developing autonomy in East Asian contexts. *Applied Linguistics, 20*(1), 71-94.

Liu, C. H., & Matthews, R. (2005). Vygotsky's philosophy: Constructivism and its criticisms examined. *International Education Journal, 6*(3), 386-399.

Liu, L. (2005). Rhetorical education through writing instruction across cultures: A comparative analysis of select online instructional materials on argumentative writing. *Journal of Second Language Writing, 14*(1), 1-18.

Matthews, B. (2001). The relationship between values and learning. *International Education Journal, 2*(4), 223-232.

Mayer, R. E. (1996). History of instructional psychology. In E. De Corte & F. E. Weinert (Eds.), *International encyclopaedia of developmental and instructional psychology* (pp. 29-33). Oxford: Pergamon Press.

Parker, S. (1997). *Reflective teaching in the postmodern world: A manifesto for education in postmodernity.* Buckingham, UK: OUP.

Piaget, J. (1953). *The origin of intelligence in the child.* London: Routledge & Kegan.

Pollard, A., & Tann, S. (1994). *Reflective teaching in the primary school: A handbook for the classroom.* London: Cassell.

Ryan, R. M. (1991). The nature of the self in autonomy and relatedness. In J. Strauss & G. R. Goethals (Eds.), *The self: Interdisciplinary approaches.* New York: Springer.

Smith, M. K. (2001). David A. Kolb on experiential learning. *The encyclopedia of informal education.* Retrieved October 28, 2005, from http://www.infed.org/b-explrn.htm

Stockhausen, L., & Kawashima, A. (2002). The introduction of reflective practice to Japanese nurses. *Reflective Practice, 3*(1), 117-129.

Ulrich, R. (1947). *Three thousand years of educational wisdom.* London: Oxford University Press.

Vygotsky, L. (1978). *Mind in society.* London: Harvard University Press.

Wright, S., & Lander, D. (2003). Collaborative group interactions of students from two ethnic backgrounds. *Higher Education Research & Development, 22*(3), 237-252.

This work was previously published in Globalizing E-Learning Cultural Challenges, edited by A. Edmundson, pp. 116-129, copyright 2007 by Information Science Publishing (an imprint of IGI Global).

Chapter 7.6
Inquisitivism:
The Evolution of a Constructivist Approach for Web–Based Instruction

Dwayne Harapnuik
University of Alberta, Canada

ABSTRACT

This chapter introduces inquisitivism as an approach for designing and delivering Web-based instruction that shares many of the same principles of minimalism and other constructivist approaches. Inquisitivism is unique in that its two primary or first principles are the removal of fear and the stimulation of an inquisitive nature. The approach evolved during the design and delivery of an online full-credit university course. The results of a quasiexperimental design-based study revealed that online students in the inquisitivism-based course scored significantly higher on their final project scores, showed no significant difference in their satisfaction with their learning experiences from their face-to-face (F2F) counterparts, and had a reduction in fear or anxiety toward technology. Finally, the results revealed that there was no significant difference in final project scores across the personality types tested. The author hopes that inquisitivism will provide a foundation for creating effective constructivist-based online learning environments.

INTRODUCTION

The purpose of this chapter is to support my claim that inquisitivism (my adaptation of minimalism) is an effective constructivist online learning approach for adult learners who are required to learn new information technologies in a Web-based setting. Inquisitivism has emerged from the author's 10 years of experiences in course development and teaching in online and distance learning environments. Since the fall of 1996, over 3,600 University of Alberta students have completed either the full-credit undergraduate online course EDIT 435 or its graduate equivalent EDIT 535. These courses have been, and are still currently, delivered exclusively online with no F2F interaction.

They are officially called The Internet: Communicating, Accessing and Providing Information (Montgomerie & Harapnuik, 1996, 1997), but are colloquially referred to as "Nethowto," which is also the Web name of the course, and subsequently, the nickname that was adopted by students and faculty. In addition, several other courses based on the inquisitive approach have been designed and delivered by the author in both the academic and professional training environment.

This presentation of inquisitivism, its development, its application, and evaluation findings presented here are not based on a single case study or a "one-off," but are based on a body of data and experiences collected over a 10-year period. The inquisitivist approach was first formalized in 1998 (Harapnuik), was updated in 2004 (Harapnuik), and has been continually revised. Inquisitivism, and its application, continues to evolve in response to the needs of the author's primary academic responsibility—his students.

CONSTRUCTIVIST APPROACHES LIKE MINIMALISM ARE EFFECTIVE FOUNDATIONS FOR DESIGNING TECHNOLOGY INSTRUCTION

There is a body of literature that calls for a change in the way we design and deliver educational material: Objectivism vs. constructivism: Do we need a new paradigm? (Jonassen, 1991), Web-based distance learning and teaching: Revolutionary invention or reaction to necessity (Rominiszowki, 1997), The Learning revolution (Dryden & Vos, 1994), Transforming learning with technology: Beyond modernism and post-modernism or whoever controls the technology creates the reality (Jonassen, 2000), and Beyond reckoning: Research priorities for redirecting American higher education (Gumport, Cappelli, Massey, Nettles, Peterson, Shavelson, & Zemsky, 2002). The authors of these works argue that traditional forms

of instruction are no longer effective. There are also claims that the deficiencies in the outcomes of learning are strongly influenced by underlying biases and assumptions in the design of instruction (Rand, Spiro, Feltovich, Jacobson, & Coulson, 1991). The systems approach to instructional design may be the primary factor contributing to the poor outcomes of instruction since it is still the predominant instructional design assumption used throughout most of education (Carroll, 1990; Dryden & Vos, 1994; Hobbs, 2002; Jonassen, 1997; Newman & Scurry, 2001; van der Meij & Carroll, 1995).

The systems approach is based on the assumption that learners are passive receptacles for information that the instructor (teacher or instructional media) relays. Educators are beginning to recognize:

...that our dominant paradigm mistakes a means for an end. It takes the means or method called 'instruction' or 'teaching' and makes it the end or purpose.... We now see that our mission is not instruction but rather that of producing learning with every student by whatever means work best. (Barr, & Tagg, 1995, p. 14)

Similarly, Carroll (1990) argued against the notion that learners are passive receptacles, and made a case against the systematic approach to learning, in his book the *Nurnberg Funnel*. The title refers to the legendary funnel of Nurnberg that was said to make people wise very quickly by simply pouring knowledge into them. The title is also a somewhat sarcastic accusation against traditional forms of instruction.

In the *Nurnberg Funnel,* Carroll presented minimalism as the culmination of 10 years of empirical research that showed that newer methods of instruction, based on constructivism and other cognitive theories or approaches, perform much better than the commonly used systems approach to instruction. Constructivists posit that knowledge is constructed, not transmitted, and that it results from activity. They also hold that knowledge is anchored in the context in which learning occurs, and that "meaning making" is in the mind of the knower, which necessitates multiple perspectives of the world (Jonassen, 1990, 1991, 1997). Meaning making is prompted by problems, questions, confusion, or even disagreement, and this meaning making is generally distributed or shared with others through our culture, tools and community (Jonassen, 1990, 1991, 1997; Kearsley, 1997; Strommen & Lincoln, 1992; Vygotsky, 1978).

Carroll's (1990) research revealed that instruction based on guided exploration (GE) was significantly more effective than the traditional systems approach. Out of a group of 12 participants at the IBM Watson research facility, 6 used (GE) cards, and the other 6 were given the traditional systems-style manual (SM). Both groups were expected to complete their respective training by working through either the drill or practice of the systems-style manual, or the 25 GE cards. Both groups were evaluated by being required to complete a real task of transcribing a one-page letter into a word processor and printing it out. The participants were asked to think aloud, and research associates recorded their thoughts. In addition, the sessions were videotaped so that all the data could be collated and taxonomized to develop a qualitative picture of how GE learning was contrasted by SM learning.

The use of guided exploration cards resulted in much faster initial learning, and more successful performance in the achievement task. The learning time for the GE participants, on average, was less than half of what it was for their SM counterparts: 3 hours and 55 minutes vs. 8 hours and 5 minutes (Carroll, 1990). Similarly, GE participants spent half as much time on the achievement task as did their SM counterparts, and the GE group achieved much greater success than the SM group. The GE group spent more time working on the actual system trying out more operations than the SM group, who spent most of their time reading about the system. Not only did the GE group work effectively with the operations needed to complete their task, they experimented with many more aspects of the system.

Carroll (1990) argued that the GE group was more successful because they worked with the system itself, and took responsibility for their own learning. They demonstrated much more initiative, and used errors as learning experiences. In contrast, the SM group often became trapped in error loops created by the systems-style manual. The problems the SM group experienced with the instructional material hindered or, in some cases, even prevented the learners from working with the system they were attempting to learn.

Carroll (1990, 1998) also argued that there is a need for a change in the way instruction is developed and delivered, and offered minimalism as a viable option for this change. An examination of the learning theory literature also reveals many theories and approaches to learning. A partial list includes structuralism, functionalism, connectionism, behaviorism, objectivism, and constructivism. When you add all the other theories that are not suffixed with an "ism" (classical conditioning, information processing model, etc.) there are over 50 learning theories and approaches (Kearsley, 1997).

Perhaps one reason that there are so many theories and approaches is that their authors have also sought out theories to substantiate or validate their research, and they, too, found that there was no single theory or approach that accurately supported or represented their work. When a suitable comprehensive theory or approach is not found, it is not uncommon for the researcher to propose new concepts, and combine elements of other theories and approaches into a new approach that could be applied specifically to a unique situation. This partially explains the creation of the inquisitivist approach.

DEVELOPMENT AND EVOLUTION OF INQUISITIVISM

Inquisitivism is a descriptive approach to designing instruction. It shares many of the same principles as minimalism, but offers two key principles, or components, that set it apart. These two principles are codependent in the sense that the second principle cannot be realized without the first. The first principle of the inquisitivist approach is the removal of the fear that many adults have when first faced with learning to use technology. Many adults who are new to technology are virtually paralyzed when placed in front of a computer. The fear of "breaking something," or perhaps the fear of looking or feeling foolish, often prevents these adults from embracing computers and technology (DeLoughry, 1993; Shul & Weiner, 2000).

The second, most significant, or dependent, principle is the stimulation of inquisitivism. By designing instruction that reduces the "hurt level" and encourages the "HHHMMM??? What does this button do?" approach/attitude to learning, adults can be encouraged to learn in a similar fashion to how children learn (Harapnuik, 1998). Exploring and discovering the power and potential of computers, and technology in general, can be an exciting and stimulating process, if the learner is confident that they "can't break the system" or that the system "won't break them." With fear reduced and the inquisitive nature stimulated, it can be argued that adults can have almost the same level of success with technological learning as children. An inquisitivist approach to learning technology is essential because technology is dynamic and is rapidly changing, forcing learners to continually adapt to these changes.

Another significant factor about inquisitivism is that the approach was developed (and continues to evolve) during the development and continued delivery of the Nethowto Web-based course. The development of the inquisitivist approach was a practical response to a need, and was the result of a search for a theoretical foundation for the design, development, and delivery of the course. As *Nethowto* evolved, it became clear that many of the principles that ultimately became foundational to inquisitivism were at work in the development of the course.

In 1997 and 1998, the third and fourth year the Nethowto course was delivered, and the second and third year it was delivered exclusively online, the minimalist approach was researched, and even though it was originally designed as an approach for document design, components of its rubric seemed very appropriate to, and were applied to, Nethowto. During this time, it became apparent that even though minimalism satisfied many of the instructional design needs of Nethowto, and had the potential of providing a sound theoretical foundation for the course, it was lacking in two key areas: fear removal and social interaction. Kearsley affirmed the "solid theoretical foundation for minimalism" (Kearsley, 1998, p. 395), but also pointed out that it does have theoretical gaps. The most significant gap in minimalism is that it does not address the social aspect of learning (Kearsley, 1998). A lesser gap is that minimalism has not been tested in a variety of media, specifically online systems. As a result, the adaptation of minimalism proceeded, and inquisitivism was formalized in 1998 (Harapnuik, 1998). Table 1 offers a comparison of inquisitivism to the constructivist learning environments (CLE) and minimalist rubric from which it ultimately evolved.

It must be noted that many of the same principles apply to all three approaches. For example, all three approaches share the need for students to work on real-world tasks in genuine settings. As would be expected of constructivist approaches, all three emphasize knowledge construction, whether it is called reasoning and improvising, or discovery learning. Since inquisitivism is an adaptation of minimalism, it shares even more of

Table 1. Comparison of constructivist learning environments, minimalism, and inquisitivism

Constructivist Learning Environments	Minimalism	Inquisitivism
Provide multiple representation of reality	Reasoning and improvising	Fear removal
Avoid oversimplification of instruction by representing the natural complexity of the real world	Getting started fast	Stimulation of inquisitiveness
	Training on real tasks	Getting started fast
Present authentic task (contextualizing rather than abstracting)	Using the situation	Using the system to learn the system
	Reading in any order	Discovery learning
Foster reflective practice		
	Supporting error recognition and recovery	Modules can be completed in any order
Focus on knowledge construction, not reproduction		
	Developing optimal training designs	Supporting error recognition and recovery
Enable context-dependent and content-dependent knowledge construction	Exploiting prior knowledge	Developing optimal training designs
Support collaborative construction of knowledge through social negotiations, not competition among learners for recognition.		Forum for discussion and exploiting prior knowledge
		Real world assignments

the same principles. Inquisitivism is continually evolving, but there are currently 10 key concepts/components that make up the approach.

Application of Inquisitivism to Nethowto

Carroll (1990) stated that taking checklists seriously is perhaps the most typical and debilitating design fallacy. Despite this strong statement, Carroll provided a rubric of minimalist principles. Similarly, inquisitivism has evolved into an approach with a rubric of principles. An early version of the following 10 principles was applied to the Nethowto course during a significant redesign of the course in the fall of 1998. It must also be noted that the course is still running, and both course and the 10 principles have continued to evolve.

Fear Removal

Dealing with the paralyzing fear that many adult learners experience must precede the stimulation of one's natural inquisitiveness. Demonstrating that the computer or any other piece of technology is not fragile, providing explanations, examples, and solutions for common errors and problems, and the application of data backup will help quell the adult learner's fear.

In an asynchronous education and Web-based environment, an instructor is not able to interact directly in person with an entire class (i.e., some students may be working in a different time zone) and to reassure the group as a whole. Nor can an instructor gauge body language, or tone and inflection of voice, to detect that fear may be an issue. Furthermore, both e-mail and Web-based conferencing interactions, which are essential to Web-based learning, are not direct forms of interaction, but are considered mediated transactions (Harasim, 1993; Lapadat, 2002). Because of these dynamics, fear or anxiety removal is perhaps one of the most challenging components to effectively facilitate, primarily because the F2F cues are missing, and students cannot be led through their anxieties. Using video or audio files to present what would be presented in a traditional F2F setting was, until recently, not a feasible option. While it is possible to use compressed video or audio to communicate with students now, there still is the issue of getting students over the initial fear or anxiety that they may have to operate this type of software for the very first time.

Because of these limitations, the asynchronous nature of the course, and the need to keep pages small to load quickly, the actual design and layout of the course's main Webpage had to be a primary factor in calming the fearful student. The main page (and the entire site for that matter), by design, is very simple and uncluttered. Students are not overwhelmed by choices on the main page, and a large "Getting Started" heading was strategically placed to be one of the first items noticed on the page.

The actual Getting Started instructions (referred to as First Steps) were broken down into four simple steps. The items in the four steps were designed to lead a student through the initial familiarization with the course. Students were not required to actually complete any assignments, but were still required to familiarize themselves with the course navigation and layout, to fill out a consent form (data was also used to create student profiles in the course administration system), to join the course conferencing system and, finally, review the introduction module.

The intention of the Getting Started page was that by following the four steps, fearful students would gain enough experience and success with the course to help them overcome or, at minimum, deal with their fear. While these four steps appear to be a linear systematic instruction (SI) type system superimposed on a minimalist structure,

students can do the steps out of sequence, or ignore them all together, and still proceed through the course, so the sequencing aspect of SI is not a factor in student progression. At some point, and in some order, students will have to fill out the consent form, join the conferences, and begin work on the introduction module. These instructions are simply presented in their most logical order. Throughout the steps, students were encouraged to contact the instructor directly if help was needed. Students had (and currently still do have) access to the course instructor via e-mail, the Web-based conferencing system called the WebBoard (WebBoard Collaboration Server, 2005), and by phone.

Stimulation of Inquisitiveness

With the fear abated, the adult learner's intrinsic (but often suppressed) inquisitive nature can be stimulated and encouraged to flourish. Nethowto students are actually encouraged to read the "HHHMMM??? What does this button do?" approach article that is linked on the main page. The article details the 10 inquisitivist principles, and makes an argument for this approach as the basis for Web-based instruction.

The design of the course forces the students to make many more decisions, and to extensively investigate and use computer programs more than they are often used to. For example, in the first formal assignment, students are asked to submit an e-mail attachment, but they are not required to use a specific e-mail client or word processor. Students are directed to resources that they can use to learn about e-mail, e-mail clients, and the sending of attachments. In addition, students are required to investigate one aspect of attaching documents that most people take for granted, the encoding format. The only way that students can be sure that they submit an attachment in the required MIME encoding format is to explore the online Help within their e-mail clients or on the Web. This starts the whole inquisitivist process.

Students quickly learn that a small amount of investigation within the programs they are currently using will reveal the results that they need. The immediate success students experience is a crucial aspect of inquisitivist design that will be further expounded in the getting started fast category.

Using the System to Learn the System

All training must take place on the actual system that is being learned. Every aspect of Nethowto is conducted online. Students are actually using the Internet, while learning about all forms of Internet communication, and accessing and sharing of information. In addition to the students conducting all aspects of the course online, the instructor of the course (the author) does not maintain an office at the University of Alberta campus, but conducts all aspects of design, development, and delivery of the courses completely online. In essence, the instructor uses the system to teach the system.

Getting Started Fast

Adult learners often have other interests than learning a new system. The learning they undertake is normally done to complement their existing work. The "welcome to the system," prefaces, and other nonessential layers in an introduction, are often ineffective uses of the learner's valuable time.

The Getting Started/First Steps sections of the course are designed to give students confidence in their initial experience with the course. The simple procedures that students are asked to follow, like joining the course conferencing system and using an online form to submit their student information, contribute positively to their learning experience. Similarly, all the information that students are required to review in the Getting Started section of the course is intended to contribute immediately and positively to their

learning experience, and ultimately, give the learner confidence in the system.

The first assignment, submitting an e-mail attachment, is relatively simple to complete, and is strategically placed and used to give students immediate success. Students usually make the e-mail submission immediately after moving through the Getting Started section. A concerted effort is made to insure that students receive an immediate reply, and have rapid confirmation of their success. Students who have difficulty with the assignment are quickly directed to the resources that they need to use to have success in the assignment. The goal of the instructor is to reply to students within 3 to 4 hours of their first assignment submission (if the assignment is submitted during regular business hours, the reply is often processed in a matter of minutes).

Discovery Learning

There is no single, correct method or procedure prescribed in the course. Allowing for self-directed reasoning and improvising, through the learning experience, requires the adult learner to take full responsibility for their learning.

Throughout all course modules and course work, students are given specific assignment requirements that specify what should be submitted or included in the portfolio. Nethowto students are also given the freedom to choose the programs they use to complete the assignments. Unlike many technology-related courses that provide step-by-step instructions on conducting a specific procedure with or within an application, students are pointed to Web-based resources that deal more with the general concept than with the specifics of a particular application. This is not to say that step-by-step instructions are not necessary. There is a section of each module that points to links for the more common applications used in the course (FTP, Telnet, Text or HTML editors, etc.) that do provide the step-by-steps instructions for those who are most comfortable with this form of instruction, or are not comfortable with learning by doing, experimenting, or exploring.

All module coursework culminates in the course portfolio in which students have to display all they have learned in a Web site (part of the learning process is learning HTML). Students are told what is to be included in the portfolio, but are not explicitly instructed on how it should be created or formatted. Instead of a rigid recipe or formula, students are given the freedom to construct their portfolio in any way they choose. Links to instructional sites on HTML, Web design, graphics utilization, and usability are provided, but students are still required to learn the application of the technical aspects of creating a Web site to create their portfolios and projects. Marking guides, (details on what markers will be looking for) and examples of previous student work are provided to offer students additional guidance on what is ultimately expected. Although many students simply copy the format of previous student work, some students embrace this freedom, and come up with innovative ways to display their portfolios. These innovative portfolios are often included in the examples, but unfortunately, most students choose the safety of copying the simple or tried and true designs.

Modules Can be Completed in any Order

Materials are designed to be read or completed in any order. Students impose their own hierarchy of knowledge, which is often born of necessity and bolstered by their previous experience. This helps to eliminate the common problems that arise from material read or completed out of sequence.

Providing a structure for openness requires a great deal of planning and structure. The course is modular, and each module, except for the portfolio, which is a compilation of all other modules, can be completed in any order. The module naming conventions do not include numbers or alphabets, to prevent any suggestion of a specific order. Despite the effort to not prescribe an order, and even though the modules can be completed in any order, most students follow the sequential listing of assignments in the course navigation structure. This, too, is part of the design. This order has been established for those students who lack confidence or experience with technology. By following the sequence of modules, students who lack technology confidence and experience can gain enough confidence and experience from the modules to successfully complete the portfolio and final project. While this sequential ordering of the modules may appear to be a linear SI type system superimposed on a minimalist structure, students can still do the modules out of order, so the sequential ordering of the modules is not as significant as it would be in a true SI system. Due to the very divergent capabilities of students in the course, the structure of the course has to serve both students with little experience, and those who may be very experienced. Students who need the order and structure can use the implied order from the navigational listing, and students who have the confidence to work on course modules in their own order have the freedom and opportunity to do so as well.

It must be acknowledged that even though there is no required order for completing the modules, the portfolio does require that the other minor assignment modules be completed first. A hierarchy of knowledge for the course is imposed by the two main course assignments. In order to complete the portfolio, students must learn HTML (hypertext mark up language) and complete the other assignments. In order to complete the final projects and earn a satisfactory grade, gaining experience in HTML development (either with a text or HTML editor) through building the portfolio is the most logical path for students to follow.

Supporting Error Recognition and Recovery

Errors must be accepted as a natural part of the learning process. Since there is such a pervasiveness of errors in most learning, it is unrealistic to imagine that errors can be ignored. Error recognition and recovery strategies need to be implemented to enable learners to learn from their mistakes, instead of being trapped by them. The use of FAQ's, Help Forums and other help strategies should be implemented to deal with the errors and problems that arise.

Once again, the asynchronous nature of Nethowto necessitates that the course itself provides support for error recovery. The Help link is strategically placed 1/3 of the way down the page and in the center (which is the area of the screen where a users eyes will first fall). The Web-based conferencing system and the Help conferences are also readily available. An online FAQ, and multiple admonitions to ask for help, are placed strategically throughout the course.

In addition to the actual design, layout, and structure of the course, the students are given immediate feedback (usually within minutes or, at most, hours) on their first assignments, and also receive detailed feedback (complete with written explanations) as to what mistakes were made on their portfolios. Students are encouraged to learn from their mistakes in the portfolios, and apply what they have learned to the final project. Students are given the option of submitting their portfolios 3 weeks prior to the end of term to receive an evaluation that will help prevent them from making the same errors on their final project that they made on the portfolios, and to give them a better of understanding of what is expected in the creation of a Web site.

When the students contact the instructor for help, they are first directed to the location in the course pages where the answer may lie. If the students report that they had reviewed the support material and were still not able to find a solution to their problems, they are then directed to additional support material where the answer could be found. If the additional support materials were not adequate, the students are then directed to even more information to help them determine the answer on their own. It is extremely important for the instructor to judge the level of frustration students may be experiencing and, if necessary, give them a direct answer sooner than later.

To insure that students help needs are met, all students are regularly queried about the course Website, and asked for suggestions on making changes to the course that would save them from having to contact the instructor, or use the Help forums for assistance.

FORUM FOR DISCUSSIONS AND EXPLOITING PRIOR KNOWLEDGE

Adult education dealing with technology is often conducted through alternative delivery methods. Distance education, Web-based instruction, and other alternative delivery methods can isolate students. Providing a conferencing system for the replacement of F2F interaction is a crucial component of any alternative delivery program. Most adult learners of technology are experts in other areas or domains. Understanding the learner's prior knowledge and motivation, and finding ways to utilize it is one of the keys to effective adult training. In addition, adult learners can share their expertise, or assist each other, and should be encouraged to use the conferencing system to facilitate social interaction.

The WebBoard conferencing system is an effective forum for enabling students to provide each other with assistance. To encourage students to assist each other (not an easy thing to do in a competitive academic environment where students strive to be at the top of departmental or faculty mandated marks distributions) students are assessed a Help participation mark based on the quantity and quality of their participation

—this mark is worth 10% of their final grade. One of the most common responses to the Help forums is how useful and helpful they are. It is not uncommon for a number of students in each session to state: "I could not have made it through the course without the Help forums." In addition to help related issues, students are required to start a topic discussion on an area that they are particularly interested in. This topic discussion is also required, and contributes toward the student's Issues participation mark.

The WebBoard forums are an example of what Vygotsky (1978) coined as social learning. In his theory, he stresses that social interaction is a critical component of situated learning because learners become involved in a "community of practice," and adopt the beliefs and behaviors of that community. Experts (experienced individuals) within the community often share the beliefs and behaviors of the community unintentionally, or model the proper conduct through their behavior. Newcomers interact with the experts, and then they themselves move into the community to become experts. This process can be referred to as legitimate peripheral participation, and occurs unintentionally (Lave & Wenger, 1990).

Some students who admit (in the WebBoard forums) to being normally reserved, or who might not even participate in a F2F setting, are encouraged by the equality they find in the WebBoard environment, and embrace this component of the course. It is not uncommon for these students to log on daily, and to participate in most (if not all) discussions. Students who may be near completion of the course often provide encouragement to students who have joined the course late or have simply started late. This exchange of information and knowledge, and sense of community, is one of the most positive aspects of this course. It is not uncommon for some students to go out of their way while traveling, to find a computer to log on and continue to participate in their special virtual community.

Despite never meeting the students F2F, it was possible for me to get familiar with the students through monitoring their e-mail and Web-based conferencing communications. In one sense, it may be easier to get a better understanding of a student's personality and needs than in a F2F setting because of monitoring all their Web-based communications. This advantage over the F2F setting is offset by the disadvantage of not being able to read students' nonverbal expression, body language, and general reactions.

Real World Assignments

"Make-work" (purposeless) projects are often not an effective use of a student's valuable time. All assignments must have a real world application.

All Nethowto assignments are genuine "real world" tasks that almost any information professional that uses the Internet as a tool would do on a daily basis. The Internet offers much more than the just the Web or e-mail, and students are required to use a variety of the Internet tools (Listserv, Usenet, Telnet, FTP, IM, HTML and Search engines) to complete their assignments, which focus on the information that can be gathered, shared, or moved using the assortment of Internet tools, rather than focusing on the tools themselves. The goal of the course is to give students experience in communicating, accessing, and providing information on the Internet. The emphasis is on the information, and not the tools used to access or provide the information. Technology is put in its place, and is relegated to its rightful role as an information access tool.

Optimal Training Designs

Feedback facilities, like online surveys or e-mail, should be used to allow learners to immediately provide feedback on any aspect of a program. Problems with instructions, assignments, wording, or other problems, should be immediately addressed and corrected. Instructional models are not deductive or prescriptive theories: they are descriptive processes. The design process should involve the actual learner through empirical analysis, so that adjustments can be made to suit the learner's needs. "Develop the best pedagogy that you can. See how well you can do. Then analyze the nature of what you did that worked." (Bruner, 1960, p. 89)

The Nethowto course has evolved to its present state because of the students who have worked through the course and provided feedback. Student feedback is immediately acknowledged, and if a particular portion of an assignment instruction (or any portion of the course for that matter) requires modification to bring clarity, this is done immediately. If the same questions are asked repeatedly, the subject of those questions is addressed, and that aspect of the course is modified to provide less confusion and to improve clarity. When significant changes are made as a result of student's feedback, announcements are made on the course News and Announcements page, to insure that all students are made aware of the change. Designing and developing an effective learning environment is a dynamic process that requires immediate responses to problems that arise. Students are encouraged to fill out detailed, online evaluation forms that provide additional information for continued improvements.

Delivery of Nethowto

Because the inquisitivist approach was developed through the delivery of the Nethowto course, it could be argued that the inquisitivist approach is not only an effective approach for the design of Web-based instruction, but it is also an effective approach for the delivery of Web-based instruction.

Another factor in the delivery of Nethowto is that the instructor (the author) does not maintain an office on the University campus, but works at a distance, and uses the same Internet tools that the students are required to use. Because the system (the Internet) is not only being used by the learners to learn the system, but also by the instructor to teach the system, the students are not asked, or required, to do anything that is not practical, or that is simply not possible with the Internet. Leading or teaching by example is often one of the most effective ways to lead and to teach. When the students learn that their instructor not only "talks-the-talk" but also "walks-the-walk" and is sensitive to the genuine problems that arise with Web-based instruction and communication (in the case of the instructor, telecommuting) because the instructor uses the same system that they do, attitudes toward the course and this approach to learning tend to become quite positive.

Necessity often breeds ingenuity. The evolution of the inquisitivist approach is tied so closely to the design, development, and delivery of Nethowto that one could argue that the approach itself evolved out of necessity. The 10 components of the inquisitivist approach are evident in the design and delivery of Nethowto (some more so than others), and while some of the components may be applied more effectively than others, they all combine to provide an approach to Web-based instruction that is practical and effective for the students and the instructor.

EVALUATION OF INQUISITIVISM

The evaluation of inquisitivism involved two phases, and employed both quantitative and qualitative measures. In the first phase, a quasiexperimental design (nonequivalent groups design) method was used to compare the grades of the final projects produced by a sample of Nethowto and comparison group students, and a comparison of the scores of the level of student satisfaction collected from both groups. The mark on the final project was used as a measure of student success in learning the concepts taught in the course, and ultimately, as a measure of the effectiveness of the instructional approach. Both the Nethowto sample and the comparison group involved undergraduate students enrolled in courses that had very similar content. Both the Nethowto and comparison group courses were designed to increase student Internet experience, knowledge, and communications skills.

To determine if students in the inquisitivist based Nethowto course had a reduction in fear of technology, students from both the groups were asked to complete three questionnaires: computer anxiety rating scales (CARS), computer thoughts survey (CTS), and general attitudes toward computers scale (GATCS) prior to the start of the course, and once again upon completion (Rosen, Sears, & Weil, 1987; Rosen & Weil, 1992).

The Nethowto sample differed from the comparison group in that they were required to take their course, while the Nethowto group chose to take the course as an elective. A second difference was that 45% of the comparison group students had taken one or two computer courses, and the rest of the comparison group had even more formal computer training (one student had a computer certificate). In contrast, 55% of the Nethowto group had no formal computer training, and the remaining students who did have formal computer training had taken only one or two courses. In addition, the comparison group was slightly younger (29 vs. 33), had a higher number of single students, with an even lesser degree of dependence (children). Another difference noted was that over half of the comparison group did not work, and the remaining portion only worked part-time. In contrast, over two thirds of the Nethowto group worked either full- or part-time. Finally, the Nethowto class was taught in conjunction with a graduate level class, which resulted in undergraduate and graduate student interaction.

The second phase of the evaluation included a student satisfaction analysis that was conducted over multiple sections of the Nethowto course, over a span of 4 years. This phase of the study also involved using the Keirsey Temperament Sorter (similar to the Myers-Briggs Type Inventory) to determine for what personality type inquisitivism is more appropriate. Both aspects of this secondary evaluation were only applied to Nethowto students.

Academic Success Comparisons

To compare the results of the final project scores for the Nethowto and the comparison group, Web sites, submitted by students from both groups, were evaluated on the same criteria. The mark on the final project was used as a measure of student success in learning the concepts taught in the course, and ultimately, as a measure of the effectiveness of the inquisitivist approach. Evaluators, who were "blind" to the group membership, used the same evaluation criteria given to students in both the Nethowto and comparison group, and scored the Web sites. The final project Web sites were scored out of 50 points, which was based on an assessment of the project's purpose, relevance, appearance, navigation, organization, level of difficulty, and content. Students were allowed to choose their own topics for the final project, to insure that motivation for the projects was high. One of the goals of the final project assignment was to demonstrate that the students could take all their newly acquired Internet skills and apply what they had learned in the course through the construction of a Web site. Assuming that this goal was met, and that students did demonstrate what they had learned in the course, the mean score of 37 (74%) on the final projects for Nethowto students demonstrated that these students had learned the course content, and were able to demonstrate their newly acquired abilities in the final project.

The first research hypothesis was whether students who learned the same course content via the Nethowto course would do better on the final project as those students who learned in a F2F model. The null hypothesis is rejected because an independent t-test (Table 2) revealed that there

Table 2. Final project scores for the Nethowto and comparison groups

	Nethowto (n = 54)	Control (n = 23)
Mean	37.27	28.96
Std. Deviation	4.69	4.32
Std. Error Mean	.64	.90
	t-test	
t	7.18	
df*	75	
Sig. (2 tailed)**	.003	
Mean Difference	8.21	
SE Difference	1.14	

*Equal variances
**$p < .05$

is a statistically significant difference between the mean final project scores for the Nethowto (M=37.27, SD=4.70) and comparison group course (M=28.96, SD=4.32), with the Nethowto students scoring higher.

Student Satisfaction Comparisons

To assess the level of satisfaction with their learning experience between the two groups, the means of the response to "Overall, this was an excellent course" were compared. Students in both the Nethowto and comparison group were given an universal student ratings of instruction (USRI) evaluation (University of Alberta Computer Network Services, 2004) form that included eight questions, near the end of the course, to assess the instruction they had received, and to assess how satisfied they were with their learning experience. The very short instrument (eight questions), the fact that students were still actively working on the course, and the comparison group's instructor having his students fill out the questionnaire during class time resulted in a high response rates for both the Nethowto and comparison groups.

Both groups indicated that they agreed that this was an excellent course: Nethowto student's average response to the question was 4.24, and the comparison group student's average response to the same question was 4.13.

An independent t-test (Table 3) demonstrates that there is no statistically significant difference between the mean final project scores for the Nethowto (M=4.24, SD=0.82) and comparison group course (M=4.13, SD=0.81), and we, therefore, fail to reject the null hypothesis. The lack of significant difference indicated that even though the Nethowto group satisfaction scores were slightly higher, the difference was not significant enough to argue that the Nethowto group was more satisfied with their learning experience.

Expanded Student Satisfaction Results

In addition to comparing the sample and comparison group results, the results of student evaluations of Nethowto undergraduate students, in multiple sections of the course, spanning a 4-year period, were examined. This supplement has been included to provide a broader perspective on the student satisfaction levels of Nethowto students over an extended period of time. It was also made possible because of the data collection instruments established when the course was originally set up, and that were unaltered in order

Table 3. Course satisfaction scores

	Nethowto (n = 54)	Control (n = 23)
Mean	4.24	4.13
Std. Deviation	.82	.81
Std. Error Mean	.11	.17

	t-test
t	.54
df*	75
Sig. (2 tailed)**	.59
Mean Difference	.11
SE Difference	.20

*Equal variances
**p < .05

to collect longitudinal data for future research. The Nethowto course remained fundamentally the same in terms of design, content, and delivery over this 4-year period. The changes or improvements made in the course during this time dealt primarily with issues of content clarity, and also reflected responses to changes in updates in software applications and systems.

Slightly more than 36% of Nethowto undergraduate students, from multiple sections of Nethowto, filled out a postcourse questionnaire over a 4-year period, resulting in a sample size of 258 for this analysis, resulting in an n of 258 for this analysis. The following six responses (Table 4) were selected and analyzed from the questionnaire because these questions dealt specifically with aspects of student satisfaction. More specifically, the questions dealt with student perceptions on the amount they learned in the course, how satisfied they were with the inquisitivist approach, and if they found the approach effective.

The course satisfaction was measured using a Likert scale (the Likert technique measures attitudes in which subjects are asked to express agreement or disagreement on a five-point scale) with one being the lowest level (strongly disagree),

and five the highest (strongly agree). While the students found they learned a lot in the Nethowto course, they were not as positive with respect to the format and structure in which the course was delivered. Students agreed or strongly agreed that they learned a lot, would be willing to take similar courses online, and perhaps most importantly, agreed that the course helped them to significantly grow in their knowledge of computers and Internet, but they did not agree that the structure was conducive to learning. In addition, a SD of 1.17 on a mean of 2.28 indicated that even though, on average, the student responses were close to neutral or leaned slightly toward disagreeing that they would have preferred to take the course via a traditional lecture/lab format, there was still a significant proportion of students that would have preferred to take the course via a traditional lecture/lab format. This observation is similar to the results of Goodwin, Miller, and Cheetham (1991) and Lake (2001). Their research confirmed that students subjected to active learning instruction would have preferred the more traditional lecture format, despite having achieved greater success.

Table 4. Student responses to questions about their satisfaction

Student response	Mean	SD	n
I learned a lot in this course	4.34	.87	258
I found the structure of the course conducive to learning.	3.85	.99	258
I would take other courses offered in this online, individualized instruction manner.	4.05	1.04	258
This course helped me grow from one level of knowledge about and familiarity with computers and the Internet to a significantly higher level.	4.36	.79	258
I found the learning theory (inquisitivism) used in this course to be effective for this type of instruction.	3.90	.91	258
I would have preferred to take this course via a traditional "lecture/laboratory" mode.	2.28	1.17	258

REDUCTION OF FEAR

The original study design included an analysis of the comparison and Nethowto groups, but because only 4 of the 23 comparison group students who completed the pretest surveys completed the posttest surveys, a comparison between the comparison group and the Nethowto was not possible. While the response rate from the Nethowto course was higher, only 11 out of 54 (20%) students completed the posttest anxiety surveys, and 10 of 54 completed the posttest thoughts and attitude surveys. The poor response rates of these posttest surveys negated any statistically useful data.

In response to this development, additional data were used to determine if there had been a change in anxiety or fear for Nethowto students as a result of the inquisitivist approach in a larger sample. Since the CARS, CTS and GATCS questionnaires, which were established when the course was originally set up, were left in place in order to collect longitudinal data, undergraduate Nethowto students from multiple sessions over a 4-year period were included in this analysis. Of the 479 undergraduate students who completed the Nethowto course during this expanded time frame, 162 students completed the posttest anxiety questionnaire, 168 completed the posttest thoughts questionnaire, and 170 students completed the posttest attitude questionnaire.

The increase in the response rate of 33% of the extended sample, compared to 20% in the original Nethowto sample, could be attributed to students being sent an additional reminder with their final project evaluations to complete the posttest questionnaires, and to an additional reminder being posted on the course conferencing system.

The anxiety levels are represented by a Likert scale with 1 (not at all) being the lowest level and 5 (very much) the highest. The attitudes toward computers are represented in a Likert scale, with 1 (strongly disagree) being the lowest level and 5 (strongly agree) the highest. The thoughts about using computer levels are represented by a Likert scale with 1 (not at all) being the lowest level and 5 (very much) the highest. Questions about thoughts and attitudes towards computers were included in two of the three surveys to help isolate the question regarding anxiety toward technology, and prevent any overlap in student responses.

Table 5 provides the mean scores for pretest and posttest attitudes and thoughts, which are virtually identical, while there is a difference between the pretest and posttest anxiety scores.

Table 6 provides ANOVA results. This analysis provides evidence of a statistically significant reduction in posttest anxiety scores ($p \leq .01$) in the expanded sample. A repeated dependent t-test would have yielded the same result as a repeated

Table 5. Means scores and standard deviations associated with pretest and posttest anxiety, attitudes, and thoughts about computers

Test		Mean	SD	N
Anxiety	Pretest	1.76	.64	162
	Posttest	1.28	.57	162
Attitude	Pre-test	3.13	.39	170
	Post-test	3.14	.35	170
Thoughts	Pretest	2.83	.39	168
	Posttest	2.87	.37	168

Table 6. Sources of variance in pretest and posttest anxiety, attitudes and thoughts about computers

Variance Source	df	MS	F	p
Pretest vs. posttest anxiety	1	2.07	14.01	.004*
Within cells error	161	.15		
Pretest vs. posttest attitudes	1	3.43	.11	NS
Within cells error	169	.25		
Pretest vs. posttest thoughts	1	1.01	.13	NS
Within cells error	167	.22		

*p < .05

measures ANOVA of the means, and could have been used, but an ANOVA was used because it reduces the chance of multiple test error, and reduces Type 1 error. There was no significant difference in the pretest and posttest scores for attitude and thoughts toward technology.

While the hypothesis that students in the inquisitivist based Nethowto course had a reduction in fear of technology is supported in the expanded sample due to the anxiety findings, this result has to be viewed in the context of there being no significant difference in the level of fear of technology in the original sample group.

Personality Type Suitability

To determine if inquisitivism is appropriate for all personality types, Nethowto students from multiple sections of Nethowto were asked, at the beginning of the course, to complete the Keirsey Temperament Sorter (KTS) II (similar to the Myers-Briggs Type Inventory). Temperament type was used as a factor in an ANOVA.

Table 7 includes the Nethowto student final project mean scores, and the standard deviations for each personality type.

Notice the similarity of mean values in the personality types. While there were significantly more artisan (147) and rational (133) than idealist (53) and guardian (40) personality types, there is very little difference in the final project mean scores. An analysis of variance showed that no significant difference exists among the mean scores of the final project for the students with the four different personality types: (F (3/369) = .303, p = .823), and have, therefore, failed to reject the null hypothesis.

These results indicate that since students from all four, personality types scored equally on the final project, the inquisitivist approach would be suitable for all four, personality types tested. Or, more specifically, the inquisitivist based Nethowto course may enable students from the four, personality types to score well in their assignments.

Not only did this study show that the online students did better on their final projects than the

Table 7. Mean scores and standard deviations of personality types of Nethowto students

Personality type	n	Mean	SD
Guardian	40	35.03	3.548
Artisan	147	34.43	4.398
Idealist	53	34.15	5.379
Rational	133	34.52	4.403
Total	373	34.49	4.459

F2F students, it also showed that there are was no significant difference in the levels of learning experience satisfaction between the online students and the students in the traditional F2F classroom. It has also been shown that there was a reduction in student anxiety, and the achievement with the inquisitivist approach did not differ (in terms of final project performance) for the four personality types measured by the Keirsey Temperament Sorter.

NETHOWTO STUDENTS EXCEEDED EXPECTATIONS

The significantly higher final project scores from the online (Nethowto) students can be corroborated by a recent meta-analysis of distance learning research (Allen, Bourhis, Burrell, & Mabry, 2002; Allen, Mabry, Mattery, Bourhis, Titsworth, & Burrel, 2004). The mean scores of the Nethowto students' final projects were 17% higher than the comparison group. This difference is especially surprising given the fact that, on average, the comparison group students had taken more computer courses, and had less work and personal responsibilities.

The difference in scores between the Nethowto and comparison groups could have been attributed to a variety of factors. It may be the case that the Nethowto students motivation to do well in the course was higher because the Nethowto group chose the course as an elective, while the comparison group was required to take their course.

Another factor affecting motivation could be related to the fact the Nethowto group was more mature, had greater marital and family responsibility, and could have been more accustomed to project work and independent learning.

Perhaps one of the most significant factors is time on task, which is a factor often not effectively controlled in quasiexperimental designs of educational research (Joy & Garcia, 2000). By its very design, inquisitivist instruction requires students to use the system while they learn the system. This translates into the Nethowto students spending virtually all their time on the actual task of learning to communicate, access, and provide information on the Internet.

In contrast, the comparison group students had traditional lectures, which meant that even though they could have been listening to Internet related topics, or even discussing these topics, they were not actually working on tasks relevant to learning how to use the Internet. Similarly, the time spent in labs for the comparison group also may not have been considered to be productive time on task, due to the systematic design of the comparison group course. With this design, students worked through lab assignments that followed the traditional step-by-step format. While this type of recipe learning does allow students to successfully complete assignments, it may not effectively foster knowledge acquisition, as minimalism would suggest.

This situation has been evident in the delivery of Nethowto. Some education students, who come into the Nethowto course having completed a prerequisite course that uses the traditional systematic approach, often have problems transferring or applying their experiences from the previous course to almost identical assignments in Nethowto. The only difference in the assignments is that Nethowto assignments do not follow the systematic recipe, and they allow the student to choose the program they should use to complete the assignment. While it must be acknowledged that this data is anecdotal, the incidents where this situation has happened have occurred enough times to warrant reporting and consideration for further investigation.

Another contributing factor that may explain the higher success of Nethowto students is that there could be significantly more direct instructor-student interaction. Direct interactions with the Nethowto instructor either fall into the category of e-mail, Web-based messages replies, or telephone conversations. Since Nethowto is conducted completely online, tracking the e-mail and Web-based conferencing interactions is very simple. On average, Nethowto students have 31 direct interactions with their instructor per session (academic term). The direct responses to student questions in the Web-based conferencing system have the advantage of being available and accessible for all other students to view at any time. Unfortunately, instructor involvement or interaction was not tested in the study, but one can assume that the number of direct interactions were much higher in the online course than they were in the F2F course.

Yet another possible success factor for the Nethowto students that was not controlled or tested was the collaborative aspect of the inquisitivist approach. Nethowto students were required to participate in a Help forum, and 10% of their final mark was also derived from this participation. Another 10% of their final mark was derived from the Issues conference participation, where students were required to start and moderate an issue of their choosing, and were required to participate in issues discussions with other students. In total, 20% of Nethowto students' final marks were from Web-based conferencing participation, so motivation to participate was quite high. While this was not controlled for and not tested, it may be speculated that the help and issues participation contributed significantly to the Nethowto students' acquisition of knowledge and final project success. Vygotsky (1978), and similar social constructivist theorists, stress the significance of social learning, and the transfer of knowledge and expertise through social interactions; therefore, it can be speculated that this dynamic applied.

A final contributing factor to the Nethowto students' success could be their involvement with graduate students in the conferencing component of the course. Since the undergraduate and graduate Nethowto students participated in the same conferencing forum, it may be the case that the graduate students attitude toward learning could have positively affected the undergraduate students.

While the author would like to posit that the inquisitivist approach was primarily responsible for the Nethowto student success, the aforementioned speculated factors need to be tested in further research. Regardless of the reason for their actual success, Nethowto students appeared to have learned the course material, and appeared to be satisfied with their learning experience.

NETHOWTO AND F2F STUDENTS LEARNING EXPERIENCE SATISFACTION

Evidence showed that there was no significant difference in the learning experience satisfaction between Nethowto students and the comparison group students. The differences between the Nethowto and comparison group satisfaction mean scores were slight, with the mean scores for the Nethowto group being slightly, but not statistically significantly higher. In addition to students being satisfied, it can be shown that Nethowto students believed that they learned a lot, and that their knowledge grew significantly. The evidence from the supplemental questionnaire given to the Nethowto students suggests that the students not only learned a lot, they agreed that the course helped them to grow from one level of knowledge and familiarity with computers and the Internet to a significantly higher level.

The only question that did not have a clearly positive response was the question of whether or not students would have preferred to take the course via a traditional lecture/laboratory mode. Even though, on average, the student responses were close to neutral, or leaned toward disagreeing that they would have preferred to take the course via a traditional lecture/lab format, there was still a significant proportion of students that agreed, and would have preferred to take the course via a traditional lecture/lab format. Similarly, the average student response, which was slightly more positive than neutral toward the online format, the wide spread, indicated by a large standard deviation (1.17), suggests that significant numbers of students would have preferred the traditional format. The slightly positive leaning toward the online format may be accounted for by the fact that approximately half the students in the course were true-distance students, and had no choice in the format of their instruction, or were accustomed to the online format. In contrast, approximately half the students in the course were nondistance students accustomed to attending traditional classes on campus. The students who indicated a preference toward the traditional lecture/lab format, may have done so because they were accustomed to this form of instruction, or they simply found traditional instruction easier, and were more comfortable following a recipe. It may also just be the case that students simply do not like active learning. These factors could be taken into account in further research.

OVERCOMING INQUISITIVIST APPROACH CHALLENGES

Even though the data reveals that students in the Nethowto course performed very well in their final projects, were as satisfied with their instruction as the comparison group, and it appears the inquisitivism is suitable for the four measured personality types, there are still challenges to the approach. For example, one of the most interesting paradoxical situations is that too many questions are asked by students who have simply not even read any of the instructions, and at the same time, not enough questions are asked students who are looking for the hidden challenge to the course. Another paradox involves encouraging student participation in the course conferencing system, while at the same time limiting excessive participation and competition. One of the most perplexing challenges is addressing the unique instructional needs of the vast diversity of students who take the course. Rather than view these issues as obstacles, these issues should be, and are, viewed as opportunities to make improvements in the design and delivery of Nethowto. Addressing these challenges, and many other challenges that have arisen in the development and delivery of Nethowto, will be addressed in future publications.

FURTHER RESEARCH AND CONCLUSION

Since the inquisitivist approach is new and an adaptation of minimalism, it could be argued that studies need to be run again (perhaps numerous times), but with much greater controls. Future investigations into the effectiveness of the inquisitivist approach would have to

- Employ true random sampling and statistically meaningful samples.
- Control for prior knowledge, ability, learning style, teacher effects, time-on-task, instructional method, and media familiarity.
- Use a comparison group for all aspects (i.e., personality).
- Use instruments with a sufficient number of items to increase reliability.
- Establish reliability scores on final projects.

- Consider using continuous data rather than discontinuous (i.e., use personality scores rather than 4-point scales).

However, even if these independent variables could be effectively controlled, their application would be artificial, calling to question the whole media comparison (Joy & Garcia, 2000).

Future research could also investigate the role of time-on-task, the impact of instructor-student and student-student interactions, and the effect of graduate and undergraduate student interactions. The affect of the instructor's personality and teaching style on the implementation and delivery of the Nethowto model could also be investigated. An even more perplexing area of future research would deal with the question of why students who demonstrated a high level of success and satisfaction with the inquisitivist approach would still have preferred a traditional form of instruction. Carroll found a similar phenomenon in his research that revealed that despite the success with minimalist documentation, people still claimed to prefer the traditional documentation (Carroll, 1990). Goodwin et al. (1991), and Lake (2001) also found that despite demonstrable improvement in achievement levels over lecture-based instruction, most students perceived active learning instruction to be ineffective, and would have preferred lecture-based instruction.

Are these claimed preferences actual preferences, or simply people's natural tendency or desire to preserve the status quo? Alternatively, does the inquisitivist approach and similar active-learning approaches expect, or require, too much of the learner? Are classes easier in the traditional systematic design format? Are inquisitivism, minimalism, active learning and many other student-centered constructivist approaches really such hard work, or are students simply more comfortable with memorization than with learning how to think? These questions are just the beginning of many more questions that would need to be effectively explored to determine why

people appear to still prefer systematic design instruction, despite demonstrable success with other instructional approaches like inquisitivism.

Inquisitivism, minimalism, and active learning can be hard work, especially for those who are not accustomed to this form of instruction. Similarly, memorization is much easier than learning how to think critically and analytically if one is accustomed to memorization. We clearly need to change student's experience and perceptions towards these forms of instruction. Lake (2001) suggested that we expand the discussion for the rational of active-learning methods, incrementally introduce active learning and, finally, change to an all-active learning curriculum. I agree with Lake, but would add that we need to move toward a much broader adoption of inquisitivist, minimalist, and other forms of constructivist approaches at the primary and secondary levels, so that when students reach the postsecondary level, they are accustomed to the challenges and benefits of these active and engaging forms of instruction.

REFERENCES

Allen, M., Bourhis, J., Burrell, N., & Mabry E. (2002). Comparing student satisfaction with distance education to traditional classrooms in higher education: A meta-analysis. *American Journal of Distance Education, 16*(2), 83-97.

Allen, M., Mabry, E., Mattery, M., Bourhis, J., Titsworth, S., & Burrel, N. (2004). Evaluating the effectiveness of distance learning: A comparison using meta-analysis. *Journal of Communication, 54*(3), 402-420.

Barr, R. B., & Tagg, J. (1995). From teaching to learning: A new paradigm for undergraduate, *Change*, (November/December), 13-25.

Bruner, J. S. (1960). *The process of education.* Cambridge, MA: Harvard University Press.

Carroll, J. M. (1990). *The Nurnberg Funnel: Designing minimalist instruction for practical computer skill.* Cambridge, MA: MIT Press.

Carroll, J. M. (1998). Reconstructing minimalism. In J. M. Carroll (Ed.), *Minimalism beyond the Nurnberg Funnel* (pp. 1-18). Cambridge, MA: MIT Press.

Carroll, J. M., & van der Meij, H. (1998). Ten misconceptions about minimalism. In J. M. Carroll (Ed.), *Minimalism beyond the Nurnberg Funnel* (pp. 55-90). Cambridge, MA: MIT Press.

DeLoughry, T. (1993). Two researchers say "technophobia" may affect millions of students. *Chronicle of Higher Education, 39*(34), 25-26.

Dryden, G., & Vos, J. (1994). *The learning revolution.* Rolling Hills Estates, CA: Jalmar Press.

Goodwin, L., Miller, J. E., & Cheetham, A. D. (1991). Teaching freshman to think: Does active learning work? *Bioscience, 41*(10), 719-722.

Gumport, P. J., Cappelli, P., Massey, W. F., Nettles, M. T., Peterson, M. W., Shavelson, R. J. et al. (2002). *Beyond reckoning: Research priorities for redirecting American higher education.* Retrieved March 18, 2005, from http://www.stanford.edu/group/ncpi/documents/pdfs/beyond_dead_reckoning.pdf.

Harapnuik, D. K. (1998). *Inquisitivism or "the HHHMMM??? What does this button do?" approach to learning: The synthesis of cognitive theories into a novel approach to adult education.* Unpublished manuscript, University of Alberta. Retrieved March 16, 2005, from http://www.quasar.ualberta.ca/edit435/theory/inquisitivism.htm.

Harapnuik, D. K. (2004). *Development and evaluation of inquisitivism as a foundational approach for Web-based instruction.* Doctoral Thesis, University of Alberta, CA.

Harasim, L. (1993). Collaborating in cyberspace: Using computer conferences as a group learning environment. *Interactive Learning Environments, 3,* 119-130.

Hobbs, D. L. (2002). A constructivist approach to Web course design: A review of the literature. *International Journal on E-Learning, 1*(2), 60-65.

Jonassen, D. H. (1990). *Computers in the classroom: Mindtools for critical thinking.* Englewood Cliffs, NJ: Prentice Hall.

Jonassen, D. H. (1991). Objectivism vs. constructivism: Do we need a new philosophical paradigm? *Educational Technology: Research and Development, 39*(3), 5-14.

Jonassen, D. H. (1997). A model for designing constructivist learning environments. In Z. Halim, T. Ottoman, & Z. Razak (Eds), *International Conference on Computers in Education* (pp. 71-80). Kuching, Sarawak, Malaysia: University Malaysia Sarawak & Asia Pacific Chapter of Association for the Advancement of Computing in Education (ACCE).

Jonassen, D. H. (2000). Transforming learning with technology: Beyond modernism and postmodernism or Whoever controls the technology creates the reality. *Educational Technology, 40*(2), 21-25.

Joy, E. J., & Garcia, F. E. (2000). Measuring learning effectiveness: A new look at no-significant-difference findings. *Journal of Asynchronous Learning Networks, 4*(1), 33-39.

Kearsley, G. (1997). *Learning & instruction: The theory into practice (TIP) database.* Retrieved March 16, 2005, from http://www.gwu.edu/~tip/

Kearsley, G. (1998). Minimalism: An agenda for research and practice. In J. M. Carroll (Ed.), *Minimalism beyond the Nurnberg Funnel* (pp. 393-406). Cambridge, MA: MIT Press.

Lake, D. A. (2001). Student performance and perceptions of a lecture-based course compared with the same course utilizing group discussion. *Physical Therapy, 81*(3), 886-902.

Lapadat, J. C. (2002). Written interaction: A key component in online learning. *Journal of Computer-Mediated Communication.* Retrieved March 16, 2005, from http://jcmc.indiana.edu/vol7/issue4/.

Lave, J., & Wenger E. (1990). *Situated learning: Legitimate peripheral participation.* Cambridge, UK: Cambridge University Press.

Montgomerie, T. C., & Harapnuik, D. K. (1996) The Internet: Communicating, accessing, & providing information [University of Alberta course Web site]. Retrieved March 16, 2005, from http://www.quasar.ualberta.ca/nethowto/.

Montgomerie, T. C., & Harapnuik, D. K. (1997). Observations on Web-based course development and delivery. *International Journal of Educational Telecommunications, 3*(2), 181-203.

Newman, F., & Scurry J. (2001). Online technology pushes pedagogy to the forefront. *The Chronicle of Higher Education, 5,* 7-11.

Rand, J., Spiro, R. J., Feltovich, M., Jacobson L., & Coulson, R. L. (1991, May). *Cognitive flexibility, constructivism, and hypertext: Random access instruction for advanced knowledge acquisition in ill-structured domains.* Retrieved March 16, 2005, from http://www.ilt.columbia.edu/ilt/papers/Spiro.html.

Romiszowski, A. J. (1997). Web-based distance learning and teaching: Revolutionary invention or reaction to necessity? In B. Khan (Ed.), *Web-based instruction* (pp. 25-37). Englewood Cliffs, NJ: Educational Technology Publications.

Rosen, L. D., Sears, D. C., & Weil, M. M. (1987). Computerphobia. *Behavior Research Methods, Instruments, & Computers, 19*(2), 167-179.

Rosen, L. D., & Weil, M. M. (1992). *Measuring technophobia: A manual for the administration and scoring of the computer anxiety rating scale (Form C), the computer thoughts survey (Form C) and the general attitudes toward computers scale (Form C).* [Manual, Version 1.1]. Dominguez Hills, Carson: California State University.

Shull, P. J., & Weiner, M. D. (2000). *Thinking inside of the box: Retention of women in engineering.* ASEE/IEEE Frontiers in Education Conference. Kansas City, MO: IEEE Education Society. Retrieved March 16, 2005, from http://fie.engrng.pitt.edu/fie2000/papers/1242.pdf

Strommen, E. F., & Lincoln, B. (1992). *Constructivism, technology, and the future of classroom learning.* Retrieved March 14, 2005, from http://www.ilt.columbia.edu/publications/papers/construct.html

University of Alberta Computer Network Services. (2004). *Universal ratings of instruction.* Retrieved March 1, 2005, from http://www.ualberta.ca/CNS/TSQS/USRI.html

van der Meij, H., & Carroll, J. M. (1995). Principles and heuristics for designing minimalist instruction. *Technical Communications, 42*(2), 243-261.

Vygotsky, L. S. (1978). *Mind in society.* Cambridge, MA: Harvard University Press.

WebBoard Collaboration Server (2005). [Computer Software]. Calsbad CA: Akiva Corporation.

Chapter 7.7
The Most Dramatic Changes in Education Since Socrates

Allen Schmieder
JDL Technologies, USA

THE CURRENT CONTEXT

This is an urgently needed topic. It is the author's conviction that, currently, there are no 21st century schools and, even worse, there is no substantive and widely held vision about what such schools should look like, and what the role and competencies of teachers in those schools should be. So, the tendency of most educators writing about needed 21st century teaching competencies will be to pretty much "rearrange the deck chairs on the Titanic." Most will be driven by another equally repugnant cliché, "Technology is only a tool," and they will try to determine how this misunderstood tool can best enhance out-of-date and fast-aging approaches to K-12 curriculum, instruction, and assessment. This is not to say that the wonderful array of traditional teaching competencies and skills that have enabled teachers to have generally done such an impressive job of teaching our children over the last century will cease to be important. The ability of teachers to understand and connect with students; to impart considerable knowledge and wisdom about their subject;

to provide them with good adult role models; to cultivate their motivation for learning; to encourage their sensitivity toward, and appreciation of, individual and cultural differences; to prepare them for post-secondary education and/or the world of work; and even, to sometimes be "the sage on the stage," will remain critical competencies as long as there is a teaching profession. *But just as technology has dramatically transformed society, the way we work, the way we live, even the way we think about things, schools must be dramatically transformed in the way they work, in the way content is processed, and maybe most importantly, in the way teachers teach and students learn.*

Given the context of this book, it should be noted that one of the major factors in any positive reform and improvement relative to the dramatic changes that are needed in teacher preparation will be the way that colleges of education respond to the challenge. They have the capacity to accelerate and lead this desperately needed reform; but, they and their host institutions (which have historically failed to give them the priority and support that they deserve) can continue to underestimate the technology-centered revolution that is taking place (albeit, much too slowly) in schools, and thus impede the inevitable changes that are needed to effectively prepare both teachers and students to thrive in the 21st century.

GENERAL 21ST CENTURY TEACHING COMPETENCIES

21st century teachers:

- must recognize and understand the rapidly increasing globalization of our world, and know how to infuse international and multi-cultural lessons and activities into their teaching;
- must have, and relentlessly reflect, a personal philosophy that all students have unlim-ited potential and the personal growth and development of every student is of critical importance;
- must help develop, and effectively use, cur-riculum and instructional approaches that customize education for every student;
- must be able to develop a classroom climate that places the highest possible emphasis on human rights, diversity, character-building, and individual responsibility;
- must be able to "take the lead" in developing a dynamic community of learners in which teachers, administrators, students, parents, and business and community persons work together to enhance the learning and growth of those involved during the regular school day;
- must advocate for, and be able to teach, a curriculum that gives the highest priority to the richness of our human and cultural heritage—one that centers on the history, growth, and present nature of the United States, and that introduces and emphasizes

the general nature and positive contributions of all of the world's peoples and cultures;

- must take the lead in developing the school as a learning center for the community—for parents, for community organizations, for local businesses that do not have adequate facilities or technologies for accomplishing their goals and objectives;
- must provide students with the knowledge and skills needed to succeed in higher edu-cation and/or in the most promising careers into which they will enter in the world of work—and for the rapidly changing world in which they will live, prosper, and eventually lead; and
- must know, keep abreast of, and be able to reflect in their teaching, the latest research on teaching and learning, and the latest standards for using cutting-edge technology, for engaging in quality professional develop-ment, and for building effective schools.

TECHNOLOGY-RELATED 21ST CENTURY TEACHING COMPETENCIES

No matter how great a teacher's command of technology-centered 21st century teaching com-petencies, they will not be able to provide 21st century teaching and learning unless their school and district (and hopefully, most student homes) have a world-class technology infrastructure that provides students and teachers with unlimited and easy access to the most powerful available personal computers; unlimited and easy access to the information highway; unlimited access to, and ability to use, a broad range of emerging technologies (e.g., handhelds, voice recognition instruments, handheld devices, GPSs, GISs, data loggers); and software and educational databases linked to the school's curriculum.

21st century, technology-savvy teachers:

- must be fully aware of the critical importance of technology in every aspect of American society, economic development, and global understanding, and must realize that unless students have a thorough knowledge of the importance of technology in future careers and a solid foundation in technology literacy and skills, they will not be well prepared to thrive in, and lead, in the 21st century;

- must be relentless advocates of the importance of technology in school policy, finance, management, instruction, and assessment. Being advocates and activists for the use of technology in teaching and learning is even more important than being able to easily use technology;

- must be aware of, and knowledgeable about, the large and rapidly growing array of new technologies that are essential to teaching effectively the basic subjects—for example, Internet I, Internet II, computers, proxy servers, handheld devices, GPSs, GISs, streaming video, and voice recognition technologies;

- must be facile with, and model the use of, appropriate technology in required reporting, classroom management, instruction, learning assessment, extra-curricular activities, and home and community communication;

- must know how to access and use the vast and growing resources of the Internet in transforming the way content is presented and processed in their assigned subject/s, and must be able to teach students how to evaluate and effectively use these same resources;

- must be aware of, and able to use, the enormous range of instructional approaches that are made possible by a technology-rich school and classroom—for example, individualized, small group, large group, distance learning, and field-based learning;

- must be aware of, recognize the value of, and know how to teach students to use the huge and dynamic databases that are now available to energize classroom instruction and make lessons more current, substantive, and "real" on almost any topic—for example, electronic libraries; information from federal, state, and local governments; Web sites of major agencies and projects; and Web sites of universities and other research centers, those of the United Nations, the World Bank, and other global information sources, those of major, validated educational programs and projects (e.g., GLOBE, GenerationY, and the Virtual High School), as well as of educational clearinghouses (e.g., ERIC, ENC, CARAT, and NEDS);

- must have a vision that reflects the needs of the high-tech and continuously changing society that supports the school and for which the school is preparing its students to live in, and lead;

- must be able to manage and utilize in strengthening the quality of instruction, a school-installed SIS (student information system) that enables appropriate and on-demand examination of a comprehensive story of the achievement, growth, and potential of individuals, groups of individuals, schools, districts; of required forms and records; of current and future schedules; and of any information considered to be "mission critical" to the school and its community of learners;

- must demand and/or take the lead in the development of in-service education programs (including substantial incentives) that will enable teachers to systematically and continuously update and improve their 21st century teaching competencies—to the extent possible, at the school, during regular school hours;

- must provide students with extensive exposure to careers in business, government, and social services, and in instructional programs, and use the most up-to-date technologies and approaches generally used in those careers;

- must have a global view with unlimited and easy access to the technologies and information that enable continuous and high-quality communication with students, teachers, communities, and organizations across the world concerned with improving understanding and economic and cultural cooperation with the United States; and
- must believe in the advantage to students in benefiting from the successes of other teachers and schools (especially in the teacher's subject or grade level), and develop a system for continuous, high-quality accessing of a broad range of developmental resources, and for sharing the experiences of successful schools and educators across the school system, state, nation, and world (Schmieder, n.d.).

REFERENCE

Schmieder, S. (Ed.). (n.d.). *Talkin' tall: Voices for "millennium" teachers* (pp. 26-29).

This work was previously published in the Encyclopedia of Distance Learning, Vol. 3, edited by C. Howard, J. Boettcher, L. Justice, K. Schenk, P. L. Rogers, and G. A. Berg, pp. 1307-1309, copyright 2005 by Idea Group Reference (an imprint of IGI Global).

Chapter 7.8
Quality Assurance and Online Higher Education

Edward D. Garten
University of Dayton, USA

Tedi Thompson
American Public University System, USA

INTRODUCTION

For higher education, the assurance of quality to others in what it does is a deeply held value. Yet, marks surrounding quality are not easily identified, clearly understood, or universally accepted. The consumer movement, among other societal factors in recent years, has nudged and in some instances pushed institutions of higher learning toward the specification of meaningful assessment measures and the subsequent reporting out to concerned parties indications of quality relative to institutional infrastructure and resources, institutional processes, and readily understood outcomes measures (Baker, 2002, p. 3).

Technology-enhanced teaching and learning has fundamental implications for quality assurance and accreditation that include:

- The reality that online learning technologies are reshaping some of the most fundamental and pervasive activities of learning and teaching.

- Digital technology will continue to change far faster than any other aspect of the academic infrastructure. Each new generation of technology calls into question fundamental values and practices with quality assurance processes, both externally and internally imposed, having roles to play in deciding what to change and what to regain.

- Computers and networked learning are being employed to broaden participation in higher education, with wider access to information and experiences. In many instances, these unfolding uses of technology are having profound effects on the identity, mission, and character of academic departments, institutions, and systems.

- Technology-enabled learning can trigger dramatic increases in costs with sometimes minimal educational payoff unless providers use careful planning, evaluation, and focused quality assurance processes.

Online higher education in multiple ways has challenged and been challenged by traditional quality assurance and accreditation processes. Online higher education alters the traditional faculty role, and it may alter many of the fundamental intellectual tasks of faculty. Moreover, many online initiatives separate curriculum design from curriculum delivery, replacing curricula designed by individual faculty or faculty teams with standardized course content. Critically, online learning can shift, in the case of some virtual university providers, responsibility for determination of academic standards from faculty to corporate leadership (Eaton, 2002, pp. 8-9). It is clear that the "continued growth of the global demand for distance education and the acceptance of the virtual university as a mainstream institution both drive the need (and also the technological capability) for more effective measurements of human and organizational performance" (Stallings, 2002, p. 53). This chapter assumes the understanding of online higher education to consist of that broad range of higher learning activities that include corporate training centers, nonprofit and governmental education activities, multi-state and international learning collaborations, and the distance learning efforts of individual institutions of higher learning both for profit and non-profit (Epper & Garn, 2004).

In this chapter we explore key elements associated with quality control and regulation of online higher education: (1) the learning outcomes movement, (2) national standards and guidelines which better ensure evidences of quality, (3) expectations of regional accreditation agencies for quality online delivery, and (4) institutionally adopted quality processes.

IMPACT OF THE LEARNING OUTCOMES MOVEMENT

Any discussion of quality control of online higher education must necessarily begin with a statement of the critical importance that the learning outcomes and learning assessment movement has had on the wider conversation regarding quality assurance. Multiple and diverse constituencies, legislative agencies, and accrediting bodies today demand improved accountability from institutions of higher learning in both online and traditionally delivered programs. These demands have resulted in a greater emphasis on learning outcomes assessment and learner-centered methodologies. Learning outcomes assessment not only assists an institution in the evaluation of the effectiveness of its programs, it provides the basis for continual quality assurance and improvement (Muirhead, 2002).

Historically, the assessment movement has its origins in the last decade. The 1990s saw a clear trend in which accountability became a critical descriptive term in higher education and, in particular, within the context of the virtual and online university (Stallings, 2002). It has been suggested that future historians of higher education are likely to observe that the latter years of the 20th century will not so much be known for educational problems solved, but rather for the intense national pressure brought by non-educators as well as accrediting and quality assurance agencies to change practice and theory in academe (Sewall, 1996). Increasingly online educators are being asked the same questions as their more traditional counterparts: "Can you provide direct measures of student outcomes? How much are students learning? And are they learning the right things?" (Erwin, 2001).

Given its nature, special consideration must be given to online learning that includes the need to address such questions as:

- What kinds of new learning and assessment opportunities are created through online learning?
- What pedagogies can be employed to support meaningful online assessment?
- What are the losses and gains of this medium for instructors and students?
- How effectively do old models and forms of assessment translate into the online environment? (Dunn, Morgan, O'Reilly & Parry, 2004, p. 39)

Critically, important questions have been raised regarding how learning communities are established and effectively assessed in the virtual higher education environment including, in particular, means through which high-quality interactions among students as well as student to instructor are nurtured (Palloff & Pratt, 1999). In short, assessment, in this context, is a manner of determining what students are acquiring in terms of general knowledge, thinking or performance-based abilities, theoretical and applied understandings, and so forth, and achieving as a result of their educational experience (Allen, 2004). The process begins with clearly articulated, measurable objectives at the institutional, program, and course levels. Those objectives can then be translated into specific goals, which can be measured through a variety of direct and indirect measures. The data collected from these measurements becomes an effective resource for measuring the overall quality of the educational experience and a powerful basis for ongoing improvement.

Whereas earlier assessment tended to focus on teaching, the focus of learning outcomes assessment in the online university increasingly has been on student learning. Prominent among those who have clarified the nuances between the teacher-centered paradigm vs. the learner-centered paradigm have been Huba and Freed (2000). They have emphasized areas of assessment that have increasingly become a focus of

concern among online educators in the online university:

1. Students are actively involved in their own learning.
2. Emphasis is on using and communicating knowledge effectively to address enduring and emerging issues and problems in real-life professional contexts.
3. The instructors' role is to coach and facilitate and, together, they evaluate learning.
4. Assessment is used to promote and diagnose learning.
5. There is an emphasis on generating better questions and learning from errors.
6. Desired learning is assessed directly through papers, projects, portfolios, and so forth.
7. The learning culture is collaborative, cooperative, and supportive,
8. Instructors and students learn together. (p. 5)

Learning outcomes assessment programs are an important component of quality assurance in any learning environment. The importance of these programs has been viewed as critical in the distance learning realm. Research in online learning indicates that interactivity is a key factor in student learning and retention in online programs. A well-structured and comprehensive learning outcomes assessment program encourages collaboration and interaction between students and faculty, ultimately enhancing the quality of the learning experience on behalf of all parties. While some critics maintain that online or distributed learning leaves external reviewers with little, apart from student achievement, on which to premise quality judgments, in the end "we will finally be forced to address student learning as a central indicator of quality" (Twigg, 2001, p. 10). Those who want to move quality assurance measures primarily in the direction of learning outcomes assessment view this as a highly desirable goal.

NATIONAL STANDARDS AND GUIDELINES

Distance Education and Training Council

With the proliferation of distance delivered options, new online universities, and changes in available technologies, national accrediting bodies have endeavored to set forth guidelines and standards for courses and programs equal to those standards to which traditional academic programs are held. The Distance Education and Training Council (DETC), established in 1926 as the Home Study Council, is one such organization; however, unlike other accrediting bodies, it deals primarily with distance learning universities and other post-secondary schools. Its mission is "to foster and preserve high quality, educationally sound and widely accepted distance education and independent learning institutions" (www.detc.org). DETC is recognized by the US Department of Education and the Council on Higher Education Accreditation; hence it holds its accredited institutions to rigid operational and academic standards. In efforts to ensure quality of instruction and service, institutions accredited by DETC are required to comply with 12 broad standards, with additional caveats set forth for institutions offering degree programs. Institutions are required to undergo reevaluation at a minimum of every five years. New programs or courses require individual approval contingent on a review by independent subject matter experts. This review entails an intensive critique of individual courses within the curriculum, the instructional methodology, and the credentials of instructors presenting courses. Instructional programs are reexamined as an integral component of re-accreditation evaluations at a minimum of every five years.

Joint Commitment of the Regional Accrediting Commissions

All six of the regional accrediting commissions have affirmed a commonly held *Statement of Commitment for the Evaluation of Electronically Offered Degree and Certificate Programs*. This statement is widely employed by online universities, as well as traditional universities offering extensive online programs, and raises quality-related questions revolving around commitment to values and principles; commitment to cooperation, consistency, and collaboration; and commitment to the support of good practice.

Council on Higher Education Accreditation

Relative to guarantees of quality, the federal government also needs assurances from the disciplinary, regional, and professional accreditation community that quality can be reviewed and promoted even in the face of significant academic changes driven by distance learning, and that student aid grants and loans will purchase a quality educational experience in a distance learning environment. The Council on Higher Education Accreditation (CHEA) in this regard has a critical coordination role to play with a broad range of accrediting agencies, assisting the focus of issues and addressing federal challenges relative to online and other technologically enabled higher education. Weighing in on these issues, the CHEA Institute for Higher Education Policy released the report "What's the Difference? A Review of Contemporary Research on the Effectiveness of Distance Learning in Higher Education" (Phipps & Merisotis, 1999). Report findings indicated some disparities in earlier research that required further investigation. Further it was noted that distance learning styles and how such styles related to the use of different technologies had not been adequately explored, nor did the research attempt to explain the higher-than-average attrition rate in distance learning courses. Available studies at the time of the report tended to focus on the learning outcomes and satisfaction with individual courses, rather than the more desirable focus on entire academic programs. Finally it was noted there was insufficient research on the effectiveness of digital libraries.

The Council for Higher Education Accreditation has also set forth *Principles for U.S. Accreditors Working Internationally*. The principles fall into four areas: (1) considerations and actions for US accreditation leaders when determining to undertake accreditation of non-US institutions and programs in another country; (2) expectations for conduct of US accreditation reviews of non-US institutions and programs in another country; (3) accreditor expectations of providers of US online and Web-based instruction and programs exporting to another country; and (4) responsibilities of US accreditors working with non-US institutions and programs for students and colleagues in another country (Eaton, 2002, pp. 36-37).

EXPECTATIONS OF REGIONAL ACCREDITING COMMISSIONS

The rapid growth of the delivery of higher education via technology has dramatically both challenged institutional accreditation agencies and altered the very meaning of the phrase *higher education institution*, given the reality that learning no longer must be anchored in a physical place. Additionally, and a stiff challenge to those who accredit, electronic access to higher learning opportunities have allowed students—on a large scale—to attend multiple institutions either serially or simultaneously (Eaton 2002, p. 6). Until recently, many regional and disciplinary accreditation leaders have chosen to assure the quality of online higher education by adapting existing standards for ground-based to online delivery.

Often these alternative standards for online and distance delivery were simply organized following traditional categories of evaluation that focused on faculty, curriculum, support services, and facilities and technologies. Typically, accreditation principals came to assume that expectations for quality for online education would be the same as for site-based education (Roberts, 1999). Higher education is experiencing a convergence of many delivery modalities, student services, and related information resources that are in many instances designed, at the front end, to serve students independent of location. Most informed quality assurance leaders today recommend against separate evaluation and accreditation criteria for technology-enabled learning in this new convergent environment, while recognizing that specific criteria initially suggested solely for online learning programs might be applied productively to all academic programs offered by a college or university.

The six regional accreditation commissions in the United States have not shied away from both their responsibilities in understanding the new realities and their obligations to modify existing standards and criteria through which to address issues of quality in this new environment. However these commissions wisely attempt to balance quality control concerns with their traditional sensitivities toward individual institutional missions (Stallings, 2002). Each of the commissions are moving away from an earlier near-exclusive emphasis on inputs, now devoting substantial attention to instructional outcomes, public disclosure, and mission focus. Critically, as of 2000, there were 26 state higher education agencies and 19 regional and disciplinary accrediting bodies that considered assessment and outcomes measures to a high level of priority (US Department of Education, 2000, pp. 11-12).

Overall, since around 1998, the language in the standards and criteria of the six regional commissions regarding expectations for electronically delivered education tends to be less prescriptive and more mission derived. More experimental and collaborative approaches to the assurance of quality are being offered, for example, The Higher Learning Commission of the North Central Association's AQIP program discussed briefly in the next section of this chapter. In the future, regional accreditation commissions will be called upon to assume more responsibility for addressing public interest in the quality of online higher education as opportunities and providers both diversify and expand. While many of these online learning providers:

...may chose to remain unaccredited, the accrediting community—including not only accreditors but also accredited colleges and universities which are still the dominant deliverers of higher education—will continue to be viewed by the public as responsible for the quality of these providers. Accreditation leaders will need to provide guidance to the public on how to reach judgments about quality in these new settings—what to examine, whom to contact, how to make comparisons. (Eaton, 2002, p. 11)

Additionally, as online learning opportunities have been expanded globally, regional accreditation commissions have been at the forefront in promulgating international quality assurance guidelines. Today, nations such as South Africa and India are heavy importers of distance learning programs and associated support services. China, Japan, and Thailand, in particular in the East, are developing their own technologies to support academic programs and degrees, while Western and Eastern European nations continue to sort out what place Internet-based education has alongside their often very traditional educational models. Internationally, the United States, the United Kingdom, and Australia are the three major exporters of technology-assisted higher education. Robust and critical international con-

versations regarding quality assurance continue and have resulted in solutions that have included bilateral agreements, generally accepted international standards, market-driven solutions, and solutions premised on agreements among World Trade Organization partners (Eaton, 2002).

Southern Regional Education Board Distance Learning Initiative

The Southern Regional Education Board (SREB) was founded in 1948 in response to business leaders' desire to improve education at all levels in the southern states. The SREB developed an electronic campus that allows students to take courses at accredited institutions from the participating 16 southern states, regardless of students' geographic location. In an effort to ensure quality of online offerings, SREB set forth *Principles of Good Practice* (retrieved April 20, 2004, from www.electroniccampus. org/student/srecinfo/publications/principles.asp). All courses offered via the SREB Electronic Campus are evaluated against the principles that, together, direct the development of new courses to ensure that components of effective teaching are not compromised, ensure institutional quality of programs and courses seeking inclusion in the Electronic Campus, and provide a consistent method of review prior to submission to the state higher education agency for inclusion. The principles fall into three broad categories: Curriculum and Instruction, Institutional Context and Commitment, and Evaluation and Assessment. In general, curriculum- and instruction-related principles provide assurance that courses offered through the SREB Electronic Campus are comparable to those offered on the traditional campus of the institution providing the course in terms of course content, faculty qualifications, appropriate levels of interaction, and academic rigor and breadth. The principles focus on institutional context and commitment address issues, including provision of appropriate student services and support, support for faculty, learning resources, and the commitment of the institution to electronically offered programs and courses. Those principles related to evaluation seek to ensure that SREB programs are systematically evaluated and learning outcomes are measured.

INSTITUTIONALLY ADOPTED QUALITY PROCESSES

Online universities and more traditional universities now evolving extensive online programs recognize that measuring inputs to the academic process are clearly insufficient measures of quality, as is an almost exclusive focus on student learning outcomes and measures, which often are narrowly defined according to data available and the subjective nature of the relative value of certain outcomes over others. Accordingly, many online universities and higher learning organizations with extensive distance delivery programs have adopted one or more non-externally mandated or driven quality processes.

Continuous Quality Improvement Processes

In the virtual university environment, continuous quality improvement (CQI) approaches are now frequently adopted from the management literature, refined, and adapted to improve learning and teaching processes. Assessment and continuous quality improvement movements are both premised on the collecting of feedback for improvement. Fundamentally, CQI techniques strive to improve the learning of both instructors and students by collecting and sharing data, and ultimately creating sustained feedback loops (Huba & Freed, 2000, p. 130). Virtual and online universities have often been in the forefront in adapting the growing range of techniques associated with CQI.

Principles and Benchmarks of Good Practice

Online universities and online programs offered by traditional universities have extensive closely followed "good practice principles" related to assessment, as promulgated by several professional associations and accreditation agencies. In the early 1990s, the American Association of Higher Education (1992) circulated nine *Principles of Good Practice for Assessing Student Learning*, followed shortly later by the North Central Association, Commission on Institutions of Higher Education's 10 "Hallmarks of Successful Programs to Assess Student Academic Achievement" (1994). Both sets of principles have been widely adopted by educational leaders of online universities and other distance delivery programs at the post-secondary level. Both sets focus on the importance of assessment for improvement and on the critical need to involve faculty and other key institutional constituents in the assessment process. Additionally, the Institute for Higher Education Policy Benchmarks were developed in 2000 with a focus on the following areas: Institutional Support, Course Development, Teaching/Learning, Course Structure, Student Support, and Evaluation and Assessment. This has been a particularly useful set of benchmarks because knowledgeable and experienced practitioners (those with concrete experiences of what works and what doesn't work) have vetted the benchmarks (Twigg, 2001, p. 6).

Baldrige Processes

The Baldrige National Quality Award Program with its associated Criteria for Performance Excellence has been a significant tool employed by thousands of US organizations to assess and improve performance on critical factors that drive success. Lessons learned from business communities increasingly have been leveraged to provide a proven course for educational organizations to pursue excellence and maintain a leadership position. Given an inherent disposition toward business models, many corporate and typically for-profit online universities, as well as some other higher learning initiatives with strong, technologically enabled delivery programs, have adapted all or many Baldrige process elements through which to assess organizational effectiveness and quality. Baldrige Educational Criteria are built upon a foundation of core values and concepts vital to high levels of organizational performance: visionary leadership; learning-centered education; organizational and personal learning; the valuing of faculty, staff, learners, and partners; agility; focus on the future; managing for innovation; management by fact; public responsibility and citizenship; focus on results and creating value; and a systems perspective. The use of Baldrige quality processes by online higher education organizations has provided:

...a different way to think about quality. Abstract and wispy words are replaced by a focus on developing discrete measures to track critical processes. Process outcomes are, in turn, expected to meet specific customer requirements. [Baldrige-type assessments in education] transform quality from a swirl of rhetoric devoid of meaning to the minimal expectations of accreditation to precise operational indices that link processes with outcomes. (Seymour, 1994, p. 26)

A more sophisticated means of assuring quality, Baldrige and Baldrige-like processes will continue to be employed by technologically enabled models of instructional delivery well into the future.

Academic Quality Improvement Program

This voluntary quality improvement program is available to higher learning organizations affili-ated with The Higher Learning Commission of the North Central Association. The program seeks to design and implement a new, alternative process for maintaining regional accreditation. The process is heavily grounded in the now established principles of continuous quality improvement and is informed by Baldrige quality principles. Increasingly, virtual universities and those with major online programs based in the North Central region have signed onto the AQIP approach. AQIP integrates contemporary quality principles—particularly their focus on institutional processes and results rather than simply on resources—and it focuses accreditation on improving teaching and learning by involving faculty more directly in all academic improvement processes. AQIP delivers practical, timely feedback and support that institutions can use to improve students' educational performance, and at the same time, it customizes accreditation to fit institutional needs and priorities, and reduces the intrusiveness, cost, and typically longer cycles of improvement associated with traditional accreditation approaches. Additionally, AQIP provides comprehensive, useful information about institutional quality to the pubic and other stakeholders, including information that recognizes and celebrates institutional distinctiveness and outstanding achievements. And, importantly, it establishes the foundation for a continuously improving system that can modify itself quickly to meet new conditions in a dynamic and highly competitive environment (www.aqip.org).

CONCLUSION

Fundamentally, the increased diversity of types of institutions of higher education, roles, missions, primary audiences, and instructional strategies demands that all quality assurance processes, both externally and internally imposed, focus on multiple measures of organizational and academic effectiveness linked specifically to institutional mission. Measuring inputs such as quality and number of faculty and financial resources is only one component in an overall quality assurance plan. Measuring student learning outcomes and institutional performance must also be linked to institutional mission clearly defined. And any quality assurance regimen must examine the effectiveness of the learning and teaching processes that have been put in place. Following her review of current issues in higher education quality assurance, Jean Mulhern offered two major observations: First, however higher education is delivered, those institutions that garner externally validated success will be attentive to quality assurance processes based on "self-monitoring against their own definitions of high standards." Second, "because the delivery of postsecondary education is not monolithic and there is a public trust involved, there always will be oversight, trying to define quality and evaluate providers against general quality criteria however defined" (2002, p. 159). The continuing growth and demand for online delivered educational programs and the growing acceptance of the university on the part of mainstream higher education continues to drive the need for more effective measurements of human and organizational performance. Accreditation commissions, partners with online higher education providers, and the expectations of governments regarding the need to guarantee quality in the most rapidly emergent sector of higher education will continue to be challenged to secure responsive means of assuring that institutions are delivering and students are receiving products of integrity. Importantly, in the dramatically expanding and increasingly more diverse online learning environment, the ability to address public interests and concerns for quality will be linked to success in the marketplace. Increasingly all who are concerned with quality in online higher education will collectively focus on issues of consumer protection; provide assurances related to the transfer of credits in the new environment; expand efforts to inform the public regarding the criticalness of external quality

reviews; and provide strong, compelling communication regarding quality issues with and among those who undertake alternative approaches to quality review (Eaton, 2002, p. 18).

The new providers of online higher education truly represent a diverse group of options including stand-along degree-granting online or virtual universities; non-degree granting online consortia; online programs not aligned with any one institution; corporate universities typically not offering degrees; and degree-granting consortia consisting of groups of degree-granting universities that offer courses online, with the actual degree granted by the consortium itself. For the foreseeable future, this array of educational providers will continue to be complicated by the highly entrepreneurial and rapidly growing presence of the for-profit higher education sector (Eaton, 2001). Moreover, academic collaborations with for-profit firms will continue to expand in continuing education, credentialing, and specialized degree programs. Technology-enabled higher education is not a passing fad. Those who do not pay close attention to the assurance of quality to present and future constituents will not survive the marketplace.

REFERENCES

Allen, J. (2004). The impact of student learning outcomes assessment on technical and professional communication programs. *Technical Communication Quarterly,* 13, 93-108.

American Association of Higher Education. (1992). *Principles of good practice for assessing student learning.* Washington, DC: AAHE Assessment Forum.

Baker, R.L. (2002). Evaluating quality and effectiveness: Regional accreditation principles and practices. *Journal of Academic Librarianship, 28*(1), 3-7.

Easton, J.S. (2001). *Distance learning: Academic and political challenges for higher education accreditation.* Council for Higher Education Accreditation Monograph Series 2001, Number 1. Washington, DC: Council for Higher Education Accreditation.

Eaton, J.S. (2002). Maintaining the delicate balance: Distance learning, higher education accreditation, and the politics of self-regulation. In *Distributed education: Challenges, choices, and a new environment.* Washington, DC: American Council on Education.

Epper, R.M. & Garn, M. (2004). Virtual universities: Real possibilities. *EDUCAUSE Review,* (March/April), 28-38.

Erwin, D. (2001). *National postsecondary education cooperative sourcebook on assessment.* Retrieved April 25, 2004, from http://nces. ed.gov/npec/evaltests

Gratch-Lindauer, B. (2002). Comparing the regional accreditation standards: Outcomes assessment and other trends. *Journal of Academic Librarianship, 28*(1), 14-25.

Huba, M.E. & Freed, J.E. (2000). *Learner-centered assessment on college campuses: Shifting the focus from teaching to learning.* Needham Heights, MA: Allyn & Bacon.

Muirhead, B. (2002). *Relevant assessment strategies for online colleges and universities.* Retrieved April 1, 2004, from http://www.usdal.org/html/journal/FEB02_Issue/artilce04.html

Mulhern, J. (2002). Current issues in higher education quality assurance. In E.D. Garten & D.E. Williams. (Eds.). *Advances in library administration and organization* (p. 19). Oxford, UK: Elsevier.

NCA Commission on Institutions of Higher Education. (1994). *Handbook of accreditation (1994-1996).* Chicago: North Central Association.

NCA Higher Learning Commission. (2001). *Best practices and protocols for electronically offered degree and certificate programs.* Chicago: North Central Association. Retrieved April 19, 2004, from www.ncahigherlearningcommission.org/resources/resources/electronicdegrees/index.html

Palloff, R.M., & Pratt, K. (1999). *Building learning communities in cyberspace.* San Francisco: Jossey-Bass.

Phipps, R. & Merisotis, J. (1999*). What's the difference? A review of contemporary research on the effectiveness of distance learning on higher education.* Washington, DC: The Institute for Higher Education Policy. Retrieved April 15, 2004, from www.ihep.com

Roberts, S.K. (1999). A survey of accrediting agency standards and guidelines for distance education. *Theological Education, 3*(20).

Sewell, A.S. (1996). From the importance of education in the 80s to accountability in the 90s. *Education, 116*(3), 325-332.

Seymour, D. (1994). The Baldrige cometh. *Change,* (January/February), 16-27.

Stallings, D. (2002). Measuring success in the virtual university. *Journal of Academic Librarianship, 28*(1), 47-53.

Statement of Commitment by the Regional Accrediting Associations for the Evaluation of Electronically Offered Degree and Certificate Programs Online. (2000). Retrieved April 18, 2004, from http://www.ncahigherlearningcommission.org/resources/electronic_degrees/index.html

Twigg, C. (2001). *Quality assurance for whom: Providers and consumers in TODAY'S distributed learning environment.* Center for Academic Transformation, Rensselaer Polytechnic Institute, USA.

US Department of Education, Office of the Inspector General. (2000, September). *Management controls for distance education at state agencies, management information report.* EDOIG/A09-90030.

KEY TERMS

Accreditation: The primary means by which colleges and universities and other higher learning programs assure academic quality to students and to the public.

Quality Assurance: Those processes that assess the relative strength of academic programs and associated student and faculty support pro-

grams, and offer strategies for both program and institutional improvements recognizable by both internal and external constituencies.

Technology-Enabled Education: The use of advanced electronic technologies for purposes of direct support and enhancement of the student learning experience, in all of its aspects and wherever it might occur.

Virtual Universities: Universities that deliver their academic programs solely or primarily in an Internet-based online mode, and which typically do not have extensive physical facilities or campuses.

This work was previously published in the Encyclopedia of Distance Learning, Vol. 3, edited by C. Howard, J. Boettcher, L. Justice, K. Schenk, P. L. Rogers, and G. A. Berg, pp. 1529-1537, copyright 2005 by Idea Group Reference (an imprint of IGI Global).

Chapter 7.9
What Factors Promote Sustained Online Discussions and Collaborative Learning in a Web–Based Course?

Xinchun Wang
California State University, Fresno, USA

ABSTRACT

Although the pedagogical advantages of online interactive learning are well known, much needs to be done in instructional design of applicable collaborative learning tasks that motivate sustained student participation and interaction. This study investigates the factors that encourage student interaction and collaboration in both process and product oriented computer mediated communication (CMC) tasks in a Web-based course that adopts interactive learning tasks as its core learning activities. The analysis of a post course survey questionnaire collected from three online classes suggest that among others, the structure of the online discussion, group size and group cohesion, strictly enforced deadlines, direct link of interactive learning activities to the assessment, and the differences in process and product driven interactive learning tasks are some of the important factors that influence participation and contribute to sustained online interaction and collaboration.

INTRODUCTION

Theoretical Framework

The pedagogical advantages of student interaction in collaborative construction of knowledge are grounded in the social constructivist perspective of learning. From the social constructivist perspective, all learning is inherently social in nature. Vygotsky's theory of the Zone of Proximal Development posits that learners benefit most from social interactions concerning tasks they cannot do alone but can do in collaboration with more knowledgeable or more experienced peers (Kern, 1995). Knowledge is discovered and constructed through negotiation, or collective sense making. Pedagogically sound tasks in an online learning environment should, therefore, reflect social learning and collaborative construction of knowledge.

In designing and implementing online collaborative learning tasks, educators also draw heavily from Bakhtin's social theories to support their models of social interaction in collaborative construction of meaning in an online learning environment (Duin & Hansen, 1994; Wang & Teles, 1998; Wu, 2003). A speaker gives voice to a thought, an utterance, this utterance, though representing the ideas of an individual, reflects a social environment that is shared. The listener interprets the utterances in a purposeful, conscious act, in terms of his or her own concept of the social context, in terms of what the words mean to him or her individually. Therefore, speech and writing are dialogical in that the meaning of an utterance is created by both the speaker/writer and listener/reader through social interaction (Duin & Hansen, 1994). Pedagogically sound online learning tasks should therefore facilitate such online interactive learning for knowledge construction.

Interactive Learning and Online Collaboration

From a student's perspective, online interaction in learning takes place at two different levels: interaction with content and interaction with instructors and between peers (Gao & Lehman, 2003). There is evidence that pedagogically well-designed interactive learning tasks actually increase rather than decrease student access to instructors; increase interactions between instructors and among students; and increase students involvement of course content as well (Lavooy & Newlin, 2003; Mouza, Kaplan, & Espinet;

2000; Wu, 2003). Interactive learning tasks also promote greater equality of participation (Mouza, Kaplan & Espinet, 2000), more extensive opinion giving and exchanges (Summer & Hostetler, 2002), empower shy students to participate, and promote more student-centered learning (Kern, 1995; Wang & Teles, 1998)

At the level of interaction with content, students benefit more from producing explanations than receiving explanations. Such proactive learning engages students in a higher level of thinking than the reactive type of learning (Gao & Lehman, 2003; Wu, 2003). To promote such proactive learning, online course instructors need to integrate more active learning tasks that require more production than reception of explanations. Therefore, tasks that require written explanations should be considered over multiple choice type of reading comprehension in interpreting learning materials. Computer Meditated Communication in both synchronous and asynchronous discussion forums is inherently supportive of tasks for exchange of such written explanations. Furthermore, the systems can also archive written explanations posted in online forums and can be easily accessed and retrieved for references.

Although CMC supports interaction and collaborative learning, it also has inherent shortcomings. Disadvantages include the time it takes to exchange messages and the increased difficulties in expressing ideas clearly in a context reduced learning environment and the difficulty in coordinating and clarifying ideas (Sumner & Hostetler, 2002). The increased time it takes to reach consensus and decisions (Kuhl, 2002; Sumner & Hostetler, 2002) and to produce a final product (Macdonald, 2003). Given all these difficulties students need to overcome in order to collaborate effectively in interactive learning environment, online instructors need to address these obstacles with careful instructional design and provide support for collaborative learning with appropriate interactive learning tasks.

Factors That Influence Student Online Interaction and Collaboration

Research has also shown that computer mediated communicative tasks require more active role of students than traditional instruction in the face-to-face environment does (Wang & Teles, 1998). Students need to be willing to send a formal written question rather than have a casual conversation with peers or with the instructor in order to have their questions answered (Kuhl, 2002). To communicate effectively with peers and the instructor, students need to create the context through written messages, which requires the writing skills to identify their problems and express them precisely in order to have the questions answered. Team work and negotiation for meaning are necessary skills in CMC that cannot be assumed. Students need to learn to be familiar with the discourse of the discipline and academic genre for an online synchronous and asynchronous forum (Kuhl, 2002; Macdonald, 2003).

In addition to negotiation skills online, previous research has identified a number of other factors that influence student participation and interaction in a Web-based learning environment. Among others, the assessment of collaborative learning tasks plays a crucial role in ensuring student participation (Kear, 2004; Kear & Heap, 1999; Macdonald, 2003). In general, assessed collaborative learning tasks attract student participation at the cost of unassessed tasks. Furthermore, grade for discussion was also positively related to students' perceived learning (Jiang & Ting, 2000).

The structure of discussion in CMC is found to be another important factor in ensuring the amount of participation and level of interaction and collaboration among the peers. Such structure includes the size of the discussion groups, the nature and types of discussion topics (Williams & Pury, 2002), and whether the collaboration emphasizes the process of learning or the end product of such collaboration, or both (Kear, 2004; Kear & Heap, 1999; Macdonald, 2003). Online collaboration can be either process or product oriented. Forum discussions regarding course contents or related issues are commonly process oriented as the sharing of ideas help learners understand the issues without necessarily leading to a final product. Students are assessed individually based on their participation and quality of their contributions. Alternatively, online interaction and collaboration may lead to a final product such as an essay, a project, or a Web page, and so forth. There can be two assessment elements to such tasks, a common grade for the group for the overall quality of the collaborative product and individual grades for the contribution of each individual to the collaborative endeavor (Kear, 2004; Kear & Heap, 1999; Macdonald, 2003).

Finally, like any other form of learning, learning collaboratively in an online course is also characterized by individual differences. Collaboration as a process of participating to the knowledge communities is not an equal process to all the members of the community (Leinonen, Järvelä & Lipponen, 2003).

To summarize, online negotiation skills, the direct link between collaborative tasks and assessment, the structure of online discussions such as the nature and types of discussion topics, the size of the group, and the differences between process and product oriented collaborative tasks are some of the factors that influence student participation, interaction, and collaboration.

It is important to note that some of the above findings are based on experiments that are not a part of an online course (Gao & Lehman, 2003). Others have based their studies on courses that integrate some collaborative tasks in mainly student-instructor/tutor interaction type courses (Kear, 2004; Kear & Heap, 1999; Leinonen, Järvelä, & Lipponen, 2003; Macdonald, 2003; Williams & Pury, 2002). Web-based courses that employ collaborative learning tasks that form the essential course syllabus are less studied. While the advantages of student interaction and collaborative learning in Web-based learning environment has long been recognized, what remains to be identified are what instructional design of course tasks and activities that promote sustained and consistent student interaction and collaboration for knowledge construction.

Moreover, there is also evidence that online interactive learning and collaboration are not always sustainable and students' participation in CMC collaborative tasks may wane after the assessed tasks that require the postings are completed (Macdonald, 2003). In a recent survey on college student's attitudes toward participation in electronic discussions, Williams and Pury (2002, p. 1) found that "contrary to much literature on electronic collaboration suggesting students enjoy online collaboration, our students did not enjoy online discussion regardless of whether the discussion was optional or mandatory." Much needs to be done to explore factors that promote sustained student interest in online interactive learning and collaboration.

The Study

Through a post course survey, this study investigates the factors that promote sustained student participation in computer-mediated discussions as the core interactive learning tasks in a small group setting in an upper division undergraduate course that was offered entirely online. It also examines students' attitudes toward process and product oriented interactive and collaborative learning. The research questions are:

1. What factors encourage sustained participation, interaction, and collaboration in asynchronous discussion forums in a Web-based course?
2. What interactive learning tasks are sustainable and what are not?
3. Are there any differences in student attitudes toward process and product oriented online collaborative learning tasks? If so, what are

the factors that influence students' different perspectives toward such tasks?
4. What pedagogical implications do the findings have?

COURSE INFORMATION AND DATA COLLECTION

Course Information

The course under study was an upper division general education course in Bilingualism and Bilingual Education delivered entirely on Blackboard in Spring and Fall 2004 at a state university in California. A total of 60 students, 22 in the Spring semester class, and 20 and 18 students in the two Fall semester classes completed the course. All were local students who took the course online because the same course offered face to face conflicted with their schedules. Some students lived over an hour of driving distance from campus (not uncommon in Central California) and chose to take the online course to avoid commute. According to student self-report, all had taken at least one Web-enhanced course and were familiar with the Blackboard interface, although most of these courses used Blackboard for downloading course materials and lecture notes rather than integrating interactive learning activities. About 20% of the students reported they had taken at least one Web-based course. It was not clear how many of them experienced interactive learning online.

Collaborative Tasks and Their Assessment

Forum discussions on course readings and related issues formed the core interactive learning activities that were 45% of the course grade. These were process oriented interactive learning tasks for which individual grades were assigned for each student based on their quantity and the

Table 1. Course activities and grading

Activities	Grading	Description
Weekly group forums	45%	Structured discussions on course readings
Weekly class forums	0%	Required postings of moderator's summaries from each weekly group forum (Spring Semester class only)
Group project	12%	Final product graded interdependently (same grade for each member of the group)
Individual assignments	8%	No interaction among students required
Three exams	35%	Online exams on course contents to assess outcome of learning

quality of postings in the forums. Small groups of 4-6 people were formed at the beginning of the semester for the weekly asynchronous group forums. During the 16 week semester, a total of 18 discussion forums were completed in each online group. For each forum, the instructor assigned a reading chapter along with comprehension questions and discussion topics to help the students to grasp the contents. Students divided the reading questions among themselves in their groups and posted the answers to each question for the first round of postings. They were also required to make comments on at least one peer's answers in the second round of postings to carry on the discussions.

To ensure participation, strict deadlines for each round of postings were enforced and each student's answers to the questions and comment messages were assessed by the instructor who assigned up to 3% of the course grade for participation of each discussion forum. After each forum was completed, the moderator of each group (in each group, students rotated as moderators) was required to summarize the discussions and post the summary messages in a class forum that was accessible to all groups. These general class forums were intended to provide the students an opportunity to learn what was going on in other group forums that they did not have access to. This way, they did not need to read the numerous messages of 3-4 other groups but could still learn the gist of other group discussions. Although the summary messages were required, they were not graded. However, the summary of group discussions in a whole class discussion forum was eliminated in the Fall semester classes because it was not popular based on the input from the Spring semester class post course survey.

For the entire semester, the mean postings of each student in group forums ranged from 62-77 messages. On average, each student posted 3.5-4.3 in each of the 18 discussion forums. Although there was some variation in number of messages posted across groups and classes, most students did more than the minimum requirement of posting two rounds of messages in each discussion forum. Messages posted in the course related forums outside the group discussion forums were not included in the calculation because they were either inquiries or socialization in nature. Moderators' postings of summary messages in the class discussions forum were not included either because these postings were not enforced in the two Fall classes.

The other major collaborative task was a product oriented group project that constituted 12% of the course grade for which all the students in the same group received a common grade based on the level of collaboration and the quality of the final written report. There was no individual assessment component for the group project. The interdependent grading (a common grade for all members of a group only) was aimed at promoting more collaboration among the peers to produce a true collaborative product with individual contributions. The group project was closely related to one of the course themes on types of bilingual education programs. Each student was required to visit a local school to interview a bilingual teacher to gain firsthand information about bilingual education programs implemented in Central California. Students then shared and synthesized the interview data to produce a group report. They were not required to meet face-to-face for the group project but exchanged information in an online forum that was mostly procedural to plan, negotiate, to reach agreement and to produce the final product. The process of planning and producing the project required negotiation, cooperation, and collaboration among peers to actually arrive at consensus to produce a report. Though not graded, the progress of each group in the online forums was closely monitored by the instructor. The deadline for submitting the group project was strictly imposed to ensure the completion of the work.

Other course activities included two individual written assignments (8%) and three online exams (35%) that assessed the learning outcomes of the course readings and group discussions. Table 1 summarizes the course activities and grading.

Data Collection: Post Course Survey Data

At the end of the semester, an online survey was administered in each class to collect information about students' learning experience and their attitudes toward the course, in particular, their experience with online collaboration in both the weekly conference discussions and the group project. The survey questionnaire, which consisted of 17 multiple choice questions and 4 open-ended questions (see Appendix) was uploaded to the survey area of the course on Blackboard. Students were able to access and complete the survey questionnaire anonymously during the week after the final exam. Blackboard automatically calculated the results of the multiple choice questions in percentage. The transcripts of the survey responses for all three classes were printed out for analysis. 16 of the 22 Spring semester students and 37 of the 38 Fall semester students completed the survey questionnaire. Therefore, the analysis of the survey data was based on the 53 completed questionnaires.

RESULTS

Students' Attitudes Towards Online Discussions and Collaborative Learning

Table 2 presents student responses to the question "what are your thoughts about the structure of the course?" Overall, 92.5 % of the students preferred the collaborative learning in the form of small group discussions to the weekly online quizzes (7.5%) if given the choices. Additionally, the first open-ended question asked the students to describe their experience with the forum discussions. Among the 47 students who answered this question, only 1 student expressed negative experience with the discussion forums. Three students commented that their experience was mixed. The majority, 43 students (91.5 %), expressed their experience with this form of learning ranged from positive to extremely positive.

What factors encouraged students to participate in this form of active and interactive learning throughout the semester? Did the students really think they learned from building on each other's insights? What were the effects of such learning as reflected by students' responses in the survey data? The survey questionnaire addressed these issues in a number of questions. Table 3 summarizes students' responses to the effectiveness of group discussions.

Table 2. Students' responses to "what are your thoughts about the structure of the course?" (N = 53)

Choices	% Reponses	Chi²
I like the way the course is structured in terms of forum discussions because we learn from each other.	92.5%	41.679*
I prefer weekly quizzes based on the readings rather than answering questions and joining the group discussions.	7.5%	

*Note: *Unless otherwise specified, the P values of the Chi² is <0.0001 in this study.*

Chi Square analyses of students' responses to the questions in Table 3 along the scale of strongly agree to strongly disagree were all significant beyond 0.0001 level. (Unless otherwise stated, Chi Square analyses reported in this study were significant beyond the level of 0.0001.) About 90% of the students agreed or strongly agreed that answering questions and participating in discussions helped them understand the readings better and that online discussion was helpful because they collaborated more and learned more from each other. Additionally, 72% of the students responded that they learned more from online discussions than they would have learned from the lectures. Furthermore, 89% of the students responded, saying group cohesion and mutual trust was an important factor in their group.

Factors That Affect Level of Participation and Sustained Interaction

Assessment

Table 4 summarizes students' responses to the level of participation in their group discussions if the postings were not required and graded. Overall, 51% of the students responded they would post some but not as many messages, 21% said they would post very few, and 8% responded they would not post any messages at all! Only 21% responded they would post the same number of

messages. One might argue that the survey data may not reflect the real level of participation in discussions if the postings were not required or assessed because all the postings in this course were actually required and assessed. Therefore, a firm claim of the effect of assessment on forum contributions must be tested with a treatment group whose postings in forums were assessed and compared with a control group whose postings in forums were optional and unassessed. Nevertheless, students' responses to this survey question still reflect the "if not" situation because they had just completed the weekly postings for the entire semester and such learning experience would certainly affect their responses. Therefore, the "if not assessed" situation was contrasted against the real situation of "assessed" postings.

There was further evidence that tasks that were not directly linked to assessment did not attract as much attention and were difficult to sustain. For example, in the first offering of the course in the Spring semester, the moderators posted summary messages of each weekly group forum in a class forum by the deadline as required. However, these messages seldom attracted voluntary comments. Table 5 summarizes the Spring semester students' responses to a survey question on the whole class forums. The results showed that while 38% of the students acknowledged that it was an important way to learn the ideas of the other groups, which may indicate these students had read summary messages from other groups, 44% of the students

Table 3. Students' views about group discussions (N = 53)

Survey Questions	% Responses				
	Strongly agree	Agree	Disagree	Strongly disagree	Chi²
My answers to the questions and comments on peers' messages help me to understand the readings better.	30%	62%	8%	0%	49.717
My peers' answers/comments helped me understand the readings better.	32%	57%	11%	0%	39.453
I learned more from online discussions than I would have learned from lectures.	25%	47%	25%	2%	21.792
The online discussion is helpful because we collaborate more and learn from each other more.	38%	55%	6%	2%	41.415
The group cohesion and mutual trust is an important factor in our group.	53%	36%	11%	0%	36.132

Table 4. Students' responses to "would you post the same number of messages as you actually did over the semester if these postings were optional, not required or graded?"

Choices	% Responses	Chi²
Yes, I will post the same number of messages	21%	21.491
I will post some messages but not as many	51%	
I will post very few messages	21%	
I will not post any messages	8%	

reported they seldom read these messages. 19% of the students responded that the class forum should be eliminated.

As the whole class forum was not popular with the majority of the students in the first offering of the course, this task was eliminated in the Fall semester classes. To investigate whether students missed the level of input from other groups, the post course survey asked Fall semester students questions about their thoughts on the input from other groups. Table 6 summarizes the responses from the two Fall semester classes.

Seventy-eight percent of the students felt that participating in their own group discussion was sufficient to learn the course contents and it would have taken too much time to read and respond to the summary messages from other groups. 14% responded that every group should have summarized their forum discussions each week and posted it to a general forum so that interested students could comment on the discussions in other groups. Very few students, 8% in all, wanted other group members to read their postings or missed the discussions in other groups. Therefore, with or without the summary postings in the whole class forums, the survey data suggest that the majority of students showed the same lack of interest in participating whole class discussions that were not graded.

Table 5. Spring semester students' responses to the tasks of "group summaries in the main message board" (N = 16)

Choices	% Responses	Chi²
Is relevant and is an important way to learn the ideas of other groups	38%	10.12*
I seldom read these summaries	44%	
Can be eliminated because I have never read the group summaries	19%	

*Note: * p < 0.0063*

Table 6. Fall semester students' attitudes toward other group discussions (N = 37)

Choices	% Responses	Chi²
I wanted other group members to read our group discussions and I also missed the discussions in other groups.	8%	90.392
Every group should have summarized their forum discussions each week and post it to a general forum so that interested students could comment on the discussions in other groups.	14%	
Participating my own group discussion is sufficient for me to understand the course contents. It would take too much time to read and respond to summary messages from other groups.	78%	

Deadlines

Table 7 presents students' responses to the importance of deadlines in the weekly postings of group forums.

Overall, 93% of the students agreed or strongly agreed that the imposed deadlines for postings had an important impact on their participation in collaborative learning and in getting the tasks done in a timely fashion.

Group Formation

Table 8 summarizes students' responses to the question on group formation.

62% of the students responded that they preferred to work with the same people for their group discussions and 30% expressed that they did not have any preferences. Only 8% responded that they wanted to work with different people because

they felt that they would also learn from other students they never interacted with in this course. Chi Square analysis yielded highly significant differences between the responses.

It appears that the group as a community for online learning established deep roots in this course. Recall that the class level discussion forums in the form of summaries from each group forum was eliminated. Except for some course related general forums in which questions regarding course activities were exchanged, students generally did not have access to the majority of the fellow students in their class. It would not have been surprising if students had expressed their desires to learn the discussions in other groups through some form of exchanges on a class level, or, through reshuffling groups. Yet, the survey responses suggest that at least two thirds of the students did not express the need to work outside their fixed groups.

Table 7. Students' attitudes toward deadlines in group discussions (N = 53)

Statement	% Responses				
	Strongly agree	Agree	Disagree	Strongly disagree	Chi²
The deadlines for the readings and postings in each forum are very important because they help to complete the readings and the course	51%	42%	7%	0%	39.755

Table 8. Fall Semester students' responses to "what is your view about group formation?" (N = 37)

Choices	% Responses	Chi²
I want to work with the same group members the way it is now because we know each other better.	62%	44.244
I want to work with different people in a group every few weeks because we will learn from other students we never meet.	8%	
It will not make a difference to me working with the same people or different people in a group.	30%	

It is important to note that the survey data reflected the student views towards their working groups that were fixed for the entire semester. If they actually had the chance to work in different groups in this online course, they might have different views. To explore the advantages and disadvantages of fixed or dynamic small groups in a Web-based course that uses weekly forum discussions, both group types need to be included in the data in future studies.

Process vs. Product Oriented Collaboration

Table 9 presents students' responses to a question that allowed for multiple choices about the group project.

Two of the choices provided in the answers were aimed at assessing whether the assignment itself was important for the course in the students' eyes because the importance of the group project may affect their overall performance. As seen in Table 9, 70% of the students from all three classes responded positively about the importance of this group project and agreed they learned a lot through doing it. However, 30% of the students felt that it could be an individual project focusing on one school rather than a group project that involved more collaboration. The Fall semester postcourse survey asked an additional question about the group project and the responses are summarized in Table 10.

While 24% of the students preferred to work with peers because they had no problems to collaborate, exactly another 24% of them did not like to depend on other peoples' schedules because some just did not get the work done on time. Similarly, although 22% of the students felt it worthwhile to collaborate for the group project despite the fact that it was difficult, 32% preferred individual work leading to a project of their own even though they would not accomplish as much.

Table 9. The group project about bilingual programs in our local schools (N = 53)

Multiple choices (choose all that apply)	% Responses
Is a good assignment and I learned a lot through doing the project.	70%
Makes the course readings more meaningful and more relevant to me.	68%
Is a good assignment but takes too much time to complete.	17%
Could be an individual assignment focusing on one school rather than a group project that involves more collaboration.	30%
Is not very important for this course.	4%

Table 10. Fall semester students' response to the group project (N = 37)

Choices	% Responses	Chi²
I prefer individual work leading to a project of my own even though I only have information about one school.	32%	1.162*
I prefer to collaborate with peers the way it is now because it is not a problem with me to collaborate.	24%	
I prefer to collaborate with others for a group projected but I do not like to depend on other people's schedule because some just do not get their work done on time.	24%	
Even though it is hard to collaborative for the group project, it is still worth doing it because we learn more about our bilingual programs in different schools through doing it together.	22%	

*Note: *P = 0.072*

Chi Square analysis failed to yield significant differences between student responses to this question. Compared to 92.5% positive responses toward collaboration in forum discussions, students' attitudes toward online collaboration in producing the group project were mixed.

Such differences were also reflected in some student comments on the group project in the open-end questions. As the open-ended questions did not address the group project directly, only 11 students expressed their views about the group project in their responses to the question about their likes and dislikes about the course (Questions 18b), and the question about any changes they wanted to recommend to improve the course (question 18d). Of these 11 students, one commented that she liked the group project the most about this course. However, 10 expressed their dislikes or frustrations about the group project. One student wrote "I think it's too inconvenient to try and get a group project together online. I also don't like having someone's performance affect my grade. I would rather do the project on my own." It appears that the end product type of collaborative tasks demands more consensus-building collaboration. When students were timed for such intensive interaction and collaboration, they became less enthusiastic about it.

DISCUSSION

A number of issues can be identified in answering the four research questions raised earlier. The following two sections discuss these questions along with the research findings.

Factors that Promote Sustained Online Small Group Discussions

Survey data suggest that a number of factors contributed to the sustained small group discussions in this course. Among others, the structure of discussions with carefully prepared discussion questions, small groups with fixed group members for interactive learning activities, the direct link between participation and assessment, and the strictly imposed deadlines for each forum were the main factors that contributed to the sustained the interactive learning in this Web course.

Previous studies suggest that topics that are not relevant to the course contents or not related to students' life experience do not attract participation and are not sustainable in online discussions (Williams & Pury, 2002). One of the factors that might have contributed to the sustained online small group discussions in this study was that students not only always had "something to say" in each forum but knew exactly what specific questions they were expected to answer in advance. These written exercises required in the first round of postings kept each individual student accountable for knowing the contents through reading. Therefore, students' interaction with the course readings, the first level of interaction with the material, was enhanced by producing written answers to be commented by peers in the group forums. Predetermined specific comprehension questions and thought provoking topic questions for each reading assignment helped students to focus on the learning contents and provided continuous discussion topics for the weekly group forums. Such proactive learning not only engaged students in a higher level of thinking (Gao &

Lehman, 2003; Wu, 2004) than the reactive type of learning but also kept the students accountable for participating in the weekly forums. The enthusiasm in group discussions never waned forum after forum because each forum focused on a new reading chapter.

Furthermore, the comment messages required students to exchange information by building on each other's ideas to negotiate for meaning and to collaboratively construction knowledge. Such interaction between peers and between students and instructors provided another level of interaction for learning. Students' positive experience with the semester long forum discussions was related to the benefits of proactive learning and learning from each other for knowledge construction. While the advantages of online interactive learning have long been proved in previous studies (Kern, 1995; Lavooy & Newlin, 2003; Mouza, Kaplan & Espinet; 2000; Summer & Hostetler, 2002; Wang & Teles, 1998; Wu, 2003), this study provided new data for the use of small group discussion as the core interactive learning tasks through the application of carefully prepared discussion questions that elicits proactive learning and through peer interaction and collaboration. When online collaborative learning tasks become main course pedagogy, such interactive learning is likely to be more sustainable and effective.

Previous studies have also indicated that collaboration as a process of participating to the knowledge communities is not an equal process to all the members of the community (Leinonen, Järvelä & Lipponen, 2003) and the size of online learning community affects the level of comforts which influences the level of participation (Williams & Pury, 2002). The current finings suggest that a group of 4-6 members can be an efficient and active learning community in which the members tend to generate sufficient responses from each other. On the other hand, the number of messages produced by each member was manageable and easy to keep track of. However, caution must be taken on this finding as the current study did not experience with other group sizes. Future studies need to test different group sizes with different learning tasks.

The survey data also indicated that students believed that group cohesion and mutual trust was the main factor of their groups. Furthermore, the majority of students said they preferred working with the same members of the group for the entire semester rather than rotating the peers. Obviously, it takes time to establish such mutual trust, even in a small group of 4-6 members. Therefore, it is very likely that the group cohesion and mutual trust comes from the semester long interaction, cooperation, and collaboration online. A small number of students expressed the desire to work with different peers and some did not show preference in working with the same peers or not. Future studies need to investigate the benefits and disadvantages of dynamic group formations in which students are given the chance to work with different online peers during the semester.

Survey data also indicated that assessment played a crucial role in motivating the students to participate in the semester-long group discussions week after week. Although over 90% of the students claimed that they learned more from reading and commenting on peers' messages, many admitted they would not have posted as many messages if the postings had not been required and assessed. The data support the previous research findings that the assessment of collaborative learning tasks plays a crucial role in ensuring student participation. Macdonald (2003) reported that students actively contributed to the discussions when the tasks were assessed but participation of discussions waned when the postings became optional. Grade for discussion was also positively related to students' perceived learning (Jiang & Ting 2000). Apparently, any optional interactive learning tasks would not have sustained for the entire semester.

Current data also suggest that required postings that were not directly linked to assessment did not attract equal amount of attention as the graded postings. This was clearly demonstrated in the lack of interest in participating the forum discussions at the whole class level in the Spring class. Williams and Pury (2002, p. 1) reported that "contrary to much literature on electronic collaboration suggesting students enjoy online collaboration, our students did not enjoy online discussion regardless of whether the discussion was optional or mandatory." It was not clear whether their "mandatory" participation of discussions was enforced by direct assessment of the actual postings in the forum discussions. This study provided further evidence that direct assessment of student interactive learning in CMC promotes sustained participation and interaction and also affects the level of participation and interaction.

Another important factor that appeared to have contributed to the completion of each discussion forum on time for the entire semester was the strictly imposed deadlines for each round of posting in each discussion forum. Student responses to survey questions suggest that required postings alone were not sufficient for guaranteed participation and interaction within a time frame. Strict deadlines seemed to be the best solution to complete the weekly forums on schedule. Therefore, the importance of imposing deadlines cannot be overemphasized for even directly assessed interactive learning activities.

Process and Product Oriented Interactive Learning

Very few studies have dealt with the differences between process and product oriented interactive learning tasks and how these differences influence peer interaction and collaboration (Kear, 2004; Kear & Heap, 1999; Macdonald, 2003). This Web-based course applied both process and product orientated interactive learning tasks that required different types and levels of interaction and collaboration. As discussed earlier, in the weekly group forums, the debate and exchange of ideas focused on the process of learning that did not lead to a final product. In contrast, the group project was a product driven collaborative task in that the interaction and collaboration among the peers through sharing and exchange of ideas and negotiation must help to reach certain consensus to produce a group report. Survey data suggest that students were more enthusiastic about process oriented group discussions than the group project even though 70% of the students agreed that the group project was a good assignment and they learned a lot through doing it.

Among others, the main reasons for students' frustration about the group project were the difficulties in reaching agreement according to a time frame, especially in the online environment. The differences in working pace and conflicts of schedules, and, perhaps more importantly, differences in level of devotion to the collaborative task in online environment made it more difficult for the peers to reach consensus in the process of doing the group project. The early birds who preferred to start and complete their parts of the work in a timely fashion conflicted with those who procrastinated in getting the work done. As peers in the same group would receive a common grade only for their project, there was pressure for them to compromise to reach agreements in completing the project.

Although the common grade can be used as a useful instructional strategy to implement end product driven collaborative tasks to encourage collaboration, the frustration and stress caused by the schedule conflicts and different levels of devotion toward such collaboration calls for more careful instructional design of such tasks. Perhaps some form of individual grading in addition to the interdependent grading are necessary to measure each individual student's efforts and contribution. In fact, Kear and Heap (1999) reported that students expressed a preference for a higher individual grade component when both common and individual grades were assigned for their group project. It is important to balance the level of collaboration among the students and the individual flexibility of online learning. Future studies need to address the pedagogical design of end product driven collaborative tasks in Web-based courses.

CONCLUSION AND RECOMMENDATIONS

This study identified some important factors that promote sustained online small group discussions as main interactive learning tasks in a Web-based course. Among other things, the structure of the online discussion, group size and group cohesion, strictly enforced deadlines, direct link of the interactive learning tasks to the assessment, and strictly imposed deadlines are some of the important factors that influence participation and motivate sustained online interaction and collaboration. The differences in process and product driven interactive learning tasks also have a different impact on student online collaboration. In general, students were more enthusiastic about process oriented than product driven collaborative tasks.

Finally, as the current data are based on one Web-based course that was mainly a reading course, the findings may not be generalized into a broad scope. Because of this limitation, the current findings may not be directly applicable to other courses that have a different online pedagogical approach. Yet, a few recommendations may be made for designing and implementing similar interactive learning activities to promote sustained and effective online collaboration.

- Although a very good tool for promoting interactive learning and collaboration, online discussion is not always sustainable if not well planned and structured. It is recommended that instructors carefully design each forum discussion with direct involvement of course contents with predetermined specific questions to engage students in a high level of thinking through providing written answers to the topics for which peer critiques are required.

- To continue to motivate the students, link the assessment with all interactive learning tasks utilizing specific grading scales.
- Impose strict deadlines for each round of postings in each discussion forum.
- Form small groups of 4-6 as learning communities for discussions so the peers will have sufficient input from each other yet still find it easy to keep track of all the postings in each new thread.
- Use process oriented interactive learning tasks to facilitate continuous online interaction and collaboration and yet still give each student sufficient amount of freedom in completing the assessed learning tasks.
- When design product oriented interactive learning tasks, much care needs to be taken in order to prepare the students to reach consensus. Give sufficient time for completing such learning assignment. Incorporate both common and individual grades in grading a group project.

ACKNOWLEDGMENT

The author thanks Sarah Maddison, Terese Thonus, and Ondine Gage-Serio for their insightful comments on earlier versions of the paper. The author also appreciates many helpful comments from the Associate Editor. Thanks are also due to Dawn Truelsen for her assistance in online course design using Blackboard.

REFERENCES

Duin, H., & Hansen, C. (1994). Reading and writing on computer networks as social construction and social interaction. In C. Selfe & S. Hilligoss (Eds.), *Literacy and computers: The complications of teaching and learning with technology* (pp. 89-112). New York: The Modern Language Association.

Gao, T., & Lehman, J. D. (2003). The effects of different levels of interaction on the achievement and motivational perceptions of college students in a Web-based learning environment. *Journal of Interactive Learning Research, 14*(4), 367-387.

Jiang, M., & Ting, E. (2000) A study of factors influencing students' perceived learning in a Web-based course environment. *International Journal of Educational Telecommunications, 6*(4), 317-338.

Kear, K. (2004). Peer learning using asynchronous discussion systems in distance education. *Open Learning, 19*(2), 151-164.

Kear, K., & Heap, N. (1999). Technology-supported group work in distance learning. *Active Learning, 10,* 21-26.

Kern, R. (1995). Restructuring classroom interaction with networked computers: Effects on quantity and characteristics of language production. *The Modern Language Journal, 79,* 457- 476.

Kuhl, D. (2002). *Investigating online learning communities.* U.S. Department of Education Office of Educational Research and Improvement (OERI).

Lavooy, M. J., & Newlin, M. H. (2003). Computer mediated communication: Online instruction and interactivity. *Journal of Interactive Learning Research, 14*(2), 157-165.

Leinonen, P., Järvelä, S., & Lipponen, L. (2003). Individual students' interpretations of their contribution to the computer-mediated discussions. *Journal of Interactive Learning Research, 14*(1), 99-122.

Macdonald, J. (2003). Assessing online collaborative learning: Process and product. *Computers and Education, 40,* 377-391.

Mouza, C., Kaplan, D., & Espinet, I. (2000). *A Web-based model for online collaboration between distance learning and campus students* (IR020521). Office of Educational Research and Improvement. U.S. Department of Education.

Sumner, M., & Hostetler, D. (2002). A comparative study of computer conferencing and face-to-face communications in systems design. *Journal of Interactive Learning Research, 13*(3), 277-291.

Wang, X., & Teles, L. (1998) Online collaboration and the role of the instructor in two university credit courses. In T. W. Chan, A. Collins, & J. Lin (Eds.), *Global Education on the Net, Proceedings of the Sixth International Conference on Computers in Education* (Vol. 1, pp. 154-161). Beijing/Heidelberg: China High Education Press and Springer-Verlag.

Williams, S., & Pury, C. (2002). Student attitudes toward participation in electronic discussions. *International Journal of Educational Technology, 3*(1), 1-15.

Wu, A. (2003). Supporting electronic discourse: Principles of design from a social constructivist perspective. *Journal of Interactive Learning Research, 14*(2), 167-184.

APPENDIX: SURVEY QUESTIONNAIRE

1.   Is this your first Web-based (entirely online) course?

    a.   ___Yes.

    b.   ___No, I already took one entirely online course before this one.

    c.   ___No, I took two or more other entirely online courses before this one.

    d.   ___I took one or more Web-enhanced course (partially online) before this Web-based (entirely online course).

    e.   ___No, I have never taken any Web-based nor Web-enhanced course.

2.   This reading course is structured on group discussions with individual and group assignments. What are your thoughts about the structure of the course?

    a.   ___I like the way the course is structured in terms of forum discussions because we learn from each other.

    b.   ___I prefer weekly quizzes based on the readings rather than answering questions and joining the group discussions.

3.   Will you post the same number of messages as you actually did over the semester if these postings were optional, not required and graded?

    a.   ___Yes, I will post the same number of messages.

    b.   ___I will post some messages but not as many.

    c.   ___I will post very few messages.

    d.   ___I will not post any messages.

4.   Please circle one answer for each of the following:

    a.   In our group forums, my answers to the questions and comments on peers' messages help me to understand the contents/readings of the course better.

    strongly agree   agree   disagree   strongly disagree

    b.   My peers' answers/comments helped me to understand the readings better.

    strongly agree   agree   disagree   strongly disagree

    c.   I learned more through online discussions than I would have learned from the lectures.

    strongly agree   agree   disagree   strongly disagree

    d.   The online discussion is helpful because we collaborate more with each other and support each other.

    strongly agree   agree   disagree   strongly disagree

    e.   The group cohesion and mutual trust is an important factor in our group forums.

    strongly agree   agree   disagree   strongly disagree

f.   I prefer individual work to group work and would have done better if I did not have to collaborate with my peers in my group for **discussions**.

strongly agree   agree   disagree   strongly disagree

g.   I prefer individual work to group work and would have done better if I did not have to collaborate with my peers in the group for the **final project**.

strongly agree   agree   disagree   strongly disagree

h.   The deadlines for the readings and postings in each forum are very important because they help me to complete the readings and the course.

strongly agree   agree   disagree   strongly disagree

i.   The overall course contents are interesting and I have learned a lot about bilingualism and bilingual education from taking this course.

strongly agree   agree   disagree   strongly disagree

5.   Choose one of the following:

a.   ____ I wanted other group members to read our group discussions and I also missed the discussions in other groups.

b.   ____ Every group should have summarized their forum discussions each week and post it to a general forum so that interested students could comment on the discussions in other groups.

c.   ____ Reading and responding to peers' messages in our own group discussions is sufficient for me to understand the course contents. It would take too much time to read and respond to summary messages from other groups.

6.   What is your view about group formation?

a.   ____I want to work with the same group members the way it is now because we know each other better.

b.   ____I want to work with different people in a group every few weeks because we will learn from other students we never meet.

c.   ____It will not make a difference to me working with the same people or different people in a group.

7.   The pace of the course, including readings and postings

a.   ____Is neither too fast nor too slow for me.

b.   ____Is too fast for me because I always try to catch up with the readings.

c.   ____Is too slow for me and we could have read more chapters.

d.   ____Should be OK for a course like this but I found it too fast for me because I work many hours a week and have limited time for course work.

8. Course documents:

    a. ____I printed out all the lecture notes and review guides (or some of them) because they are helpful.
    b. ____I read the lecture notes and the review guides online but did not print them all.
    c. ____I never printed out nor read the lecture notes and the review guides because they are not essential for me.

9. The videos on reserve in the music library are used in all other face to face sessions of the same course. I found these videos

    a. ____worth seeing because they are informative and very relevant to the course content.
    b. ____relevant to the course content, but it is hard for me to make special trips to the university to watch them all.
    c. ____are not relevant to the course content and can be omitted.

10. You took all the three exams online in this semester. Do you think the online exam should be kept the way they are now, or do you prefer to take these exams in a classroom on a certain date?

    a. ____I prefer online exams the way they are now.
    b. ____I prefer to come to a classroom to write the exams.
    c. ____I have no preference.

11. Exam format:

    a. ____ I prefer multiple choice exams.
    b. ____ I prefer essay question type of exams.
    c. ____ It does not make a difference for me.

12. The group project about bilingual programs in our local schools (circle all the answers that apply to you)

    a. ____Is a good assignment and I learned a lot through doing the project.
    b. ____Makes the course readings more meaningful and more relevant to me.
    c. ____Is a good assignment but takes too much time to complete.
    d. ____Could be an individual assignment focusing on one school rather than a group project that involves more collaboration.
    e. ____Is not very important for this course.

13. For the group project:

    a. ____I prefer individual work leading to a project of my own even though I only have information about one school.
    b. ____I prefer to collaborate with peers the way it is now because it is not a problem with me to collaborate.
    c. ____I prefer to collaborate with others for a group projected but I do not like to depend on other people's schedule because some just do not get their work done on time.
    d. ____Even though it is hard to collaborative for the group project, it is still worth doing it because we learn more about our bilingual programs in different schools through doing it together.

14. Overall, my experience with this Web-based course

 a. ____Is very positive.
 b. ____Is positive.
 c. ____Is negative.
 d. ____Is very negative.

15. Experience with the Blackboard and the online forums: circle all apply to you.

 a. ____I found it challenging at the beginning but quickly picked up and like it now.
 b. ____The interface is straightforward and easy to learn, although I was not very experienced with any online courses.
 c. ____It was never a problem for me because I am good at technology.
 d. ____It was a plus because I learned the technology as well as the course contents.

16. If I have the choice in future,

 a. ____I will take a similar Web-based course.
 b. ____I will not choose to take a similar Web-based course.
 c. ____It will not make a difference, Web-based or face-to-face version.

17. Would you recommend a friend to take this Web-based course?

 a. ____Yes
 b. ____No
 c. ____Not sure

18. Please take some time to answer the following questions:

 a. Please describe your experience with the forum discussion part of the course. (positive, negative, expectation, effect on learning, etc. anything you think is relevant)

 b. What do you like the most, or dislike the most about this course?

 c. In your opinion, what are the most important elements for a Web-based course like this to be successful?

 d. To improve the course for future students, what changes do you recommend?

This work was previously published in the International Journal of Web-Based Learning and Teaching Technologies, Vol. 2, Issue 1, edited by L. Esnault, pp. 17-38, copyright 2007 by Idea Group Publishing (an imprint of IGI Global).

Chapter 7.10
Challenges to Implementing E-Learning in Lesser-Developed Countries

Bolanle Olaniran
Texas Tech University, USA

ABSTRACT

The integration of communication technologies and the Internet has created an explosion in the use of e-learning both locally and globally. The beneficiaries of this new media integration are organizations at large, in both developed and lesser-developed countries. For instance, globalized organizations have been able to develop training programs that serve their needs. However, global e-learning raises some implications, which include communication, culture, and technology, that must be addressed before successful implementation and outcome can occur.

CHALLENGES TO IMPLEMENTING E-LEARNING AND LESSER-DEVELOPED COUNTRIES

As communication technologies and the Internet continue to merge, organizations continue to integrate them within their activities and corporate practices. One of the key benefits of such integration includes learning and curriculum development, which is otherwise referred to as e-learning, and more appropriately referred to as global e-learning. Because of the trend toward globalization of research and development (R&D), there exists the need for uniform and customized training. On a more comprehensive scale is the need for employees' continued training, which circumvents traditional college training and requires participants to be in a specific location in order to access and participate in learning. It is not surprising that online universities (e.g., AIU Online, Capella University, Devry University, Kaplan University, University of Phoenix, Walden University, and Westwood College Online) are thriving and attractive to corporate travelers and expatriates. Notwithstanding, as corporate e-learning solutions continue to explode and gain popularity in the sphere of global e-learning, challenges exist from cultural standardization rather than differentiation. Standardization creates problems for learners who are culturally different from the culture that developed the learning content.

The major advantage of e-learning remains cost savings. However, for e-learning to produce desired results, there should be some kind of accounting for effectiveness of the learning program. Effectiveness of e-learning cannot be assessed outside of its cultural underpinnings. To this end, the current chapter examines cultural implications of global e-learning and education. It explores significant challenges created by learning preferences and adoption of innovation using the Hofstede's REF (1983) dimensions of cultural variability. Finally, the chapter provides recommendations for implementing successful global e-learning programs.

As the Internet goes global, so does e-learning (Van Dam & Rogers, 2002). However, the e-learning programs mostly emphasize organizational goals in terms of how and what organizational leadership intends to accomplish in their respective region and employees. In essence, users of e-learning often ignore cultural implications and insights that employees or customers have in controlling how they learn and the learning process as a whole. Specifically, consumers of e-learning (e.g., purchasers, instructors, students, and end-users) are expected to work with curriculum designed in and for another culture.

Internet Usage and E-Learning

There is a correlation between Internet usage and global learning penetration and adoption in any given society. In Asia, Internet usage was expected to increase from 64 million in 2001 to 173 million by 2004, but the most recent data on internet usage in Asia show an actual figure that reflects a jump to 323.76 million—a 405% increase from the 2001 figure, and a 87% increase above projected figure (Internetworldstats, n.d.). A 65.8% increase is reported for Japanese Internet use while a 357.8% jump was recorded for Chinese between 2000 and 2005, which represents a jump from 22.5 million to 103 million users (Internetworldstats, n.d.). In spite of the tremendous increase in the number of users, the overwhelming predominance of English is considered a major deterrent that limits Internet usefulness for most countries and regions (Barron, 2000). For instance, when the number of Internet users is compared with actual population figure, it reveals that there are still a lot of potential users to reach. The Internet world statistics, for example, puts the total Internet usage for China at 7.9% of the country's population figure.

CULTURE AND E-LEARNING

In order to realize the aim of e-learning as an educational tool, it is essential to accommodate the learning needs of different cultures in order to promote equitable learning outcomes for targeted students, and to promote education and technological literacy that improve socio-economic opportunities in developing nations (Dede, 2000; Henning, 2003; Selinger, 2004). Attention to geographic cultures and implications for diffusion of technological innovation is warranted. But first it is important to look at the dimensions of cultural differences. One useful model in exploring cultural differences includes Hofstede's dimensions of cultural variability (Hofstede, 1980). The four dimensions of cultural variability are power distance, uncertainty avoidance, individualism, and masculinity (Hofstede, 1980, 1983, 2001; see also Dunn & Marinetti, 2002, overview of cultural value orientations and cultural dimensions). These four categories result from data collected from 50 countries and three world regions (Hofstede, 1980, 1983). Past research uses these four dimensions to operationalize cultural differences and their effects on uncertainty reduction in intercultural communication encounters (Gudykunst, Chua, & Gray, 1987; Olaniran, 1996; Olaniran & Roach, 1996; Roach & Olaniran, 2001, 2004; see also www.worldvaluessurveys.org). A brief description of the four dimensions includes the following.

Power distance is "the extent to which the less powerful members of institutions and organizations accept that power is distributed unequally" (Hofstede & Bond, 1984, p. 418). Uncertainty avoidance describes "the extent to which people feel threatened by ambiguous situations and have created beliefs and institutions that try to avoid these" (Hofstede & Bond, 1984, p. 419). Individualism-collectivism acknowledges the fact that in individualistic cultures, "people are supposed to look after themselves and their family only," while in collectivistic cultures, "people belong to in-groups or collectivities which are supposed to look after them in exchange for loyalty" (Hofstede & Bond, 1984, p. 419). Masculinity-femininity refers to cultures "in which dominant values in society are success, money, and things," while femininity refers to cultures "in which dominant values are caring for others and quality of life" (Hofstede & Bond, 1984, pp. 419-420). One of the challenges to dimensions of cultural variability is that comparisons are "relative" and "restricted" to two cultures or regions. Notwithstanding, these dimensions can still serve as a starting point for educational providers in global e-learning contexts. It is noted that cultural drivers of people play a significant role in learning, while representing the foundation for which global e-learning platforms must be based (Henning, 2003; Van Dam & Rogers, 2002).

Van Dam and Rogers (2002), using Hofstede's dimensions of cultural variability, consider design elements and actions for adaptation of e-learning. Within uncertainty avoidance dimension and e-learning, issues of security and risk are of primary concern. For instance, the e-learning is expected to be seen in a high-risk environment (i.e., low uncertainty avoidance culture) as something intriguing and potentially fun, motivational and interesting; while in a low security environment (i.e., high uncertainty avoidance culture), it can be perceived as dangerous and risky. Power distance, which is a measure of inequality in a given culture, suggests that in high equality culture (low power distance), the expectation is that knowledge is shared or distributed equally across an organizational structure. In a high status culture (high power distance), however, the expectation calls for "telling" strategies where the knowledgeable are required to teach whatever needs to be learned. Individualism then suggests that in high independence culture, there is a sense of controlling one's destiny as far as career and work choices go (i.e., freedom to choose). But in collectivistic culture (group oriented), the success of the group is more important. The masculine-feminine dimension describes the idea of work-life balance where work-focused countries require achievement and recognition, that is, people "live to work" (masculine culture), whereas in a "life-focused" culture, work-related issues, including learning, must be performed within the context of life, in essence, people "work to live" (feminine culture).

Value Preferences and Technology

Despite decentralization and opportunity for increased participation, facilitated by technology in certain cultures, most cultures still remain high context and power distant (e.g., African countries, Japan, South East Asian countries). In a high-context culture, information is internalized in the person or situations, while power-distant cultures recognize or accept the fact that power is not evenly distributed (Hall, 1976; Hofstede, 1980). These cultural categories have implications for implicit and explicit communication tendencies and the general propensity to use technology in global education and e-learning. Therefore, cultural factors tend to influence how individuals use or view communication technologies, and interpretations drawn from messages through them. Specifically, Devereaux and Johansen (1994) argue that it might be difficult to get people to use certain technology such as the computer-mediated communication (CMC) systems in power-distant cultures where status dictates every aspect of interpersonal communication. Others (Ess, 2002), however, have argued that the "soft deterministic" effect of technology, implying that every culture tends to find ways to adapt technologies to their cultural communication patterns. For instance, in the African culture, where significant emphasis is put on relationships, it was found that when e-mail was used for local communications, organizations habitually followed the e-mail with the telephone as a back-up medium to ensure that the message had been received (McConnell, 1998), and had the desired effect (Olaniran, 2001). Following through with a more traditional medium may be less cultural than McConnell suggests, however. No scholars disputed the fact that cultural differences affect technology adoption and use. Japanese designers acknowledge the effect of culture that not all types of communication can be supported by communication technology such as the CMC systems (Heaton, 1998).

Heaton (1998) contends that if communication technologies are to be useful in Japan, it is important that a familiar sense of atmosphere or feeling must be conveyed through the system. Her research on computer-supported collaborative work (CSCW) systems in Japan suggests that it is problematic for groups to use computers without first meeting face-to-face to establish a trust environment (see also Barron, 2000; Mason, 1998).

The challenge in what a technology innovation such as e-learning can offer, and the hindrance by

traditional (local cultural) approach, is not to be taken lightly. People fear new things despite the fact that change itself is a constant in human life. Moreover, in high power-distance cultures, people tend to see a technology system as threatening to their traditional learning methods. The perceived threat creates anxiety about technologies, and consequently, the ensuing negative reactions to using these technologies. For instance, Henning (2003) provides a synopsis of sample effects of culture in response to online courses and information. The encounter synopsis looks at interaction with technology between two teachers in South Africa. She found a high level of anxiety among e-learners. In addition to the physical distance to cyber café, and not having computer facilities at their schools, home, or village, the students' struggle with technology is vividly apparent in the following statements:

When I wrote my first discussion posting I was so afraid. Would this get to others? Will they laugh, what will Prof say? ... I feel I have not the same control as before. I type and I read and I am scared to click because when I do that I feel I am falling down. (Henning, 2003, p. 308)

From this example, one can see the terrifying feelings that emerge from the feeling of loss of control emanating from the attempt to adapt to the new mode of learning. The feelings of anxiety are real to the people from this culture, because they have certain expectations of how learning ought to occur. Henning interprets this information and concludes that these individuals face confusion with who is in charge of the learning environment (i.e., the teacher or the student). Furthermore, the scenario also points to another problem that needs to be addressed by e-learning content providers who are usually from industrialized and economically-developed countries (EDCs) and low power-distance culture: These providers often stress the convenience of online learning and, more importantly, the freedom to put learners in control of learning. Yet in the above scenario, participants point to the need for instructors to perform their job and teach by telling. This case illustrates an example of what we can expect from learners in high power-distance cultures, especially when using a technology system that focuses on a different (i.e., low power distance) learning preference.

Attention to differences in the oral tradition of certain collectivistic cultures and the non-oral tradition's emphasis among individualistic cultures has different implications for e-learning. For instance, e-learning in oral tradition cultures might be better to allow for more interpersonal interactions where students and instructors get to explain ideas to one another. On the other hand, the concept of self-paced independent focus for e-learning might succeed in non-oral tradition cultures. Cultural differences can influence the use or choice of technologies in e-learning. The use of PowerPoint in both nonoral and oral cultures provide supports for the above recommendation. For example, it was reported that some instructors from the United Kingdom (i.e., non-oral culture) succeeded in e-learning contexts when using PowerPoint presentations, while their counterparts from the United Arab Emirates (UAE) saw PowerPoint presentations as extra work that did not fit into oral cultures where students explained ideas to one another about what they read (Selinger, 2004). Specifically, Selinger (2004) reports one student claimed that he does not like e-learning because he prefers to work from a book and from talking with people.

The fact that most of the teachers' toolboxes, provided by the content provider and based on cultural differences, are considered irrelevant by instructors from cultures different from those of the content providers; this adds support to the argument that the needs of the end users should be incorporated into the e-learning course design. This prompts one to ask, "What use is an e-learning toolbox when information provided is not useful to those who need it to do their duties?" Selinger

(2003) finds that instructors from about half of the countries participating in the evaluation of the Cisco global e-learning program hardly use the materials available on the teachers' site, and have to develop or adapt the contents to their own eclectic cultures. A more effective strategy would be to resolve problems before the course, by asking the instructors about their requirements and students' preferences, instead of after the fact.

Language

Central to cultural challenges in globalized or e-world learning is the issue of language, since the majority of Internet content is in the English language (Barron, 2000; Van Dam & Rogers, 2003; Wilborn, 1999). Non-English speaking individuals may feel that technology has nothing to offer them since they cannot understand the content. Even in situations where people speak or understand English, its use is limited to certain contexts such as e-mail and entertainment. For instance, it is not uncommon for people in other cultures to restrict their English usage to work, school, or formal business settings, and using the local language and dialects for most of their daily communication encounters. Furthermore, scholars acknowledged that learning a second language in school is quite different from simply learning a foreign language itself (Collier, 1995). Moreover, as much as 36% of online users indicated that they would prefer a language that differs from English (Van Dam & Rogers, 2003). In a study of the global e-learning program offered by Cisco in the English language, students who use English as a second language indicated that they prefer their instructors to first overview the contents of the chapter before readings were assigned (Selinger, 2004). In a global e-learning curriculum, it may also help to note that even when curriculum is made available in languages other than English, there are differences within languages. For example, in the Cisco scenario mentioned above, French and Spanish versions of its e-learning course were provided. Unfortunately, the French version was the Canadian French, and the Spanish version was the South American Spanish, both of which differ from their European versions, thus creating problems for students from France and Spain, and necessitating the need to provide contents and services in localized language.

A scholar addresses the non-participation by the Japanese in online education by attributing it to their language (Kawachi, 1999). Specifically, the Japanese language, which is developed early in life, is conducive to right brain learning modality (i.e., visual and memorization skills) when compared to left brain (i.e., analytic and argumentation skills) which is required by online content (Kawachi, 1999). The limited English proficiency in Japan is attributed to why the Internet is used primarily for searching and printing-out information for reading or translating off-line, and for entertainment and games (Kawachi, 1999).

In relation to the difficulty in language, it is suggested that the potential for information overload exists because non-native speakers read at slower speed than native speakers read. For instance, Chinese-English bilinguals read English at 255 words per minute, compared to Chinese at 380 words per minute (Chambers, 1994; Wang, Inhoff, & Chen, 1999). Kawachi (1999) speculates that the English reading rate for Japanese is slower than the figure for Chinese, given the Japanese English proficiency and learning style. In Europe, the language barrier is seen as a hindrance to the rapid adoption of e-learning. The language barrier results in an increased call for "native-language" content development for local companies who are unwilling to adopt English (Barron, 2000). Unfortunately, language barriers often result in national or cultural pride, which further put U.S. companies at a disadvantage in competing with home-grown developers and content providers.

The need to be part of a group rather than an independent person is imminent when different cultures view or use e-learning. For instance, people from collectivistic cultures tend to seek the connections or look for signs or symbols that provide them a general sense that they connect with others. Henning (2003) found that when some of her participants view Web pages for information, the participants claim that all they see are words and graphics. The participants had no sense of someone else being on the other side to "invite them into their homes", and they could not physically interact with them. Simply put, the personal feel and connection with other learners is not present, and participants' lack of interaction affects willingness to participate in e-learning environments. Although theories such as "tie strength" and "transactional distance" are suggested as solutions to the obvious deficit in e-learning, the apprehension about participation persists and is reflected in instructors' willingness to not have learners' initial e-learning course experience graded or evaluated. Learners express the need for trial run and assurance to know that they are doing well, but more importantly, someone to encourage them that everything will be all right as noted in the next statement from a teacher participating in an e-learning environment:

I have learned something, but not a lot. What I think they [instructors] should do is to teach us how to behave in this sort of set-up before we do a course. I mean not just computers, but the real e-learning thing. By the time I got used to it, it was too late and I think I will fail this one. I still dream of a book and a neat study guide and I am not happy with professor...she thinks we are Americans who breathe through the lungs of the Web. (Henning, 2003, p. 310)

In another example, Henning noted a participant longing for his friend to come into the discussion, just like coming into a house so that they can work on the project together. Henning reports that the learner did not want his peer to complete the project, but to provide a rescue from a very bad experience or "oppression." This analogy clearly illustrates the fact that there are different expectations among learners in individualistic and collectivistic cultures about learning, and these expectations are transferred to e-learning settings.

In general, the challenge is that there does not appear to be a technological and cultural fit in the diffusion of some Westernized technology (Green & Ruhledder, 1995; Mesdag, 2000). Specifically, from the global e-learning standpoint, the learning content needs to match the needs of users. Thus, the key to resolving cultural problems with technology use, especially in any global e-learning environment or curriculum, is to recognize cultural differences and associate technology use with the prevailing cultural values, structures, and activities within these different environments.

Access and Propensity to Use Technology

The propensity to use any communication technology starts with access to technology and the willingness to use it (Olaniran, 1993), and e-learning is no different. A study on attitude and perceptions of e-learning finds easy accessibility as one of the top motivating factors. Forty-seven percent report accessibility of participants, while course relevance to future career and user friendliness follow at 29% and 24% respectively (Vaughan & MacVicar, 2004). A factor contributing to technology access in less economically-developed countries (LEDCs) is "technology transfer." Bozeman (2000) argues that technology transfer is based on cost and benefits, and usually the transfer exists only when the benefits outweigh the costs. The rate of technology transfer is significant in the race to bridge the digital divide between developed countries and LEDCs. Since major communication technology advances occur in economically-developed countries (EDCs), there is disparity in access to this technology in LEDCs when compared with EDCs. With global e-learning, the implications extend beyond mere access attributable to lack of financial capabilities to pay the cost, especially when organizations

provide technologies to employees. The access problems create challenges with frequency of use and lack of comprehension of basic commands and protocols to be successful and to facilitate adoption that result in renewal—continued use of the technology (Olaniran, 1993). Furthermore, due to lack of adequate infrastructure and failure to transfer technology to LEDCs, most organizations subcontract their e-learning needs to third-party vendors. Some employees may have to travel several miles to access required e-learning curriculum, which does not bode well in motivating potential users to adopt the technology.

The lack of technological infrastructure could derail any e-learning program regardless of how lofty its goal might be. In addition to how different cultures use or react to different technology media, certain infrastructure, such as high-speed Internet access, are simply not commonplace elsewhere around the globe. Selinger (2004) alludes to this problem with the level of frustration expressed by some e-learning participants in South Africa. One participant expressed:

Did they not know that I had no electricity? Did they not know that I had no telephone line? Did they not know I could hardly type? ... First, we stayed over in Joburg for an extra day, when we could, because that was the only way we could get to computers. (p. 311)

The above scenario is not unique; there are significant portions of the world population that have never made a phone call or used the Internet. Therefore, lack of critical technological infrastructure will hinder the deployment of e-learning. Consequently, lack of access to computers, transportation, and convenience in accessing technological infrastructure are conditions that must be considered before any successful deployment of e-learning can materialize.

When technology is available, the propensity or motivation to use it is an internal one. There has to be a willingness or motivation on the part of potential users to use the technology (Olaniran, 1993; Storck & Hill, 2000; Vaughan & MacVicar, 2004). However, the motivation is also tied to cultural norms. Some LEDC cultures adopt technology only as long as it does not conflict with their cultural norms (Heaton, 2001), while others simply adopt technology without considering the impact. Also, one research study reported that older generations may have the tendency to resist new technology because of complacency with old ways of doing things (Wheeler, 2001). In general, the reliability and effectiveness of the communication taking place over the information technology medium has been called into question (Daft, Lengel, & Trevino, 1987; Mitra, 1997; Vaughan & MacVicar, 2004).

CULTURE AND IMPLICATIONS FOR IMPLEMENTING GLOBAL E-LEARNING

While the move toward globalizing e-learning is driven by the need to reduce educational cost and the drive to improve efficiency (Pargman, 1998; Sproull & Kiesler, 1991), it has been found that efficiency and cost advantage showed no significant predictive value on usage or social or cultural effects once technologies have been deployed (Pargman, 1998). Nevertheless, it is possible to implement effective and successful global e-learning, such that individuals from LEDCs are receptive to adopting the technology systems and implied structural changes. Successful adoption process would be facilitated only when it is done in a way that takes the idiosyncrasies of cultural factors into consideration both when designing and implementing the technology (Ess, 2002). By doing this, the implementation strategy would enhance the propensity to adopt communication technologies, leading to embracing them for global e-learning and curriculum dissemination. Therefore, attempts to develop a successful global

e-learning program that encourages people to use information technologies accompanying e-learning must incorporate the cultural characteristics of the given nation into the design. Another advantage to this approach is that consideration for the cultural characteristics can be a persuasive mechanism for motivating people, and, consequently, speeding up the adoption process.

Adoption Concerns

According to Sitkin, Sutcliffe, and Barrios-Choplin (1992), CMC technology has symbol-carrying capacity such that the users often are presumed to have specific status in using it. This argument appears to have validity, as computers and telephones in certain societies put people who own or use them in the elite category of the society. Few cultural implications are presented to help in this regard. First, people in harmonious or collectivistic cultures are members of a "social network," where conflicts are handled in a non-confrontational manner despite the contexts. Second, the hierarchical structure of a culture makes the use of technology a status symbol. Third, oral tradition in certain cultures use planned and organized face-to-face meetings, visits, and the telephone as primary modes of interaction, while e-learning (or e-mail based on written tradition) are problematic and are hardly ever used (Barron, 2000; Nulens & Audenhove, 1999).

By themselves, the above criteria do not necessarily imply that these technologies will be rejected automatically by a culture. However, they convey the fact that a re-orientation that is built around the adaptation of technology within the cherished values and societal norms of a culture is a necessity for organizational vendors and e-learning providers. For example, the need for individuals in a collectivistic culture to maintain close contact with families and loved ones is a common thread that can serve as a selling point to get the people in collectivistic cultures to adopt technology. Perhaps preference for close contact is one of the reasons that the need for personal contact with teachers and students during learning was stressed by users in e-learning environment (Henning, 2003; Vaughan & MacVicar, 2004).

Blending Needs

While the needs of organization and e-learning content providers are essential, the needs of end users (i.e., cultural perspectives and learning styles) must be taken into consideration in the design and development of the technology. To such an end, Vaughan and MacVicar (2004) indicate that e-learning, as in any learning, is doomed to fail when it fails to focus on learners. For example, developers need to take into consideration access to infrastructure and accruing costs. High Internet access charges often hinder students from accessing e-learning curricula outside class. When this is the case, provisions must be made for learners to access the Internet at work or campus sites. Access and usage must be allowed even when it means spending significant amounts of class time on assignments so that students can familiarize themselves with the curriculum and the technology. Similarly, in situations where students need to work together in groups in order to resolve problems, it would be ideal for the e-learning instructors to allow learners to have each other's contact information (e.g., instant messenger where available), and also those of their instructors (if appropriate). Increased access to their peers and their instructors would help students to avoid the feeling of being lost from the start.

Change Agents and Their Roles

E-learning content providers (in-house or third-party) need to realize that they are cultural change agents and, thus, need to make the change process for e-learners as painless as possible. In essence, providers must make sure that the change aligns with specific cultures in order to successfully ac-

complish their goals. One way of accomplishing this goal is to recognize that, while the curriculum contents may be universal in their goals, the process for accomplishing those goals must be particularistic. The need for modification in how exercises and assignments are structured for end users is warranted. For instance, different instructor toolboxes can be made available depending on countries, languages, and culture types.

Furthermore, teachers of foreign languages can also serve as leaders who often facilitate the change process and help by using the e-learning curriculum as a tool for teaching foreign languages (e.g., English). After all, end users planning on working in the information technology area usually are in organizations where English is spoken, and there is little need to learn another language (Selinger, 2004). The advantage of this method is "redundancy reinforcement," a principle considered necessary for successful diffusion of innovation in order for novel users to become continued users (Olaniran, 1993). However, foreign language acquisition should never be considered as a substitute for making e-learning contents available in local cultures' languages.

The use of translation software has been suggested (Selinger, 2004), but this software is not readily available across contexts and, more importantly, it is still lacking in precision and accuracy. A valuable approach is to collaborate with other change agents who can help organizations from EDCs with their e-learning projects to convey information and persuade end users that the use of technology (i.e., e-learning) would help to achieve other valuable goals such as learning English or other foreign languages (which could also be necessary to advance students' respective careers). The goal of education and partnering in reducing the digital divide should be to prepare students and teachers to master new skills that current programs may not address. One of these skills includes the ability to collaborate with a diverse team of people in both face-to-face and distance environments in order to achieve different tasks and goals. Making technology available to students and teachers would help them acquire and develop the technological skills and knowledge (Dede, 2000), that are necessary in today's globalized economy.

Facilitating Technology Use

Another proposed benefit to getting users to use technology that could be adapted to e-learning environment is the need to allow students to use technology for reasons other than the designated purpose (Olaniran, 1993, 1994, 1995), as long as it does not violate legal or ethical uses. For instance, if users are allow to send and receive personal e-mails, or browse the Internet for personal reasons, the time spent using the technology to access information could indirectly help students overcome their fears in using e-learning technologies. Research indicates that up to 70% of job-related learning is informal, and driven by individuals rather than acquired from instructors in a structured environment or class (Center for Workforce Development, 1998; see also Brussee, Grootveld, & Mulder, 2003).

User Support

Critical and readily available supports must be a component of any e-learning program. In the adoption process, it is frustrating for individuals to feel flustered, with no help during moments of need. Consequently, e-learning will be better served, especially in culturally-diverse environments, to provide communication tools and social settings, such as virtual classroom for peer supports, as well as fostering competent management in planning, implementation, learner tracking, and certification issues (Brussee et al., 2003). Provision of support is important in the sense that it can improve and perhaps facilitate knowledge transfer even when learners or participants are spread across the globe. The reason for good technological support is that the social setting, or virtual classroom, provides a support group where learners can attempt to discuss or resolve complex problems on their own. Therefore, the more readily available the user support, the easier a user can figure-out the technology and help other students in doing the same.

Instructors' Roles

Teachers need to understand their specific and changing roles in the e-learning environment. First of all, the "one size fits all" teaching is not going to be successful with different cultures and different learning styles. In individualistic cultures where students tend to seek greater control in their learning, the instructor's role would have to shift towards that of a learning facilitator and a coach (Brussee et al., 2003; Selinger, 2004), meaning that instructors only come in as needed. On the other hand, in a collectivistic and a high power-distance culture, instructors must help provide the initial structure, and take a more active role in explaining e-learning contents, and only after that can they use group collaborations where students are able to work on complex problems. Even then, the instructors in a collectivistic culture would have to be available to guide individuals or groups along tasks toward goal accomplishment.

Learning Environment and the Choice of Media

Perhaps one of the greatest struggles in e-learning across cultural contexts that one needs to be prepared for is the need to cultivate learning in a cultural environment, with the provision of a stronger blend of the familiar social environment and the virtual technology environment. In order to do this, the physical world of learners needs to coincide with the tools, signs, and symbols of the e-learning world. To this end, it is suggested that simple visual materials such as icons, sounds, and menus can be replaced with localized words or symbols. In addition, the discussion tool does not have to adhere to a strict structure, but to offer an alternalte and innovative way of presenting learning content (Selinger, 2004; Van der Westhuizen & Henning, 2001). In essence, students should be allowed to have inputs on the structure and format of e-learning as much as possible. Similarly, the social environment in which learning takes place contributes to the students' motivation to learn. The choice of communication media represents another area where significant deliberation must be given. For example, while videoconferencing provides more cues and offers participants opportunities to see other learners (through synchronous interaction), it is more difficult to implement in contexts where high-speed broadband access are not available. At the same time, videoconferencing activities have been found to reduce active participation. Videoconferencing has a tendency to be more formal than the classroom because interactions are stifled, and students are more reluctant to ask questions (Brussee et al., 2003; Malpani & Rowe, 1997). Mailing lists or bulletin boards, on the other hand, offer a low bandwidth alternative but are difficult to set up for synchronous interaction, and are considered lacking in a "personal feel," which may reduce participation level.

The choice and selection of the technology medium ought to be done with significant consideration for different cultures. For instance, it has been shown that in Korea and Japan, e-mail usage is common in peer interaction, but not in superior-subordinate interactions (i.e., power distance). Thus, learner preference is shown for alternative media such as the phone, the fax, and the face-to-face when communicating with superiors, in order to acknowledge and convey respect (Lee, 2002; Olaniran, 2004). Western cultures do not share similar perceptions of respect, and thus do not perceive the use of e-mail between subordinates and superiors to be rude. In essence, the role of culture and the complexity it can create in e-learning and other virtual collaboration work cannot be over-emphasized.

At the same time, the choice of technology medium for disseminating e-learning curricula points to the fact that "technology for the sake of technology" is not a sufficient criterion for motivating learners' and end users' interest in acceptance of, and satisfaction with, e-learning. Rather, it is better when the technological innovation in learning context supports communication and interaction between learners, and builds a social climate that fosters knowledge exchange (Brussee, 2003) and retention of learning, in order to secure the commitment of and the acceptance from users. Gallagher (2003) echoes a similar claim when he argues that people need to take the center stage, while other issues, such as content and technology medium, should take on supportive roles. Similarly, the SMART model of technology planning for delivery of management education suggests the need to assess user needs against the available technology infrastructure to implant or to provide e-learning (Hamlin, Griffy-Brown, & Goodrich, 2003).

Beyond internal organizational technology needs, technology planning for management education and e-learning should reflect global environments and contexts. In essence, the challenge is not whether e-learning is potentially useful or that the trend to use e-learning will continue to grow or to have wider acceptance, but how e-learning fits into specific organizations' strategies, and whether e-learning providers are using the technologies in a way that is feasible and can help them and their participants (employees, students, and target audience) accomplish organizational goals. On this note, the recommendation is that the implementation of e-learning and accompanying technologies be instituted with a long-term view where acceptance and use are based on ongoing simultaneous process (Cummings & Buzzard, 2002; Wankel & DePhillippi, 2003).

CONCLUSION

Certainly e-learning is evolving, whether it is intended to supplement traditional face-to-face learning or be used to replace traditional learning altogether within organizations. Regardless of the aim or purpose, it is paramount that providers tailor e-learning use to the client, and pay attention to differences in cultural learning styles and the preferences of its end users. Yes, we live in a global world where product standardization seems to be the norm. Notwithstanding, the idea of a "one size fits all" learners' approach across different cultural contexts is not feasible. Moreover, the lack of attention to different cultures could be the ultimate deciding factor between success and failure. This chapter argues that cultural contexts matter, that the dimensions of cultural variability have implications on the decision to introduce e-learning technology, and stresses the need for e-learning providers to incorporate these cultural differences into their planning. E-learning technologies must be about the learner. Attempting to seek the learner's commitment during planning is crucial to user acceptance, to user satisfaction, and to continued use of the innovation object. This chapter adds to the scholarly dialogue the role of culture values in the adoption process of technological innovations, especially those developed in a different culture. The arguments explore and provide specific problems and challenges facing potential users of e-learning technology. The chapter offers some recommendations for organizations using e-learning, and porviders of e-learning technologies, as well as ideas for facilitating successful adoption.

Finally, e-learning is not a cure-all; e-learning is a means to an end. Therefore, while the potential for e-learning is enormous among emerging markets and developing economies, careful planning and attention to the idiosyncrasies of cultures is necessary in order to realize the potential benefits of e-learning, and this information must be communicated to any e-learning content providers, whether organizational in-house or subcontracted third-party vendors and platforms.

REFERENCES

Barron, T. (2000, September). E-learning's global migration. *Learning Circuits*. Retrieved August 26, 2005, from http://www.learningcircuits. org/2000/Sep2000/barron.html

Brussee, R., Grootveld, M., & Mulder, I. (2003). Educating managers, managing education: Trends and impacts of tomorrow's technologies. In C. Wankel & R. DePhillippi (Eds.), *Educating managers, with tomorrow's technologies* (pp. 1-16). Greenwich, CT: Information Age Publishing.

Center for Workforce Development (1998). The teaching firm: Where productive work and learning converge. *Report on research findings and implications*. Newton, MA: Education Development Center.

Collier, V. P. (1995). Acquiring a second language for school. *Directions in Language & Education, 1*(4). Retrieved May 4, 2005, from http://www. ncela.qwu.edu/pubs/directions/04.htm

Cummings, D., & Buzzard, C. (2002). Technology, students, and faculty: How to make it happen! *Techniques, 77*(8), 30-33.

Daft, R. L., Lengel, R. H., & Trevino, L. K. (1987). Message equivocality, media selection, and manager performance: Implications for information systems. *MIS Quarterly, 11*, 355-366.

Dede, C. (2000). A new century demand new ways of learning. In D. T. Gordon (Ed.), *The digital classroom* (pp. 171-174). Cambridge, MA: Harvard Education Letter.

Devereaux, M. O., & Johansen, R. (1994). *Global work: Bridging distance, culture, & time*. San Francisco: Jossey-Bass.

Dunn, P., & Marinetti, A. (2002). *Cultural adaptation: Necessity for global e-learning.* Retrieved May 4, 2005, from http://www.linezine.com

Ess, C. (2002). Cultures in collision philosophical lessons from computer-mediated communication. *Metaphilosophy, 33*(1-2), 229-253.

Gallagher, J. (2003). The place and space model of distributed learning: Enriching the corporate-learning model. In C. Wankel & R. DePhillippi (Eds.), *Educating managers, with tomorrow's technologies* (pp. 131-148). Greenwich, CT: Information Age Publishing.

Green, C., & Ruhleder, K. (1995). Globalization, borderless worlds, and the tower of Babel: Metaphors gone awry. *Journal of Organizational Change Management, 8*(4), 55-68.

Gudykunst, W. B., Chua, E., & Gray, A. J. (1987). Cultural dissimilarities and uncertainty reduction processes. In M. McLaughlin (Ed.), *Communication yearbook: Vol. 10* (pp. 457-469). Beverly Hills, CA: Sage.

Hall, E. T. (1976). *Beyond culture.* New York: Doubleday.

Hamlin, M. D., Griffy-Brown, C., & Goodrich, J. (2003). From vision to reality: A model for bringing real-world technology to the management education classroom. In C. Wankel & R. DePhillippi (Eds.), *Educating managers, with tomorrow's technologies* (pp. 211-238). Greenwich, CT: Information Age Publishing.

Heaton, L. (1998). Preserving communication context: Virtual workspace and interpersonal space in Japanese CSCW. In C. Ess & F. Sudweeks (Eds.), *Cultural attitudes towards communication and technology* (pp. 163-186). Sydney, Australia: University of Sydney.

Heaton, L. (2001). Preserving communication context: Virtual workspace and interpersonal space in Japanese CSCW. In C. Ess (Ed.), *Culture, technology, communication: Towards an intercultural global village* (pp. 213-240). Albany: State University of New York Press.

Henning, E. (2003). I click therefore I am (not): Is cognition "distributed" or is it "contained" in borderless e-learning programmes? *International Journal of Training and Development, 7*(4), 303-317.

Hofstede, G. (1980). *Culture's consequences.* Beverly Hills, CA: Sage.

Hofstede, G. (1983). Dimensions of national cultures in fifty countries and three regions. In J. Deregkowski, S. Dziurawiec, & R. Annis (Eds.), *Expiscations in cross-cultural psychology* (pp. 335-355). Lisse, Netherlands: Swets & Zeitlinger.

Hofstede, G. H. (2001). *Culture's consequences: Comparing values, behaviors, institutions, and organizations across nations.* Thousand Oaks, CA: Sage.

Hofstede, G., & Bond, M. (1984). Hofstede's culture dimensions: An independent validation using Rokeach's value survey. *Journal of Cross-Cultural Psychology, 15*, 417-433.

Internet Usage in Asia. (n.d.). Retrieved December 31, 2005 from http://www.internetworldstats.com/stats3.htm

Kawachi, P. (1999, April 19-21). *Language curriculum change for globalisation* (mimeograph). Paper presented at the 34th Annual RELC Seminar—Language in the Global Context: Implications for the Language Classroom. Singapore: SEAMEO RELC. Retrieved from ouhk..edu.hk/cridal/gdenet/Teaching/Design/EATL11A.html

Lee, O. (2002). Cultural differences in email use of virtual teams a critical social theory perspective. *Cyberpsychology & Behavior, 5*(3), 227-232.

Malpani, R., & Rowe, L. A. (1997). Floor control for large-scale MBone seminars. *ACM Multimedia*, 97. Retrieved May 4, 2005, from http://bmrc. berkeley.edu/research /publications/1997/137/qs-bmm97.html

Mason, R. (1998). *Globalising education: Trends and applications*. London: Routledge.

McConnell, S. (1998). NGOs and Internet use in Uganda: Who benefits?. In C. Ess & F. Sudweeks (Eds.), *Cultural attitudes towards communication and technology* (pp. 104-124). Sydney, Australia: University of Sydney.

Mesdag, M. V. (2000). Culture-sensitive adaptation or global standardization: The duration of usage hypothesis. *International Marketing Review, 17*, 74-84.

Mitra, A. (1997). Virtual commonality: Looking for India on the Internet. In. S. G. Jones (Ed.), *Virtual culture* (pp. 55-79). London: Sage.

Nulens, G., & Audenhove, L. (1999). The African information society: An analysis of the information and communication technology policy of the World Bank, ITU, and ECA. *Communicatio: South African Journal of Research and Theory, 25*(1-2), 28-41.

Olaniran, B. A. (1993). An integrative approach for managing successful computer-mediated communication technological innovation. *Ohio Speech Journal, 31*, 37-52.

Olaniran, B. A. (1995). Perceived communication outcomes in computer-mediated communication: An analysis of three systems among new users. *Information Processing & Management, 31*, 525-541.

Olaniran, B. A. (1996). Social skills acquisition: A closer look at foreign students on college campuses and factors influencing their level of social difficulty in social situations. *Communication Studies, 22*, 72-88.

Olaniran, B. A. (2001). The effects of computer-mediated communication on transculturalism. In V. Milhouse, M. Asante, & P. Nwosu (Eds.), *Transcultural realities* (pp. 83-105). Thousand Oaks, CA: Sage.

Olaniran, B. A. (2004). Computer-mediated communication in cross-cultural virtual groups. In G. M. Chen & W. J. Starosta (Ed.), *Dialogue among diversities* (pp. 142-166). Washington, DC: National Communication Association.

Pargman, D. (1998). Reflections on cultural bias and adaptation. In C. Ess & F. Sudweeks (Eds.), *Cultural attitudes towards communication and technology* (pp. 73-91). Sydney, Australia: University of Sydney.

Roach, K. D., & Olaniran, B. A. (2001). Intercultural willingness to communicate and communication anxiety in international teaching assistants. *Communication Research Reports, 18*, 26-35.

Sitkin, S. B., Sutcliffe, K. M., & Barrios-Choplin, J. R. (1992). A dual capacity model of communication media choice in organizations. *Human Communication Research, 18*, 563-598.

Sproull, L., & Kiesler, S. (1991). *Connections: New ways of working in the networked organization*. Cambridge, MA: MIT Press.

Storck, J., & Hill, P. A. (2000). Knowledge diffusion through strategic communities. *Sloan Management Review, 41*(2), 63-74.

Van Dam, N., & Rogers, F. (2002, May). E-Learning cultures around the world: Make your globalized strategy transparent. *E-Learning*, 28-33.

Vaughan, K., & MacVicar, A. (2004). Employees' pre-implementation attitudes and perceptions to e-learning: A banking case study analysis. *Journal of European Industrial Training, 28*(5), 400-413.

Wang, J., Inhoff, A. W., & Chen, H. (1999). *Reading Chinese script: A cognitive analysis*. Marwah, NJ: LEA Publishers.

Wankel, C., & DePhillippi, R. (2003). Introduction: Emerging technological contexts of management learning. In C. Wankel & R. DePhillippi (Eds.), *Educating managers, with tomorrow's technologies* (pp. vii-ix). Greenwich, CT: Information Age Publishing.

Wheeler, D. (2001). New technologies, old culture: A look at women, gender, and Internet in Kuwait.' In C. Ess (Ed.), *Culture, technology, communication: Towards an intercultural global village* (pp. 187-212). Albany: State University of New York Press.

Wilborn, J. (1999). The Internet: An out-group perspective. *Communicatio: South African Journal of Research and Theory, 25*(1-2), 53-57.

Chapter 7.11
Information Technology Certification:
A Student Perspective

Tanya McGill
Murdoch University, Australia

Michael Dixon
Murdoch University, Australia

ABSTRACT

Certification has become a popular adjunct to traditional means of acquiring information technology skills and employers increasingly specify a preference for those holding certifications. This chapter reports on a study designed to investigate student perceptions of both the benefits and risks of certification and its importance in obtaining employment. Certification was perceived as an important factor in achieving employment and students undertaking it anticipate that it will lead to substantial financial benefits. Yet higher salaries are not seen as the most important benefit of certification. The potential benefits that students believe are most important relate to real-world experience. The respondents were aware of the possible risks of certification but did not appear to be overly concerned about them.

INTRODUCTION

Certification has become a popular adjunct to traditional means of acquiring information technology (IT) skills and increasing numbers of job advertisements specify a preference for those holding certifications. Certification intends to establish a standard of competency in defined areas. Unlike traditional academic degrees, certifications tend to be specific to narrow fields or even to individual products. They are designed to provide targeted skills that have immediate applicability in the workplace.

Vendors such as Microsoft and Cisco Systems dominate the vendor specific certification market worldwide with qualifications such as the Microsoft Certified Systems Engineer (MCSE), Cisco Certified Network Associate (CCNA) and Cisco Certified Internetwork Expert (CCIE). Vendor neutral certifications such as those provided by the Institute for Certification of Computing Professionals (ICCP), the Computer Technology Industry Association and the Disaster Recovery Institute also play a role. It has been reported that there are more than 300 IT certifications available and that approximately 1.6 million people have earned approximately 2.4 million certifications (Nelson & Rice, 2001), and no doubt these figures have already increased dramatically. Gabelhouse (2000) quoted an IDC Inc. report that found that the IT training and testing industries had revenues of $2.5 billion in 1999 and were expected to reach $4.1 billion by 2003.

Vendors create certifications as a way of promoting widespread adoption of their products and technologies, but they have also become important for educational institutions in attracting students and placing graduates (Brookshire, 2000). This chapter explores the perceptions of students who are undertaking courses of study that can lead to certification. It reports on a study designed to investigate student perceptions of both the benefits and risks of certification and its importance in obtaining employment.

Benefits of Certification

Numerous benefits have been proposed to result from IT certification. As Nelson and Rice (2001) note, many of the claims of benefits have originated in the brochures and Web sites of certification agencies; however, there seems also to be a wider recognition of their importance. The major benefits that have been claimed can be categorized as relating to employers, educational institutions and students (i.e., potential employees). The major benefit for employers is believed to be the provision of more capable employees (Ray & McCoy, 2000), and one in eight IT job advertisements have been found to mention certifications (Clyne, 2001; Nelson & Rice, 2001). Some support for the benefit of employee certification to employers is provided by a study by IDC Inc. (1999), which found that 92% of managers surveyed said that they

realized all or some of the benefits they expected from their certified employees. The major benefits to employers accruing from certified employees were found to be:

- Greater knowledge and increased productivity
- A certain level of expertise and skill
- Improved support quality
- Reduced training costs
- Higher morale and commitment

The major benefit proposed for educational institutions is the opportunity to extend program content and to have an increased assessment capability (Ray & McCoy, 2000). Institutions that successfully offer certifications can become known for their expertise in these areas and attract more students and employers for their graduates (Brookshire, 2000). Student performance on certification exams also provides additional and generalizable measures of student competencies.

The greatest benefits of certification are believed to accrue to students (Ray & McCoy, 2000). Marketability is proposed as a major benefit. Students are marketable if their programs of study contain content considered valuable by employers. For example, holders of Cisco certifications should have substantial experience as network administrators, designers and troubleshooters on real networks. Higher salaries are also commonly cited as a benefit, and there is evidence to support this. A survey conducted by *Certification Magazine* (Gabelhouse, 2000) reported that on average certification resulted in a 12% increase in income. This study also reported varying values for different certifications. For example, a MCSE led to an average increase in income of 12.6%, a Cisco CCNA to a 16.7% increase and a Novell CNA to a 13.3% increase. However, Alexander (1999) speculates that increased supply of people with the most popular certifications (such as MCSE) means diminished value in the marketplace. Other proposed benefits that are associated with increases in marketability and salary include increased self-confidence and increased credibility (Karr, 2001).

Risks Associated with IT Certification

Despite these benefits, various concerns have been expressed about the current popularity of IT certification. Ray and McCoy (2000) identify the heavy involvement of vendors as an issue for concern, citing the absence of unbiased neutral groups for determining content, creating exams and authorizing examiners. They also recognize that the rapidly changing knowledge base might mean that certification is not of lasting value. Wilde (2000) also highlights the fact that some certifications do not require practical or real-world experience, thus limiting the claims of usefulness.

As IT certifications are increasingly offered by universities and colleges, concerns have been raised that academics might be uncomfortable with the loss of control over content that arises when certification exams determine the content of courses and academic programs (Nelson & Rice, 2001; Ray & McCoy, 2000). Academics might also be uncomfortable with the pressure to maintain their own proficiency levels and certification status.

Given the increasing pervasiveness of certification in the IT profession, more research is needed to verify the benefits of IT certification and to determine the importance of the proposed risks.

RESEARCH PROJECT

The exploratory study reported in this chapter contributes to the need for further research on the risks and benefits of IT certification by investigating student perceptions of both the benefits and risks of certification, focusing particularly on Cisco certification. This research was conducted by survey. Participants in the study were students enrolled in several electronic commerce, telecommunications management and information technology courses at an Australian university. Students who have successfully completed these particular courses can also pursue Cisco certification, as the courses make use of the Cisco curriculum. Participants were recruited during class and completed a questionnaire on the spot. It was stressed that the completion of the questionnaire was voluntary and that it formed no part of their assessment in the course.

The questionnaire was designed to be easy to read and understand and to require no more than 10 minutes to complete. The questionnaire contained four main groups of items. The first section asked about:

- Age
- Gender
- Amount of previous work experience (both total and IT experience)
- Whether the skills provided by their degree are those employers require

The second group of questions related to the perceptions of the participants about the importance of industry certification for employment. Those participants who were not currently working in the IT industry were firstly asked to rate the importance of industry certification for obtaining their initial IT employment. This item was measured on a five-point scale ranging from (1) "Not Important" to (5) "Vital". They were then asked to indicate how much higher (as a percentage) than the average graduate starting salary they believed their starting salary would be if they obtained various certifications. The list of certifications included those currently available to the participants and several other popular certifications (see Table 2 for a list of the certifications included).

Those participants who were currently working in the IT industry were instead asked to rate the importance of industry certification for getting ahead in their current employment. This question also used a five-point scale ranging from (1) "Not Important" to (5) "Vital". They were then asked to indicate how much they thought their salary (as a percentage) would increase if they obtained the various IT certifications.

The third group of questions related to the participants' perceptions of the importance of various proposed benefits of seeking certification. A list of 11 benefits proposed for IT certification was developed from the literature on IT certification (e.g. Alexander, 1999; IDC Inc., 1999; Karr, 2001; Nelson & Rice, 2001; Otterbourg, 1999). Each potential benefit was rated for importance on a five-point scale ranging from (1) "Not Important" to (5) "Very Important" (see Table 3 for a list of the proposed benefits included).

The fourth group of questions related to the participants' perceptions of the importance of various concerns about certification. A list of potential risks of reliance on IT certification was drawn from the literature on certification (e.g., Nelson & Rice, 2001; Ray & McCoy, 2000; Wilde, 2000). Participants rated each potential risk for impor-

tance on a five-point scale ranging from (1) "Not Important" to (5) "Very Important" (see Table 5 for a list of the proposed risks included).

The participants in the study were 145 students with an average age of 23.4 years (with a range from 18 to 48). Twenty-one of the participants (14.5%) were female and 124 (85.5%) were male. The gender proportions in this study are consistent with the low representation of females in IT courses around the world (Downes & Hobbs, 2000; Fitzsimmons, 2000; Klawe & Leveson, 1995). The majority of the participants were at undergraduate level (89.7%), with approximately 10% at postgraduate level. The participants who had previously been employed had on average 5.8 years of work experience, of which 3.4 years were in the IT industry. Table 1 summarizes some of the background information about the participants in this study.

RESULTS AND DISCUSSION

Benefits

IT certification was perceived as very important both for obtaining initial IT employment and for

Table 1. Background information about participants

		Number	Percentage
Gender			
	Male	124	85.5
	Female	21	14.5
Degree level			
	Undergraduate	130	89.7
	Postgraduate	15	10.3
Work experience* (mean = 5.8 years)			
	No	75	51.7
	Yes	70	48.3
IT work experience (mean = 3.4 years)			
	No	97	66.9
	Yes	48	33.1

*Note: * Work experience includes both IT and non-IT experience*

getting ahead if currently employed in the IT industry. The average importance rating given to IT certification by those not currently employed in IT was 4.09 (out of five) and it was 3.75 (out of five) for those currently employed in the IT industry (see Table 2). These perceptions of students who were not yet certified are consistent with results of a survey of 470 IT contractors described by Alexander (1999). In that study, 83% of the contractors believed that IT certifications were either "very important" or "somewhat important" to their prospects for career advancement. Thus, student perceptions of the importance of certification appear to be consistent with industry perceptions. The majority of student participants in the current study also believed that the studies they were undertaking would provide the skills required by employers (yes: 64.5%, not sure: 32.4%, no: 2.8%).

In general, participants perceived that obtaining IT certification would lead to clear financial benefits. The average increases that students who were not currently working in the IT industry believed they would receive from obtaining certification ranged from a high of 25.27% for CCNP certification down to 18.87% for CCNA certification (Table 2). The range of increases anticipated by participants was very large, with some suggesting that no increase would result, up to a maximum of 100% for all of the certifications. This wide range of responses suggests that this group of participants did not have a good sense of the value of these certifications in the marketplace. It would be reasonable to expect this result for those certifications that are not currently available to participants as part of their program of study, but it is more surprising for the CCNA and CCNP, as these certifications are readily available to the participants. The potential financial benefit resulting from them could be assumed to have influenced their decisions to undertake the courses being surveyed. This lack of knowledge about the financial value of certification is also reflected in the large number of participants who did not provide answers to these items (around 20% did not respond to at least one of the ques-

Table 2. Perceived importance of certification

	N	Mean	SD	Min.	Max.
Importance of certification for initial job	119	4.09	1.00	1	5
Importance of certification for current job	24	3.75	1.26	1	5
Anticipated percentage increase in starting salary (if not currently in IT employment)					
CCNA certification	99	18.87	21.46	0	100
CCNP certification	97	25.27	23.85	0	100
Security certification	96	22.88	21.61	0	100
Wireless certification	95	21.27	21.95	0	100
Unix certification	96	21.47	20.55	0	100
MCSE certification	98	19.77	21.26	0	100
Anticipated percentage increase in salary if currently in IT employment					
CCNA certification	19	6.32	4.96	0	15
CCNP certification	20	16.60	14.30	0	50
Security certification	20	19.20	23.31	0	100
Wireless certification	19	10.68	10.19	0	30
Unix certification	18	16.72	23.37	0	100
MCSE certification	18	9.28	14.90	0	60

tions about salary). Instructors have a major role to play in providing up to date information about employers' needs and likely outcomes of obtaining certification. They need to be highly accessible and to ensure that their knowledge of the marketplace that graduates will enter remains current so that they can help guide their students (McGill & Dixon, 2003).

Those participants who were currently working in the IT industry also anticipated financial gains from certification, but the average percentage gains they suggested were lower than those anticipated by students not working in the IT industry. The percentage increases anticipated by those who were currently employed were consistent with the figures available from surveys such as the one conducted by *Certification Magazine* (Gabelhouse, 2000), suggesting that employed students have realistic expectations. There was also a narrower range of responses provided, suggesting less confusion about likely financial outcomes resulting from certification. Presumably those working in the IT industry would have received better quality information, as they would

have had access to IT work colleagues; whereas those without IT work experience would have been receiving information from a pool of people with perhaps limited direct IT industry experience (McGill & Dixon, 2003).

Table 3 presents the average perceived importance of each potential benefit of IT certification. The ratings of benefits are ranked by perceived importance. All benefits were ranked relatively highly, with averages above the midpoint of the scale. The two most highly ranked benefits were practical experience with real networking tasks, and experience with real equipment. Almost 97% of the respondents considered practical experience with real networking tasks to be important or very important. This finding reflects that fact that the participants were primarily undertaking Cisco certifications. Wilde (2000) comments that Cisco Systems has the most "realistic" certification program, requiring those undertaking certification to perform real tasks, using real equipment. Wilde also raises concerns that some certifications do not emphasize practical skills.

Table 3. Benefits of certification

Rank	Benefits	N	Mean	SD	Min.	Max.
1	Practical experience with real networking tasks	143	4.57	0.60	2	5
2	Experience with real equipment	143	4.55	0.62	2	5
3	Widely recognized qualification	142	4.39	0.71	1	5
4	Greater knowledge/skill	143	4.29	0.64	2	5
5	Able to apply for the increasing number of jobs that require certification	143	4.19	0.75	1	5
6	Obtaining a formal marketable qualification	143	4.11	0.85	1	5
7	Academics that teach certifications must be certified, so you can be confident of their knowledge	143	4.08	0.84	1	5
8	Higher salaries	143	4.00	0.88	1	5
9	Increased credibility	143	3.99	0.77	1	5
10	Increased self-confidence	143	3.89	0.97	1	5
11	Flexibility of study because of online curriculum	142	3.62	0.97	1	5

The third ranked perceived benefit in terms of importance was having a widely recognized qualification. IT certifications are global and enable those who have them great flexibility in terms of obtaining employment around the world. The fourth ranked benefit was greater knowledge and skill. It appears that the intrinsic value of the knowledge and skill obtained during certification is perceived as important beyond the job related benefits that can result.

The fifth and sixth ranked proposed benefits relate to the role of certification in improving opportunities to obtain jobs. The ability to apply for the increasing number of jobs that require certification was ranked fifth, and obtaining a formal marketable qualification was ranked sixth. Improving employment opportunities is clearly important to those who undertake certification, but the higher rankings of practical experience and improving knowledge and skill suggest that employment is not the sole motivation for undertaking certification. The perceived importance of practical experience obtained goes beyond just improving marketability.

Confidence in the knowledge of those who teach certification programs was ranked as the seventh most important benefit. Whilst having knowledgeable instructors is clearly important (with an average of 4.08 out of five), the relative ranking perhaps suggests that students perceive those who teach them to be well qualified for the job regardless of whether the unit of study involves a certification and hence requires instructor certification.

Higher salaries were ranked eighth in terms of importance. Whilst potential salaries perhaps receive the most publicity in terms of benefits to holders of certifications, this ranking suggests that salary is not the major driving factor for students. The ninth and 10th ranked benefits relate to the importance of certification for how students see themselves. Increased credibility and self-confidence did not appear to be major reasons for undertaking certification. The lowest ranked of

the proposed benefits was the flexibility of study enabled because of online curriculums. While certification providers such as the Cisco Networking Academies pioneered delivery of quality e-learning material, online materials are now routinely available to IT students whether or not they are attempting certifications (McCormick, 2000; Peffers & Bloom, 1999), thus reducing the perceived importance of this benefit.

Several themes appear to emerge from the examples of benefits that have been proposed. To determine the number and nature of factors underlying the various benefits identified from the literature, a principal components factor analysis with varimax rotation was performed in SPSS 11.5 using the data from the 145 respondents. Three factors with eigenvalues of greater than one emerged, indicating the existence of three underlying dimensions (see Table 4 for factor loadings).

Examination of the benefits associated with each factor led to naming the factors as follows:

- **Marketability benefits:** Which relate to desirability in the eyes of employers
- **Personal benefits:** Which relate to the impact of the certification on the way in which students perceive themselves and to the ease of their study
- **Learning benefits:** Which relate to intrinsic fulfillment from the type of learning

These factors summarize the major types of benefits that students anticipate will accrue from certification.

Risks

Table 5 presents the average perceived importance of each of the potential risks of, or concerns about, IT certification. The ratings of risks are ranked by perceived importance. The average importance ratings for the risks are mostly well below those

Table 4. Factor loadings of benefits

	Marketability Benefits	Personal Benefits	Learning Benefits
Experience with real equipment	0.10	0.18	**0.90**
Practical experience with real networking tasks	0.21	0.18	**0.89**
Obtaining a formal marketable qualification	**0.75**	0.13	0.26
Greater knowledge/skill	0.48	0.13	**0.52**
Higher salaries	**0.76**	0.28	0.06
Widely recognized qualification	**0.71**	0.27	0.20
Flexibility of study because of online curriculum	0.12	**0.78**	0.16
Increased credibility	0.34	**0.78**	0.16
Increased self-confidence	0.17	**0.80**	0.19
Able to apply for the increasing number of jobs that require certification	**0.60**	0.46	0.11
Academics that teach the Cisco curriculum must be certified, so you can be confident of their knowledge	0.43	**0.53**	0.06
Percent of variance explained	24.04%	23.20%	18.99%

of the benefits discussed previously. Therefore, whilst the participants were conscious of the potential risks, they did not appear to be overly concerned about them.

The highest ranked risk was that the rapidly changing knowledge base might mean that certification is not of lasting value. IT has been changing rapidly over a long period and this rate of change is likely to continue or increase (Benamati & Lederer, 2001; Fordham, 2001). Organizations find it difficult to obtain personnel with the appropriate knowledge and skills in order to meet the growing demands for IT services (Doke, 1999), and this has contributed to the desirability of certified employees, as they provide a way for employers to obtain a pool of employees with up to date skills. However, the rapidly changing knowledge base also means that certification may not be of enduring value, and means that recertification is necessary. Gabelhouse (2000) found that 75% of certification holders shoulder some of the costs of certification, with 45% paying for everything themselves. If regular recertification is required,

the costs and investments of time can become prohibitive.

The middle ranked group of risks relate to the potential for bias in certification. The second ranked risk was the absence of an unbiased neutral group for creating exams and approving examiners, and the third ranked risk was the absence of an unbiased neutral group for determining content. The fourth ranked risk was heavy involvement of vendors. Whilst vendor neutral certifications do exist, most certifications are linked to vendors and this has been raised as an issue of concern (Ray & McCoy, 2000). Again, students appear to be aware of the issue, but not overly concerned about it. They appear to accept the central role of vendors in the IT industry.

Concerns have been raised by several authors (Nelson & Rice, 2001; Ray & McCoy, 2000) that academics might not be comfortable with the loss of control over content that occurs because of the role of certification exams. They might also be uncomfortable with the pressure to maintain their own proficiency levels and certification sta-

Table 5. Risks of certification

Rank	Risks	N	Mean	SD	Min.	Max.
1	The rapidly changing knowledge base might mean that the certification is not of lasting value	138	3.91	0.82	1	5
2	The absence of an unbiased neutral group for creating exams and approving examiners	139	3.58	0.78	1	5
3	The absence of an unbiased neutral group for determining content	140	3.49	0.81	1	5
4	Heavy involvement of vendors	139	3.47	0.81	1	5
5	Academics might be uncomfortable with the pressure to maintain their own proficiency levels and certification status	138	3.34	0.79	1	5
6	Academics might be uncomfortable with the thought that certification exams determine content of courses and academic programs	138	3.30	0.79	1	5

tus. Not surprisingly, these concerns are the two lowest ranked concerns of the students surveyed in this study.

CONCLUSION

IT certifications are a popular adjunct to traditional means of preparing for a career in IT. Many educational institutions offer a range of IT certifications. This study explored the perceptions of students currently undertaking courses of study that could lead to IT certification. Certification was perceived as an important factor in achieving employment and students undertaking it anticipate that it will lead to substantial financial benefits. Yet higher salaries are not seen as the most important benefit of certification. The potential benefits that students believe are most important relate to the real-world experience that is part of some certifications. They also value the potential improvements in knowledge and skill to which certification should lead.

Those respondents who were currently working in the IT industry had realistic perceptions of the likely salary increases available once cer-

tification was obtained, but those students with no IT experience appeared to overestimate the potential financial benefits. Instructors should ensure that they have current information about salaries and employers' skill requirements so that they can help guide their students.

The respondents were aware of the possible risks of certification but did not appear to be overly concerned about them. The issue considered most important was the potential for the rapidly changing knowledge base to mean that certification is not of enduring value.

Obtaining IT certification has become an important consideration for the IT profession. More research is needed to understand the benefits of IT certification and to determine the importance of the proposed risks. The study reported on in this chapter has provided a starting point, but future research should extend it to holders of IT certifications and to employers.

REFERENCES

Alexander, S. (1999, December 13). Sorting out certifications. *Computerworld.*

Benamati, J., & Lederer, A.L. (2001). Coping with rapid changes in IT. *Communications of the ACM, 44*(8), 83-88.

Brookshire, R.G. (2000). Information technology certification: Is this your mission? *Infor. Technology, Learning, and Performance J., 18*(2), 1-2.

Clyne, M. (2001). Employee recruitment & retention – Certification's role. *Professional Certification Magazine.*

Doke, E.R. (1999). Knowledge and skill requirements for information systems professionals: An exploratory study. *Journal of IS Education, 10*(1), 10-18.

Downes, S., & Hobbs, V. (2000). An exploratory study of the representation and performance of females in Information Technology at Murdoch University. *Proceedings of the International Information Systems Education Conference* (IS-ECON), Philadelphia.

Fitzsimmons, C. (2000, April). Doing IT for themselves. *Information Age.*

Fordham, D.R. (2001). Forecasting technology trends. *Strategic Finance, 83*(3), 50-54.

Gabelhouse, G. (2000). *CertMag's salary survey.* Retrieved July 9, 2003: http://www.certmag.com/issues/dec00/feature_gabelhouse.cfm

IDC Inc. (1999). *Benefits and productivity gains realized through IT certification.* Retrieved July 9, 2003: http://www.ecertifications.com/idcrep_it-cert.html

Karr, S.S. (2001, December). IT certification pays off. *Financial Executive,* 60-61.

Klawe, M., & Leveson, N. (1995). Women in computing where are we now? *Communications of the ACM, 38*(1), 29-35.

McCormick, J. (2000). The new school. *Newsweek, 135*(17), 60-62.

McGill, T., & Dixon, M. (2003). How do IT students stay up to date with employer skill requirements. In T. McGill (Ed.), *Current Issues in IT Education* (pp. 144-152). Hershey, PA: IRM Press.

Nelson, M.L., & Rice, D. (2001). Integrating third party-certification with traditional computer education. *The Journal of Computing in Small Colleges, 17*(2), 280-287.

Otterbourg, S.D. (1999). Cisco systems and Hewlett-Packard prepare the workforce for the future. *Education + Training, 41*(3), 144-145.

Peffers, K., & Bloom, S. (1999). Internet-based innovations for teaching IS courses: The state of adoption, 1998-2000. *Journal of Information Technology Theory and Application, 1*(1), 1-6.

Ray, C.M., & McCoy, R. (2000). Why certification in information systems? *Information Technology, Learning, and Performance Journal, 18*(1), 1-4.

Wilde, C. (2000, September 25). Demand for IT pros drives vendor certification growth—but multiple-choice tests aren't always a true measure of skills and experience. *Information Week,* 214.

This work was previously published in the International Journal of Information and Communication Technology Education Vol. 1, No. 1, pp. 19-30, copyright 2005 by Idea Group Publishing (an imprint of IGI Global).

Chapter 7.12
Electronic Paralanguage:
Interfacing with the International

Katherine Watson
Coast Community College District, USA

ABSTRACT

Psychologists and linguists agree that communicative elements other than words alone transmit more than 65% of the meaning of any linguistic message. New messages in new languages can be learned quickly and in their cultural context if instructional materials are sheathed in the L2 ("foreign") "electronic paralanguage" rather than in the students' native "L1" language. That is, L2 acquisition can take place at an extraordinarily rapid pace if the Netiquette and interfaces, page layouts, buttons, and alternative correspondence styles of the L2 mode of expression are employed. Exemplary adult students of French as a Second Language have demonstrated achievement of unusually high-level reading, writing, and cultural competence skills quickly in an online environment that immerses them in their new L2. Indeed, these students' success demonstrates at least two things: First, learning a new language may be at least as effective, and is clearly more complete, in an online environment than it is in a traditional classroom, and second, that educators online should attend to all features of the electronic environment, rather than simply to the subject matter that it transmits.

INTRODUCTION

Psychologists and linguists agree that communicative elements other than words alone transmit more than 65% of the meaning of any linguistic message (Birdwhistell, 1952; Collier & DiCarlo, 1985). And it is not just kinesics, our "body language," which affects the way in which our words are understood. Intonation and pitch, loudness, and the use of hesitations or pauses all comprise "paralanguage," the influential vocal but non-verbal noise that sheathes every human utterance.

In cyberspace, it is page layout, background color, graphics-to-text ratio, arrangement of words and pictures on a page, and even typeface which act as a kind of "electronic paralanguage," enveloping the electronic texts and offering cyberspatial "suprasegmentals" that augment or detract from meaning. Web designers have begun to attend to page layout, line length, font, and page color, as these features affect cognition (Hudson et al., 2005), and non-English language users of the Internet have noticed how variations in these features affect understanding (Hudson et al., 2005, Vasquez, 2000). These "beyond-words" features of electronic data delivery can affect comprehensibility, if not comprehension, and they transmit ineffable cultural information.

Learning new subject matter online, with the aid of data delivered in the parlance that characterizes that subject matter and that envelopes it in the most pertinent electronic paralanguage, is different from learning in conversational contexts or in the traditional classroom. That is, just as the harmony of figure and ground can augment the message transmitted by a work of art, so can the concord of language and electronic paralanguage expedite understanding online.

LANGUAGE LEARNING AND LINGUISTIC COMPETENCIES

Favored with a complicated cerebrum with which he can coordinate, communicate, and comprehend data transmitted aurally, visually, tactilely, or in writing, the human adult exploits his experience each time he takes on a new learning task, profiting from techniques honed throughout his life (Dobrovolny, 2003). If his task is to learn a new language, the human adult will automatically try to compare/contrast/discern patterns in the new mode of expression that might relate to those of the language(s) he already knows (Singhal, 1998). Because babies learn their mother tongues while they are expanding their understanding of color, texture, sound, sight, and movement, the flowering of their first language, their "L1," happens in parallel to development of cognition. Moreover,

L1 acquisition flows smoothly in a sea of L1 paralanguage.

But the learning of secondary languages, "L2s," does not occur in like manner (Cook, 2005). Rather, it is self-analysis, self-criticism, and self-correction, all resulting from interaction with native speakers, that influence this process. Interaction is key, especially for the comprehension of the non-verbal, and yet novelty is a defining feature of all human expression. Indeed, it is the development of at least three sorts of competencies, explained below, that underlies human linguistic understanding (Thanasoulas, 2000).

Three Competencies Necessary to Understanding

Each sentence that each human being produces in any language is novel, never having being uttered before in exactly the same way or in exactly the same context, but people understand one another if they share basic *linguistic competence* in the same language. And perhaps even more interestingly, people know how to react to one another's utterances, based upon their shared *communicative competence*, their ability not necessarily to create grammatically-perfect structures, but rather to say things that are contextually fitting in an appropriate way. Thus, if we humans understand one another's grammatical structures, if we generally recognize that linguistic interaction is an amicable social act, then we are on the way to communicative competence.

As James (1969) has said, in defining this second sort of competence, "(we) can tell whether our interlocutor is speaking seriously or in jest, we can use information we have about the interlocutor to interpret his utterances (e.g., the political party he belongs to and whether he had a domineering mother) and our knowledge of whether the interlocutor is a stranger, a friend, or a professional foe (will) undoubtedly affect the inferences drawn about what his sentences imply."

And finally, the person who would learn a new mode of expression should also attend to the need to attain a tertiary *cultural competence,* defined as an ensemble of congruent behaviors, attitudes, and strategies comprising a system designed to facilitate the integration of human behavioral patterns. Cultural competence embraces an understanding of thoughts, communications, actions, customs, beliefs, values, and institutions of alternative racial, ethnic, religious, or social groups, as Kalyampur and Harry have noted (1999). He who is culturally competent has moved beyond the "cultural destructiveness" characterizing those who would mock or denigrate another point of view, he has passed beyond the ignorance of "cultural blindness," and he has attained a degree of masterful proficiency in his interactions with others.

A Novel Exploration in Second Language Learning Online

At Coastline Community College, in Fountain Valley, California (http://coastline.edu), students in an online French course have been able to attain high levels of competence in their new language quickly and extraordinarily well because their course materials have taught them to place language in a cultural context. This has been possible because the Internet provides them French language materials swathed in an electronic paralanguage incorporating French Netiquette and interfaces, layouts, design/array, colors, and buttons that augment and expedite the learning of words, grammar, and meaning. Rather than feeling their eyes float toward "home page" or "next" or "back," and rather than being directed to the typically "low-context" (Hall, 1976) or quick and straightforward systems typical of American Web-delivered data, these learners are immersed immediately into a "high-context" (Hall, 1976), unhurried, frequently elliptical and text-heavy set of materials; moreover, with "page d'accueil," "suivant," and "précédent" replacing

the aforementioned Anglophone buttons, the eyes have no escape from the French context. Indeed, francophone patterns of communicative behavior are quickly integrated online.

Online learners of French at Coastline begin within weeks, often without realizing it, to accept that the francophone world prefers the implicit, the highly contextual, the nuanced, the creative; French Web materials tend to harmonious colors and shades (e.g., http://www.academie-fran-caise.fr/), often with images sliding gently or dancing rhythmically (cf. http://www.louvre.fr http://www.cite-sciences.fr/); sites tend to be text-heavy as well (cf. http://www.herodote.net/). Thus, online language learners find right away that skill in writing, often imbued with a poetic style (even the prime minister is lauded as a poet: http://www.abidjan.net/qui/profil.asp?id=541), is not only valued, but required as a precursor to understanding.

Coastline Community College, in Orange County, California (http://coastline.edu), offers a linguistic/paralinguistic portal to students of French language and culture in the form of four semester-length courses bristling with links to authentic, francophone-produced data. Course homepages retain as little of the College's boilerplate template as possible, so that the aforementioned French-named buttons demand to be clicked upon for clarification, and the course *matière,* or content, is made up of Web resources from throughout the French-speaking world, sites brought together in the interest of a few general topics. For instance, the question of language and society brings together sites about the nature of Breton, the history of the French language, the essence of dialect versus language, the imposition of American-language films in French movie theatres, and the artistic need for dubbing as opposed to sub-titling in film. Students' question topics comprise small Web portals, lists of lists, amassing French-sourced information in a single place concerning a general topic area. Likewise, course *ressources* pages facilitate improvement in grammar and vocabulary acquisition and in media awareness, with each *resources* page offering URLs from everywhere in the interest of news or language.

Hyperlinks and links from around the world, combined with free-of-charge and easily translatable interfaces, have become the heart of this novel educational experience. And these have been demonstrated to promote steady, enduring progress in writing improvement and critical thinking, as well as in international cultural awareness. Coastline student writers begin their work in French as most second-language learners do, seeking a correct answer to each question and, in the American tradition, reasoning in a linear way from thesis to conclusion. They produce short, direct, present-tense utterances. Gradually, however, sentences become longer and more complex, graced with metaphors and linguistic nuance (e.g., long student papers responding to short questions at http://www.classebranchee. com/classe/motpasse.asp?noclasse=10975653, mot de passe : etudiant).

Because Coastline has been offering completely-online French courses for more than thirteen years, the institution has been able to remain in the marketing eye of international promoters of linguistic/cultural materials based in the francophone Net; these promoters have used the college to beta-test tools and software, communications curricula and services, stylistics improvement programs, and vocabulary data bases. Free applets and interfaces have borne a fruitful array of international connections and resources for Coastliners' use, along with electronic pen pals and news-rich communiqués from Asia, Southeast Asia, Africa, Haiti, Australia, and India, not to mention Europe and the Americas. Students enrolled in Coastline's "French Topics Online" have been able to improve language and cultural fluency as well as international knowledge while learning to précis-write in alternative styles and to synthesize alternative perspectives without

recourse to any textbooks or written materials in a completely virtual, freely-accessible space.

During the 13-plus years that Coastline has offered French language and culture courses online, most enrollees have ranged in age from 18 to 85. Unlike children, these post-pubescents arrive at their French course competently able to form and discern all the significant sounds (*phonemes*) of their native languages as well as the basic structures (word formation units, or *morphemes*), and meaning units (*sememes*) thereof. Anyone older than twelve or thirteen has attained all three of the aforementioned types of language competence defining fluency, a comfortable use of these *emic* elements, in his native language. This fact, along with the obvious actuality that adults are capable of making decisions on their own, is important for secondary language learning. That is, after having spent years honing communication strategies, adults have developed a high degree of competence in their native language's cognitive style. They are also good at blocking out what they find to be irrelevancies, they can discern and obey pattern restrictions, and they can analyze; in essence, fluent adult speakers of any L1 language have learned how to learn.

At Coastline, many enrollees arrive with the complaint that they have already studied some subject matter without success and are giving it a not-entirely optimistic new effort. Online French language students often report that they have studied the language "for years," but "don't know how to talk." Many of them suffer from the fact that, although they have become excellent analytical thinkers, largely due to Western educational practices, "Analytical thinkers... sometimes never acquire communicative fluency in a second language because their left-brain, sequential processing slows them down" (SIL International, 1998).

But it seems that in international, transnational, non-national cyberspace, adult analytical thinkers, especially those who have been trained in the American system of searching for a single right answer in a linear way, find themselves having to use all their skills in a new manner. Online-delivered course materials comprise text and images, colors, and movement; Coastline's online French courses offer francophone country-sourced materials that necessarily demand a view of the world that is different from the American one. Just as it has been reported that cultural differences in thought processes may be related to variations in what people focus on as they view images (Nisbett, 2005), so Coastline's students of French online have reported orally and in writing that they "feel the eyes moving in a different way" and "notice different stuff" when surfing the Web *à la française*. The Coastline French onliners' experience in "French Topics" may have yielded such high success (their native francophone pen pals report being impressed at their fluency, and their reading and writing skills rank them well beyond the "intermediate" level of American Council on the Teaching of Foreign Languages (ACTFL) proficiency) because they have been able to alter their learning patterns so as to harmonize with the French mentality.

Alternative Modes of Thinking, Viewing, and Analyzing the World

The Coastline online French students have come to find that many non-American cultures' worldviews are suffusing the Americanocentric Internet through a distinctively "field-sensitive" or "field-dependent" presence. That is, as has been noted by Nisbett (2005), among others, Web-based materials generated outside the USA tend to incorporate background into foreground, field into ground, so that the entire mass of material can be considered as a whole; by contrast, American-sourced materials place the figure against the ground, emphasizing the foreground as it is superimposed against a background in a "field-independent" way (Watanabe, 2003).

Thus, particular, ethnic, or even national cultural identities are seen in many non-American countries as being parts of a global, integrated whole (Griggs & Dunn, 1996); people of those countries consider themselves to be part of an international, or as modern American business-people say, "transnational," field. For instance, French children's textbooks generally contain timelines of international discoveries and events, artistic products, and political progress, reminding youths that human advancement is an intellectual, worldwide thing, that humans make progress thanks to one another's creativity, all in the interest of each.

Many European and Asian classrooms emphasize the overarching, the deductive, with the general rule presented, the overall picture revealed, before each detail of it is analyzed; this sort of "field sensitivity" differs from the American inductive approach, in which particular instances of a phenomenon are laid out so that students may arrive at generalizations about them on their own (Baudry, 2003; Watanabe, 2003) In American science classes, numerous experiments are carried out that are meant to lead students to generalizations; many other countries' scientific education begins with theories that are subsequently clarified through demonstration or experimentation.

The distinctive nature of field sensitivity/field dependence has led business psychologist Pascal Baudry (2003) to note that communication for the French in particular is situated in a context of abstract relationships and of relativism more than it is in a context of concrete content. The French make decisions of almost every kind on an ethno-relative, rather than an ethnocentric basis; this perspective holds that each ethnic group develops only in relation to others, since we all live in the same "field," on the same Earth. A "worldview," to this way of thinking, is literally just that, a truly global perspective. The complexity and beauty of life are viewed as they are exhibited in every production of every living being; even online productions are ultimately the fruits of beautiful human minds. Web sites are, therefore, to be designed as syntheses, representations of our artistic side, our "right brain," rather than simply as mechanical, "left brain" miscellany; as Fischer (2002) notes, the French categorization of *le Webdesign* as an art like painting or drama or sculpture must entail a "transversality of thought," a cross-pollination of mathematics/ engineering with literature and the visual arts. And as Tice-Deering (2002) has pointed out, with respect to the sort of intercultural integration resulting from the necessary mental interchange that has emerged among the *Weltanschauungen* of the francophone world as they have mingled with those of Coastline's online language learners, an underlying anthropological assumption is being revealed here: As our experiences with cultural difference become more complex, our potential competence in intercultural interaction increases.

Melding Modes of Thinking into Long-Term Learning Online

It appears that the most successful Coastline Community College online learners of French have achieved what Vasquez (2000) has called a *bicognitive* or "field-mixed" manner of thinking. That is, these onliners are no longer oriented solely toward the independent, inductive thought process encouraged in American society; rather, their learning styles have come to share certain features of the dependent, the deductive. For example, Coastline's French onliners express in their essays generalizations that are inclusive, incorporating geology and geography, economics, demographics, and language as bases for political actions, rather than hoping to pinpoint a single cause for what nations do. They look for the science in art and the art in science, the physics in football and the chemistry in athletic performance, executing the kind of transversality

that Fischer (2002) would have thinkers of the future exploit. Coastline's French learners seek out student colleagues with whom to do joint, interdisciplinary research projects; group work and frequent discussion like this characterize the field-dependent thinker, while individual efforts are the product of the field-independent mind. Coastline's onliners thus end up delivering Web-based research with multi-faceted subject matter examined from multifarious perspectives.

High levels of intercultural awareness and consequent linguistic competence have developed through interactivity with the non-American Net unconstrained by temporal or physical boundaries. Indeed, although Coastline's Internet-delivered courses, like traditional ones, have been scheduled to extend for a single 16-week semester each time they are offered, a dozen students have remained continuously enrolled in online French activities with the college for more than thirteen years, whether school is in session or not. Coastline's adult French language learners have reported during year-round electronic live chat sessions and in e-mail and spoken conversations that their modes of thinking, reasoning, and writing have taken on a new cultural context awareness, a field sensitivity, through participation in French language and cultural fluency courses, requiring them to read, communicate, and interact exclusively in French online. For example, a woman interested in French feminism has discovered through the francophone 'Net that much of Simone de Beauvoir's strength emanated from aspects of the existentialist philosophy that she shared and discussed with her partner Jean-Paul Sartre, a young physicist has found that the University of Geneva is performing experiments of a kind he had imagined impossible, and a dog-lover has discovered veterinary advice at http://www.chiensderace.com, where Swiss, Belgian, Canadian, and Luxembourgeois reports accompany those of French DVMs.

Indeed, online learning offers excellent, if not unparalleled, opportunities to enhance erudition through cultural, linguistic, argumentative exchanges across time and space (Stevens, 2002), and Coastline College's online French language students have thus been able to enrich their understanding of a new mode of idea exchange without ever reading hard-copy texts or meeting one another or their professor face-to-face in a physical classroom. In addition, their course materials have remained dynamic, authentic, and tailored to each student's goals at his level of competence; they select their own topics of research and progress at their own pace, working and re-working with one another for feedback, for editing, and for advice.

Altered modes of self-expression and unsolicited remarks provide evidence that the French onliners are learning in a new way. For example, instead of producing sentences with an average of seven words, in the Anglophone style (Watanabe, 2003), they have begun to generate regular sequences of ten-to-fifteen-word constructions in the francophone mode. Instead of writing short essays with more than 50% simple sentence structures, typical of American students in traditional second-language classrooms, these onliners are producing work abounding in complex and compound utterances, often using subjunctives and pluperfects, abstract adjectives, and adverbs. And instead of presenting arguments in the standard American, single-thesis, single-antithesis, single-conclusion manner (Watanabe, 2003), they are entertaining ancillary commentaries, exploring possible rebuttals to their claims, and even researching supportive documentation, in the manner of the European learner (Watanabe, 2003; Baudry, 2003). As Hall (1976) has remarked, proof of acculturation is often apparent in language use; mimicking a pattern typical among native French speakers, Coastline's French language learners begin after less than 25% of their first semester online to write in a francophone-style high-context mode (Baudry, 2003), referring explicitly to the question they are answering and embedding responses to any queries in complete sentences expressing the who, the why, and the how, rather

than using two-to-five word e-mail notes in the low-context manner typifying American communication (Hall, 1976).

Since the first semester of online course offerings in French at Coastline, more than 50% of the students in each single-semester class have attained "advanced" or "advanced plus" reading and writing skills in French online, as these proficiency levels are defined by the American Council on the Teaching of Foreign Languages (ACTFL) (1986); more than 25% have attained "superior" writing skills, making themselves understood by native speakers and arguing in a francophone field-mixed style (Vasquez, 2000). But even more notably, the onliners report spending more "time on task" than ordinary classroom students do overall; that is, with 8-10 hours of live chat per week, nearly daily e-mail, and asynchronous messaging to accompany their four expository-essay-based, ordinary assignments, these learners are putting themselves in a pool of French on a far more frequent basis than they would do traditionally. In foreign language learning, perhaps more than in other areas of study, total immersion (in online learning of a new tongue, this implies interactivity and idea interchange) is necessary at least once daily if progress is to continue and frustration is to be reduced (Magny, 2002).

SERENDIPITOUSLY CREATIVE COURSE DEVELOPMENT TO INCITE LEARNING

All this linguistic and cultural development among Coastline's onliners has occurred rather serendipitously. The College's Distance Learning Department has saved money by avoiding "courses in a box"; neither Blackboard nor Web CT, for instance, is promoted as a template, and each instructor is encouraged to work with departmental Web experts to set up sites. French students are therefore able to immerse themselves instantly upon arriving at their course *page d'accueil*, designed for them with a French ambiance by their instructor, and aimed to pique the interest of the analytical thinker while tantalizing the intuitive one with immediate entrée into the francophone 'Net. The *ressources* link clickable from the course homepage is an example of one that depends on English-French cognates, at http://dl.ccc.cccd.edu/classes/internet/french198/ressourcessensass.htm. Rather than exhibiting a course-in-a-box Anglophone layout, pages are conceived with materials made freely available by such organizations as the francophone http://thot.cursus.edu and http://www.classebranchee.com with the former evaluating software and services and the latter providing course-building tools under the auspices of the Canadian *Office de la langue française*. Netscape Composer and Microsoft Front Page have permitted additional page creation tools; hyperlinks are simply listed/incorporated from freely-available sources, along with *gratuiciel*, that Coastline's French instructor has found, and often translated, on the international 'Net. New assignment variations, comments, and so forth, are sent to students via e-mail using AOL France and France Telecom's free Voilà service. These mail interfaces' use of *expéditeur, destinataire, daté le..., connexion* and *déconnexion* helps to keep learners' eyes from wandering to English-language words while using their new mode of expression. Indeed, on the rare occasions when College-based Outlook Express e-mail has been sent, students have remarked that the English is "obvious" and "makes me look at it." Anglophone chat interfaces with Anglophone advertising attract the eyes of learners who should be attending to things French; this is avoided as much as possible.

With very little recourse to English available, Coastline French language students are offered general, topical areas of study/Websurfing as their course content (e.g., *l'éthique* (ethics, either in business, medicine, the workplace, or elsewhere), *le voyage* (travel), etc.). *Travaux pratiques*, or course content questions, depend upon "surf

reports" submitted in French via e-mail; these comprise pieces somewhat like film or book reviews enhanced by expository writing. Carnegie Mellon University's free Lycos Course Builder, which has permitted translation into French, houses many of the course's content questions in an external area safe from occasional local power outages or glitches: Even short quizzes, also written by the instructor and posted at the course home page for instant interactivity, are embedded with francophone links.

International experts in distance education (CRPUQ, 2002; Magny, 2000) have cited freedom from constraints and a desire to take advantage of opportunities of all kinds whenever they arise as features uniting effective learning objects to successful learners. Coastline's online French course material is always "up," permitting students to work whenever they wish, at their own pace, and at the intensity they wish, free from ordinary institutional limitations. The aforementioned dozen students who have continued to enroll in online French courses or to have kept in touch with synchronous chat and/or asynchronous bulletin boards throughout the thirteen years that the College has offered online French study say they have become "addicted" to their virtual *francophonie*. Their course materials, freely produced and freely accessible from anywhere, remain ready for access all the time. If a student's "learnable moment" does not come until late in the semester, he may wait until then to begin his work; no "due dates" exist.

Although projects without due dates require self-discipline to complete, it seems that Coastline online French students have been able to raise their "engagement rate" in schedule-free francophone cyberspace. That is, Coastline students' online French course activity demonstrates something that researchers have observed elsewhere: The number of minutes of active participation in coursework is frequently higher online than it is in traditional classes (Han, 1999). Coastline French onliners frequently report that they spend "way more time on this class than I do on my other ones" or "get lost in those Web sites and end up spending hours surfing around and reading stuff" with French online. These students' intrinsic motivation, their engagement in their subject matter simply because of curiosity, interest, or enjoyment, seems to be greater online than it is in other contexts. And as a result of this, their learning is more efficient and broader-based than it might be otherwise. As Brewster and Fager (2000) have noted, the intrinsically-motivated learner exploits strategies that demand greater effort and which facilitate fast information processing. This sort of learner is also more likely than his peers to feel confident about his ability to learn new material, and he is more likely to retain information and concepts longer. It is clear that retention arising from something other than external prodding is an essential feature of learning language in cyberspace. Ultimately, as a result of increased motivation to learn, it is perhaps useful to note that the aforementioned traits of the intrinsically-interested individual tend to grant him a higher likelihood of being a lifelong learner, continuing to educate himself beyond the formal institutional classroom.

An Electronic Interface Promoting Deeper Learning at a Faster Pace

It has become clear in the thirteen years during which Coastline Community College has been offering current-events French courses online that an extraordinary quality of learning has been taking place. Evidently, as the University of Quebec's Estelle Magny (2002), professor of Distance Learning, has said, the pedagogical possibilities of online learning have only just begun to be probed and the quality of their results analyzed. Cyberspace can and will permit learners to redefine their relationship with information, Magny claims, enabling them to profit from a new nexus of knowledge that will enable them to teach themselves new things however and

whenever they desire, often without realizing it. Coastline's online learners of French have profited from linguistic contextual freedom in online chat and from the dynamic realism of electronically-delivered news arriving to their computers from everywhere; this has enabled them to broaden their vocabulary and their sophistication in understanding the francophone worldview while they are immersed in authentic linguistic realia in all social stylistic registers. Live chat sessions and asynchronous bulletin board postings include subjects ranging from new films and books of interest to the political difficulties of France's prime minister and the comparison of immigration concerns in France and California. Students report that they appreciate francophone page layout and content, calling it "classy" and "neatly organized, almost like a French garden," they have begun to notice how the affective domain penetrates the interactive. They have altered their field-independent cognitive style to have accommodated field dependence, resulting in a useful, bicognitive field mix.

It seems that continuously-available contact with language in all its contexts, in all its authenticity, in all its social, stylistic registers, can be realized online in a way that can force language learners to "sharpen the intellect," as Crystal (2004) has said. Indeed, as Crystal (2004), Baudry (2003), and Magny (2002), and Caron (1999) have all noted, human interaction and modes of communication are changing as a result of synchronous and asynchronous electronic systems. Happily enough for Coastline's students, the French government has embraced Internet communication. Indeed, the fact that their course materials are authentic and dynamic is a fortunate result of Prime Minister Lionel Jospin's 1997 initiative in favor of electronic connectivity: All schools and government-associated institutions were to receive high-speed connections and *soutien informatique* before 2002; government-sponsored news services, schools for children and adults, and all information sources were to be housed in cyberspace; Web development was to be granted university departmental status. Fortunately for Coastline's French onliners, their course development was to benefit. Free materials easy to download were available, usually without the busy *frénésie* characterizing Anglophone 'Net materials, and this ended up being one reason cited by Coastline students for their interest in the course. And fortunately for the Coastline students, it may be the case that the pictorial, subtly-colored, artistic nature of many francophone Web sites strengthens the message-learning process, a process that, as Crystal (2004) remarks, demands deeper study for its uniqueness. It is perhaps true that, since pictorial data are generally stored in, and decoded by, the brain's right hemisphere, while linguistic data are decoded by the left (Nisbett, 2005), written text overlying or accompanying static images may offer increased opportunities for the brain to apprehend.

CONCLUSION

For Coastline's non-traditional online learners of French, who have comprised a range of age from post-adolescence to octogenarian, and who never have to see any college buildings or parking lots, electronic media have enhanced their learning of a new language in multifarious contexts, permitting them to travel well beyond the bounds of any classroom and across time zones, from the American "monochronic" state of assuming that everything is tangible, happening in measurable seconds or minutes or days or years (Hall, 1976), into the polychronic francophone one (Baudry, 2003) where a multiplicity of tasks, ideas, or actions may occur simultaneously, just as they do in international, pluricultural cyberspace.

REFERENCES

American Council on the Teaching of Foreign Languages. (1986). *ACTFL proficiency guidelines* (Rev. ed.). Hastings-on-Hudson, New York: ACTFL Materials Center.

Baudry, P. (2003). *Français et Américains: L'autre rive.* Paris: Village Mondial.

Birdwhistell, R. (1952). *Introduction to kinesics.* Louisville, KY: University of Louisville Press.

Brewster, C., & Fager, J. (2000). *Increasing student engagement and motivation: From time-on-task to homework.* Portland, OR: Northwest Regional Educational Laboratory.

Caron, F. (1999). *Les deux révolutions industrielles du XXe siècle.* Paris: Pocket.

Collier, G., & DiCarlo, D. (1985). *Emotional expression.* Hillsdale, NJ: Lawrence Erlbaum Associates.

Conférence des recteurs et des principaux des universités du Québec (CRPUQ). (2002, September). Les normes et standards de la formation en ligne. *Profetic, revue internationale des technologies.* Rubrique ressources, normes et standards. Retrieved August 31, 2005, from http://profetic.org/file/norm-0210-d-RAPPORT.pdf

Cook, V. J. (2005). First and second language learning. In G. E. Perren (Ed.), *The mother tongue and other languages in education, CILTR* (pp. 7-22). London: Center for Information on Language Teaching and Research.

Crystal, D. (2004). *Language revolution (Themes for the 21st century).* London: Polity.

Dobrovolny, J. (2003, October). Learning strategies. *Learning Circuits, 4*(10), 27. Retrieved May 5, from http://www.learningcircuits.org/2003/oct2003/dobro volny.htm

Fischer, H. (2002). Technologies – en attendant le huitième art, le rêve de l'œuvre totale. *Le Devoir.com*, edition of Monday, 18 November, 2002, Rubrique Technologies. Retrieved September 12, 2005, from http://www.ledevoir. com/2002/11/18/13590.html

Griggs, S., & Dunn, R. (1996). Hispanic-American students and learning style. *Clearinghouse on Early Childhood and Parenting* (393607). Urbana-Champaign, Illinois: College of Education, Early Childhood and Parenting Collaborative, University of Illinois.

Hall, E. (1976). *Beyond culture*. New York: Anchor Press.

Han, X. (1999, November 15). Exploring an effective and efficient online course management model. *Teaching with Technology Today, 5*(2), 3. Retrieved May 7, 2006, from http://www.uwsa. edu/ttt/han.htm

Hudson, R., Firminger, P., & Weakly, R. (2005). Developing sites for users with cognitive disabilities and learning difficulties. Retrieved October 5, 2005, from http://juicystudio.com/article/cognitive-impairment.php

James, L. (1969). Prolegomena to a theory of communicative competence. *Extreme Psychology, Extreme Research*, Fall 1998 (p.499). Retrieved September 10, 2005, from http://www.soc.hawaii. edu/leonj/499f98/libed/competence/titlepage. html

Kalyampur, M., & Harry, B. (1999). *Culture in special education*. New York: Paul H. Brooks.

Magny, E. (2002). *Les représentations reliées à la langue seconde et à son enseignement/apprentissage chez les formateurs universitaires et scolaires*. University of Québec papers.

Nisbett, R. (2005, September). Culture and point of view: Eye movements may betray your culture.

In *Proceedings of the National Academy of Sciences, 100*(19), 11163-11170.

SIL International. (1998, September). Your brain dominance and language. *Ethnologue.com, LinguaLinks* library resources, SIL International Linguistics Center, Dallas, TX. Retrieved September 13, 2005, from http://www.sil.org/LinguaLinks/ LanguageLearn ing/OtherResources/YorLrnng-StylAndLnggLrnng/YourBrainDominanceAnd LanguageL.htm

Singhal, M. (1998, October). A comparison of L1 and L2 reading: Cultural differences and schema. *TESL Journal, 4*(10). Retrieved May 7, 2006, from http: //iteslj.org/Articles /Singhal-ReadingL1L2.html

Stevens, V. (2002). Rationale for chat in language learning. *WWW.Study.com*, Writing for Webheads, V. Stevens files. Retrieved September 14, 2005, from http://www.homestead.com/prosites-vstevens/files/efi/why_chat.htm

Thanasoulas, D. (2000). Language and culture: A thesis. *Developing Teachers.com*. Teacher Training/Culture. Retrieved September 16, 2005, from http://www. developingteachers.com/articles_tchtraining/culture1_dimitrios.htm

Tice-Deering, B. (2002). Inquiry and research. *Ontario Library Association Database of Expertise*. Retrieved March 7, 2005, from http://www. accessola.org/action/positions/info_studies/html/ research.html

Vasquez, J. (2000, June). Difference is not deficiency. *IN CONTEXT: A quarterly of humane sustainable culture, 4*(27), 30.

Watanabe, M. (2003). *Comparisons of cooperative learning in the U.S., Japan, and France*. Paper presented at the Japan Society of Educational Sociology Annual Meeting, Meiji Gakuin University, Tokyo, Japan.

Chapter 7.13
Enhancing Inclusion in Computer Science Education

Donald D. Davis
Old Dominion University, USA

Debra A. Major
Old Dominion University, USA

Janis V. Sanchez-Hucles
Old Dominion University, USA

Sandra J. DeLoatch
Norfolk State University, USA

Katherine A. Selgrade
Old Dominion University, USA

Shannon K. Meert
Old Dominion University, USA

Nikki L. Jackson
Norfolk State University, USA

Heather J. Downey
Old Dominion University, USA

Katherine M. Fodchuk
Old Dominion University, USA

INTRODUCTION AND BACKGROUND[1]

We describe an intervention that uses computer science (CS) faculty and students to create an inclusive learning environment. Our intervention model assumes that persistence and retention are the result of a match between student motivation and abilities and the university's social and academic characteristics. This match in turn influences the effective integration of students with the university and, as a result, their persistence and retention (Cabrera, Castaneda, Nora, & Hengstler, 1992; Tinto, 1993). We are currently implementing and evaluating this intervention at Old Dominion University, a research intensive urban university with a culturally diverse student body, and Norfolk State University, an urban and historically black university (HBCU) that primarily emphasizes teaching.

A MODEL FOR CREATING INCLUSIVE LEARNING ENVIRONMENTS

Organizational Support for Faculty and Students

This portion of our model depicts external resources available to support change in faculty and students (see Figure 1). Support for faculty includes peers, teaching assistants, and external consultants who provide assistance in areas targeted for change, for example, pair programming practices. Support for students includes academic resources such as tutoring, advising, and mentoring. Professional organizations, such as the Association for Women in Computing, may provide support to both faculty and students.

Changing Faculty

We focus on faculty because they influence student outcomes. The intervention concentrates on faculty who teach introductory programming classes because these classes represent the first and largest barrier to success in CS. We build on

Figure 1. Intervention model for creating inclusive learning environments

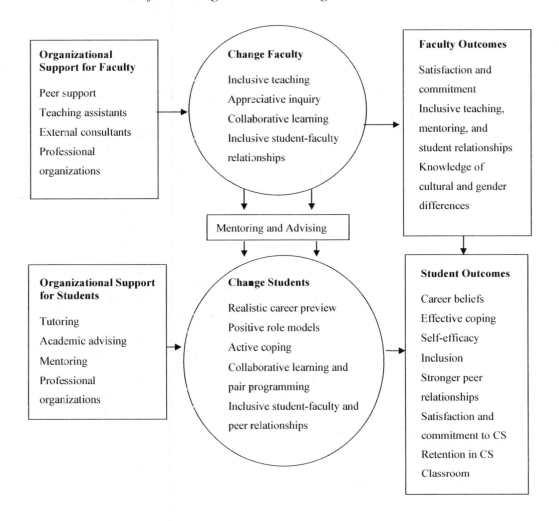

existing talents and strengths of faculty by emphasizing four skill areas: (1) inclusive teaching practices, (2) appreciative inquiry, (3) collaborative learning, and (4) inclusive student-faculty relationships.

Inclusive Teaching

Teaching style strongly influences retention in science, mathematics, and engineering (Seymour & Hewitt, 1997). Men and women and members of different ethnic groups experience IT work environments and learning environments differently (Major, Davis, Sanchez-Hucles, & Mann, 2003; Steele, 1997; Steele & Aronson, 1995). Inclusive teaching encourages acceptance of diverse styles of learning and avoids actions based on stereotypes concerning group membership.

Appreciative Inquiry

Mentoring and tutoring typically focus on deficits and remediation. Appreciative inquiry, in contrast, emphasizes optimism, positive expectations, and challenge (Srivastva & Cooperrider, 1990). People tend to act in ways to fulfill expectations of powerful others, such as professors. Moreover, use of positive expectations energizes and directs behavior in a manner to fulfill expectations. Positive expectations may be especially powerful for those for whom low expectations are held (McNatt, 2000) or for those for whom gender or racial stereotypes influence expectations (Jussim & Eccles, 1992).

Collaborative Learning

Faculty members experience positive outcomes when they learn and share their experiences together. Collaboration may lead faculty to learn more deeply, encourage and support one another, explore new methods of teaching, increase colleagueship, discover and appreciate student differences, and develop a shared vision for teaching

inclusiveness (Cox, 2002). Faculty who share learning experiences report increased interest in the teaching process, effectiveness as a teacher, and awareness and understanding of how differences may influence and enhance teaching and learning (Cox, 2002).

Inclusive Student-Faculty Relationships

The quality of exchange in relationships influences important outcomes. Inclusive student-faculty relationships enhance mentoring and advising and increase the retention of women and minority IT students (Cohoon, 2001, 2002; Gürer & Camp, 2002). Mentoring has more positive effects when it is not remedial in nature. Inclusive mentoring may be critically important because *not actively encouraging* women and minorities may have the same effect as *actively discouraging* them (Leggon, 2003).

Mentoring and Advising

Mentoring and advising link together faculty change and student change. Mentoring represents an attempt to provide information required for successful socialization and social support needed for effective stress management. Women and minorities may experience greater barriers to mentoring due to restricted access to mentors, unwillingness of mentors to mentor them, or disapproval by others of mentoring activities. Some women may also worry that a request for mentorship may be perceived to be a sexual advance (Ragins & Cotton, 1991). Women and minority students may have more difficulty establishing mentoring relationships because mentors often choose protégés with whom they can identify; this choice is often influenced by gender, race, and social class. Because women and minorities are underrepresented in CS, they may face greater challenges in seeking mentorship, particularly from mentors of the same gender or ethnicity (Wright

& Wright, 1987). Fortunately, mixed-ethnic and mixed-gender mentorships can be as beneficial as same-ethnic and same-gender mentorships (Atkinson, Neville, & Casas, 1991).

It is important that mentoring and advising be inclusive and appreciative. Well-meaning support such as mentoring may elicit "stereotype threats" and lead to unintended negative consequences (Steele, 1997). Stereotype threat can be particularly insidious when combined with mentoring that emphasizes deficits and remediation. This may explain, in part, the positive relationship between mentoring and attrition in CS students reported by some researchers (Cohoon, Cohoon, & Turner, 2003).

Changing Students

Freshman and sophomore students in CS provide the second focus for change. Emphasis is placed on student attitudes, beliefs, and behaviors that are likely to improve socialization into the learning environment that exists in CS education and in the IT workplace. We have implemented a class required of new CS majors that has five components: (1) realistic career preview, (2) positive role models, (3) active coping, (4) collaborative learning and pair programming, and (5) developing inclusive relationships.

Realistic Career Preview

Our conception of a realistic career preview (RCP) is based on a practice developed in industry called the realistic job preview (RJP). The RJP provides potential new hires realistic—that is, both favorable and unfavorable—information that describes their likely work experience in the hiring organization. This method contrasts with traditional approaches to hiring and socialization that emphasize only favorable information. The logic of the RJP is that accurate information improves the fit between individual and organizational expectations and hence reduces later

disappointment and dissatisfaction. Job candidates who receive RJPs have higher performance, job satisfaction, and organization commitment, and are less likely to quit (Philips, 1998; Premack & Wanous, 1985). RJP recipients are also more committed to training and profit more from it (Hicks & Klimoski, 1987). We expect to get the same results from the RCP.

The RCP presents to students a realistic portrayal of computer science education and careers. We use research findings to debunk common myths about working in IT, for example, we show that women in IT are frequently happy with their workplace and career in IT. We also "inoculate" students with previews of potentially difficult relationships and situations to prepare them for future challenges.

Positive Role Models

We use diverse IT professionals to serve as role models to show students the variety of career opportunities available to them with a degree in CS. Female students have less clearly defined career goals than their male counterparts when they begin studying science; they are disproportionately likely to abandon their plans to major in science after their first undergraduate year as a result (Seymour, 1999). Efforts to improve socialization during the first year of study, therefore, may have pivotal importance for improving retention among women. Margolis and Fisher (2002) recommend that early in their careers female students learn the context for technology and that they learn to understand why they are being asked to do what they do. This includes being presented with role models and career information that will help them to believe that they can be successful IT professionals.

Active Coping

Ethnic and gender minority status may interfere with the socialization process by reducing access to information and mentors and by increasing anxiety and stress (Jackson, Stone, & Alvarez, 1992). The extent to which newcomers to CS have a cultural or gender background that differs from most others in their department may reduce the rate of socialization and increase the likelihood of failure and quitting (Bauer, Morrison, & Callster, 1998). Women and minority students in CS report greater stress than other students (Davis, Gregerman, & Hathaway, 2003). The manner in which students deal with stress influences their level of social integration, institutional commitment, and intention to remain in university (Bray, Braxton, & Sullivan, 1999). We teach practices to cope with stress and anxiety that result from outsider status in order to strengthen academic and social integration. Enhanced academic and social integration increase commitment to academic goals and the academic institution and, as a result, enhance persistence and retention (Cabrera et al., 1992).

Active coping is the process of taking steps to remove or circumvent stressors or to reduce their impact (Carver, Scheier, & Weintraub, 1989). Active coping is positively associated with optimism, control, self-esteem, and hardiness, and negatively associated with anxiety (Carver et al., 1989). Moreover, when students experience stressful events, active coping is more likely to lead to efforts to: reduce stress, suppress competing activities, seek instrumental and emotional social support, and reduce maladaptive behaviors such as denial and behavioral and mental disengagement (Carver et al., 1989). Active coping also leads to increased information seeking, relationship building, positive cognitive framing of events, more successful socialization results and, as a result, increased satisfaction and intention to remain in the organization (Ashford & Black, 1996; Major & Kozlowski, 1997; Major, Kozlowski, Chao, & Gardner, 1995).

Collaborative Learning and Pair Programming

Pair programming provides an important means of collaborative learning for CS students (Williams & Kessler, 2003). Pair programming directs one partner to act as the "driver" who is responsible for writing code, while the "non-driver" partner is directed to continuously observe the work of the driver, watch for errors, and recommend alternatives. Programmers who cooperate with one another in this way work faster, detect more errors, create higher quality programs, express more satisfaction with their work, and report more confidence in their programs than do programmers working in isolation (Williams, 1999; Williams & Kessler, 2000a, 2000b; Williams & Kessler, 2001). Students who program together in pairs achieve higher grades and may be more likely to remain in CS (McDowell, Werner, Bullock, & Fernald, 2002; Williams, Wiebe, Yang, Ferzli, & Miller, 2002).

Inclusive Student: Faculty and Peer Relationships

As IT has permeated all organizational functions, employers of IT workers have encouraged development and use of "soft skills," such as communication, interpersonal relationships, teamwork, and working with diverse coworkers. Interviews with senior managers in IT reveal that such soft skills are better predictors of on-the-job success among IT workers than technical experience or skills (U.S. Department of Commerce, 2003), yet soft skills are not taught in conventional undergraduate CS programs (Bevan, Werner, & McDowell, 2002). Moreover, the global nature of IT requires the ability to work successfully with people from different cultural backgrounds; such knowledge should be part of the CS and IT curriculum (Little et al., 2000). Working effectively in teams and developing inclusive relationships with others from different cultures requires special knowledge and skills (Davis & Bryant, 2003). We highlight for CS students the importance of cultural differences and the need to consider these when developing relationships with faculty, graduate teaching assistants, and other students.

CONCLUSION AND FUTURE DIRECTIONS

With continued decline in enrollments, CS departments must pay attention to the learning environment if the percentage of women and minorities in CS is to increase. We describe a comprehensive, research-based intervention that is designed to enhance inclusiveness for all students—including women and minority students—by simultaneously addressing change in faculty and students and, as a consequence, increasing student commitment and retention.

Components of this intervention are likely to exert a synergistic impact on retention. Because there are multiple causes of women and minorities quitting IT study and careers, interventions intended to address these causes must be similarly complex. Moreover, interventions must be sufficiently flexible to allow their diffusion to educational institutions with different missions. The components of this intervention are broadly applicable and can be used to enhance inclusiveness and student retention in all areas of education.

REFERENCES

Ashford, S. J., & Black, J. S. (1996). Proactivity during organizational entry: The role of desire for control. *Journal of Applied Psychology, 81,* 199-214.

Atkinson, D. R., Neville, H., & Casas, A. (1991). The mentorship of ethnic minorities in professional psychology. *Professional Psychology, 22,* 336-338.

Bauer, T. N., Morrison, E. W., & Callster, R. R. (1998). Organizational socialization: A review and directions for future research. In G. R. Ferris (Ed.), *Research in personnel and human resources management* (Vol. 16, pp. 149-214). Stanford, CT: JAI Press.

Bevan, J., Werner, L., & McDowell, C. (2002). *Guidelines for the use of pair programming in a freshman programming class.* Paper presented at the 15th Conference on Software Engineering Education and Training, Los Alamitos, CA.

Bray, N. J., Braxton, J. M., & Sullivan, A. S. (1999). The influence of stress-related coping strategies on college student departure decisions. *Journal of College Student Development, 40,* 645-657.

Cabrera, A. F., Castaneda, M. B., Nora, A., & Hengstler, D. (1992). The convergence between two theories of college persistence. *The Journal of Higher Education, 63*(2), 143-164.

Carver, C. S., Scheier, M. F., & Weintraub, J. K. (1989). Assessing coping strategies: A theoretically based approach. *Journal of Personality and Social Psychology, 56,* 267-283.

Cohoon, J. M. (2001). Toward improving female retention in the computer science major. *Communications of the ACM, 44*(5), 108-114.

Cohoon, J. M. (2002). Recruiting and retaining women in undergraduate computing. *SIGCSE Bulletin, 34,* 48-52.

Cohoon, J. M., Cohoon, J. P., & Turner, S. E. (2003). Departmental factors in gendered attrition from undergraduate IT majors. In *Proceedings of the National Science Foundation's ITWF & ITR/EWF Principal Investigator Conference* (pp. 126-130). Albuquerque: The University of New Mexico.

Cox, M. (2002). *Designing and implementing staff learning communities: An effective approach to educational development.* Paper presented at the 4th World Conference on International Consortium for Educational Development in Higher Education, Perth, Australia.

Davis, C. S., Gregerman, S. R., & Hathaway, R. S. (2003). Information technology pathways in the academy: Identifying barriers for women and underrepresented minorities. *Proceedings of the National Science Foundation's ITWF & ITR/EWF Principal Investigator Conference* (pp. 91-95). Albuquerque: The University of New Mexico.

Davis, D. D., & Bryant, J. (2003). Leadership in global virtual teams. In W. H. Mobley & P. W. Dorfman (Eds.), *Advances in global leadership* (Vol. 3, pp. 303-340). Amsterdam: JAI.

Gürer, D., & Camp, T. (2002). An ACM-W literature review of women in computing. *SIGCSE Bulletin, 34,* 121-127.

Hicks, W. D., & Klimoski, R. J. (1987). Entry into training programs and its effects on training outcomes: A field experiment. *Academy of Management Journal, 30,* 542-552.

Jackson, S. E., Stone, V. K., & Alvarez, E. B. (1992). Socialization amidst diversity: The impact of demographics on work team oldtimers and newcomers. In L. L. Cummings & B. M. Staw (Eds.), *Research in organizational behavior* (Vol. 15, pp. 45-109). Greenwich, CT: JAI Press.

Jussim, L., & Eccles, J. S. (1992). Teacher expectations II: Construction and reflection of student achievement. *Journal of Personality and Social Psychology, 63,* 947-961.

Leggon, C. B. (2003). Women of color in IT: Degree trends and policy implications. *IEEE Technology and Society, 22*(3), 36-42.

Little, J. C., Granger, M., Adams, E. S., Holvikivi, J., Lippert, S. K., Walker, H. M., & Young, A.

(2001). Integrating cultural issues into the computer and information technology curriculum. *ACM SIGCSE Bulletin, 33*(2), 136-154.

Major, D. A., Davis, D. D., Sanchez-Hucles, J., & Mann, J. (2003). Climate for opportunity and inclusion: Improving the recruitment, retention, and advancement of women and minorities in IT. In *Proceedings of the National Science Foundation's ITWF & ITR/EWF Principal Investigator Conference* (pp. 167-171). Albuquerque: The University of New Mexico.

Major, D. A., & Kozlowski, S.W.J. (1997). Newcomer information seeking: Individual and contextual influences. *International Journal of Selection and Assessment, 5,* 16-28.

Major, D. A., Kozlowski, S. W. J., Chao, G. T., & Gardner, P. (1995). A longitudinal investigation of newcomer expectations, early socialization outcomes, and the moderating effects of role development factors. *Journal of Applied Psychology, 80,* 418-431.

Margolis, J., & Fisher, A. (2002). *Unlocking the clubhouse: Women in computing.* Cambridge, MA: MIT Press.

McDowell, C., Werner, L., Bullock, H., & Fernald, J. (2002). The effects of pair programming on performance in an introductory programming course. *SIGCSE Bulletin, 34,* 38-42.

McNatt, D. B. (2000). Ancient Pygmalion joins contemporary management: A meta-analysis of the result. *Journal of Applied Psychology, 85,* 314-322.

Philips, J. M. (1998). Effects of realistic job previews on multiple organizational outcomes: A meta-analysis. *Academy of Management Journal, 41,* 673-690.

Premack, S. L., & Wanous, J. P. (1985). A meta-analysis of realistic job preview experiments. *Journal of Applied Psychology, 70,* 706-719.

Ragins, B. R., & Cotton, J. (1991). Easier said than done: Gender differences in perceived barriers to gaining a mentor. *Academy of Management Journal, 34*, 939-951.

Seymour, E. (1999). The role of socialization in shaping the career-related choices of undergraduate women in science, mathematics, and engineering majors. *Annals of New York Academy of Science, 869*, 118-126.

Seymour, E., & Hewitt, N. (1997). *Talking about leaving*. Boulder, CO: Westview Press.

Srivastva, S., & Cooperrider. (1990). *Appreciative management and leadership: The power of positive thought and action in organizations*. San Francisco: Jossey-Bass.

Steele, C. M. (1997). A threat in the air: How stereotypes shape intellectual identity and performance. *American Psychologist, 52*(6), 613-629.

Steele, C. M., & Aronson, J. (1995). Stereotype threat and the intellectual test performance of African Americans. *Journal of Personality and Social Psychology, 69*, 797-811.

Tinto, V. (1993) *Leaving college: Rethinking the causes and cures of student attrition* (2nd ed.). Chicago: University of Chicago Press.

U.S. Department of Commerce. (2003). *Education and training for the information technology workforce: Report to Congress from the Secretary of Commerce*. Washington, DC: Author.

Williams, L. (1999). *But, isn't that cheating? Collaborative programming*. Paper presented at the 29th Annual Frontiers in Education Conference, Champaign, IL.

Williams, L. A., & Kessler, R. R. (2000a). *The effects of "pair-pressure" and "pair-learning" on software engineering education*. Paper presented at the 13th Conference of Software Engineering Education and Training, Los Alamitos, CA.

Williams, L. A., & Kessler, R. R. (2000b). All I really need to know about pair programming I learned in kindergarten. *Communications of the ACM, 43*, 108-114.

Williams, L. A., & Kessler, R. R. (2001). Experimenting with industry's "pair programming" model in the computer science classroom. *Computer Science Education, 11*, 7-20.

Williams, L. A., & Kessler, R. R. (2003). *Pair programming illuminated*. Boston: Addison-Wesley.

Williams, L., Wiebe, E., Yang, K., Ferzli, M., & Miller, C. (2002, September). In support of pair programming in the introductory computer science course. *Computer Science Education, 12*(13), 197-212.

Wright, C. A., & Wright, S. D. (1987). The role of mentors in the career development of young professionals. *Family Relations, 36*, 204-208.

KEY TERMS

Active Coping: Process of taking initiative to remove or circumvent stressors or to reduce their impact.

Appreciative Inquiry: An approach to change that emphasizes optimism and challenge rather than deficits.

Historically Black College or University (HBCU): About one-hundred private and public colleges and universities created to educate African Americans starting in 1837 with the founding of Cheney University (PA).

Inclusiveness: Work and learning practices that emphasize involving rather than excluding others. This includes acceptance of diverse styles of learning and avoids actions based upon stereotypes concerning group membership.

Pair Programming: A programming practice involving two programmers working together; one partner acts as the "driver" and is responsible for writing code, while the "non-driver" partner is instructed to observe the work of the driver and watch for errors, consider alternatives, and provide feedback.

Realistic Career Preview: An individual change and socialization practice in which one is provided a realistic preview of an occupational future that includes both positive and negative features.

Stereotype Threat: The extra pressure that members of underrepresented groups such as women and minorities may feel caused by fear of confirming and reinforcing negative stereotypes held about their group, for example, they are not skilled in mathematics and science. This extra pressure can reduce performance and persistence.

ENDNOTE

[1]   This material is based upon work supported by the National Science Foundation under Grant 0420365.

This work was previously published in the Encyclopedia of Gender and Information Technology, edited by E. Trauth, pp. 269-275, copyright 2006 by Idea Group Reference (an imprint of IGI Global).

Chapter 7.14
Technology–Mediated Progressive Inquiry in Higher Education

Hanni Muukkonen
University of Helsinki, Finland

Minna Lakkala
University of Helsinki, Finland

Kai Hakkarainen
University of Helsinki, Finland

INTRODUCTION

In higher education, students are often asked to demonstrate critical thinking, academic literacy (Geisler, 1994), expert-like use of knowledge, and creation of knowledge artifacts without ever having been guided or scaffolded in learning the relevant skills. Too frequently, universities teach the content, and it is assumed that the metaskills of taking part in expert-like activities are somehow acquired along the way. Several researchers have proposed that in order to facilitate higher level processes of inquiry in education, cultures of education and schooling should more closely correspond to cultures of scientific inquiry (Carey & Smith, 1995; Perkins, Crismond, Simmons & Under, 1995). Points of correspondence include contributing to collaborative processes of asking questions, producing theories and explanations, and using information sources critically to deepen one's own conceptual understanding. In this way, students can adopt scientific ways of thinking and practices of producing new knowledge, not just exploit and assimilate given knowledge.

BACKGROUND

The best practices in the computer-supported collaborative learning (CSCL) paradigm have several features in common: consideration, in an interrelated manner, of the development of technological applications, use of timely pedagogical models, and attention to the social and cognitive aspects of learning. Emphasis is placed on creating a collaborative community that shares goals, tools, and practices for taking part in an inquiry process.

Synthesizing these demands, Kai Hakkarainen and his colleagues at the University of Helsinki have developed a model of *progressive inquiry* as a pedagogical and epistemological framework. It is designed to facilitate expert-like working with knowledge in the context of computer-supported collaborative learning. It is primarily based on Carl Bereiter and Marlene Scardamalia's (Scardamalia & Bereiter, 1994) theory of knowledge building, on the interrogative model of scientific inquiry (Hakkarainen & Sintonen, 2002; Hintikka, 1999), and on the idea of distributed expertise in a community of learners (Brown & Campione, 1994). The model has also been implemented and studied in various educational settings from elementary to higher education (see e.g., Hakkarainen, Järvelä, Lipponen & Lehtinen, 1998; Lakkala, Ilomäki, Lallimo & Hakkarainen, 2002; Lipponen, 2000; Veermans & Järvelä, in press).

THE PROGRESSIVE INQUIRY MODEL

In progressive inquiry, students' own, genuine questions and their previous knowledge of the phenomena in question are a starting point for the process, and attention is drawn to the main concepts and deep principles of the domain. From a cognitive point of view, inquiry can be characterized as a question-driven process of understanding; without research questions, there cannot be a genuine process of inquiry, although in education, information is frequently conveyed or compiled without any guiding questions. The aim is to explain the phenomena in a deepening question-explanation process, in which students and teachers share their expertise and build new

knowledge collaboratively with the support of information sources and technology.

The progressive inquiry model specifies certain epistemologically essential processes that a learning community needs to go through, although the relative importance of these elements, their order, and actual contents may involve a great deal of variation from one setting to another. As depicted in Figure 1, the following elements have been placed in a cyclic, but not step-wise succession to describe the progressive inquiry process (Hakkarainen, 2003; Muukkonen, Hakkarainen, & Lakkala, 1999; 2004).

a. *Distributed expertise* is a central concept in the model. Progressive inquiry intends to engage the community in a shared process of knowledge advancement, and to convey, simultaneously, the cognitive goals for collaboration. Diversity in expertise among participants, and interaction with expert cultures, promotes knowledge advancement (Brown et al., 1993; Dunbar, 1995). Acting as a member of the community includes sharing cognitive responsibility for the success of its inquiry. This responsibility essentially involves not only completing tasks or delivering productions on time, but also learners' taking responsibility for discovering what needs to be known, goal-setting, planning, and monitoring the inquiry process (Scardamalia, 2002). There should be development of students' (and experts') social metacognition (Salomon & Perkins, 1998)—students learning to understand the cognitive value of social collaboration and gaining the capacity to utilize socially distributed cognitive resources.

b. The process begins by *creating the context* to anchor the inquiry to central conceptual principles of the domain or complex real-world problems. The learning community is established by joint planning and setting up common goals. It is important to create

Figure 1. Elements of progressive inquiry (Reprinted by permission from Muukkonen et al., 2004)

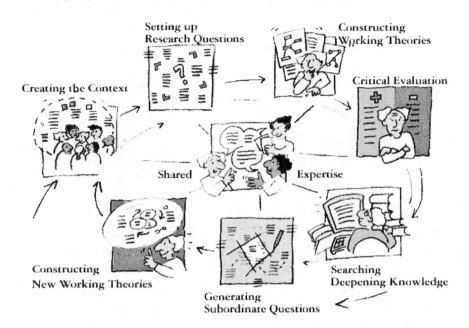

a social culture that supports collaborative sharing of knowledge and ideas that are in the process of being formulated and improved.

c. An essential element of progressive inquiry is *setting up research questions* generated by students themselves to direct the inquiry. Explanation-seeking questions (Why? How? What?) are especially valuable. The learning community should be encouraged to focus on questions that are knowledge-driven and based on results of students' own cognitive efforts and the need to understand (Bereiter, 2002; Scardamalia & Bereiter, 1994). It is crucial that students come to treat studying as a problem-solving process that includes addressing problems in understanding the theoretical constructs, methods, and practices of scientific culture.

d. It is also important that students explain phenomena under study with their own existing background knowledge by *constructing working theories* before using information sources. This serves a number of goals: first, to make visible the prior (intuitive) conceptions of the issues at hand; second, in trying to explain to others, students effectively test the coherence of their own understanding, and make the gaps and contradictions in their own knowledge more apparent (e.g., Hatano & Inakagi, 1992; Perkins et al., 1995); third, it serves to create a culture in which knowledge is treated as essentially evolving objects and artifacts (Bereiter, 2002). Thoughts and ideas presented are not final and unchangeable, but rather utterances in an ongoing discourse (Wells, 1999).

e. *Critical evaluation* addresses the need to assess strengths and weaknesses of theories and explanations that are produced, in order to direct and regulate the community's joint cognitive efforts. In part, it focuses on the inquiry process itself, placing the process as the center of evaluation, not only the end result. Rather than focusing on individual students' productions, it is more fruitful to evaluate the community's productions and efforts, and give the student participants a main role in this evaluation process. Critical evaluation is a way of helping the community to rise above its earlier achievements, creating a higher level synthesis of the results of inquiry processes.

f. Students are also guided to engage in *searching deepening knowledge* in order to find answers to their questions. Looking for and working with explanatory scientific knowledge is necessary for deepening one's understanding (Chi, Bassok, Lewis, Reiman & Glaser, 1989). A comparison between intuitive working theories produced and well-established scientific theories tends to show up the weaknesses and limitations of the community's conceptions (Scardamalia & Bereiter, 1994). The teacher of a course must decide how much of the materials should be offered to the students and how much they should actually have to search out for themselves. Questions stemming from true wonderment on the part of the students can easily extend the scope of materials beyond what a teacher can foresee or suggest. Furthermore, searching for relevant materials provides an excellent opportunity for self-directed inquiry and hands-on practice in struggling to grasp the differences between various concepts and theories.

g. *Generating subordinate questions* is part of the process of advancing inquiry; learners transform the initial big and unspecified questions into subordinate and more specific questions, based on their evaluation of produced new knowledge. This transformation helps to refocus the inquiry (Hakkarainen & Sintonen, 2002; Hintikka 1999). Directing students to return to previously stated problems, to make more subordinate questions, and answer them are ways to scaffold the inquiry.

h. *Developing new working theories* arises out of the fresh questions and scientific knowledge that the participants attain. The process includes publication of the summaries and conclusions of the community's inquiry. If all productions to the shared database in a collaborative environment have been meaningfully organized, participants should have an easy access to prior productions and theories, making the development of conceptions and artifacts a visible process.

CASES OF PROGRESSIVE INQUIRY IN HIGHER EDUCATION

Progressive Inquiry in a Cognitive Psychology Course

In a study reported by Muukkonen, Lakkala, and Hakkarainen (2001), the progressive inquiry model was implemented in a cognitive psychology course with the use of the Future Learning Environment (FLE). The FLE-environment (http://fle3.uiah.fi) is an open-source collaborative tool that has the progressive inquiry model embedded in its design and functionality (Muukkonen, Hakkarainen, & Lakkala, 1999). All the students in the course were guided, during the first two lectures, to formulate research problems. In the beginning, they individually produced these formulations. They continued by discussing their research problems with a peer and, finally, within a small group, selected the most interesting questions to pursue. These questions were then presented to all the participants in the lecture. After this initial problem setting, the technology-mediated groups (three groups of four to seven volunteers) were instructed to continue their inquiry processes between the weekly lectures in the FLE-environment. The tutor-facilitators took part in the FLE-environment, whereas the teacher conducted the weekly lectures without participating in the database discourse. The rest of the students also formed groups based on their questions, but continued their inquiry process by writing learning logs and commenting on the logs produced by other members of their group without collaborative technology.

A comparative analysis of the knowledge produced by the students in the two conditions provided evidence that the technology-mediated groups were more engaged in problem-setting and redefining practices. Further, they reflected on the process they had undertaken, with respect to the collaboration and their individual efforts. In the productions of the groups who had not used collaboration tools, the social and communal aspects of inquiry and knowledge building were not evident at all in their learning logs, although they were engaged in collaboration during the lectures. The type of the comments they provided to two of the learning logs written by other members of their group were very general, and they concentrated mainly on evaluating the level of writing, not on advancement of ideas. However, many of the learning logs were conceptually well-developed and integrated. Discourse interaction within the FLE environment was different in that the participants sometimes engaged in extensive dialogues about ideas presented by the fellow students.

Progressive Inquiry in a Design Course

Two studies carried out by Seitamaa-Hakkarainen and her colleagues (Seitamaa-Hakkarainen, Lahti, Muukkonen & Hakkarainen, 2000; Seitamaa-Hakkarainen, Raunio, Raami, Muukkonen & Hakkarainen, 2001), analyzed a collaborative design process as it occurred in the complex and authentic design task of designing clothing for premature babies. The framework of the studies was based on evidence from cognitive research on expertise that indicated that novices in design tend to generate problem solutions without engaging in extensive problem structuring; experts, by

contrast, focus on structuring and restructuring the problem space before proposing solutions (Glaser & Chi, 1988). The studies described in this case were designed to examine whether an expert-like engagement in design process would be supported in the FLE-environment. Features of the environment were used to encourage the users to engage in expert-like designing and to enable graphic presentation of the knowledge artifacts in the form of importing students' drafts and prototypes into the collaborative environment and developing multiple versions of the designs.

During the collaborative design course, the students were first guided to find out information about the constraints of their design task, such as the size of the babies, special needs for the usability of the clothing, and about the materials. Then they were asked to produce their own sketches and work in small groups to share design ideas and develop their designs. Following this development, each group produced a prototype, which was tested by actual end-users in hospital. Feedback and suggestions were then used to develop advanced design ideas.

In these studies of designing with the support of a networked collaborative environment, Seitamaa-Hakkarainen and her colleagues (2001) found that a key aspect of these environments is its provision of tools for progressive discourse between the designers and users of the future products. Further, the environments offer shared spaces and tools to elaborate conceptual knowledge related to the design problem. The collaborative technology made design thinking more explicit and accessible to the fellow designers and enabled participants to share their ideas and construct a joint understanding of design problems and solutions.

TUTOR'S ROLE AND ACTIVITY

A special question in implementing progressive inquiry and knowledge building practices in higher education is the teacher's or tutor's role in supporting and guiding students' collaborative inquiry. In progressive inquiry, the traditional role of a teacher as an expert who delivers the essential information by lecturing is radically changed. The important roles of the teacher and the facilitators of collaboration are to create the context for collaboration, and provide anchors between the theoretical representations, world knowledge, and the real-life experiences that students report (Muukkonen et al., 2004). It is also necessary for the teacher to structure and scaffold the process, keep it active and in focus during the progression of the course, and to help students to gradually take upon themselves the responsibility for the higher level cognitive processes (Scardamalia, 2002).

FUTURE TRENDS

Productive changes in educational systems towards establishing inquiry-based approaches in studying and teaching call for an alignment of epistemic, pedagogical, and institutional goals and actions (Muukkonen et al., 2004). Learning technologies also need to be critically viewed for their role in fostering expert-like skills in advancing knowledge. For instance, availability of scaffolding, support for multiple forms of collaboration, and shared development of knowledge objects are challenges for designing learning technologies.

CONCLUSION

The progressive inquiry model may be utilized in a variety of educational settings to provide a heuristic framework for the key activities and epistemic goals of a knowledge-building community. The community may provide multiple levels of expertise and, equally important, social support for engaging in a strenuous quest for learning and advancing knowledge. In higher education, a progressive inquiry approach may support the development of academic literacy, scientific thinking, and epistemic agency, particularly when integrated with the use of appropriate collaborative technology and supportive arrangements in curriculum design.

REFERENCES

Bereiter, C. (2002). *Education and mind in the knowledge age*. Hillsdale, NJ: Erlbaum.

Brown, A.L. & Campione, J.C. (1994). Guided discovery in a community of learners. In K. McGilly (Ed.), *Classroom lessons: Integrating cognitive theory and classroom practice* (pp. 229-287). Cambridge, MA: MIT Press.

Brown, A.L., Ash, D., Rutherford, M., Nakagawa, K., Gordon, A., & Campione, J. (1993). Distributed expertise in the classroom. In G. Salomon (Ed.), *Distributed cognitions: Psychological and educational considerations* (pp. 188-228). Cambridge, UK: Cambridge University Press.

Carey, S. & Smith, C. (1995). On understanding scientific knowledge. In D.N. Perkins, J.L. Schwartz, M.M. West, & M.S. Wiske (Eds.), *Software goes to school* (pp. 39-55). Oxford, UK: Oxford University Press.

Chi, M.T.H., Bassok, M., Lewis, M.W., Reiman, P., & Glaser, R (1989). Self-explanations: How students study and use examples in learning to solve problems. *Cognitive Science, 13*, 145-182.

Dunbar, K. (1995). How scientist really reason: Scientific reasoning in real-world laboratories. In R.J. Sternberg, & J. Davidson (Eds.), *Mechanisms of insight* (pp. 365-395). Cambridge, MA: MIT Press.

Geisler, C. (1994). *Academic literacy and the nature of expertise*. Hillsdale, NJ: Erlbaum.

Glaser, R., & Chi, H. T. M. (1988). Overview. In H.T.M. Chi, R. Glaser, & M. Farr (Eds.), *The nature of expertise* (pp. xv-xxviii). Hillsdale, NJ: Erlbaum.

Hakkarainen, K. (2003). Emergence of progressive inquiry culture in computer-supported collaborative learning. *Learning Environments Research, 6*, 199-220.

Hakkarainen, K., & Sintonen, M. (2002). Interrogative model of inquiry and computer-supported collaborative learning. *Science & Education, 11*, 25-43.

Hakkarainen, K., Järvelä, S., Lipponen, L. & Lehtinen, E. (1998). Culture of collaboration in computer-supported learning: Finnish perspectives. *Journal of Interactive Learning Research, 9*, 271-287.

Hatano, G., & Inagaki, K. (1992). Desituating cognition through the construction of conceptual knowledge. In P. Light & G. Butterworth (Eds.), *Context and cognition: Ways of knowing and learning* (pp. 115-133). New York: Harvester.

Hintikka, J. (1999). Inquiry as inquiry: A logic of scientific discovery. *Selected papers of Jaakko Hintikka*, Volume 5. Dordrecht, The Netherlands: Kluwer.

Lakkala, M., Ilomäki, L., Lallimo, J. & Hakkarainen, K. (2002). *Virtual communication in middle school students' and teachers' inquiry.* In G. Stahl (Ed.), Computer support for collaborative learning: Foundations for a CSCL community. *Proceedings of CSCL 2002* (pp. 443-452). Hillsdale, NJ: Erlbaum. Available online: http://newmedia.colorado.edu/cscl/97.html

Lipponen, L. (2000). Towards knowledge building discourse: From facts to explanations in primary students' computer mediated discourse. *Learning Environments Research, 3*, 179-199.

Muukkonen, H., Hakkarainen, K., & Lakkala, M. (1999). Collaborative technology for facilitating progressive inquiry: future learning environment tools. In C. Hoadley & J. Roschelle (Eds.), *Proceedings of the Computer Support for Collaborative Learning (CSCL) 1999 Conference* (pp. 406-415). Mahwah, NJ: Erlbaum.

Muukkonen, H., Hakkarainen, K., & Lakkala, M. (2004). Computer-mediated progressive inquiry in higher education. In T.S. Roberts (Ed.), *Online collaborative learning: Theory and practice* (pp 28-53). Hershey, PA: Information Science Publishing.

Muukkonen, H., Lakkala, M., & Hakkarainen, K. (2001). Characteristics of university students' inquiry in individual and computer-supported collaborative study process. In P. Dillenbourg, A. Eurelings & K. Hakkarainen (Eds.), *European perspectives on computer-supported collaborative learning. Proceedings of the first European conference on CSCL* (pp. 462-469). Maastricht, The Netherlands: Maastricht McLuhan Institute.

Perkins, D.A., Crismond, D., Simmons, R., & Under, C. (1995). Inside understanding. In D.N. Perkins, J.L. Schwartz, M.M. West, & M.S. Wiske (Eds.), *Software goes to school* (pp. 70-87). Oxford, UK: Oxford University Press.

Salomon, G. & Perkins, D. N. (1998). Individual and social aspects of learning. *Review of Research of Education 23*, 1-24.

Scardamalia, M. (2002). Collective cognitive responsibility for the advancement of knowledge. In B. Smith (Ed.), *Liberal education in a knowledge society* (pp. 67-98). Chicago, IL: Open Court.

Scardamalia, M. & Bereiter, C. (1994). Computer support for knowledge-building communities. *The Journal of the Learning Sciences, 3*, 265-283.

Seitamaa-Hakkarainen, P., Lahti, H., Muukkonen, H., & Hakkarainen, K. (2000). Collaborative designing in a networked learning environment. In S.A.R. Scrivener, L.J. Ball, & A. Woodcock

(Eds.), *Collaborative design: The proceedings of CoDesigning 2000* (pp. 411-420). London, UK: Springer.

Seitamaa-Hakkarainen, P., Raunio, A.M., Raami, A., Muukkonen, H., & Hakkarainen, K. (2001). Computer support for collaborative designing. *International Journal of Technology and Design Education, 11*, 181-202.

Veermans, M. & Järvelä, S. (in press). Generalized achievement goals and situational coping in inquiry learning. *Instructional Science.*

Wells, G. (1999). *Dialogic inquiry: Towards a sociocultural practice and theory of education.* Cambridge, UK: Cambridge University Press.

KEY TERMS

Distributed Expertise: Cognition and knowing are distributed over individuals, their tools, environments, and networks.

Epistemic Agency: Taking responsibility for one's own learning efforts and advancement of understanding.

Knowledge Building: A framework for collective knowledge advancement and development of knowledge artifacts.

Learning Community/Community of Learners: All participants in a learning process (students, teachers, tutors, and experts) have valuable expertise and skills, which benefit collective efforts.

Metaskills: Skills involved in academic literacy as well as metacognitive skills related to planning, monitoring, and regulating comprehension-related activities.

Progressive Inquiry: A pedagogical model for structuring and supporting a group of learners in a deepening question-explanation process.

Scaffolding: Providing support that enables a learner to carry out a task that would not be possible without that support, and enabling the learner gradually to master that task without support.

This work was previously published in the Encyclopedia of Information Science and Technology, Vol. 5, edited by M. Khosrow-Pour, pp. 2771-2776, copyright 2005 by Idea Group Reference (an imprint of IGI Global).

Chapter 7.15
Enhancing Phronesis:
Bridging Communities
Through Technology

Anders D. Olofsson
Umeå University, Sweden

J. Ola Lindberg
Mid Sweden University, Sweden

ABSTRACT

In this chapter, the possibilities to use technology in order to improve the contextual and value-based dimensions in online distance-based teacher training in Sweden are explored. Aristotle's (1980) concept of phronesis is used as a starting point for raising questions whether the Internet, and the establishing of educational online learning communities, can be used to enhance the teacher trainees' skills of making moral decisions in unpredictable situations. It is argued that active participation, collaboration, and dialogue are vital in order to foster common moral and societal values among the teacher trainees, but that there is a need for rethinking how technology could be used in order to accommodate such processes. This chapter suggests that the development of a shared teacher identity is possible by expanding the scope of online community, and bridging teacher-training practices to teacher practices, thus including already practicing teachers, teacher trainers, and teacher trainees in a joint educational community.

INTRODUCTION

Society is changing. We have become part of a society characterised by multiplicity and globalisation. Teacher training is an institution buffeted by these changes. It must foster societal values such as democracy, freedom, multiculturalism, and equity, as well as ideas about teaching and learning, education, and instruction. However, can it do this within the ICT frameworks that have emerged in the learning society?

Enhancing learning through technology, in this chapter, is therefore conceptualised as a matter of being able to "walk the walk," in other words, of building bridges between the text-based learning environments of Web-based conference systems and situated practices relevant to teacher training. This can be done, we believe, by using an understanding of knowing in action based on the Aristotelian concept of phronesis, which captures the idea that engaged social practice — doing something "well," — has both contextual and moral dimensions (Aristotle, 1980; Gadamer, 1989).

In this chapter, then, we explore two issues. First, we comment on the work of an ICT-supported distance-based teacher training programme. Secondly, we consider whether online learning communities (OLC) are a valid basis for fostering a practice built around common societal values. Overall, our aim is to problematise the current relationship between online learning and teacher training. In short, can online "talk" be converted into classroom "walk"?

FOSTERING VALUES AS PHRONESIS

Aristotle (1980) used phronesis to denote an aspect of knowledge that he claimed had both practical and moral implications, and he saw it as practiced rather than possessed, knowing rather than knowledge. Phronesis embraces prudence: the moral considerations in doing, the deliberative quest for the wisdom of the chosen actions. Phronesis, for Aristotle, was the knowledge linked to practicing morality.

Following Gadamer (1989), action is linked to practical wisdom and, as phronesis, always interlaced with the application of understanding. Application provides understanding with a direction. It defines the moral, in a specific case, in relation to collective understandings of right and wrong. Application is always present as an open opportunity for seeing things differently. Through the embodied aspects of morality (Merleau-Ponty, 1962), we build on a different rationale

of phronesis (Lindberg & Olofsson, 2005), or practical wisdom, than seems to be present within the reflective practitioner paradigm (Clandinin, 2002; Clandinin & Connelly, 1999; Noel, 1999). For us, understanding is always focused by its application, and by values that frame moral actions, like teaching, and not only by critical reflection and reason.

Transferring this understanding of phronesis to the distance-based teacher training programme in question, the work of the teacher trainers becomes more than, or perhaps different from, ensuring that the teacher trainees develop a theoretical understanding of teaching and learning. If teacher training becomes a question of developing the skill of making moral decisions in unpredictable situations within the classroom, and since this particular teacher training is provided on a distance basis, questions arise concerning fostering and sharing notions of societal values. What are the possibilities for a communication technology to become an educational technology through which teacher trainees can develop these skills?

EDUCATION AND TECHNOLOGY

The use of technology for educational purposes has undergone major changes during the last 50 years. From being primarily concerned with transmission or delivery, the changes have passed a cognitive focus on representation, and later on construction, and have gone towards a focus on social theories of learning and collaboration (Koschmann, 1996). In particular, the Internet has changed the way technology is used for educational purposes (Palloff & Pratt, 2003; Stephenson, 2001).

The idea of being part of an online education marks a shift from earlier views of the learning process. Focus has shifted from outcomes, in terms of students' performances as a product, and more to the view that today emphasises democratic learning, often in terms of participation, engagement, motivation, and ownership (Imel, 2001; Ó. Murchú & Sorensen, 2003) and presence (Anderson, Rourke, Garrison, & Archer, 2001; Garrison & Anderson, 2003).

Distance-based tertiary education has also experienced this shift (Keegan, 1996). It has gone from being an educational form used to overcoming structural constraints such as time and place, to being an educational form characterised by two-way communication among a network of learners (Garrison, 2000). Distance education is no longer equivalent to correspondence education. Rather, it should be seen as an interactive learning experience supported by the use of ICT (Vrasidas & Glass, 2002).

Students that have had difficulties attending educational programmes are newly given the opportunity to participate (Brown, 2001). Teacher training is no exception to this trend. Students are recruited from all walks of life, making them more heterogeneous than previously. Accordingly, new values and beliefs are constantly being added to the melting pot of teacher training. A different practice is in the making whenever the traditional universities, with their emphasis on theoretical knowledge, offer professional programmes at a distance. One issue immediately raised is how do such courses address the normative, fostering, and value-based elements of education? How do teacher trainees learn that teaching has a moral aspect, as well as being a technical activity? These issues underlie the pedagogical rationale of teacher training in Sweden.

Teacher Training in Sweden

Teacher training has been a vocational programme since 1842. However, it was not until 1977 that it became a part of the university system. In the same year, distance education also became an integrated part of the university system as universities reached out to sections of Swedish society that had been denied university access. Although the possibility to study via distance-based teacher training has been available for nearly 20 years, most teacher training has been on campus and not online.

The 1999 proposal for teacher training reform in Sweden stressed the importance of teachers developing an understanding of the norms and values of the Swedish society as a means of establishing their personal ethic as a schoolteacher. Thus, they need to appreciate the value/framework based, for example, on the democracy, freedom, multiculturalism, and equity that is central to Swedish life.

Likewise, the parallel proposals on distance education identify the value of ICT-based flexible learning unconstrained by place or time. Insofar as such forms of distance education are adopted in teacher training, it must be assumed that the norms and values of Swedish society should necessarily occupy a central position. It can be assumed that the same values that permeate Swedish society should also permeate Swedish teacher training.

Education and Community

Sergiovanni (1999) claims that communities arise from shared values, sentiments, and beliefs. Humans are bonded together in an oneness, a set of shared values and ideas. In this sense, both universities and schools can be considered social organisations. It aims to create a we, rather than an I culture. Humans are joined together by being part of webs of meaning that tie humans together. They are no longer fulfilled by their individuality, but rather, by their commonality.

According to Bauman (2001), the word "community" has a double meaning. On the one hand, it gives an image of security and belonging; on the other hand, it offers a sense of place. Wenger (1998) suggested that communities share a history, and develop a shared repertoire of practices through the negotiation and fostering of such meanings.

Palloff and Pratt (2003), like Wenger, suggest that community entails developing personal identity and shared values.

The concept of community is often used in connection with online learning. The community is considered to be located on the Internet, and is frequently described in terms of an online learning community (Carlén & Jobring, 2005; Haythornthwaite, 2002; Olofsson & Lindberg, 2005; Seufert, Lechner, & Stanoevska, 2002), or a virtual learning community (VLC) (Daniel, Schwier, & McCalla, 2003; Lewis & Allan, 2005; Schwier, 2002). Carlén and Jobring (2005) describe three main types of online learning community: educational, professional, and interest related. In this chapter, we focus on an educational online learning community. To qualify, certain characteristics are required, such as a curriculum, or other kind of steering document, related to its activities. Further, it can be assumed that students involved in those activities have the intention to learn (Nolan & Weiss, 2002). Additionally, within an educational OLC, students demonstrate their commitment to community by their joint participation, collaboration, and dialogue.

Participation, Collaboration, and Dialogue in an Educational OLC

In shaping the practice of online learning, teachers and students participate primarily through written text (Lock, 2002), organised around computer-mediated communication (CMC) (Garrison & Anderson, 2003). Different Web-based conference systems build online learning environments, and participation is possible both asynchronously and synchronously (Chong, 1998; Schwier & Balbar, 2002).

The definition of participation often excludes passive participation conducted by, for example, lurkers. Those are only reading or eavesdropping and not taking an active part in the ongoing dialogue. Many guidelines, design principles, or theoretical elaborations attempt to find methods that actively contribute to reducing the possibility of lurking or eavesdropping (see Sorensen & Takle, 2004).

Two potential strategies for dealing with the question of lurkers are, for example, ropes courses (Lowell & Persichitte, 2000) and initial bonding (Haythornthwaite, Kazmer, Robins, & Shoemaker, 2000). Ropes courses and initial bonding provide the members with opportunities to come together at the start of the education, and to build up a kind of social capital (Schwier, Campell, & Kenny, 2004) that holds the community together. This creates feelings of belonging, safety, and trust, and works as a foundation for future participation and activity within the educational OLC (Haythornthwaite, 2002; Hossain & Wigand, 2004).

This definition of participation, in terms of exclusion of lurkers, is closely aligned with a view of learning as an active process, where different aspects of knowledge building (Sorensen & Takle, 2002) are in focus, and where both active participation and close collaboration is considered to lead to learning (Dennen, 2000; Ingram & Hathorn, 2004; Mitchell & Sackney, 2001) and meaning making (Stahl, 2003).

To be able to both participate and collaborate in an educational OLC, with the aim of negotiating meaning and fostering a shared understanding among the members, some kind of asynchronic or synchronic dialogue is often used. This dialogue becomes a tool for learning, construction of knowledge, and creation of understanding (see Muukkonen, Hakkarainen, & Lakkala, 2004). This means that understanding built on dialogue provides the member in an educational OLC with an opportunity to contribute and be part of a process in which learning becomes a joint and collaborative venture.

Understanding through dialogue aligns us again with Gadamer (1989). By using dialogue, the possibilities for understanding are open for further questions, instead of answering questions with reproductive and predetermined content. Gadamer's approach brings up other issues concerning, for example, examination and normative features of education. It opens for a transformation of the university's more traditional role as producer and custodian of knowledge (Grundy, 1999). It also enables for an understanding of an educational OLC in which joint participation, as being, is central (Olofsson & Lindberg, 2006).

In Heidegger (1962), we find emphases on the unavoidable participation and presence in the world, which is always a shared world, a kind of being-together. Being together enables a common ground of existence, it enables a joint participation. This view of participation thereby includes the kind of passive participation conducted by, for example, lurkers within an educational OLC, and participation with other humans is considered a part of life as important as breathing.

In the sense of community, humans share history, and being is constructed and created as identities with certain and specific meanings in relation to shared repertoires negotiated and fostered in practice (Wenger, 1998). To participate with others raises questions about being together with others in a moral sense, and how to negotiate shared values. In the following, we will try to elaborate upon these questions by using the concept of phronesis in relation to data from an ICT-supported distance-based teacher-training programme.

COLLECTING AND ANALYSING DATA: CONSTRUCTING THE PEDAGOGIES OF TODAY

The empirical part of this chapter is based on an analysis of interviews with university teachers, all active teacher trainers within the ICT-supported distance teacher-training programme in question.

Programme and Participants

The programme was three-and-a-half to four-and-a-half years long, depending on the level of exam (which ranged from kindergarten teacher to upper secondary school teacher). The programme was distance-based, with compulsory gatherings two to four times each term at the university campus, or at the local learning centre nearest the teacher trainees' homes. These gatherings ranged from two consecutive days up to a week. Between gatherings, teacher trainees worked in study groups together, with a university teacher acting as a tutor.

The work between gatherings was conducted using a Web-based conference-system, WebCT, including both asynchronic and synchronic functions such as e-mail, chats, notice boards, and so forth. This Web-based conference-system formed a Web-based platform to which the teacher trainees had continuous access. The courses in the programme had their own web pages within the platform, and the study groups created their own educational OLC within the platform. The study group was intended to provide the teacher trainees with the social context for their studies between gatherings, and the Web-based platform was intended to provide the teacher trainees with an online learning environment.

The teacher trainers continuously encouraged the teacher trainees to actively use the Web-based conference system. This was, for example, stressed in the study guides handed over to the teacher trainees. Further, the tasks, designed by the teacher trainers, often required the teacher trainees to use ICT (chats, e-mail, and so on) in order to be solved (for example, comment upon each others' electronic portfolio and participation in online seminars).

In the programme, several activities were built around an idea of a social dimension permeating the programme, and that the Web-based conference-system or ICT should provide for the creation of a social dimension within this distance-based teacher training programme. In other words, the pedagogical rationale, or model, behind the programme included an idea of supporting a collaborative and social learning process using ICT.

The teacher trainers interviewed were chosen to represent as many university departments and courses as possible within the programme. The courses given were in the areas of social science, natural science, and the humanities. In this particular programme, the teacher trainees met one or more of the teacher trainers for at least half of the programmes' duration. Of interest might be that several of the teacher trainers were more or less unfamiliar with how to organise teacher training at a distance with the support of ICT. Since this was known in advance, it was addressed with courses and mentoring in distance education techniques for teacher trainers, as a way to make sure that the teacher trainers were well prepared prior to the start of the programme.

Interviews and Analysis

In the interviews, focus was on two major themes. Main interest was of issues concerning what the teacher trainers regarded as important aspects of teacher training and distance education, and issues regarding how they worked with the teacher trainees. The teacher trainers were given the interview guide in advance, and they had time to reflect over the questions before the interview (one teacher trainer, however, received the questions at the time of the interview). The interviews lasted for about 30 to 45 minutes, and were recorded. The interviews were thereafter transcribed, and given back to the teacher trainers for comments and correction. The commented transcripts are the data used for analyses.

In the analysis, we have built upon the concepts of affordances and constraints in line with Hutchby (2001), where the empirical use, and the possibilities for developing an understanding of how technological artefacts become involved in everyday conduct, are central. In other words, the

technology affords certain forms of interactions and constraints others.

The descriptions below are abstractions based on the data collected. The descriptions should not be seen as an objective account of what occurs or does not occur within the programme: they are constructions of the pedagogies the teacher trainers talk about and use in the programme on different occasions and in different courses. They are also accounts of the aspects the teacher trainers mention as important for distance-based teacher training, but appear to carry out without the support of technology. In short, technologically afforded pedagogies include the talk about what teacher trainers do using technology, and technologically constrained pedagogies include the talk about what teacher trainers do without the use of technology.

Pedagogies that Appear to be Afforded by Technology

The following is a description of the constructed pedagogies used by the teacher trainers when applying technology to ensure active participation, required collaboration, and necessary dialogue.

Enabling Participation

Participation in the program is structured through the use of technology, for example, using different ways of distributing information. Including accessible study guides in the Web-based conference system, providing calendars with timetables for the courses, and by thinking about issues of information and access, the teacher trainers try to design for a supportive structure:

...when I design Web-based courses I always include a lot of contact points into the course, so that the teacher trainees have to send in an email or a memo or have been part of a discussion group... (Teacher trainer, humanities course)

The structure provided by technology also seems useful to the teacher trainers in providing clarifying information, and it helps them when responding to questions from single teacher trainees, as well as giving answers to the group as a whole. It could be used to provide direct answers in certain Web-based conferences, or to possibly build an FAQ (frequently asked questions) of teacher trainee difficulties:

...it was a good conference-system and the best thing was that you could reach everyone very easy. We could add questions from every person... (Teacher trainer, social science course)

The teacher trainers also emphasise the quality of the text-based information. The teacher trainers seem to see the need for clear and informative study guides, and a need to think about differences in text-based communication, so as to provide some kind of internal structure for the communication:

...because the writing is so special it is rather difficult to handle actually, and sometimes I even check, could I write this? Partly, so that the content is factual but also if it feels like a friendly answer. For that might matter I assume... (Teacher trainer, social science course)

The structure given in the Web-based conferences is well thought through. For several teacher trainers, the intention of the programme is perceived to provide for groups of people living in remote and sparsely populated areas with varying home conditions, and to whom tertiary education has traditionally been a distant alternative. The teacher trainers have these aspects in mind when designing support for individual teacher trainees:

...at the same time, you have these people, with a lot in their baggage, and at the same time as there is some kind of alienation towards academic studies ... and it is important I think how you adapt or design the content to fit these categories of teacher trainees... (Teacher trainer, social science course)

The use of technology provides a structure for the programme, enabling these teacher trainees the possibility to attain a university degree and a vocational education. Although the opportunity to reach the teacher trainees as a group over a considerable geographical distance is considered as an advantage of using technology, it is sometimes not so easily achieved, since it is also associated with a lot of planning and coordination:

...we were supposed to have lectures and I was trying to arrange, coordinate schedules, it was terrible the way we had to carry on... (Teacher trainer, social science course)

At the same time, the emphasis on participation and dialogue also raises questions. There are several aspects of the teacher trainers' role that differs within a technology-supported distance-based programme. Perhaps aspects of the professionalism of the teacher trainers themselves come into focus:

...and then, I know a couple of teachers who think like, when it comes to teaching at campus they prepare very little, and not that they give bad lectures, but they kind of keep their knowledge in their heads, they can stand there and lecture, but how do you do that at a distance... (Teacher trainer, humanities course)

This opens up for an emphasis on the teacher trainer as facilitator and mentor of learning processes, and for a structure of the work within the programme to be designed in line with a student-centred approach focused on collaboration.

Working with Collaboration

Between gatherings at the university, teacher trainers use technology to enable collaboration in study groups. This is done, for instance, by providing the teacher trainees with Web-based conferences of their own:

...we create rooms for the teacher trainees where they can talk, it is important right, although it is rather a different matter whether they work or not, but the possibility is there, and in some groups, there can suddenly be added like hundreds of little comments... (Teacher trainer, natural science course)

The work carried out in the study-groups is adapted to a group work mode, and is expected to enhance learning at an individual level. The general feeling seems to be that group work is beneficial to all, but that it requires a different focus: a focus on the discussions and not on the products of the work:

...it is not the same course as a campus course, in that you have to think about the study-groups, to create tasks and assignments that fit the discussion forum... (Teacher trainer, humanities course)

This collaborative working mode is, therefore, not always chosen for assessment by the teacher trainers. Since the working mode is based on discussions, and there is less focus on product, teacher trainers exclude it from the ordinary forms of examination. However, the group work mode leads to questions concerning the form and role of assessment and examination:

...if it is as I have thought that the learning aspect of a certain examination form is the most dominant or important then you could ease up on the control aspect and it gives a lot in return, I give for example different kinds of home assignments and group work, and there too that you have to

trust the teacher trainees that they do this because they are interested... (Teacher trainer, natural science course)

The earlier stated clarity that is needed in study guides and information seems not to be as apparent in the group work. The support and guidelines on group work seems to be less emphasised. Only some of the teacher trainers mention strategies for working in groups as being part of the overall structure, for instance, by establishing contracts within the study group to regulate the individual efforts, by circulating roles within the study group, or being a role model for discussions. What seems to be a general opinion is that dialogue has to be established.

Establishing Dialogue

The teacher trainers see a need to establish a dialogue with, and among the teacher trainees. An issue several teacher trainers mention but express mixed, if not sceptical, opinions about is the use of technology for giving lectures. Opinions voiced claim both technical and social difficulties in relation to lecturing:

...that is [give lectures] at the same time simultaneously in another [local city] so that we could link by videoconference and it didn't work... (Teacher trainer, social science course)

Technical difficulties are not the only problem, since other aspects connected to lecturing through videoconferencing are also mentioned. For example, the troublesome nature of the videoconference makes it more appropriate for monologues and as an information channel, rather than for dialogues among the participants:

...I could say from what I have seen of the Web-based that this with videoconferences is in many cases poor lecturing, and then I don't think we use the advantages of the computer, I mean we

could just as easily have sent out a videotape since there is no dialogue it is still a monologue and we might as well have handed it out in writing... (Teacher trainer, natural science course)

Teacher trainers also use technology to establish dialogue with individual teacher trainees concerning content matters. In the courses, the teacher trainees frequently hand in assignments and use e-mail to ask questions about the theoretical content in the courses. Using this practice, teacher trainers both provide content matters as such, and didactical aspects of the content matter taught:

...with this email communication, you get opportunities in a totally different way or I don't know who is given the opportunity, the teacher trainees above all I assume, to show what they cannot and thereby get help... (Teacher trainer, humanities course)

Sometimes, in discussions concerned with aspects of the content, the discussions end up with issues or aspects that the teacher trainers had not themselves thought of:

...we have tried different models with chat models and some other stuff and it is obvious that sometimes you see the discussion taking turns you would never have anticipated... (Teacher trainer, natural science course)

Within the Web-based conference, dialogue is ongoing in the discussion forums, and the teacher trainees have access to these on a study group basis. However, in relation to teacher training as a governmental concern, with aims that must be reached and assessed, teacher trainers sometimes feel themselves forced into situations where they see themselves as having the responsibility to initiate and guide both the discussions and the content in chats, forums, or in seminars. The dialogue between teacher trainees and the dialogue with

the teacher trainers is considered very important for reaching desired goals, but is, at the same time, paradoxical since, as mentioned above, the discussion might take unexpected turns:

...sometimes we join one of these discussion-groups we have created in the conference-system and try to steer them towards the goals... (Teacher trainer, natural science course)

In these ways, teacher trainers enable the teacher trainees to participate, using collaboration among the teacher trainees as a means, and ensuring that a dialogue is possible between teacher trainer and teacher trainee and between teacher trainees themselves. Thereby, the teacher trainees are regarded as being connected to an educational OLC, centred in the practice of teacher training. It seems, though, to be a practice primarily directed at the individual, and conducted mainly using different kinds of text-based communications. In short, the pedagogies afforded by the use of technology are focused on the theoretical perspective of teacher training; the content matter.

WHAT ARE THE AFFORDED PEDAGOGIES?

The afforded pedagogies seem to be more in-line with the use of technology in the first three paradigms of Koschmann (1996). Teacher trainees receive information through the technology, and then they elaborate and reflect on the content, and produce signs of these elaborations and reflections in text. The educational OLC seems to be problematic when it comes to assessing qualities and aspects that are not content matter. We find this to be the case in the theoretical aspects of OLC literature, for example when passivity and lurking are unwanted forms of participation. If active participation is desired, in-line with a constructivist assumption about learning (or perhaps mainly since activity is more easily assessable when it comes to learning), then the passive aspects of learning a practice, of moving from the peripheral participation (Wenger, 1998) towards the centre of the practice, is overlooked.

Further, collaboration is seldom the basis for assessment, but is a mean of learning. This restricts the possibility for using the collaborative features and dialogue for educational intentions. A technologically afforded pedagogy, though not always articulated in a positive manner, involves the use of text. Text is often assessed from an individual perspective as a cognitively produced and reified thought (Anderson, Greeno, Reder, & Simon, 2000). If social aspects of collaboration are to have a chance of making their way into the teacher training programmes in practice, then this division ought to be considered. In the next section on pedagogies constrained, we elaborate further on this issue.

Pedagogies that Appear to be Constrained by Technology

The following is a description of the pedagogies used by the teacher trainers during the gatherings at the university, but for which they do not use technology.

Enabling Socialisation

When participating at the gatherings, teacher trainers' intentions are that the teacher trainees develop social skills and interpersonal capacities, as well as a readiness and capacity to handle complex and unpredictable situations in the classroom:

...they learn very much at the gatherings, and perhaps not primarily by sitting for eight hours in class at lectures, but by the total interaction with the others in the group, which they cannot have when they are at home... (Teacher trainer, natural science course)

The socialisation that is present in group participation seems difficult for the teacher trainers to re-create without the gatherings, likewise, certain aspects of assuming a teacher's role. It seems that teacher trainers view this process as something that the teacher trainees have to experience in a face-to-face situation that the teacher training programme has to provide insights into:

...I'm not totally convinced, that you could replace the physical meetings altogether with ICT, I think that, and above all in a vocational program where you actually educate people to work with socialisation actually is in need of its physical meetings which would mean meeting between people in both an informal as well as an formal meaning, that this is actually socialising for the people who are going into a profession... (Teacher trainer, social science course)

It seems that teacher trainers regard learning as a social enterprise that is important for future teachers, and that discussion in study groups, physical meetings, and seminars are vehicles of learning a view of life:

...all people are part of all, in larger situations these situations have roots back and it is always a question of a cultural situation in some way and this, is about really, we humans can have a deeper dimension of being understanding of how things in the world are really connected... (Teacher trainer, humanities course)

This is an understanding in which values become part of everyday life and practice, and as such, they become part of the teacher-training programme, and thereby under the teacher trainers' control. The teacher trainers express an apparent need for careful consideration before deciding if these aspects are to be incorporated into the teacher training without being part of the gatherings at the university.

Working with Values and Beliefs

When considering the aims and goals of participating at the gatherings, the teacher trainers express several values, beliefs, and views that the teacher trainees should embrace ; a view of themselves as competent, active, and self-reliant learners. This implies a view of individuals as responsible for their own situation and actions, and it seems to the teacher trainers very much to be a question of forming the teacher trainees as persons:

...we also want to be clear in both the study-guide and when we introduce courses of our view of learning and our view of knowledge and our view of humans and so, right, and in that I also feel that we are saying, it usually amounts to a bit of a discussion and questions of what it means and so on that the responsibility lies very much with the learner... (Teacher trainer, social science course)

This is apparently an ongoing process that is not only concerned with the work conducted by the teacher trainees when they are studying at a distance, but also with the work carried out during the gatherings. The ongoing fostering of the teacher trainee as independent and self-reliant is also built into the work at the gatherings:

...the problem may also concern the gatherings having to be a bit different to strengthen the issue of the independent studies, to strengthen this community so that there is a parallel learning process rather than just being about the content we try to pass on... (Teacher trainer, social science courses)

Establishing Teacher Trainees as Teachers

Teacher trainers see their role as that of catalysts for challenging assumptions about both learning and subject matters that influence the teacher trainees at an individual level. This could be a way of looking beyond the 'here and now" of the programme towards the teacher trainees' futures as teachers, or as a way of questioning the experiences teacher trainees might have from periods of student teaching and reflecting on their chosen actions. It could raise the possibility of developing a deeper understanding of the content matter, and a view of potential difficulties future pupils may have, as well as the assumptions of the teacher trainees themselves:

...then connecting to the experiences met during periods of student teaching, having the time to stop and listen, how did you do then, why is that do you think... (Teacher trainer, social science course)

When working as a teacher trainer, there follows a responsibility to educate the teacher trainees to be good persons and good teachers. The teacher trainers might feel, for example, that the teacher training as an institution has this responsibility, or they might find themselves deciding who is to be given enough help to manage their studies. There even seems to be an underlying assumption that not all teacher trainees are suitable for a job as a teacher:

...well, it is a vocation with responsibilities to be a teacher and then I feel that it is important questions to keep in mind who we educate on an individual level too. It is a responsibility that we at the university have, that at the same time as I say yes that this will be very good pedagogues, well others you might feel less certain about and then you have to think about how in which way they should come prepared and then you look more to the individual level... (Teacher trainer, social science course)

These are issues the teacher trainers see a need for. Collaboratively, with the teacher trainees, creating a teacher-training programme that is connected to the expected practice of teaching, where the teacher trainers can assume the responsibilities of the university to educate the teacher trainees with both a capacity to teach as a skill, as well as fostering values aligned with those of the national curriculum.

What are the Pedagogies Constrained?

In the description of the pedagogies that are included at the gatherings, we find aspects of teacher training that are based on working with humans. It seems as if fostering values is possible only during the gatherings at the university. The forms of working seem to be more collaborative, and more in-line with a student-centred approach based on social theories of learning. Sharing the characteristics of the forth paradigm of Koschmann (1996), the one he sees as emerging based on collaboration and social theories of learning, these pedagogies seem to contain more of those aspects of becoming a teacher that are in-line with the knowledge of phronesis.

The concept of phronesis is concerned with being good, being able to choose and to deliberate upon different means and ends in relation to life as a whole. Using phronesis as the knowledge base that is needed to handle the complex and unpredictable teaching situation, and thereby also linking actions to moral deliberations, seems to be difficult for the teacher trainers using technology. Above all, it seems to be a question of creating a way for the teacher trainers to be present with the teacher trainees when they have experiences that ought to be challenged, and to be able to point out the assumptions they might have that should be questioned. This is both in relation to the teacher trainees' own assumptions, and in relation to their future pupils' assumptions and conceptions of the world.

Teacher trainers also seem to need to have the possibility to identify those teacher trainees who ought not to become teachers, and they feel a need to do so by challenging and questioning the

teacher trainees in person. They also see a need for the teacher trainees to become socialised into the profession, to be able to take on the teachers' role and act as teachers in the future classroom. Above all, socialising teacher trainees to work themselves with socialisation requires the teacher trainees acting as their own role models. It seems that the teacher trainers have difficulties in realising these aspects of the teacher training in ways other than using the gatherings at the university.

We believe that the teacher trainees, by being part of a teacher training programme, are moving from being novices in the community of teachers, that is, being only peripherally involved, towards becoming full members (Lave & Wenger, 1991) and being initialised in the discourse of the practice (Wenger, 1998). The teacher trainees are learning to talk as professionals, and also learning to embrace the values that underlie the practice of teaching, but they do this as it is constructed within teacher training.

It seems as though the fourth paradigm of Koschmann (1996) is more present in the pedagogies that are technologically constrained. Working with humans is based on assumptions and values, and being part of a community is to share those beliefs. Even though the teacher trainers seem to include few aspects of working with the values common to Swedish society through the use of technology, the teacher trainees are still becoming members of a community. In the next section, we give some possible solutions and recommendations on how to include these aspects, by connecting the recommendations to the knowledge in action of phronesis.

BEYOND THE PEDAGOGIES OF TODAY

We believe that the work carried out at the gatherings could also take place between the gatherings, but that it involves using a pedagogy afforded not only by the text-based conference systems, but also by other kinds of technology than are used today. This means moving teacher training towards being a more student-centred rather than teacher-centred practice, and a belief in the use of more collaborative approaches. We can, by posing the question whether teacher trainers and teacher trainees necessarily have to meet physically if there is to be a fostering of common values, formulate the following afforded uses of technology, where meetings can be arranged through the Internet.

- Firstly, we suggest that teacher trainers and teacher trainees have access to the teacher trainees' experiences when teaching in classrooms, through the use of Web-cams and the common use of software for video conferencing.

The idea is to situate the Web-cams within different classrooms, and broadcasting live through a streaming server on to which the community members are logged into. This could open new possibilities for moving beyond the "talk" towards the "walk." It could create possibilities for a collaborative dialogue around different values that should embrace the future work for the teacher trainees. This could further create an opportunity to see how a future college deals with the different situations that emerges in classrooms, and also provides an opportunity for discussing the solutions from different aspects or perspectives. This opens the door to the future classroom and the fostering of phronesis, the knowledge of a practived moral.

- Secondly, we suggest that the Web-cams could be used to conduct "live role-play." To set up and act out different dilemmas in a classroom, related to the teaching profession, and to use educated teachers and pupils as "actors."

The teacher trainees could watch the play online, and thereafter discuss, either through a videoconference or a text-based chat, the actions taken by, for example, the teacher. This creates a dialogue between the teacher trainees and the "actors" (the teacher and the teacher trainee). It would also be possible to include the teacher trainers in the discussion. Phronesis is always interlaced with the application of understanding, and by conducting a reflective and critical dialogue around different scenarios played out, and by the actions taken by the "actors," it could provide an understanding of the morality involved in teaching.

- Thirdly, we suggest that digital cameras could be used by the teacher trainees to take snapshots of the practice they experience, and use them for reflection and discussion.

Organising snapshots from different schools and different contexts, and publishing them within the conference system used in the teacher training, can provide a sense of understanding for the diversities involved in teaching. Furthermore, it could display the differences of conditions framing the teaching, in terms of learning environments and pupils, and make this the common ground for a negotiated meaning among the members of the educational OLC. The teacher trainees could reflect differently depending on the meanings invested in the current situation displayed on the snapshot. In the dialogue, the teacher trainees have the chance to collaboratively negotiate a shared meaning, built on phronesis, of what the snapshot is all about.

- Fourthly, we suggest that digital video cameras could be used by teacher trainees to document certain ethical, as well as teaching, dilemmas, and later discuss them with their teacher trainers or with their study-group.

If certain ethical, as well as teaching, dilemmas are organised thematically, and highlighted in relation to aspects of the content matter in the university courses, different aspects of the knowledge base behind teaching could be brought to life within the collection of video clips available on the Internet. These might include issues regarding gender in relation to classroom discourse in social science, or related to discourse in natural science. Gender issues could also generate questions about other ways of defining social justice, and place those questions in a context of real life. Such questions that touch upon the application of understanding, the embodied aspects of morality, foster a readiness to work as a teacher.

- Fifthly, we suggest that digital video clips could make up a directory of good teaching experiences, as well as a directory of dilemmas associated with different subject areas.

If the video clips are organised on the basis of the content matter, it could help teacher trainees to identify whether the pupils have difficulties in certain areas. Collaboratively in the study-group and with the teacher trainers, teacher trainees can design learning strategies and teaching approaches towards difficulties that are (near to) authentic. This way of using technology allows an interlacing of theory and practice, and to work towards best practices of those solutions. This method of working opens for moral considerations in teaching, with trajectories towards the future work with pupils with special needs, or just those simply in need of support.

- Sixthly, we suggest that virtual reality (VR) would enable situating the teacher trainees in digital classrooms, and the provision of (near to) authentic teaching experiences.

The teacher trainees are provided with complex, real life dilemmas by using VR. For example, the teacher trainees could be placed in different classroom situations where different conditions are inherent due to the pupils' social background, the actions taken by the teacher trainees, the facilities in the classroom, and so on. The teacher trainees could be given the task of giving a lecture for the pupils, and the teacher trainers could change the circumstances of what happens in the classroom from one time to another. This provides a kind of social presence, and in a way, forces the teacher trainees to use knowledge in action. The technology thereby contributes to the fostering of phronesis. This integration of theory and practice is conducted in a safe and trusted learning environment, where the teacher trainees also have the possibility to always rethink their actions taken, and over and over again, step into the VR-created classroom to face new pupils and new real life dilemmas.

Ethical Considerations

Establishing these kinds of pedagogies causes a number of ethical problems. It is, for instance, difficult to obtain a licence to publish real-life situations and stories on the Internet, especially when there are children involved. The issues of integrity and ethical aspects of distributing individuals and their learning difficulties worldwide on the Internet is a troublesome aspect. But on the other hand, this could also incorporate the possibilities of involving the pupils themselves in the reflections and deliberations of teacher trainers and teacher trainees in a more active and democratic way. It could be a way to avoid student teaching periods becoming a practice *on* children, but rather a practice *with* children. We do not intend to provide solutions to these problems in this chapter (since we believe that there are no easy solutions), but one point we still want to make is that the issue is better handled if it is part of a democratic process of dialogue, and not silenced.

Any Problems Left?

Could these recommendations provide teacher trainers with possibilities to foster aspects of teaching aligned with both knowledge in action such as phronesis, and reaching the aims of teacher training associated with the values common to Swedish society? What remains to be a legitimate question for further thought is whether the practice of teacher training is different from the practice of teaching itself? In other words, are there any aspects of teacher training that involve learning to talk the talk, to which there is a different walk than teaching? If so, what are the legitimate grounds for teacher training? To avoid yet other questions, in the final part of this chapter we intend to propose another way to conceptualise teacher training through the use of technology that is more radical, and perhaps involves a more thorough change.

FUTURE TRENDS: BRIDGING COMMUNITIES

In a future perspective, technology affords teaching in tertiary education to include aspects that are common in vocational programmes. Vocations are basically practices, and as such, practices are fostering their participants into, not only a discourse of a professional (a talk), but an embodied knowledge, and an approach to the vocation of a professional (a walk). In teacher training, much of the focus in the national steering documents is concerned with ensuring that future teachers are aware of the values common to Swedish society. Teacher training is a moral arena, and as such, a state-controlled moral formation of the prospective teachers. Here, technology can afford the teacher trainers to develop practices where this formation could take place. The values in focus here are, for example, democracy and freedom, multiculturalism, and equity.

We can envisage a development of the educational OLC to include within the practice of teaching, more professionals with a relation to the teacher profession. Our idea is to widen the perspectives, and to put forth a more complex picture of the values that will embrace the teacher trainees' future work. Connected through asynchronic, as well as synchronic participation, collaboration, and dialogue, we can envisage the development of a shared teacher identity, made possible, if we expand the scope of the community from teacher training to teacher practice, by including already practicing teachers, teacher trainers, and teacher trainees in an (for all parts equally) educational OLC. Using Web-cams and software for video-conferencing, this bridged community could be realised, if sustained membership could be attained over time as a trajectory of participation into the practice of teaching.

With the help of technology, then, there is a real chance that the walks of life might affect the talks of academy. That the trajectory of participation of a teacher, the life of a professional among professionals, starts at teacher training, includes teacher training and the academic society and the workplaces in a true community of practice (Wenger, 1998), sustained and upheld in the virtual teaching of online learning, and thereby dissolving the notion of separate communities.

CONCLUSION

One main issue that hereby is addressed, is "Which community teacher trainees belong to?" Is it the university community, the teaching professionals' community or a community of teacher trainees? Who has the power to decide which belongings should have priority, and which communities should count? Since belonging to a community of a shared practice is the foundation to learning in a meaningful way, the question arises "Where is meaningfulness created"? Is it academically meaningful or future professional meaningfulness? Is it the practice of teacher training, that is being able to complete university studies, or is it the practice of teaching? Is the practice of teacher training the same as the practice of teaching? These questions all have one answer in a bridged community of teaching.

Our belief is that the use of Web-cams, software for videoconferencing, and the possibility to connect members of an educational OLC to each other, are technologies that could enhance the learning not just for the teacher trainees. It would enhance the practicing teachers' own teaching practices, and work as an in-service teacher training by including aspects and interests of the same community. Being a teacher in Sweden, in a future perspective, then becomes an aspect of a shared history of participation and a shared negotiated meaning. To make this happen, technology, and how it is used in teacher training, is more crucial than ever.

REFERENCES

Anderson, J. R., Greeno, J. G., Reder, L. M., & Simon, H. A. (2000). Perspectives on learning, thinking and activity. *Educational Researcher, 29*(4), 11-13.

Anderson, T., Rourke, L., Garrison, D.R., & Archer, W. (2001). Assessing teaching presence in a computer conferencing context. *Journal of Asynchronic Learning Networks, 5*(2), 1-17.

Aristotle (1980). *The Niceomachean ethics.* (Sir David Ross, Trans.). Oxford: Oxford University Press. (Original work published 1925)

Bauman, Z. (2001). *Community: Seeking safety in an insecure world.* Cambridge: Polity Press.

Brown, R. E. (2001). The process of community-building in distance learning classes. *Journal of Asynchronic Learning Networks, 5*(2), 18-35.

Carlén, U., & Jobring, O. (2005). The rationale of online learning communities. *International Journal of Web Based Communities, 1*(3), 272-295. Retrieved May 13, 2005, from http://www.inderscience.com/search/index.php?action=record&rec_id=6927&prevQuery=&ps=10&m=or.

Chong, S. M. (1998). Models of asynchronic computer conferencing for collaborative learning in large college classes. In C. J. Bonk & K. S. King (Eds.), *Electronic collaborators: Learner-centered technologies for literacy, apprenticeship, and discourse* (pp. 157-182). Mahwah, NJ: Lawrence Erlbaum Associates.

Clandinin, D. J. (2002). Storied lives on storied landscapes. Ten years later. *Curriculum and Teaching Dialogue, 4*(1), 1-4.

Clandinin, D. J., & Connelly, F. M. (1999). *Shaping a professional identity: Stories of educational practice.* New York: Teachers College Press.

Daniel, B., Schwier, R. A., & McCalla, G. (2003). Social capital in virtual learning communities and distributed communities of practice. *Canadian Journal of Learning and Technology, 29*(3), 113-139.

Dennen, V. P. (2000). Task structuring for online problem based learning: A case study. *Educational Technology & Society, 3*(3), 329-336.

Gadamer H. G. (1989). *Truth and method.* London: Sheed and Ward.

Garrison, D. R. (2000). Theoretical challenges for distance education in the 21st century: A shift from structural to transactional issues. *International Review of Research in Open and Distance Learning, 1*(1), 1-17.

Garrison, D. R., & Anderson, T. (2003). *E-learning of the twenty-first century. A framework for research and practice.* London: Routledge Falmer.

Grundy, S. (1999). Partners in learning. School-based and university-based communities of learning. In J. Retallick, B. Cocklin, & K. Coombe (Eds.), *Learning Communities in Education* (pp. 44-59). London: Routledge.

Haythornthwaite, C. (2002). Building social networks via computer networks: Creating and sustaining distributed learning communities. In K. A. Renninger & W. Shumar (Eds.), *Building virtual communities—Learning and change in cyberspace* (pp. 159-190). Cambridge: Cambridge University Press.

Haythornthwaite, C., Kazmer, M. M., Shoemaker, S., & Robins, J. (2000, September). Community development among distance learners: Temporal and technological dimensions. *Journal of Computer-Mediated Communication, 6*(1). Retrieved January 25, 2005, from http://www.ascusc.org/jcmc/vol6/issue1/haythornthwaite.html.

Heidegger, M. (1962). *Being and time*. Oxford: Blackwell.

Hossain, L., & Wigand, R. T. (2004, November). ICT enabled virtual collaboration through trust. *Journal of Computer-Mediated Communication, 10*(1). Retrieved January 25, 2005, from http://jcmc.indiana.edu/vol10/issue1/hossain_wigand.html.

Hutchby, I. (2001). *Conversation and technology. From telephone to the internet*. Cambridge: Polity Press.

Imel, S. (2001). Learning technologies in adult education. *Myths and realities. ERIC Clearinghouse on Adult, Career, and Vocational Education,* (17). Retrieved February 8, 2005, from http://www.cete.org/acve/docs/mr00032.pdf.

Ingram, A. L., & Hathorn, L. G. (2004). Methods for analyzing collaboration in online communications. In T.S. Roberts (Ed.), *Online collaborative learning: Theory and practice* (pp. 215-241). London: Information Science Publishing.

Keegan, D. (1996). *Foundations of distance education*. London: Routledge.

Koschmann, T. D. (1996). Paradigm shifts and instructional technology. In T. D. Koschmann (Ed.), *CSCL: Theory and practice of an emerging paradigm* (pp. 1-23). Mahwah, NJ: Lawrence Erlbaum Associates.

Lave, J., & Wenger, E. (1991). *Situated learning: Legitimate peripheral participation*. Cambridge: Cambridge University Press.

Lewis, D., & Allan, B. (2005). *Virtual learning communities. A guide for practitioners*. Berkshire: Open University Press.

Lindberg, J. O., & Olofsson, A. D. (2005). Phronesis: On teachers' knowing in practice. Towards teaching as embodied moral. *Journal of Research in Teacher Education, 12*(3), 148-162.

Lock, J. V. (2002). Laying the groundwork for the development of learning communities within online courses. *The Quarterly Review of Distance Education, 3*(4), 395-408.

Lowell, N. O., & Persichitte, K. A. (2000, October). A virtual course: Creating online community. *Journal of Asynchronic Learning Networks, 4*(1). Retrieved January 25, 2005, from http://www.aln.org/publications/magazine/v4n1/lowell.asp.

Merleau-Ponty, M. (1962). *Phenomenology of perception*. (C. Smith, Trans.). London: Routledge & Kegan Paul.

Mitchell, C., & Sackney, L. (2001, February 24). Building capacity for a learning community. *Canadian Journal of Educational Administration and Policy, 19*. Retrieved February 1, 2005, from http://www.umanitoba.ca /publications/cjeap/articles /mitchellandsackney.html.

Muukkonen, H., Hakkarainen, K., & Lakkala, M. (2003). Computer-mediated progressive inquiry in higher education. In T. S. Roberts (Ed.), *Online collaborative learning: Theory and practice* (pp.

28-53). London: Information Science Publishing.

Noel, J. (1999). On the varieties of phronesis. *Educational Philosophy and Theory, 31*(3), 273-289.

Nolan, D. J., & Weiss, J. (2002). Learning in cyberspace: An educational view of virtual community. In K. A. Renninger & W. Shumar (Eds.), *Building virtual communities: Learning and change in cyberspace* (pp. 293-320). Cambridge: Cambridge University Press.

Olofsson, A. D., & Lindberg, J. O. (2005). Assumptions about participating in teacher education through the use of ICT. *Campus Wide Information Systems, 22*(3), 154-161.

Olofsson, A. D., & Lindberg, J. O. (2006). "Whatever happened to the social dimension?" Aspects of learning in a distance-based teacher education programme. *Education and Information Technologies, 11,* 7-20.

Ó. Murchú, D., & Korsgaard Sorensen, E. (2003, September 23-26). "Mastering" communities of practice across cultures and national borders. In *Proceedings of the 10th Cambridge International Conference on Open and Distance Learning,* Cambridge, UK. In A. Gaskell, & A. Tait (Eds.), *Collected Conference Papers.* The Open University in the East of England Cintra House. Retrieved February 23, 2005, from http://www2.open.ac.uk/r06/conference/Papers.pdf.

Palloff, R. M., & Pratt, K. (2003). *The virtual student: A profile and guide to working with online learners.* San Francisco: Jossey-Bass.

Schwier, Richard A. (2002, June 1). *Shaping the metaphor of community in online learning environments.* Paper presented at the International Symposium on Educational Conferencing. The Banff Centre, Banff, Alberta. Retrieved January 9, 2005, from http://cde.athabascau.ca/ISEC2002/papers/schwier.pdf.

Schwier, R. A., & Balbar, S. (2002). The interplay of content and community in synchronous and asynchronous communication: Virtual communication in graduate seminar. *Canadian Journal of Learning and Technology, 28*(2), 21-30.

Schwier, R. A., Campbell, K., & Kenny, R. (2004). Instructional designers' observations about identity, communities of practice and change agency. *Australasian Journal of Educational Technology, 20*(1), 69-100.

Sergiovanni, T. (1999). The story of community. In J. Retallick, B. Cocklin, & K. Coombe (Eds.), *Learning communities in education* (pp. 9-25). London: Routledge.

Seufert, S., Lechner, U., & Stanoevska, K. (2002). A reference model for online learning communities. *International Journal on E-Learning, 1*(1), 43-55.

Sorensen, E. K., & Takle, E. S. (2002). Collaborative knowledge building in Web-based learning: Assessing the quality of dialogue. *International Journal of E-learning, 1*(1), 28-32.

Sorensen, E. K., & Takle, E. S. (2004). A cross-cultural cadence. Knowledge building with networked communities across disciplines and cultures. In A. Brown & N. Davis (Eds.), *World yearbook of education 2004. Digital technology, communities & education* (pp. 251-263). London: Routledge Falmer.

Stahl, G. (2003). *Meaning and interpretation in collaboration.* Paper presented at Computer Support for Collaborative Learning (CSCL 2003), Bergen, Norway. Retrieved January 18, 2005 from http://www.cis.drexel.edu/faculty/gerry/cscl/papers/ch20.pdf .

Stephenson, J. (Ed.). (2001). *Teaching & learning online. Pedagogies for new technologies.* London: Kogan Page.

Vrasidas, C., & Glass, V. S (Eds.). (2002). *Current perspectives on applied information technologies:*

Distance education and distributed learning. Greenwich, CT: Information Age Publishing.

Wenger, E. (1998). *Communities of practice. Learning, meaning and identity.* Cambridge: Cambridge University Press.

Chapter 7.16
Female Retention in Post–Secondary IT Education

Jeria L. Quesenberry
The Pennsylvania State University, USA

INTRODUCTION

The historical gender stratification in technical disciplines has been an area of study for many years and researchers have concluded that women are alarmingly under-enrolled in post-secondary information technology (IT) education (e.g., Camp, 1997; Teague, 2002; von Hellens, Nielsen, Greenhill, & Pringle, 1997). One challenge facing the IT gender gap discourse is the application of theories that focus on a variety of levels of analysis (Korpela, Mursu, & Soriyan, 2001; Walsham, 2000). Recently, the Individual Differences Theory of Gender and IT has been proposed by Trauth (Trauth, 2002; Trauth, Huang, Morgan, Quesenberry, & Yeo, 2006; Trauth & Quesenberry, 2006, 2005; Trauth, Quesenberry, & Morgan, 2004) to explain the underrepresentation of women in the IT workforce at both the societal *and* individual levels of analysis. To date, the majority of the Individual Differences Theory of Gender and IT research has focused on improving our understanding of the underrepresentation of women in the IT workforce.[1] Hence, in an attempt

to build on the theoretical foundation, this chapter reports on a literature survey of the influences on American women's retention in post-secondary IT education.

BACKGROUND

Walsham (2000) stresses the importance of research agendas that help to improve the understanding of IT in the contemporary world. Hence, researchers should investigate IT that enables connectivity, but supports diversity by studying particular levels of analysis in detail and context including: the individual, group, organization, inter-organization, and society levels. Korpela et. al. (2001) also support the level of analysis concept by constructing a 2x4 + History framework, which contains four integrative levels of analysis: the individual, group/activity, organizational, and societal levels.

The Individual Differences Theory of Gender and IT articulated by Trauth (Trauth, 2002; Trauth & Quesenberry, 2006, 2005; Trauth et al., 2004,

2006) answers the call for research at multiple levels of analysis stressed by Walsham (2000) and Korpela et al. (2001). First, the theory focuses on women as individuals, having distinct personalities, experiencing a range of socio-cultural influences, and thus exhibiting a range of responses to the social construction of IT. As a result, the theory focuses on an individual level of analysis while acknowledging gender group and societal influences. Secondly, the Individual Differences Theory of Gender and IT takes into account the role of gender group and societal shaping and the importance of individual critical life events (Trauth, 2002; Trauth & Quesenberry, 2005, 2006; Trauth et al., 2004, 2005a).

This theory accounts for the differences among women in the ways they experience and respond to the IT workforce using three constructs: personal data, shaping, and influencing factors and environmental context (Trauth et al., 2004). The personal data construct includes: demographic data (e.g., age, race, and ethnicity), lifestyle data (e.g., socio-economic class and parenting status), and workplace data (e.g., job title and technical level). The shaping and influencing factors construct includes: personal characteristics (e.g., educational background, personality traits, and abilities), and personal influences (e.g., mentors, role models, experiences with computing, and other significant life experiences). The environmental context construct includes: cultural attitudes and values (e.g., attitudes about IT and/or women), geographic data (about the location of work), and economic and policy data (about the region in which a woman works).

MAIN THRUST OF THE CHAPTER

A literature survey of gender and IT research was conducted to understand how individual attributes contribute to the retention of females in post-secondary IT education in support of the Individual Differences Theory of Gender and IT

(Trauth, 2002; Trauth & Quesenberry, 2005, 2006; Trauth et al., 2004, 2006). This analysis resulted in the identification of several research themes that influence women in IT education. These themes include: personal attributes, learning experiences, and responses to support structures and will be discussed in more detail in the remainder of this section.

Personal Attributes of Students

Researchers have found that self-confidence has a large influence on female retention in post-secondary education (Alper, 1993; Ambrose, Lazarus & Nair, 1998; Beckwith, Burnett, Wiedenbeck, Cook, Sorte, & Hastings, 2005; Vest & Kemp, 1999). Shashaani (1994) argues that there is a direct connection between informal computing experiences and high levels of self-confidence in using and understanding IT. This notion is articulated by Margolis and Fisher (2002) in their argument that men tend to have more informal computing experiences than women at the post-secondary level. As a result, male students generally have higher levels of computing self-confidence than female students. Over time, this causes female students to question their technical knowledge and personal fit with an IT related degree, eventually leading to lower female retention numbers.

Research has also demonstrated that negative images of the IT workforce have an influence on the retention of women in post-secondary IT education (Ahuja, Robinson, Herring, & Ogan, 2004; Balcita, Carver & Soffa, 2002; Camp, 1997; Nielsen, von Hellens & Wong, 2000; Vest & Kemp, 1999). Joshi, Schmidt, and Kuhn (2003) and Camp (1997) explain that both men and women acknowledge the negative stereotypes of IT work (e.g., IT work is solitary and a male domain). Furthermore, these stereotypes persist despite receiving accurate information about IT careers (Joshi et al., 2003). Balcita et al. (2002) argues that these negative stereotypes are shaped by the media, and society, which frequently present the IT domain as a place that does not welcome feminine characteristics or traits. For instance, television shows such as *Bill Nye the Science Guy* and *Mr. Wizard* demonstrate strong male presence in science, but lack female role models with whom women can identify. As a result, women, more than men, leave post-secondary IT education programs because they cannot imagine their roles within the field.

Student Learning Experiences

Barker, Garvin-Doxas, and Jackson (2002) and Katz, Aronis, Allbritton, Wilson, and Soffa (2003) argue that experiences in learning environments also have an influence (positive and negative) on female retention in post-secondary IT education.[2] These authors found that the influences of impersonal environment and guarded behavior, informal student hierarchy, and the creation of a defensive social climate are the most important factors in student retention. For instance, Barker et al. (2002) observed that technical courses in post-secondary IT education have the tendency to be impersonal social environments in which it is easy for students to remain relatively anonymous and socially distant. The authors found that it was rare to hear the names of students or the sharing of personal information among students and professors in technical classes. Students in these technical courses were typically referred to as "the woman in the red shirt" or by desk location such as "F1." In addition, informal hierarchies were created by the attainment and display of status in social structure of the classroom. In these technical courses, status was afforded to those students who displayed the highest level of technical skills. Students with programming experience were frequently referred to as "smart" and it was implied that they had intellectual superiority over the other students. Eventually, students became aware of whether they belonged in the group and their places in the social hierarchy. Unfortunately, the female students felt as outsiders and did not share social hierarchy with their male counterparts.

Support Structures of Students

Researchers have found that role models and mentors constitute a powerful support structure that has a positive influence on the retention of females in post-secondary IT education (e.g., Beise et al., 2000; Symonds, 2000; Vest & Kemp, 1999). Role models provide confirmation that women can succeed in technical disciplines and encourage their persistence in the program (Cohoon, 2001, 2002). Mentors are also a powerful support structure that influence the retention of women in IT education. Mentors provide encouragement in a genuine and sincere manner that transforms the way a woman views her connectedness to a technical discipline (Townsend, 2002).

Having faculty who enjoy teaching is also an important influence on the retention of women in post-secondary IT education. Departments with faculty reporting high degrees of personal satisfaction tend to have lower student attrition rates (Cohoon, 2000, 2001, 2002). Quality teaching helps overcome the disproportionately negative effects of unfavorable environments. Female students can also be influenced by accessible and motivated teachers. Faculty who promote interactions among classmates and develop learning communities and other forms of peer support for students also have an influence as they help to foster friendship and support networks that provide effective retention of women in technical programs (Ahuja et al., 2004; Cohoon, 2002).

Institutional and community resource support also has an influence on the retention of females in post-secondary IT education. Educational programs with adequate resources and student organizations, the use of the local job market, and volunteer opportunities all contribute to increased retention rates. The more support women receive from institutional and community resources the higher the female retention level of IT educational programs. Likewise, programs with strong institutional support and adequate resources have a positive influence on the retention of female students (Cohoon, 2002; Frieze & Blum, 2002).

IT educational programs with active student organizations also have a positive influence on the retention of female students in post-secondary IT education (Frieze & Blum, 2002; Gabbert & Meeker, 2002). For example, at Carnegie Mellon University a student leadership advisory council called the Women@SCS Advisory Council was formed to build a community to act as the driving force behind proactive efforts to improve the academic and social climate for women in the computer science department. The Council holds various mentoring events and activities such as freshmen orientation, Big Sisters/Little Sisters, school help sessions, dessert study breaks, invited speaker series, conferences, and outreach programs. As a result, the Council fosters a supportive community that promotes the academic success of women in computer science by improving the academic and social climates (Frieze & Blum, 2002).

FUTURE RESEARCH

This chapter has discussed several factors that influence the retention of women in post-secondary IT education in order to build on the Individual Difference Theory of Gender and IT. This analysis has attempted to go beyond the identification of societal messages that operate at the group level of analysis to also understand the variances at an individual level of analysis. In doing so, several sub category constructs have been added to the conceptual framework of the Individual Differences Theory of Gender and IT (see table one).[3] Future research plans include the continued building of the Individual Differences Theory of Gender and IT to more fully account for gender and IT education issues in order to conduct empirical investigations.

CONCLUSION

The concerns of IT worker shortages and the importance of diversity as a key component of the American economy highlight the utmost importance of attracting more women to IT careers. An initial step in this process is to recruit and retain more female students in IT educational programs and in the IT workforce. In order to do so, it is important for researchers, policy makers, and practitioners to understand the factors that influence women's choices to pursue IT degrees. The factors that have been elucidated in this chapter can be leveraged when recruiting and marketing to potential students as a way to encourage more women to consider a technical curriculum. They can also be used to reshape educational institutions and the IT workforce to better accommodate female recruitment and retention.

Table 1. Gender and IT education sub-category constructs

High Level Construct	Sub Category Construct
Personal Data	
School Place Data	Student Learning Experiences (Environment and Classroom) Student Support Structures (Faculty, Role Models and Mentors for Students, Institutional and Community Response)
Shaping and Influencing Factors	
Personal Characteristics	Personal Attributes (Student Self Confidence, Student Images and Stereotypes of IT)

REFERENCES

Ahuja, M., Robinson, J., Herring, S., & Ogan, C. (2004). Gender issues in IT organizations: Exploring antecedents of gender equitable outcomes in higher education. In *Proceedings of the 2004 SIGMIS Conference on Computer Personnel Research* (pp. 120-123).

Alper, J. (1993). The pipeline is leaking women all the way along. *Science, 260*, 409-411.

Ambrose, S., Lazarus, B., & Nair, I. (1998). No universal constants: Journey of women in engineering and computer science. *Journal of Engineering Education, 87*(4), 363-368.

Balcita, A. M., Carver, D. L., & Soffa, M. L. (2002). Shortchanging the future of information technology: The untapped resource. *ACM SIGCSE Bulletin, 34*(2), 32-35.

Barker, L. J., Garvin-Doxas, K., & Jackson, M. (2002). Defensive climate in the computer science classroom. In *Proceedings of the ACM SIGCSE 2002 Conference* (pp. 43-47), Covington, Kentucky.

Beckwith, L., Burnett, M., Wiedenbeck, S., Cook, C., Sorte, S., & Hastings, M. (2005). Educational issues: Effectiveness of end-user debugging software features: Are these gender issues? In *Proceedings of the SIGCHI Conference on Human Factors in Computing Systems* (pp. 869-878), Portland, OR.

Beise, C., Chevli-Saroq, N., Andersen, S., & Myers, M. (2000). A model for examination of underrepresented groups in the IT workforce. In *Proceedings of the 2002 ACM SIGCPR Computer Personnel Research Conference* (pp. 106-110), Kristiansand, Norway.

Camp, T. (1997). The incredible shrinking pipeline. *Communications of the ACM, 40*(10), 103-110.

Cohoon, J. M. (2000). *Non-parallel processing: Gendered attrition in academic computer Science.* Dissertation. University of Virginia.

Cohoon, J. M. (2001). Toward improving female retention in the computer science major. *Communications of the ACM, 44*(5), 108-114.

Cohoon, J. M. (2002). Recruiting and retaining women in undergraduate computing majors. *ACM SIGCSE Bulletin, 34*(2), 48-52.

Frieze, C., & Blum, L. (2002). Building an effective computer science student organization: The Carnegie Mellon Women@SCS Action Plan. *ACM SIGCSE Bulletin, 34*(2), 74-78.

Gabbert, P., & Meeker, P. H. (2002). Support communities for women in computing. *ACM SIGCSE Bulletin, 34*(2), 62-65.

Joshi, K. D., Schmidt, N. L., & Kuhn, K. M. (2003). Is the information systems profession gendered? Characterization of IS professionals and IS careers. In *Proceedings of the ACM SIGMIS/CPR 2003 Conference* (pp. 1-9), Philadelphia.

Katz, S., Aronis, J., Allbritton, D., Wilson, C., & Soffa, M. L. (2003). A study to identify predictors of achievements in an introductory computer science course. In *Proceedings of the ACM SIGMIS/CPR 2003 Conference* (pp. 157-161), Philadelphia, PA.

Korpela, M., Mursu, A., & Soriyan, H. A. (2001). Two times four integrative levels of analysis: A framework. In N. Russo, B. Fitzgerald, & J. I. DeGross (Eds.), *Realigning research and practice in information systems development: The social and organizational perspective* (pp. 367-377). Boston: Kluwer Academic Publishing.

Margolis, J., & Fisher, A. (2002). *Unlocking the clubhouse: Women in computing.* Cambridge, MA: MIT Press.

Morgan, A. J., Quesenberry, J. L., & Trauth, E. M. (2004). Exploring the importance of social networks in the IT workforce: Experiences with the "Boy's Club." In E. Stohr & C. Bullen (Eds.), *Proceedings of the 10th Americas Conference on Information Systems* (pp. 1313-1320). New York.

Nielsen, S., von Hellens, L., & Wong, S. (2000). The game of social constructs: We're going to winIT! In *Proceedings of the International Conference on Information Systems (ICIS 2000)* (pp. 1-13). Brisbane, Australia.

Quesenberry, J. L., Morgan, A. J., & Trauth, E. M. (2004). Understanding the "Mommy Tracks": A framework for analyzing work-family issues in the IT workforce. In M. Khosrow-Pour (Ed.), *Proceedings of the Information Resources Management Association Conference* (pp. 135-138). Hershey, PA: Idea Group Publishing.

Quesenberry, J. L., & Trauth, E. M. (2005). The role of ubiquitous computing in maintaining work-life balance: Perspectives from women in the IT workforce. In C. Sorensen, Y. Yoo, K. Lyytinen, & J. I. DeGross (Eds.), *Designing ubiquitous information environments: Socio-technical issues and challenges* (pp. 43-55). New York: Springer.

Quesenberry, J. L., Trauth, E. M., & Morgan, A. J. (2006). Understanding the "Mommy Tracks": A framework for analyzing work family balance in the IT workforce. *Information Resource Management Journal, 19*(2), 37-53.

Shashaani, L. (1994). Gender-differences in computer experience and its influence on computer attitudes. *Journal of Educational Computing Research, 11*(4), 374-367.

Symonds, J. (2000). Why IT doesn't appeal to young women. In E. Balka & R. Smith (Eds.), *Women, work and computerization: Charting a course to the future.* Boston: Kluwer Academic Publishers.

Teague, J. (2002). Women in computing: What brings them to it, what keeps them in it? *ACM SIGCSE Bulletin, 34*(2), 147-158.

Townsend, G. C. (2002). People who make a difference: Mentors and role models. *ACM SIGCSE Bulletin, 34*(2), 57-61.

Trauth, E. M. (2002). Odd girl out: An individual differences perspective on women in the IT profession [Special Issue on Gender and Information Systems]. *Information Technology and People, 15*(2), 98-118.

Trauth, E. M., Huang, H., Morgan, A. J., Quesenberry, J. L., & Yeo, B. (2006). Investigating the existence and value of diversity in the global IT workforce: An analytical framework. In F. Niederman & T. Ferratt (Eds.), *Managing information technology human resources.* Greenwich, CT: Information Age Publishing.

Trauth, E. M., & Quesenberry, J. L. (2005). Individual inequality: Women's responses in the IT profession. In G. Whitehouse (Ed.), *Proceedings of the Women, Work and IT Forum*, Brisbane, Queensland, Australia.

Trauth, E. M., & Quesenberry, J. L. (2006). Gender and the information technology workforce: Issues of theory and practice. In P. Yoong & S. Huff (Eds.), *Managing IT professional in the Internet age.* Hershey, PA: Idea Group Reference.

Trauth, E. M., Quesenberry, J. L., & Morgan, A. J. (2004). Understanding the under representation of women in IT: Toward a theory of individual differences. In M. Tanniru & S. Weisband (Eds.), *Proceedings of the 2004 ACM SIGMIS Conference on Computer Personal Research* (pp. 114-119). New York: ACM Press.

Trauth, E. M., Quesenberry, J. L., & Yeo, B. (2005). The influence of environmental context on women in the IT workforce. In M. Gallivan, J. E. Moore, & S. Yager (Eds.), *Proceedings of the 2005 ACM SIGMIS CPR Conference on Computer Personnel Research* (pp. 24-31). New York: ACM Press.

Vest, S. N., & Kemp, J. J. (1999). The retention of women in the computing sciences. *1999 ACM Southeast Regional Conference.*

von Hellens, L. A., Nielsen, S. H., Greenhill, A., & Pringle, R. (1997). Gender and cultural influences in IT education. *PACIS* (pp. 389-397).

Walsham, G. (2000). Globalization and IT: Agenda for research. In R. Baskerville, J. Stage, & J. I. DeGross (Eds.), *Organizational and social perspectives on information technology* (pp. 195-210). Boston: Kluwer Academic Publishing.

KEY TERMS

Conceptual Framework: The basic structure of concepts of a given theoretical perspective.

Individual Differences Theory of Gender and IT: A social theory developed by Trauth (2002; Trauth et al., 2004) that focuses on within-group rather than between-group differences to explain differences in male and female relationships with information technology and IT careers. This theory posits that the underrepresentation of women in IT can best be explained by considering individual characteristics and individual influences that result in individual and varied responses to generalized environmental influences on women.

IT Education: Educational programs including, but not limited to computer science (CS), management information systems (MIS), information sciences and technology (IST), and computer engineering.

Level of Analysis: Research agendas and projects in various domain areas of individuals, groups, organizations, inter-organizations, and societal levels (Walsham, 2000).

Post-Secondary Education: University or college educational experiences, which follows secondary education such as high school.

ENDNOTES

[1]   For instance, the influence of environmental context in the underrepresentation of women in the IT workforce (Trauth et al., 2005), the role of parenthood (Quesenberry et al., 2004, 2006) and ubiquitous computing in work-life balance (Quesenberry & Trauth, 2005), and the role of social networking in the IT profession (Morgan et al., 2004).

[2]   The learning environments include the physical surroundings, psychosocial conditions, emotional conditions, and social or cultural influences present in a learning situation (Barker et al., 2002).

[3]   The full conceptual framework for the Individual Differences Framework can be found in Trauth et al. (2004).

Chapter 7.17
Contextualized Learning:
Supporting Learning in Context

Marcus Specht
Fraaunhofer FIT-ICON, Denmark

ABSTRACT

This chapter presents an overview of research work for contextualized learning, integrating the background of adaptive hypermedia, ubiquitous computing, and current research on mobile learning systems that enable support for contextualized learning. Several examples for new learning paradigms are analyzed on their potential for mobile learning and contextualization. In the second part, examples for systems that integrate mobile learning solutions in existing learning systems for schools and working context are presented. The RAFT project realizes application for computer-based field trip support and shows an integration of m-learning tools in an established teaching method of school field trips. The SMILES prototype shows the integration of e-learning services and its stakeholders with mobile learning technology.

INTRODUCTION

New technology develops fast, and the reality of information and learning delivery everywhere is changing monthly. Nearly every week new devices and gadgets appear on the market and enable new ways of mobile access to information, mobile games, and online applications. Currently, the new research field of m-learning and a community working on that topic is establishing, and a variety of research groups work on new approaches supporting mobile learning. Those approaches mainly come from the background of collaborative learning, mobile information systems, adaptive hypermedia, and context-aware computing. From our point of view, adaptive educational hypermedia plays a central role in new models for m-learning and contextualized learning support. Applications from this area include a range of examples from personalized guiding systems for cities, art exhibition guides, and adaptive learning management systems.

In the field of adaptive hypermedia, several approaches have been doing work on the adaptation of interfaces and contextualized user interaction to specific devices and interaction modalities. Adaptations mostly have been based on the constraints of the devices used (mostly small screens) or network constraints like low bandwidth for mobile devices. From our point of view, there is more to contextualized computing than delivering content to small screens or converting it to new technical formats. By the variety of devices and the new possibilities of ubiquitous computing, information access gets embedded in the environment and gets contextualized to the current context of use (Oppermann & Specht, 2000).

For educational applications, this enables new possibilities for learning in context and understanding artifacts in the real world with the help of computers that can support the learning process in the current situation by adapting to a variety of context parameters. The underlying theoretical background of situated cognition and situated learning (Wenger & Lave, 1991) clearly states the target and motivation for contextualized learning support. Furthermore, it demonstrates the benefits for learners and authors that can be achieved by having information available in context. Mobile learning seems to be one of the fields where new paradigms for mobile cooperation and the integration of mobile and stationary activities are analyzed in most detail up to date. Most of the empirical studies currently looking at the usage of mobile devices in learning come from the classroom and learning situations related to field trips. The classroom in this sense seems to be a highly adequate field to introduce tools and services that allow a new way of learning and handling digital media for contextualized experiences.

This chapter will try to connect the theoretical foundations of situated learning and cognition and how those relate to new forms of computer supported contextualized learning.

In the first part, we will describe our view on adaptive hypermedia and the variations of adaptive methods. Based on this, we will give examples of how contextualized learning extends current approaches for personalization of learning processes and content delivery by taking into account additional parameters of the current context (environment, location, time, social context). On this background, some applications realized in Fraunhofer FIT and the European project RAFT [1] and the prototype SMILES will be presented and their usage of context information to adapt to users will be demonstrated. It will be shown that especially the combination of different context parameters with classical learner modeling approaches allows for information delivery tailored to the individual learner and his/her current situation in a very effective way.

SITUATED LEARNING AND ADAPTIVE METHODS

In the following section, we will give the motivation and background for situated learning and describe current scenarios for contextualized learning applications and use of mobile devices in classroom learning.

Situated Learning and Blended Learning

Situated learning as introduced by Wenger and Lave (1991) states the importance of knowledge acquisition in a cultural context and the integration in a community of practice. Learning in this sense must not only be planned structured by a curriculum, but also by the tasks and learning situations and the interaction with social environment of the learner. This is often contrasted with the classroom-based learning where most knowledge is out of context and presented de-contextualized. On the one hand, the process of contextualization and de-contextualization might be important for abstraction and generalization of knowledge on the other hand in the sense of cognitive apprenticeship (Collins, Brown, & Newman, 1989) it is reasonable to guide the learner towards appropriate levels and context of knowledge coming from an authentic learning situation.

From a constructivist point of view, not only knowledge is always contextualized but also the construction of knowledge, for example, learning is always situated within its application and the community of practice (Mandl, Gruber, & Renkl, 1995). Stein (1998) defines four central elements of situated learning where the content emphasizes higher order thinking rather than acquisition of facts; the context for embedding the learning process in the social, psychological, and material environment in which the learner is situated; the community of practice that enables reflection and knowledge construction; and the participation in a process of reflecting, interpreting and negotiating meaning. From the perspective of situated learning, several requirements for new learning tools can be stated, such as: use authentic problems, allow multiple perspectives, enable learning with peers and social interaction within communities, enable active construction and reflection about knowledge. A shift towards a new tradition of online learning is described by Herrington et al. (2002).

Moreover, the idea of situated learning also is closely related to the ideas of "blended learning" and "learning on demand", especially in educational systems for adults and at the workplace. An important point that is not taken into account by a lot of new approaches for delivering learning on demand is the aspect that the need (demand) for knowledge and learning arises in a working context with the motivation for solving specific problems or understanding problem situations. This notion of "learning on demand" in the workplace exemplifies the potential of contextualized learning in the workplace. Learners who identify a problem in a certain working situation are highly motivated for learning and acquiring knowledge for problem solving. They have a complex problem situation as a demand, which can be used for delivering learning content adapted to their situation.

The contextualization of learning on demand can not only be seen from the point of view of an actual problem or learning situation, but also in a longer lasting process of learning activities that are integrated. Different learning activities are combined in blended learning approaches where the preparation for a task updates on base knowledge, and then the application in an actual working situation and the documentation of problem solutions and the reflection about one's activities evaluates that process. An example of a blended learning process with situated learning components and different activities in such phases is shown in Figure 1.

From this perspective, e-learning brought the possibility of delivering learning content quickly and easily to immense numbers of learners and enabled them to cooperate with computer-based tools. Adaptive educational hypermedia now gives the possibilities to adapt those curricula and learning process to the individual and his/her strengths and weaknesses. M-learning and new technologies will bring a closer integration in the life long learning process and the disappearing computer will allow the learners to access information and content in every day life where planned learning but also accidental learning can take place.

A Need for Mobile Learning Support: Scenarios

In recent years, several initiatives researched scenarios for learning and mobile information support at the workplace and in the classroom. According to Kling (2003), the classroom and research in the classroom might be one of the key drivers for a next generation of social software. The classroom gives a variety of scenarios and situations where

Figure 1. Integrating a blended learning solution with situated learning components

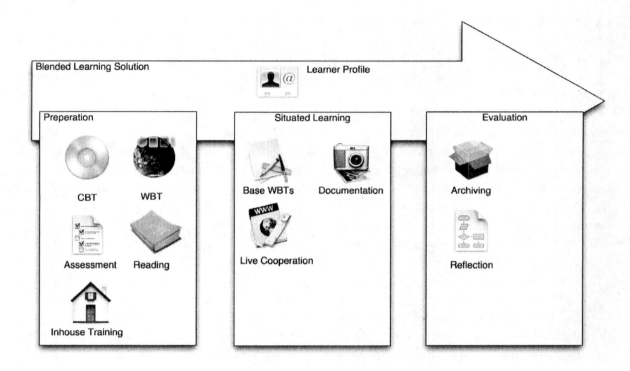

ad hoc collaboration and the contextualization of information play an important role.

The PEP program (Tatar et al., 2002) looked especially at classroom-based learning and how mobile devices can give new possibilities for classroom based learning. In a study conducted in the PEP program, 84% of teachers strongly agreed that the quality of teaching was improved by handheld devices in the classroom. New possibilities were seen in the live interaction about data and the reflection about easily exchangeable and copied data sets. The teacher could collect feedback and get anonymous assessment for the current understanding of topics in the class. Recommendations for system designers coming out of PEP are focused application designs and a clear structure for blending of computer-based collaboration and information usage and teacher based instruction periods.

Curtis et al. (2002) identified the top five scenarios and applications on PDAs for sing the handheld in the school. These included sketching, focused simulations with curriculum integration, picture chat, concept mapping, word processing, online information research, and beaming class notes. The eSchoolbag System clusters the functionality supported in the classroom and in ad hoc learning scenarios in several functional modules like the scheduler system, a broadcasting system, voice and image transmission, text transmission, real time examination, notebook, contacts, reporting, and others (Chang & Sheu, 2002). In the context of the m-learn project user studies analyzed the different scenarios being relevant at the working context for learning (Curtis et al., 2002).

Based on a pattern approach DiGiano and colleagues (2002) are working on a systematic approach to structure and identify patterns for collaborative mobile work and learning. In that sense, most of the current research works can be clustered according to their contribution to certain patterns in collaborative learning scenarios.

Several empirical findings stress the opportunity of using mobile technologies for training and education of young adults and expect high acceptance rates for educational applications for that target group (Atewell & Savoll-Smith, 2003; Eldridge & Grinter, 2001). Usability studies about activities when interacting with mobile Web content and important findings can be seen in the Electronic GuideBook System (Hsi, 2002).

Most of the studies reported here show a high potential and acceptance for supporting new forms of mobile and contextualized learning approaches in the classroom. From our point of view, the integration of focused applications with specialized interfaces and their integration in more complex task contexts is crucial for the design of contextualized learning. The methods of adaptive hypermedia play an important role because the generation and selection of personalized views on shared data and cooperative tools is an essential aspect of the more complex cooperative applications used by individuals. In the following section, we will present some extensions for an adaptive methods classification which we perceive as important for building contextualized learning tools.

Adaptive Methods and Extensions

Adaptive educational hypermedia gives a variety of research work about questions on how to adapt curricula and learning content to individuals and groups of learners. Brusilovsky (1996) gives a comprehensive overview of adaptive methods and techniques in general. From our point of view, the application of adaptive methods to educational hypermedia applications can mainly be structured according to four main questions (Specht, 1998):

What parts or components of the learning process are adapted? This question focuses on the part of the application that is adapted by the adaptive method. Examples can be the pace of the instruction (Leutner, 1992; Tennyson & Christensen, 1988) that can be modified based on diagnostic modules embedded in the learning process or adaptation of content presentations, the sequencing of contents and others. Extensions with new forms of information delivery allow the distribution of learning materials to different learning contexts relevant to the individual user or groups of users.

What information does the system use for adaptation? In most adaptive educational hypermedia applications, a learner model is the basis for the adaptation of the previously given parameters of the learning process. Nevertheless, there are several examples where the adaptation takes place not only to the learner's knowledge, preferences, interests, and cognitive capabilities, but also to tasks and learner goals. In contextualized learning, the information used for adaptation is extended by the environmental parameters. The inference methods of the adaptive system can gain precision from the additional information of environmental sensors. A variety of sensors available from the area of ubiquitous computing can for example be seen in Schmidt (2002).

How does the system gather the information to adapt to? There are a variety of methods to collect information about learners to adapt to. Mainly implicit and explicit methods like those described in works from user modeling can be distinguished. An overview can be found in Jameson, Konstan, and Riedl (2002). Sensors play an important role in extending existing hypermedia approaches to contextualized learning. There are several works in the literature to create context sensor middleware allowing for higher-level contexts based on sensor data from different sources (Schmidt, 2002). As a simple example in learning, a tracking system in physical space can enrich the information from a user questionnaire for getting more valid assumptions about the user's preferences.

Why does the system adapt? This question mainly focuses on the pedagogical models behind the adaptation. Classical educational hypermedia system mainly adapted according for compensation of knowledge deficits, ergonomic reasons, or adaptations to learning styles for an easier introduction into a topic. Location-based services are an example for the type of adaptation we want to discuss in the chapter situated learning and cognition. In those applications, the individual possibility for encoding and decoding information is one interesting aspect for better understanding artifacts "in context". Furthermore, the authentic collaboration on a topic is another example where co-learners can be selected according to the learning task of an individual.

Several empirical findings show that adaptation to learner models and parameters of the learning situation lead to more efficiency, effectiveness, and motivation for learning. Basically, our understanding of contextualization comes

from an extension of the adaptive educational hypermedia approach. Adaptive educational hypermedia systems collect information from user assessment, feedback, the current task or user goal, and other implicit and explicit acquisition methods. Additionally, in a contextualized learning system, the current user context with a variety of environmental parameters can be taken into account. Those environmental sensors enable the application to collect much more information about the behavior of a user. Even more, by the integration of environmental sensors and user sensors, new applications can collect direct feedback from user sensors dependent on the variation of environmental parameters measured by environmental sensors like those shown in Figure 2. A good example for such an application can be a training system that monitors the users' moves while handling a complex machine and giving direct feedback for training purposes. Such systems are already used today in medical training applications like echo tutor (Grunst et al., 1995).

Table 1. A classification schema for adaptive methods

Adaptive Educational Hypermedia		
What is adapted?	To which features?	Why?
Learning goal • Content • Teaching method • Content Teaching style • Media selection • Sequence • Time constraints • Help Presentation • Hiding • Dimming • Annotation	Learner • Preferences • Usage • Previous knowledge, professional background • Knowledge • Interests • Goals • Task • Complexity • ...	Didactical reasons • Preference model • Compensation of deficits • Reduction of deficits Ergonomic reasons • Efficiency • Effectivness • Acceptance
Extensions in Contextualized Learning		
Presentation • 3D Sound • Augmented Reality displays • Distribution to different contexts	Context Sensors • User Location • Time • Lighting, Noise • Other User's Locations	• Authenticity of Learning Situations • Situated Collaboration • Active Construction of Knowledge

For the adaptation to individual users, the system in that sense can have shorter feedback cycles and adapt not only to the individual learner model and explicit user feedback, but also to implicit feedback loops from a variety of contextual parameters. First, simple examples for new adaptive methods in content delivery are location-based services and museum information systems like hippie (Oppermann & Specht, 2000). Besides new adaptive methods, this additionally can have an important impact on the interaction with the learning system. New forms of augmented reality training systems in this sense are not restricted to the request of information, but also enable the learner to explore the learning subject and its artifacts either in virtual reality training simulations (Rickel & Johnson, 1997) or in a tracked real training environment (Fox, 2001). From our point of view, this is not only a different way of accessing learning materials, but can be seen as support for constructivist learning approaches in combination with an adaptive intelligent system that tracks the users' learning activities and responds to them.

Another example for the extension of adaptive methods comes from the field of adaptive augmented reality systems. The LISTEN (Goßmann & Specht, 2001) system tracks the user with a resolution of 5 cm and 5 degrees, which allows to identify if a user looks onto a detail of an artwork in a gallery or just on the frame or beside the artwork. Additionally, the system can present information to the user embedded in the physical environment with 3D audio technology. So the user experiences the sound of presentations coming from the environment or from specific objects in the environment. Based on those location tracking sensors and the presentation possibilities, new adaptive methods can be realized. Some examples are:

Figure 2. User sensors and environmental sensors for more valid inferences and validation of implicit user tracking methods

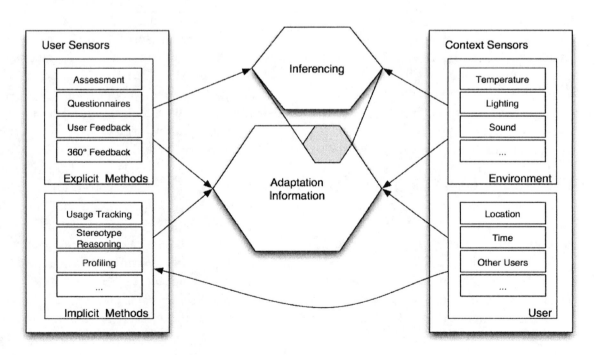

- **Adaptation of presentation to position and object distance:** The user's position in space relative to an object in the physical space is mapped onto the direction and the volume of the sound for the presentation of the information.

- **Selection of presentation style based on position:** If a user moves in a room, s/he will get different presentation styles from the system. If the user moves into the center of a room, more general information about the room as a whole is presented and a sound collage for the single objects in the room is generated with directed sound sources coming from the objects. If a user moves close towards an objects and focuses on that object, the volume of that piece of the collage is turned up and more detailed information of the object will be selected.

- **Adaptation to movement and reception styles:** Several kinds of common behavior can be identified with people walking through the environment (e.g., clockwise in museums). By using the fine-grained tracking technology, the system can learn about preferred user movement and perception styles. The information about the time of listening to object descriptions can be combined with the movement. The selection and dynamic adaptation of tour recommendations can be adapted to the stereotypical type of movement and his/her preferred perception style.

- **Adaptation to time and lighting conditions and position of user** is a complex adaptive method taking into account environmental factors, the time, and the user position for the explanation of artworks in LISTEN. The sound presentation can be adapted to the changing lighting conditions during the day (based on sensor data) for explaining certain details that can only be visible during a certain time period or from a certain position in the room.

The contextualization of learning experiences and information is not only important on a level of presenting single contents, but is additionally important on the level of integrating and synchronizing learning activities in blended learning like those described in in the section Situated Learning and Blended Learning.

In this sense, we perceive m-learning as a natural evolution of e-learning: new technologies allow for a better support of learning than classical ways of e-learning, where the textbook often was just replaced by a computer screen. While in the current discussion about e-learning, blended learning approaches are often mentioned as a solution for the integration of e-learning in existing educational scenarios, we see m-learning and contextualized learning as a good chance to develop e-learning one step further. Often in blended learning scenarios intermixing computer-based and face-to-face learning in the learning process describe the way towards a certain educational goal. Nevertheless, this often neglects the problem of synchronization of learning steps. How should an intelligent learning environment get aware of the users' progress? How should remote peers support a user when he is in an actual learning/working context? Many of those questions can be answered when the computer disappears in the environment or gets mobile in a first step. Learners could use contextualized learning tools just like a mobile telephone where they not only could call an expert for advice but also could use a variety of other learning tools for helping in an actual situation. In this sense, we understand m-learning and the contextualization as a natural way of integrating learning technology in the learning process on demand. That this does not only work with planned instruction can be seen with examples of system that use more accidental learning like in museum environments (Oppermann & Specht, 1999).

We do not see a major question of m-learning in the conversion of learning materials into PDA formats. The idea of a mobile book in our understanding is mainly an issue of technology but not of pedagogy. Nevertheless, the use of PDAs in mobile learning scenarios can be very fruitful and get a new quality into learning. As soon as there is communication between learners new mobile devices even without a contextualization of materials and learning activities can be very helpful as seen in studies reported from the PEP program (Tatar et al., 2002).

Nevertheless, we just do the first steps towards more integrated learning tools that allow for a natural learning process with embedded intelligent learning, tutoring, and collaboration systems. From our point of view, paradigm changes can be triggered by positive experiences and integrated solutions that allow the users to use new pedagogical approaches and systems within existing infrastructures and content networks. Therefore, we want to present some examples in the following section that we think can be interesting starting points for the integration of contextualized learning into today's learning infrastructures.

SOLUTIONS AND RECOMMENDATIONS

We want to introduce two examples of systems that are used in two different contexts at Fraunhofer FIT. The first system shows the integration of mobile learning into the school scenario of field trips that is done in the project RAFT. The second application shows the integration of learning tools in a working scenario where different stakeholders structure and reuse information for different purposes.

Mobile Learning for Field Trips and Collaboration

In the context of the European-funded project RAFT (Remotely Accessible Field Trips), the consortium creates a learning tool for field trips in schools. The system should support a variety of learners with different tasks either in the classroom or in the field. The main objectives of the RAFT project are:

- To demonstrate the educational benefits and technical feasibility of remote field trips, with a view to promoting a market for products and to prompting best practice to support this learning activity.
- To establish extensions on current learning material standards and exchange formats for contextualization of learning material. This is combined with the embedding of learning and teaching activities in an authentic real world context.
- To establish new forms of contextualized learners' collaboration with real time video conferencing and audio communication in authentic contexts.

An additional emerging objective is to give students the opportunity to experience vocational domains for themselves before being committed to a particular course of studies for a future career.

RAFT envisions to facilitate field trips for schools and to enable international collaboration of schools. Instead of managing a trip for 30 students, small groups from the RAFT partner schools go out to the field, while the other students and classes from remote schools participate interactively from their classrooms via the Internet. The groups going to the field will be equipped with data gathering devices (photographic, video, audio, measuring), wireless communication, and a video conferencing system for direct interaction between the field and the classroom.

In the first year of the project, the different phases and functional requirements for supporting live collaboration and information access during field trips were worked out. Field trips with school kids were held in Scotland, Slovakia, Canada, and Germany in order to identify different activities in the field and in the classroom and to draw first evaluations of critical factors. Through these trials, different phases for preparing the field trip, experiencing the field trip in the classroom and in the field, and the evaluation after the field trip were identified. In those phases a variety of stakeholders and participants contribute to the field trip and take an active role in it.

Field trips are an ideal example for an established pedagogical method that can be enhanced with computer-based tools for new ways of collaboration and individual active knowledge construction. The learners in the field can collect information and contextualize it with their own experiences and in the same time work on tasks with their peers and detect new perspectives and solutions to given problems. To foster the variety of perspectives and activities in the field trip process RAFT develops tools for the focused

support of different activities in the field and in the classroom. A basic schema of some roles in the field, the used devices, and their activities can be seen in Table 2.

Based on these roles, the RAFT project develops focused applications that also integrate the collaboration with other team members. The interaction flow between classroom and field site on different channels can be seen in Figure 3.

The RAFT applications enable different participants in a synchronous collaborative learning situation to solve common tasks and learn with different activities about a topic. In the RAFT scenario, students should change their role either from field trip to field trip or sometimes even within one fieldtrip to learn about the different activities how to learn about a topic and also take different perspectives to the same topic. This

Table 2. Overview of field roles and their activities

User Role	Device features	Description of activity and related use cases
Scout	Gotive(WLAN), PDA-CAM(WLAN), Walkie-talkie	To look around the field trip site to identify appropriate locations to gather the data required by field trip tasks
Data Gatherer	PDA-CAM(WLAN), Gotive(WLAN), Sensors, Digital camera, Digital video camera	To gather data from the field in response to a field trip task. He/She collects raw data of type video, picture, audio and sensor data and tells sensor values to annotators for form filling.
Annotator	Tablet-PC(WLAN), PDA(WLAN) Gotive(WLAN)	To gather the raw data being generated by the Data Gatherer and to add initial meta data prior to the material being placed within a collection.
Communicator	Camcorder, Tablet-PC in Backpack with CTM2.0 Laptop PC(WLAN) Web camera	To work in partnership with a Reporter to follow the activity taking place in the field so that classroom participants can watch the activity
Reporter	PDA(WLAN)	To work in partnership with a communicator to comment on the activity taking place in the field and to interview remote experts in the field
Field Coordinator	Laptop PC(WLAN), Tablet PC (WLAN) Walkie-talkie	Overview Field Trip
Teacher	PDA(WLAN)	Overview Students. Accept/Reject task completion/deletion/creation etc
Observer	No device	Observing other field members

ensures the integration of different pedagogical approaches in a "blended" learning situation, where different learning activities are distributed in a team of learners.

From the prototyping and usage of the RAFT applications by end users, we see the following main activities as new qualities of contextualized learning approaches:

- **Cooperative task work:** The distributed work on a task focuses the interaction and communication between the learners, and technology moves into the background when the curiosity about the given task and its exploration in physical and knowledge space become the main interest. The context in this

sense is an enabling mean that allows the learners to immerse in the learning subject at hand.

- **Active construction of knowledge and learning materials:** Users are much more motivated when "self made" learning materials get integrated into the curriculum and they have the possibility to extend existing structures for learning.

- **Clear task structures are helpful in the school context:** In schoolwork in the field, it is highly recommendable to have structured tasks and different roles for taking of the tasks. This can be seen as giving a structured task context for individuals to contribute to a group work on a shared basis.

Figure 3. The roles in the RAFT Interactive Field trip System and the interaction flows between them

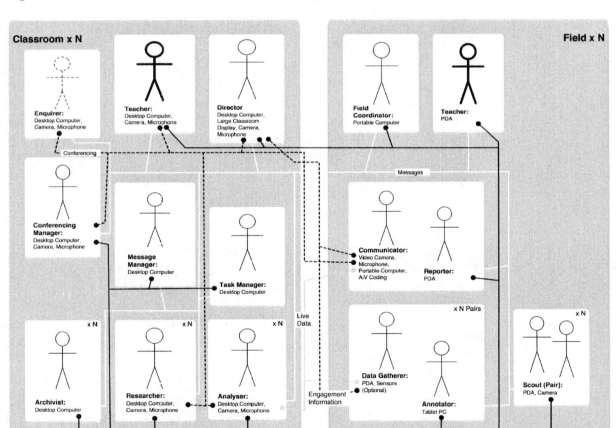

The RAFT applications are based on adaptive methods on different levels. The system supports the user with different tools depending on his/her current phase in the field trip process in general: preparation, field trip activity, or evaluation. During the field trip, the selection of information and collaboration tools is based on the position and current user task of a user. Based on the experiences made in the prototyping phase of the project, the implementation of different user roles and interfaces is not based on a software solution for intelligent rendering of interface components. Instead, it is developed with specialized applications for the different roles and role-specific devices for fulfilling the tasks.

Another focus was the development of contextualized learning materials in RAFT. Besides the classical learning object metadata (LOM, SCORM) attached to materials used in preparation, field trip activity and evaluation additional metadata was required for contextualized learning objects. For learners who collected materials, it is essential to be able to store information about the location where the materials where collected. For learners exploring a field trip site, it is crucial to get information that fits with the current time of the year and the position—or maybe even the weather conditions on that day. Therefore, we developed a specialized framework for collecting context sensor data in real time together with the learning materials and used the context metadata to make the collected information accessible to other participants of a field trip. As an example, a scout can collect small pictures or audio annotations and tag them with the location information (sensor metadata) from a GPS device. This tagging and the information instantly appear on the task lists of other team members and are highlighted in the user interface. Classical learning object metadata can be helpful for adaptive methods on sequencing and selecting the appropriate learning objects for a learner. Context metadata allowed for new approaches for structuring and accessing shared assets and learning objects and in RAFT.

As one example, learners could browse a database of pictures in a biology field trip filtered by the location and the time of the year. Using this approach students could explore and learn about simple questions like "Which flowers grow here at what time of the year?" Additionally, metadata such as the precise time when the picture was taken and the weather conditions on that day can give interesting materials for exploring and learning about important factors of flower growth.

Situated Mobile Learning Support (SMILES)

The prototype SMILES was developed at Fraunhofer FIT for supporting the different actors in a working situation with contextualized tools that allow to record learning material and access learning materials by context parameters.

For applying contextualized learning principles to working life and learning on demand in working contexts, we have built the SMILES prototype. We have chosen a mobile maintenance scenario where a user has to work outside in the field to repair some complex machinery. Additionally, to the mobile worker there are a number of other stakeholders with different tasks, which are all integrated, in a complex training and documentation cycle.

How the different stakeholders work together with the SMILES system is shown in Figure 4.

As seen in Figure 4, one main point of the SMILES prototype was the integration of heterogeneous resources that can be used for learning like technical documentation, courseware authoring, quality assurance, maintenance experts, and others. Those resources are integrated in the learning network based on SMILES and cooperate on the same database and information. Coming from the typical maintenance scenario, mobile workers in the field can request information about certain problems from the SMILES system and search in the case base for problems that are similar to their current situation. Additionally, they can document a new case if they cannot find useful information in the database, thus integrating a newly documented case in the learning environment. Technical documentation can insert content via classical authoring tools for Web-based courses and so insert live materials in the learning contexts. The courseware authoring or human resources department can use the basic documentation to integrate it with didactically structured e-learning lessons and WBT. The quality assurance takes care that the resources structured from the human resources department are technically correct and feeds back usage from maintenance people into the technical documentation. Training on the job and on demand feeds back live experiences to the quality assurance and into the system, thereby ensuring consistent learning materials. In addition, the content can of course be used for classical training and WBT.

As a base for implementation we used the ALE (Specht et al., 2002) system and extended its functionality with several frameworks. First, we needed to create specialized learning objects for collecting and structuring experiences in a different way than in learning units and learning elements and created content templates for cases, problem descriptions, and solutions similar to problem base learning approaches. Additionally, we created different applications for accessing the LCMS with a different task focus. On the one hand, we used a PDA for the maintenance person being in the field and searching for cases and problems. For extended usage and documentation in the field, we used a tablet PC that allowed the users to access the case base, search it, and contact experts for different problems via a live conferencing link. Examples for those applications can be seen in Figure 5 and Figure 6.

Based on a backend LCMS ALE and the authoring tool author42 (bureau42), the SMILES system allowed us to produce live content and synchronize it with a case base and also export the contents to different target formats that allow the usage as SCORM based courseware.

The SMILES prototype shows different examples for context adaptive methods applied in a collaborative learning and documentation application. Depending on the task and the stakeholder

Figure 4. The different activities and stakeholder in the SMILES scenario

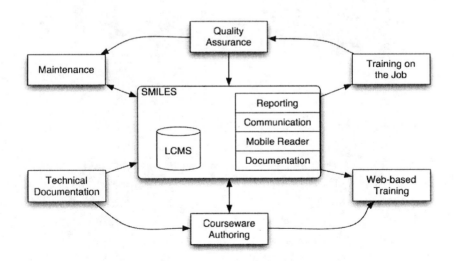

Figure 5. A case base searching and documentation interface

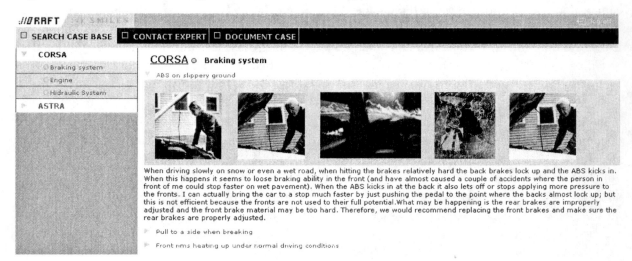

Figure 6. Tablet PC live conferencing tool

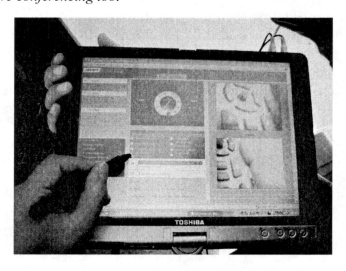

accessing the system, different functionalities are highlighted in the user interface. Nevertheless, all information goes into a shared repository and is instantly available to the other users. As a special need for structuring the materials collected in the field, we developed an easy way of adding metadata to collected materials with the semantic concepts of the learning domain. Based on the described software backend, additional sensor data would allow for fast and implicit access to cases in the field and in cooperation with remote experts. As an essential functionality for supporting mobile workers we identified the possibility to get adaptive recommendation of experts based on their availability and their expertise focus. This information about experts connected to learning objects in the LCMS could be used for most stakeholders involved in the SMILES prototype. For learning on the job, simple extensions of the approach gave promising feedback in first user workshops. The easy access to manual and training information in the working context for example driven by barcodes was seen as a main advantage compared to classical de-contextualized e-learning in a computer room or printed manuals on the job.

CONCLUSION

Contextualized learning appears to be a chance to explore new paradigms for computer-based learning embedded in authentic learning situations. In the examples presented, we have looked at two target groups of current e-learning approaches and have shown how to integrate existing solutions like authoring tools and LCMS with new interfaces and tools for contextualized learning tools. In most current systems we just see that as a first step and the current drawbacks of devices often become obvious soon after starting to use a mobile tool. One of the central insights of our work is that the tools for mobile and contextualized learning must be highly focused and adaptable to the actual task and situation. On the one hand often the characteristics of the situation in which the mobile devices are used give critical constraints to the design of hardware and software. In some cases this can even lead to the conclusion that an adaptation of a tool supporting contextualized learning must be done by the hardware and software designers and not during runtime. This can lead to less flexible tools today and sometimes the adaptation can be "hardcoded" in the system. On the other hand only this procedure makes the tool usable in real mobile learning situations.

Mobile learning in that sense can also be the missing link between learning from a computer screen and the learner grasping the idea by in-

teracting with a physical object. It offers the possibilities of interacting with virtual and physical objects and learning from the response of the objects. This is clearly linked to inquiry-based learning approaches like WISE (Slotta & Britte Cheng, 2001). Even more adaptive instructional methods can get important for learners to constrain the available information at the right time and especially on the current task at hand.

The construction of learning materials like demonstrated in the mobile collector of the RAFT project has positive effects on the learning and understanding. Moreover, this active construction of knowledge does not only have positive effects on the individual learning but also in cooperative scenarios like shown in the context of RAFT. For the application in the working context, it seems extremely important that learning can be delivered on demand in an actual working situation. In classical e-learning approaches, learners identify a problem, go back to the computer, and learn about possible solutions. This break in the learning context often has a bad impact on motivation and knowledge acquisition as such.

It should be mentioned that the chapter explicitly has taken a more pedagogically oriented viewpoint to the question of adaptive instruction and contextualization. Important technical issues like the definition and usage of metadata for storing and retrieving contextualized learning materials or the relations to semantic Web technologies have only been touched. Nevertheless, we perceive it as crucial to engineer educational systems with clear educational motivation and organizational constraints in mind.

REFERENCES

Atewell, J., & Savoll-Smith, C. (2003). M-learning and social inclusion - Focusing on learner and learning. *MLEARN 2003*. London: Learning and Skills Development Agency.

Brusilovsky, P. (1996). Methods and techniques of adaptive hypermedia. *User Models and User Adapted Interaction, 6*(6), 87-129.

Chang, C. Y., & Sheu, J. P. (2002). Design and implementation of ad hoc classroom and eSchoolbag systems for ubiquitous learning. *Proceedings of the IEEE International Workshop on Wireless and Mobile Technologies in Education.* Växjö, Sweden. IEEE Computer Society.

Collins, A., Brown, J. S., & Newman, S. E. (1989). Cognitive apprenticeship: Teaching the craft of reading, writing, and mathematics. In L. B. Resnick (Ed.), *Knowing, learning and instruction* (pp. 453-494). Hillsdale, NJ: Lawrence Erlbaum Associates.

Curtis, M., et al. (2002). Handheld use in K-12: A descriptive account. *Proceedings of the IEEE International Workshop on Wireless and Mobile Technologies in Education,* Växjö, Sweden. IEEE Computer Society.

DiGiano, C., et al. (2002). Collaboration design patterns: Conceptual tools for planning for the wireless classroom. *Proceedings of the IEEE International Workshop on Wireless and Mobile Technologies in Education,* Växjö, Sweden. IEEE Computer Society.

Eldridge, M., & Grinter, R. (2001). Studying text messaging in teenagers. *Human factors in computing systems CHI.* Seattle, WA.

Fox, T. (2001). *Präsentation Neuer Interaktiver Lehrmedien in der Sonographie.* In *Dreiländertreffen DEGUM –SGUM – ÖGUM.* Nürnberg.

Goßmann, J., & Specht, M. (2001). Location models for augmented environments. *Proceedings of Ubicomp 2001, Workshop on Location Modelling for Ubiquitous Computing,* Atlanta, GA.

Grunst, G., et al. (1995). *Szenische Enablingsysteme - Trainingsumgebungen in der Echokardiographie.* In *Ulrich Glowalla/Erhard Engelmann/*

Arnould de Kemp/Gerhard Rosbach/Eric Schoop (Hrsg.): Deutscher Multimedia Kongreß '95. Auffahrt zum Information Highway, S. 174 - 178.

Herrington, J., et al. (2002). Towards a new tradition of online instruction: Using situated learning theory to design Web-based units. *Proceedings of the 17th Annual ASCILITE Conference.* Lismore: Southern Cross University Press.

Hsi, S. (2002). The electronic guidebook: A study of user experiences using mobile Web content in a museum setting. *Proceedings of the IEEE International Workshop on Wireless and Mobile Technologies in Education,.* Växjö, Sweden. IEEE Computer Society.

Jameson, A., Konstan, J, & Riedl, J. (2002). AI techniques for personalized recommendation. *Proceedings of AAAI 2002, The 18th National Conference on Artificial Intelligence,* Edmonton, Alberta, Canada.

Kling, A. (2003). *Social software.* Tech Central Station.

Leutner, D. (1992). *Adaptive Lehrsysteme; Instruktionspsychologische Grundlagen und experimentelle Analysen.* Fortschritte der psychologischen Forschung. Winheim: Beltz. 246.

Mandl, H., Gruber, H., & Renkl, A. (1995). *Situiertes lernen in multimedialen lernumgebungen.* In L.J. Issing & P. Klimsa (Ed.), *Information und lernen mit multimedia* (pp. 167-178). Weinheim: Psychologie Verlags Union.

Oppermann, R., & Specht, M. (1999). A nomadic information system for adaptive exhibition guidance. *Proceedings of ICHIM99, International Cultural Heritage Meeting,* Washington, DC.

Oppermann, R. & Specht, M. (2000). A context-sensitive nomadic exhibition guide. *HUC2K, Second Symposium on Handheld and Ubiquituous Computing.* Bristol, UK: Springer.

Rickel, J., & Johnson, L.W. (1997). Integrating paedagogical agents in a virtual environment for training. *To appear in the journal Presence.*

Schmidt, A. (2002). Ubiquitous computing Computing in context. In *Computing department.* Lancaster University, UK: Lancaster.

Slotta, J. D., & Cheng, B. (2001). *Integrating Palm technology into WISE inquiry curriculum: Two school district partnerships.*

Specht, M. (1998). Adaptive Methoden in computerbasierten Lehr/Lernsystemen. Psychology, ed. G.R. Series. Vol. 1. Trier: University of Trier. 150.

Specht, M., et al. (2002). Adaptive learning environment for teaching and learning in WINDS. *Proceedings of the 2nd International conference on Adaptive Hypermedia and Adaptive Web-based Systems.* Malaga.

Stein, D. (1998). *Situated learning in adult education.* Educational Resources Information Center.

Tatar, D., et al. (2002). *Handhelds go to school: Lessons learned.*

Tennyson, R. D., & Christensen, D. L. (1988). MAIS: An intelligent learning system. In D. H. Jonassen, (Ed.), *Instructional designs for microcomputer courseware.* Hillsdale, NJ: Erlbaum.

Wenger, E., & Lave, J. (1991). *Situated learning: Legitimate peripheral participation.* Cambridge; New York: Cambridge University Press.

ENDNOTE

[1] The project RAFT (Remote Access to Field Trips) is funded by the European Commission under #IST-2001-34273. Information can be found on www.raft-project.net

This work was previously published in Advances in Web-Based Education: Personalized Lerning Environments, edited by G.D. Magoulas and S.Y. Chen, pp. 331-352, copyright 2006 by Information Science Publishing (an imprint of IGI Global).

Chapter 7.18
Elecronic Portfolios and Education:
A Different Way to Assess Academic Success

Stephenie M. Hewett

The Citadel, The Miltary College of South Carolina, USA

ABSTRACT

The use of electronic portfolios for students as an assessment tool is explored in this chapter. Portfolios have expanded from use in the arts and humanities to the field of education. Teachers, administrators, and students understand the benefits of portfolio assessment. The age of technology has improved the use of portfolio assessment by allowing the portfolio information to be transmitted and shared worldwide. No longer are portfolios limited to the single assessment of one person. Based on the current literature on electronic portfolios, the simplicity of creating electronic portfolios, the efficiency of collecting and organizing massive amounts of work, the ease of worldwide transmission of portfolio material, and the promotion of candidate-centered (student-, teacher-, professor-centered) assessment through the use of e-portfolios, the author hopes to promote the electronic portfolio as a beneficial way for the student, teacher, and professor to highlight their achievements for assessment.

INTRODUCTION

Portfolios have been used in a variety of careers including art, architecture, photography, and modeling. The portfolios are used to display a person's skills and talents. Portfolios have opened the doors to many opportunities for the person who has a professionally organized display of their finest works. Portfolios are strong representations of the identity of the person. Portfolios can help describe the person and his/her talents. Looking through a person's portfolio provides insights into the person's thinking and personality. The architect's portfolio allows the prospective builder to look at what the architect has designed and determine if the designs match the building ideas wanted by the builder. The artist can demonstrate types of art that he/she has produced to get commissions from buyers. Models get jobs with their best pictures and poses placed in a portfolio. Advertisers search for certain looks to sell their products to a target audience. The model's portfolio projects the different looks of the model so that advertisers can match their products with the appropriate model. Portfolios display the best products of the person creating the portfolio.

Education has been behind the times in the use of the portfolios. For many years, teachers have had only one way for students to show their knowledge. The one way typically used by teachers to find out what a student knows is through a test. Tests can be standardized or informal but only provide one way to show knowledge. A student must be able to read the questions and be able to write the answers to show what they know on a test. This type of assessment is a linguistic approach according to Gardner (1994), who has written many articles and books on the theory of multiple intelligences. Gardner believes that a person is born with not one, but several intelligences. The intelligences include intrapersonal, interpersonal, musical, linguistic, logical/mathematical, spatial, and artistic. Gardner states that a person has a dominant intelligence that is the best way to demonstrate a person's knowledge. For example, a student may be studying about crustaceans in biology. A typical test may not be the best way for a person to show what all they know about crustaceans. The person may be able to draw and label the crustaceans to visually show what they know or show his/her knowledge through some form of music.

The learner-centered philosophy of education recognizes the need to provide choices for students to show their knowledge. They may not be able to linguistically present their knowledge. Typical tests require the learner to show what they have learned through the linguistic intelligence.

Recently, educators have begun using portfolios to allow students to show off their best works as well as show what knowledge they have gained. Educational portfolios give students choices in the way to present their knowledge. An educational portfolio is "a purposeful collection of student work that exhibits the student's efforts, progress and achievements in one or more areas" (Paulson, Paulson, & Meyer, 1991). A portfolio gives a broader picture of what a student has achieved than typical assessments. Herman and Michael (1999) argued that portfolios shift the balance from teacher-centered learning to student-centered learning. The student takes the responsibility of selecting what products best display their learning. The students also decide how to professionally present their materials.

The K-12 setting is ideal for the introduction of portfolios for assessment. In the early years, students love to show off their work to anyone who will listen! As a first-grade teacher, I endlessly sought ways to keep and display students' artwork and important writings and math assignments to show parents and students the progress that they made throughout the year. The introduction of portfolios at this level enables that collection and display of work, as well as provides a forum for students to present their portfolios to parents and other students. As the student grows and matures, he/she wants to show off his/her work, but does not know how to do it without looking childish. For elementary and middle school children, the portfolio offers the opportunity to display their work in a professional manner. High school students are able to add their creativity and computer knowledge as they develop their portfolios and even display those portfolios electronically. The opportunities for portfolios do not end in high school. Many colleges are requiring students to have writing portfolios to demonstrate their writing proficiencies.

Students are not the only people who can use portfolios to exhibit their achievements. Teachers and professors can create portfolios to highlight their educational careers. Teaching portfolios have been used over the years for teacher candidates to show their skills in planning and assessment. The portfolios are typically used to show growth in planning and professional knowledge. Professors use portfolios to collect and show their professional growth in teaching, scholarly activities, and service to the field. The quantity of materials for professors is massive and can be displayed in numerous file boxes, which are difficult to handle and maneuver to different locations.

The problem with the typical portfolio for K-12 students, teachers, and professors is the cumbersome nature of portfolios. The student portfolios can contain videos, artwork, essays, dioramas, and many other creative presentations and products. The teacher and professor portfolios can contain journals, conference programs, books, lesson plans, syllabi, and evidence of service activities. Keeping all of these materials together in a neat and organized fashion is a challenge within itself. The rise of technology has solved many of the problems associated with the collection and presentation of products. The electronic portfolio, sometimes called an e-portfolio, simplifies the collection and storage of the products. Written assignments, artwork, videos, music, and even tests can be saved to a disk and linked to a Web page to provide a simple way to store and organize the products.

Based on the current literature on electronic portfolios, the simplicity of creating electronic portfolios, the efficiency of collecting and organizing massive amounts of work, the ease of worldwide transmission of portfolio material, and the promotion of candidate-centered (student-, teacher-, professor-centered) assessment through the use of e-portfolios, the electronic portfolio offers a beneficial way for the student, teacher, and professor to highlight their achievements for assessment.

BEST PRACTICES IN ASSESSMENT

Positive effects on students' learning have occurred through the use of portfolio assessment (Santos, 1997; Sweet, 1993; Tierney, Carter, & Desai, 1991; Wolf & Siu-Runyun, 1996). As an assessment, a portfolio:

...matches assessment to teaching, has clear goals, gives a profile of learner abilities, is a tool for assessing a variety of skills, develops awareness of own learning, caters to individuals in the heterogeneous class, develops social skills, develops independent and active learning, can improve motivation for learning and thus achievement, is an efficient tool for demonstrating learning, and for student-teacher provides opportunity dialogue. (http://www.etni.org.il/ministry/portfolio/default. html)

Educators utilize portfolios to get the most effect from assessments by:

...encouraging self-directed learning, enlarging the view of what is learned, fostering learning about learning, demonstrating progress toward identified outcomes, creating an intersection for instruction and assessment, providing a way for students to value themselves as learners, and offering opportunities for peer-supported growth. (http://www.pgcps.pg.k12. md.us/~elc/portfolio. html)

The research points to performance assessment as the most commonly used assessments.

Performance assessment is a dynamic process calling for students to be active participants, who are learning even while they are being assessed. No longer is assessment perceived as a single event...The purpose of assessment is to find out what each student is able to do, with knowledge, in context. (Wiggins, 1997, p. 20)

*Performance is an umbrella term that embraces both **alternative assessment** and **authentic assessment**. The term alternative assessment was coined to distinguish it from what it was not: traditional paper-and-pencil testing. There are even now distinctions within performance assessment, a distinction which refers to the fact that some assessments are meaningful in an academic context whereas others have meaning and value in the context of the real world, hence they are called 'authentic'.*

*Performance assessment is **a continuum of assessment formats** which allows teachers to observe student behavior ranging from simple responses to demonstrations to work collected over time. Performance assessments have two parts: a clearly defined task and a list of explicit criteria for assessing student performance or product.* (Rudner & Boston, n.d.)

Astin and others (2005) discuss the characteristics of best practices in assessments which include the following concepts regarding assessment. Assessment:

- Begins with education values
- Is most effective when the assessment reflects an understanding of learning as multidimensional, integrated, and revealed in performance over time
- Must have clear, explicitly stated purposes
- Looks at outcomes and experiences that lead to those outcomes
- Is ongoing and not sporadic
- Is enhanced when representatives from across the educational community are involved
- Illuminates questions that people really care about
- Promotes change by being a part of a larger set of conditions

As many states have turned to standardized tests to assess educational progress, the educational community recognizes the pitfalls of having a single assessment to evaluate educational progress. Suskie (2000) disputed the thought that one simple assessment could fairly assess educational progress. In the May 2000 issue of the *American Association for Higher Education Bulletin,* Suskie wrote an article for the Fair Assessment Practice Column entitled "Giving Students Equitable Opportunities to Demonstrate Learning." She stated:

An assessment score should not dictate decisions to us; we should make them based on our professional judgment as educators, after taking into consideration information from a broad variety of assessments.

The characteristics of best practice assessments match the characteristics of portfolios. The research literature on portfolios follows and demonstrates the connection of a best practice assessment and electronic portfolios.

RESEARCH LITERATURE ON ELECTRONIC PORTFOLIOS

The research shows that people generally see assessment as "something that is done to them" (Sweet, 1993). People have little knowledge in what is actually involved in the evaluation process. The authentic and performance-based measures of the portfolio clarify the assessment process. Lankes (1995) identified six different types of portfolios:

1. Developmental portfolios to document improvements and growth over an extended period of time.
2. Planning portfolios to identify weaknesses and develop an action plan to address those weaknesses.

3. Proficiency portfolios to demonstrate competencies and performances in a variety of areas.
4. Showcase portfolios to document a person's best work.
5. Skills portfolios to demonstrate proficiency of skills required to accomplish a variety of specific tasks.
6. Admissions/employment portfolios to show a person's capability to perform at an expected level in the specific setting.

Regardless of the type of portfolio assessment, it is a multi-faceted process with the following qualities:

- It is continuous and ongoing, providing both formative (i.e., ongoing) and summative (i.e., culminating) opportunities for monitoring progress toward achieving essential outcomes.
- It is multidimensional, i.e., reflecting a wide variety of artifacts and processes reflecting various aspects of the learning process(es).
- It provides for collaborative reflection, including ways for people to reflect about their own thinking processes and metacognitive introspection as they monitor their own comprehension, reflect upon their approaches to problem solving and decision making, and observe their emerging understanding of subjects and skills. (George, 1995)

The major characteristics of effective portfolio assessments are that they:

1. Reflect clearly stated outcomes
2. Focus on performance-based experiences and the acquisition of key knowledge, skills, and attitudes
3. Contain samples of work over an extended period of time
4. Contain works that represent a variety of different assessment tools
5. Contain a variety of work samples and evaluations based on different sets of audiences. (George, 1995)

Kemp and Toperoff (1998) prepared a set of guidelines for portfolio assessment. The guidelines include a list of reasons that a person should use a portfolio to organize and display their works. Portfolios should be used to:

1. Match assessment to the purpose for the collection of products
2. Clarify goals
3. Give a profile of the individual, including depth of quality of work available because of lack of time restraints, breadth of a wide range of products, and growth over time
4. Assess a variety of different skills
5. Develop awareness of one's own learning and growth
6. Cater to different learning styles and allow expression of different strengths
7. Develop active independent learning
8. Improve motivation and thus achievement
9. Demonstrate different types of learning and growth
10. Provide opportunities for dialog with the evaluator

Electronic portfolios are a collection of works made available on the Internet. Electronic portfolios are technology based. Electronic portfolios:

1. Are used to foster active learning
2. Are used to motivate
3. Are instruments of feedback
4. Are instruments of discussion
5. Demonstrate benchmark performances
6. Are accessible
7. Can store multiple media, are easy to upgrade, and allow cross-referencing. (Creating and using..., 2003)

Bull, Montgomery, Overton, and Kimball (2000) state that electronic portfolios promote self-evaluation and maximize the use of a variety of independent learning strategies. In addition, electronic portfolios serve as an excellent activity to enhance problem-solving skills (Barrett, 1994). The person takes responsibility for the compilation and organization of their work, therefore having a degree of control over the learning process (Campbell, Cignetti, Melenyzer, Nettles, & Wyman, 1997). The creation of the portfolio requires that the person become an assessor of his/her own products. The self-evaluation involved in selecting the most important and best representation of the person's work is a major instructional benefit. The final selection and display of the person's accomplishment demonstrates the process of learning that the person underwent to reach the present point. The key benefit of this type of assessment is that the portfolio enables the process of learning to be assessed as well as the products. The multiple sources of evaluation, in combination with the self-evaluation required in portfolio development, aids in the recognition of the person's strengths and weaknesses (Barrett, 2000).

The purposes, qualities, and benefits of electronic portfolios have been well documented in the research literature. Electronic portfolios have been portrayed in the literature as a best practice for authentic performance assessments. As a best practice, it is important that it is easy to create, or it will become an underutilized assessment tool.

CREATING ELECTRONIC PORTFOLIOS

Before beginning an electronic portfolio, one must understand and have working knowledge of a portfolio. In order to create an effective portfolio, one must first identify the goal of the portfolio. If the goal is to assess growth, then specific goals and criteria should be set so that everyone is clear as to what they are trying to attain. The goals will guide the selection of the inclusion of work in the portfolio. The required portfolio contents should also be identified. For a student, the portfolio may have examples of writings, auditory presentations, drawings, and reflections. Teachers may include sample lesson plans, pictures of classes, videos of teaching, examples of assessments, and reflections of teaching experiences. Professors would include evidence of teaching excellence, scholarly activities, and service to the field. Everyone must understand the purpose of collecting the work so that their selection process would meet the assessment goal.

Barrett (2002) identified five steps inherent in the development of effective electronic portfolios:

1. **Selection:** The development of criteria for choosing items to include in the portfolio based on established learning objectives.
2. **Collection:** The gathering of items based on the portfolio's purpose, audience, and future use.
3. **Reflection:** Statements about the significance of each item and of the collection as a whole.
4. **Direction:** A review of the reflections that looks ahead and sets future goals.
5. **Connection:** The creation of hypertext links and publication, providing the opportunity for feedback. The power of a digital portfolio is that it allows different access to different artifacts. The user can modify the contents of the digital portfolio to meet specific goals. As a student progresses from a working portfolio to a display or assessment portfolio, he or she can emphasize different portions of the content by creating pertinent hyperlinks. For example, a student can link a piece of work to a statement describing a particular curriculum standard and to an explanation of why the piece of work meets that standard. That reflection on the work turns the item

into evidence that the standard has been met. (Barrett, 2002)

Once the goal of the portfolio is established along with the standards stating the criteria for success, the organization and planning of the portfolio must occur. The major task in the planning and organization of portfolios is deciding on what types of authentic products need to be included and how the works will be organized. Mandatory performance measures including specific evidence of knowledge, skills, and activities should be identified. A timeframe of works should also be developed. The collection of the samples over a period of time shapes the depth of the quality of the sample collection. The ability to collect work over a period of time makes portfolio assessment a needed alternative in a world where people recognize that knowledge can be shown in a variety of ways.

The next step is the actual collection of the work samples. The selection process is of as much benefit to the portfolio creator as the preparation of the works themselves. The collection of the works is dependent on the type of portfolio being established. The collection and selection of the products for the portfolio offer the creator a chance to look back and reflect on his/her own learning. As the person selects the work samples to be included in the portfolio reflecting on individual growth, he/she completes a self-assessment that identifies strengths and weaknesses. The weaknesses then become improvement goals. As the collection continues, the person strives to strengthen the weak areas and attain the improvement goals. This process deepens the person's understanding of his/her own growth and learning process. The reflections of the person's achievements and process to attain those achievements lead to the major self-evaluations, creating further growth.

The reflections become a crucial part of the work collection in the portfolio. The reflections may include learning logs, reflective journals, experience logs, and descriptions of the thinking processes employed in the development of the work samples. As the reflections continue while the portfolio is developed, the person's overall assessment of his/her progress is developed. People gradually recognize where they began and the progress that they have made. In any portfolio development, the recognition of growth and the overall self-assessment are major benefits of portfolio assessment.

As the development of a portfolio continues, problems also can emerge. The enormous amount of paperwork that is collected as work samples within the portfolio creates organization and storage issues. Large numbers of portfolios cannot be easily handled and maintained. Paperwork can easily be misplaced or filed incorrectly. Multimedia presentations, videos, and auditory tapes require specialized equipment to view and hear. With the increase in the use of technology, it became evident that a portfolio that could be electronically developed and maintained would solve many of the storage and maintenance issues.

Electronic portfolios have the capability of storing a large amount of works. Pictures, artwork, writing samples, videos, and auditory samples can be easily linked to the Web page. Work can be scanned that normally would not be computer based. In creating an e-portfolio, one must have knowledge of technology and the appropriate terms. The use of computers has necessitated a whole new language. Literacy of technology is essential to insure that the intended message is communicated when speaking to the expert or the novice. Key technology terms that need to be defined before the discussion of e-portfolios continues are:

- **Web Page:** A starting point for the electronic portfolio; a page designed to lead the viewer through the different products.
- **Hot Spot:** A blue underlined word or phrase that when clicked on links/opens the product the person is presenting.
- **Links:** The products that are part of the

portfolio and must be opened by clicking on hot spots.

The electronic portfolio utilizes a Web page to link to the sample products. Most Internet service providers have free space that can be used for a personal Web page. There are many programs available to assist in Web page design and development. Netscape Communicator offers a free and easy program to develop a Web page. Creation of a Web page does not require a high level of computer skill and/or literacy; however, the level of technology skill and knowledge will increase as the person creates the e-portfolio. A trial-and-error creation process will result in the most professional e-portfolio design. The electronic portfolios allow the work samples to be posted on the Internet, where it can be easily accessed. The Web page is the start of the electronic portfolio. Hot spots are created with hyperlinks so that a hot spot is clicked on and it links the viewer with the work sample. Artwork and pictures can be scanned. Video and audio clips can be linked to the page. Electronic portfolios can be more comprehensive with the inclusion of multimedia presentations. The e-portfolio can be as basic as a Web page with links to Microsoft Word documents, to a more sophisticated page with multimedia presentations.

COLLECTION AND ORGANIZATION OF WORK SAMPLES

When collecting work samples, the portfolio should include examples with:

1. Evidence of reflection and productive thinking processes
2. Growth and development in relation to the purpose
3. Understanding and application of key processes

4. Completeness, correctness, and appropriateness of work samples and reflections
5. A variety of formats to demonstrate performance and growth. (http://www. pgcps. pg.k12.md.us/~elc/portfolio5. html)

DeFina (1992, pp. 13-16) lists the following assumptions about portfolio assessment:

- "Portfolios are systematic, purposeful, and meaningful collections of students' works in one or more subject areas.
- Students of any age or grade level can learn not only to select pieces to be placed into their portfolios but can also learn to establish criteria for their selections.
- Portfolio collections may include input by teachers, parents, peers, and school administrators.
- In all cases, portfolios should reflect the actual day-to-day learning activities of students.
- Portfolios should be ongoing so that they show the students' efforts, progress, and achievements over a period of time.
- Portfolios may contain several compartments, or subfolders.
- Selected works in portfolios may be in a variety of media and may be multidimensional."

The actual collection of the work samples is not thought provoking. All work samples can be saved on a flash drive or other storage device to be accessed at any time. To link to the Web page, all documents should be saved on the same storage device (CD, flash drive, zip disk, or DVD). Saving documents on different devices is inefficient and limits the ease of linking to the documents from the Web page. Although collecting the work samples does not require much thought, the organization of the Web page and e-portfolio is extremely thought provoking. The selection of the products to be included in the portfolio is one of the best learning tools that an assessment can offer. While selecting the products, students not only review what they have done over a period of time, but can see the progress and growth that has been made. The review of the material is actually a wonderful study guide to help students recall what was studied and to review major concepts. The selection of material is the self-assessment component that is missing in most assessment devices. The self-assessment creates opportunities for the person to reflect on what he/she learned and the mistakes that have been made. People typically learn from their mistakes, making the process of creating an electronic portfolio an important learning process.

A frequently missed learning tool in creating an electronic portfolio is the increase in technology skills that go along with development of the Web page and electronic portfolio components. Presenting the material in the most professional manner is more difficult with an electronic portfolio than with a regular portfolio. An electronic portfolio offers so many different options. Creative fonts and word art, along with clip art and animations, can be added to the electronic portfolio to enhance its appearance. Instead of boxes of work samples, the electronic portfolio is a blank screen with the opportunity for creative integration of technology. The electronic portfolio is eye catching and provides an additional insight into the technology skills of the creator. For the typical young person, the technology skills are a given. According to *USA Today* (2003), 83% of college students regularly use information technology. With basic computer knowledge, the person can easily create and transmit his/her professional portfolio across the world.

TRANSMISSION OF THE PORTFOLIO

Electronic portfolios can be easily transmitted worldwide on the Internet. Once the Web page has been developed, the Internet service provider may offer a free Web page to upload the portfolio to the Internet. The Web page, all pictures, scanned items, Microsoft Word documents, video clips, and audio clips must be uploaded. For ease of uploading, save all items in one folder marked "portfolio." The opportunity to publish the portfolio on the Internet has its advantages. The Internet address of the Web page can be e-mailed and distributed to interested people worldwide. The assessment process can be enhanced with electronic portfolios by inviting educators from around the world to look at the electronic portfolio and assess it. The diversity of assessments gives a fair indication of the overall quality of the work.

If a person does not decide to make the electronic portfolio available to all on the Internet, he/she can save the portfolio on a CD or disk, make copies, and share with other interested people. When exploring the possibilities of information sharing through technology, it is noteworthy that electronic portfolios:

- Increase opportunities for peer review
- Provide flexibility in the overall assessment process
- Serve as excellent introductions of the professionals who create the portfolios
- Encourage feedback from people outside of the education profession
- Eliminate barriers to parent participation in the schools
- Ensure fair assessments based on a variety of assessors

The purpose of the electronic portfolio dictates whether it will be saved on a disk or transmitted worldwide. For student assessments, saving on a disk to provide a type of scrapbook of learning does not require transmission of the portfolio on the Internet. In order for students to have peer review and share information with other students around the world would require transmission through the Internet. Teacher electronic portfolios can be used for awards and employment opportunities. The transmission of the documents through the Internet is beneficial for both the teacher and the future employer and award committees. An e-mail with a link to the electronic portfolio is a quick and easy way to share the professional documents with the awards committee members and future employers. The electronic portfolio also offers a unique way for professors to present their documentation for employment and promotion. The ease of transmission without the massive paperwork makes the electronic portfolio the best way to present professional documents.

Using electronic portfolios can change the way that people think about assessments. Instead of the typical dread that a person feels when undergoing an evaluation, the electronic portfolio offers an opportunity to "show off" his/her works. He/she has ownership in the collection and selection of works to be included, as well as the option of transmitting the portfolio to whomever he/she chooses.

CANDIDATE-CENTERED ASSESSMENTS

The Citadel School of Education has adopted a learner-centered philosophy of teaching. As written by Reilly (2000):

Learner-centered education is defined by Mc-Combs and Whisler (1997, p. 9) as: the perspective that couples a focus on individual learners (their heredity, experiences, perspectives, backgrounds, talents, interests, capacities, and needs) with a focus on learning (the best available knowledge about learning and how it occurs and about teaching practices that are most effective in promoting the highest levels of motivation, learning, and achievement for all learners). This dual focus, then, informs and drives educational decision-making. Learner-centered education in this perspective embodies the learner and learning in

the programs, policies and teaching that support effective learning for all students.

Administrators are responsible for developing, maintaining, and enhancing a school environment that enhances effective learning. They are also responsible for assuring teachers are knowledgeable about their students and how learning best occurs. Teachers are responsible for having classrooms that promote effective learning for all, as well as being familiar with the instructional techniques that promote effective learning for all. School counselors are concerned with improving both the conditions for learning (parent education, classroom environment, teacher attitude), as well as assisting each learner develop his/her fullest potential. The following five premises support these assertions:

1.   Learners have distinctive perspectives or frames of reference, contributed to by their history, the environment, their interests and goals, their beliefs, their ways of thinking, and the like. These must be attended to and respected if learners are to become more actively involved in the learning process and to ultimately become independent thinkers.

2.   Learners have unique differences, including emotional states of mind, learning rates, and learning styles, stages of development, abilities, talents, feelings of efficacy, and other needs. These must be taken into account if all learners are to learn more effectively and efficiently.

3.   Learning is a process that occurs best when what is being learned is relevant and meaningful to the learner, and when the learner is actively engaged in creating his or her own knowledge and understanding by connecting what is being learned with prior knowledge and experience.

4.   Learning occurs best in an environment that contains positive interpersonal relationships and interactions, and in which the learner

feels appreciated, acknowledged, respected, and validated.

5. Learning is seen as a fundamentally natural process; learners are viewed as naturally curious and are basically interested in learning about and mastering their world. (Reilly, 2002)

With those premises in place, the electronic portfolio becomes the assessment tool of choice. Learners are encouraged to become actively involved in the learning process as they collect, select, design, and create their electronic portfolios. Their different perspectives of what they have learned are incorporated in the e-portfolio, as these include works that address a history of their learning, their beliefs, and their way of thinking. The electronic portfolio gives learners a chance to embrace their unique differences, including their abilities and talents. As learners create electronic portfolios, they maximize their learning potential. The products of the portfolio are relevant and meaningful to the learner, and actively engage him/her in creating and understanding the knowledge gained and experiences that lead to the knowledge gain. The portfolio serves as a tool in which the learner begins to feel appreciated, acknowledged, respected, and validated through sharing his/her portfolio with others. Learners' curiosities are challenged as they learn the technology skills required to produce an electronic portfolio. Learners begin to see the assessment process as meaningful and fun!

E-PORTFOLIO RESOURCES

The Internet offers a wide range of resources in the development, implementation, and assessment of electronic portfolios. Interesting Internet sites that are useful in the creation of electronic portfolios include:

- http://www.essdack.org/port/—Tammy Worcester (2005) from Soderstrom Elementary School in Lindsberg, Kansas, presents information on why to use electronic portfolios, what to include in electronic portfolios, an assessment of electronic portfolios, and how to create electronic portfolios. She also includes examples and resources for electronic portfolio preparation.

- http://electronicportfolios.com/—Dr. Helen Barrett is an internationally known expert on electronic portfolio development of all ages and has sponsored *electronicportfolios. org.* The site includes listservs, blogs, wikis, resources, and special topics on electronic portfolios.

- http://www.educationworld.com/a_tech/ tech/tech111.shtml—Education World at Ashland University has a site entitled "Electronic Portfolios in the K-12 Classroom" (2005), which informs about what electronic portfolios are and how they can help the teacher and benefit the student including guidelines for developing personal portfolio.

- http://www.uvm.edu/~jmorris/portresources.html—"Electronic Portfolio Resources" provides examples of online portfolios, selecting electronic portfolio programs, and links to electronic software resources and electronic articles.

- http://eduscapes.com/tap/topic82.htm —"Electronic Portfolios: Students, Teachers and Life Long Learners" provides information on electronic portfolios, including what a digital or electronic portfolio is, how to develop an electronic portfolio, and how to integrate text, photos, diagrams, audio, video, and other multimedia into the electronic portfolio, along with examples of electronic portfolios.

Many people use rubrics, a criteria-rating scale, giving the teachers a tool that allows them to track student performance to assess the quality of the work. Rubrics describe the expectations of the portfolio and the rating system. The Pearson/Prentice-Hall Publishing Company includes a rubric for electronic portfolios (Rubrics for electronic portfolios, 2005). Not only does the rubric evaluate the content choice, organization, and personal reflections, but the rubric also assesses the creative use of technology. The rubric encourages the use of varied technology. There are other excellent examples of rubrics on the World Wide Web, but this template provides the flexibility for evaluators to develop the scoring to their needs and purposes.

The rubric template shows evaluation of the following elements:

1. The creative use of technology
2. The content choice
3. The organization and mechanics
4. The personal reflections

Rubrics guide the students as they create their portfolios. When creating e-portfolio directions and rubrics, explore the World Wide Web to determine if there are existing directions and rubrics which will be useful in the planning for portfolio creation.

CONCLUSION

As computer technology progresses, so will the uses and benefits of electronic portfolios. As students and teachers continue to refine electronic portfolios, the assessment benefits will also continue to emerge. The research literature supports the use and benefits of electronic portfolios; and based on the ease of creating electronic portfolios, the efficiency of collecting and organizing massive amounts of work, the possibilities of worldwide transmission of portfolio material, and the promotion of candidate-centered (student-, teacher-, professor-centered) assessment through the use of e-portfolios, the electronic portfolio is becoming the most effective and efficient way to showcase and assess K-12 students', college students', teachers', and professors' academic growth and progress.

REFERENCES

Astin, A. W., Banta, T. W., Cross, K. P., El-Khawas, E., Ewell, P. T., Hutchings, P., et al. (2005). *9 principles of good practice for assessing learning*. [online]. American Association for Higher Education. Available at http://www.aahe.org/assessment/principl.htm

Barrett, H. C. (1994). Technology-supported portfolio assessment. *The Computing Teacher, 21*(6), 9-12.

Barrett, H. (2000). Electronic teaching portfolios: Multimedia skills + portfolio development = powerful professional development. *Proceedings of the Society for Information Technology and Teacher Education* (pp. 1111-1115). Charlottesville, VA: Association for the Advancement of Computing in Education.

Barrett, H.C. (2002). Retrieved April 1, 2005, from http://electronicportfolios.com

Bull, K. S., Montgomery, D., Overton, R., & Kimball, S. (2000). *Developing teaching portfolio quality university instruction online: A teaching effectiveness training program*. Retrieved February 9, 2002, from http://home.okstate.edu/homepsages.nsaf/toc

Campbell, D. M., Cignetti, P. B., Melenyzer, B. J., Nettles, D. H., & Wyman, R. M. (1997). *How to develop a professional portfolio: A manual for teachers*. Boston: Allyn and Bacon.

Chriest, A., & Maher, J. (Eds.). (2005a). *Why use a portfolio?* Retrieved March 31, 2005, from http://www.pgcps.pg.k12.md.us/~elc/portfolio.html

Chriest, A., & Maher, J. (Eds.). (2005). *How can portfolios be evaluated?* Retrieved March 31, 2005, From http://www.pgcps.pg.k12.md.us/~elc/portfolio5.html

Corbett-Perez, S., & Dorman, S. M. (1999). Technology briefs. *Journal of School Health, 6*(69), 247.

Creating and using portfolios on the alphabet superhighways. (2003). Retrieved October 15, 2003, from http://www.ash.udel.edu/ash/teacher/portfolio.html

DeFina, A. (1992). *Portfolio assessment: Getting started.* New York: Scholastic Professional Books.

Electronic portfolios in the K-12 classroom. (2005) Retrieved April 1, 2005, from http://www.educationworld.com/atech/tech/tech111.shtml

Gardner, H., & Boix-Mansilla, V. (1994). Teaching for understanding: Within and across disciplines. *Educational Leadership, 51,* 14-18.

George, P. S. (1995). *What is portfolio assessment really and how can I use it in my classroom?* Gainesville, FL: Teacher Education Resources.

Herman, L. P., & Morrell, M. (1999). Educational progressions: Electronic portfolios in a virtual classroom. *T.H.E. Journal, 26*(11), 86.

Kemp, J., & Toperoff, D. (1998). *Guidelines for portfolio assessment in teaching English.* Retrieved March 31, 2003, from http://www.etni.org.il/ministry/portfolio/default.html.

Lamb, A. (2002). *Electronic portfolios: Students, teachers, and life-long learners.* Retrieved March 30, 2005, from http://www.eduscapes.com/tap/topic82.htm

Lankes, A. M. (1995). *Electronic portfolios: A new idea in assessment.* ERIC Digest, ED390377.

Marklein, M. B. (2003). Students aren't using info technology responsibly. *USA Today,* (November 9).

McCombs, B. L., & Whisler, J. S. (1997). *The learner centered classroom and school: Strategies for enhancing student motivation and achievement.* San Francisco: Jossey-Bass.

Morris, J. (2005). *Portfolio resources.* Retrieved March 28, 2005, from http://www.uvm.edu/~jmorris/portresources.html

Paulson, L. F., Paulson, P. R., & Meyer, C. (1991). What makes a portfolio a portfolio? *Educational Leadership, 49*(5), 60-63.

Reilly, D. H. (2000). The learner-centered high school: Prescription for adolescents' success. *Education, 121*(2), 219.

Rubrics for electronic portfolios. (2005). Retrieved April 12, 2005, from http://www.phschool.com/professional_development/assessment/rub_electronic_portfolio.html

Rudner, L., & Boston, C. (n.d.). *A long overview on alternative assessment.* Halifax, Nova Scotia: Norwood Publishing Company.

Santos, M. G. (1997). Portfolio assessment and the role of learner reflection. *Forum, 35*(2), 10-16.

Suskie, L. (2000). Fair assessment practices: Giving students equitable opportunities to demonstrate learning. *American Association of Higher Education Bulletin,* (May).

Sweet, D. (1993). Student portfolios: Classroom uses. *Office of Educational Research: Consumer Guide, 8.*

Tierney, R. J., Carter, M. A., & Desai, L. E. (1991). *Portfolio assessment in the reading-writing classroom.* Halifax, Nova Scotia: Norwood Publishing Company.

Wiggins, G. (1996). Practicing what we preach in designing authentic assessments. *Educational Leadership, 55*(1), 18-25.

Wolf, K., & Siu-Runyan, Y. (1996). Portfolio purposes and possibilities. *Journal of Adolescent & Adolescent Literacy, 40*(1), 30-36.

Worcester, T. (2005). *Electronic portfolios*. Retrieved April 13, 2005, from http://www.essdack.org/Port

This work was previously published in the Handbook of Research on Literacy in Technology at the K-12 Level, edited by L. Tan and R. Subramaniam, pp. 437-451, copyright 2006 by Idea Group Reference (an imprint of IGI Global).

Chapter 7.19
Behaviour Analysis for Web–Mediated Active Learning

Claus Pahl
Dublin City University, Ireland

ABSTRACT

Software-mediated learning requires adjustments in the teaching and learning process. In particular active learning facilitated through interactive learning software differs from traditional instructor-oriented, classroom-based teaching. We present behaviour analysis techniques for Web-mediated learning. Motivation, acceptance of the learning approach and technology, learning organisation and actual tool usage are aspects of behaviour that require different analysis techniques to be used. A combination of survey methods and Web usage mining techniques can provide accurate and comprehensive analysis results. These techniques allow us to evaluate active learning approaches implemented in form of Web tutorials.

INTRODUCTION

Since its inception, the Web has been widely and successfully used as a platform for teaching and learning. Technology-mediated teaching and learning, however, requires adjustments in the teaching and learning process for both instructors and students. The complexity of the symbiotic relationship between learning and instructional design on one hand, and technology and tool mediation on the other needs to be understood. Rose (1999) observes that the words "interactive" and "interactivity" proliferate in texts on educational computing, despite their apparent lack of denotative value. However, it seems to be understood widely that interactive instruction is learner-controlled, an opportunity for students to engage in active, hands-on exploration (Northrup, 2001). Interactive tools can enable active learning in a constructivist style if they create a representation of reality in which learning is relevant. According to Ravenscroft, Tait, and Hughes (1998), students integrate the use of computer-based learning resources into their study habits in an incremental fashion. Instructors need to carefully analyse the learning behaviour with new educational technologies in order to support new student learning processes through an incremental instructional design approach.

The Web-mediated interactive tutorial system that we are going to analyse is part of an undergraduate course in computing. This tutorial allows students to construct programming knowledge and acquire programming skills in the database language SQL through engaging and interactive exercises based on meaningful problems, (Pahl, Barrett, & Kenny, 2004). At the core of the tutorial is an interactive submission feature that allows students to execute programs and that gives feedback on those submissions. Engagement in the learning process is, according to Northrup (2001), a key objective in interactive instruction. In self-controlled environments, students actively construct meaning to determine how to proceed in the learning activity.

The goal of this investigation is the behaviour analysis of tool-mediated active learning. We demonstrate novel analysis techniques for the evaluation of learning behaviour in tool-mediated, interactive environments that combines classical survey-based techniques with Web usage mining technology. The motivation to analyse and evaluate the students' learning behaviour and learning processes is to gain an understanding of student learning in interactive learner-controlled environments. This is a prerequisite for the successful and effective implementation of instructional design for active learning and for the empirical evaluation of the implementations.

THE INTERACTIVE TUTORIAL

An interactive tutorial is a software tool that facilitates active learning in a guided learning process. Learners learn to solve problems in a dialogue with the tool. The interactive tutorial we analysed is part of an undergraduate courseware system for a database course, part of a computing degree, with online lectures, tutorials, and labs that is implemented using Web technologies and accessed through Web browsers and plug-ins and that supports active and autonomous learning (see Figure 1). This environment is the target of our experimental and empirical study of learning behaviour.

Solutions to programming problems, which are presented as a guided tour through the material, can be submitted through a Web interface to a remote database server, which executes the input and replies with data from a database, or error messages (right-hand side of Figure 1). Scaffolding in form of feedback, self-assessment functionality, and links to background material is available (bottom and left-hand side of Figure 1). The tutorial prepares the student for coursework, such as lab tests and projects, and final exams. The courseware system aims at providing the student with a realistic learning context by integrating features and problems into a learning environment that are similar to tools and tasks that would be faced by a database engineer in a real development scenario.

METHODS

Our research goal is the analysis and evaluation of student learning behaviour in tool-mediated active learning environments. We define tool-mediated active learning as a software-supported approach to learning where a learner creates knowledge, in other words, a meaningful representation of some part of reality, within the software environment. Behaviour in learner-controlled environments is determined by the learners' motivation, their acceptance of pedagogical approach and technical environment, their learning organisation, and their activities in the environment (i.e. tool usage). Consequently, the instruments for the behaviour

Figure 1. The interactive tutorial: With lecture material in the background

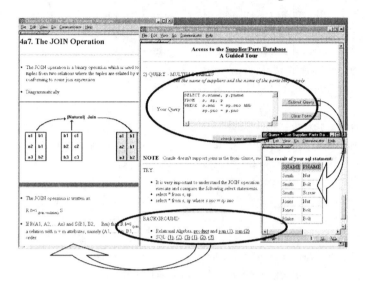

analysis include two instrument types: survey methods to address motivation and acceptance and Web usage mining techniques (Pahl, 2004) to capture organisation and usage in a Web environment. This combination provides a more complete and accurate picture than surveys and student observation alone (Kinshuk, Patel, & Russell, 2000) or student tracking features available in various learning technology systems. We propose a novel, mainly quantitative method that combines classical survey methods with computational techniques for data mining and analysis. The survey design is addressed at the end of this and in the next section. We will focus here on mining techniques and the overall design of the analysis framework.

Web mining is a technology that discovers and extracts knowledge from structured Web data—usually access logs that record requests from a Web browser. A Web log record—the basis for statistical analysis and data mining—contains a user and/or machine ID, the time of the request, and the requested resource. To derive learning activities from navigation and interaction in Web-

based systems is not always straightforward. Web logs record accesses to resources, which can be associated with activities.

Web mining has the advantage of being non-intrusive and useable at all times. Web log data can give a precise and objective account of student activities in Web-based systems. In addition to classical Web-usage statistics such as number of hits in a period of time, Web usage mining allows a more targeted analysis of Web log data for educational purposes, (Zaiane & Luo, 2001). Our analysis is based on two mining techniques developed for the educational context, (Pahl, 2004):

- Session classification. A Web log is a chronologically ordered list of Web requests. The first task is to identify learning sessions, which are defined as periods of uninterrupted usage of an individual user. The classification tries to identify purposes or activities of a session, for example interactive learning, attending a virtual lecture, or downloading resources.

- Behavioural pattern discovery. The Web log, if sorted by user, provides a sequential list of learner requests representing the learner activities in the system. The first task is to find sequential patterns (i.e. recurring sequences of requests). The second step is the identification of behavioural patterns such as repetition or the parallel use of features in these sequences and sequential patterns.

We have complemented a standard Web usage mining product with a research prototype for the education-specific features to implement Web usage mining. We recorded only information logged by standard Web servers – a fact that students were aware of and that should not have impacted their behaviour.

We have adopted complementary instruments—student surveys and observation-based Web usage mining—that allow us to address the different aspects of behaviour. Adding Web usage mining gives us an improved interpretative strength over classical methods for our behaviour analysis, as we have demonstrated in (Pahl, 2004). A benefit of the combination is the validation of behaviour-specific survey results and addition of preciseness through usage mining.

The behaviour of students in computer-based teaching and learning environments is influenced by the motivation to use the system and the acceptance of the approach. These two behaviour aspects relate concrete learning behaviour with the objectives and state-of-mind that have led to that behaviour. A learning activity is an engagement towards a learning objective. We distinguish two aspects of the student's concrete behaviour, which defines the learning activity. Firstly, the learning organisation addresses the study habits and captures how students organise their studies over a longer period of time. This includes how they plan to learn and work on coursework, and how they prepare for exams. Secondly, the usage of the system captures single learning activities and embraces how the student works with and behaves

in the system in a single study session. Overall, we have identified four aspects of behaviour:

- **Motivation:** The reason to do something —causes the learner to act in some planned and organised way, giving the activities a purpose.
- **Acceptance:** To follow the learning approach and use the system willingly—is crucial for the introduction of new educational technology.
- **Organisation:** The way the learning activities are planned and put into logical order —reflects the study habits and is guided by the purpose.
- **Usage:** The way the tool is actually used —reflects the actual learning activities.

Both the pedagogical approach and the Web-based system need to support the objectives that form the students' motivation in order to be accepted. The organisation is determined by the motivation – the objectives determine how activities are organised and executed. The usage follows the organisational plan to achieve the objectives. Motivation and acceptance are necessary to interpret organisation and usage. Except the motivation, we have analysed all aspects using both instruments for each category.

An iterative process of instructional Web design, based on a formative evaluation, facilitates feedback and exploration of new technologies. Formative evaluations are vital for identifying key design issues and for improving our understanding of pedagogical issues, (Kinshuk, Patel, & Russell, 2000). Our analysis techniques combine mostly quantitative, but also qualitative aspects, leading to a more comprehensive picture of learning behaviour. This will result in a better understanding of how to develop new, effective types of learning environments.

We have surveyed and analysed the behaviour of two classes in two successive years with 79 and 112 students in each year, respectively. Of these

37 and 69, respectively, took part in the survey. Both classes were comparable with respect to age, sex, and also performance in coursework and exams. Both classes have used the tutorial system in the same version. Since no significant differences between the two classes emerged, their respective evaluation results shall not be distinguished. Tables 1, 2, and 3 below detail the questions and results of the questionnaire used in the student survey. All questions provided an open-ended part in order to record qualitative answers. Table 1 contains questions for which a number of alternative answers were given; the students were asked to rank these answers. Response categories for questions in Table 1 were decided based on standard categories from the literature and a pilot survey with open-ended questions. The questions in Tables 2 and 3 were presented in a Likert scale style—a 5-point scale (strongly disagree, disagree, undecided, agree, strongly agree – see Table 2) for a number of statements that students were asked to classify and a 5-point scale (traditional, rather traditional, undecided, rather virtual, virtual – see Table 3) where students were asked to compare delivery approaches. Table 4 presents session classification results. All reported results are statistically significant at the 95 percent confidence level. Web mining was deployed constantly throughout the term.

RESULTS

Motivation

According to question Q1—see Table 1—here is a clear preference for practical course elements, i.e. coursework preparation, as the main motivation. A Web log analysis shows that the tutorial is mainly used during term to support coursework (about 2/3) and to a lesser extent (about 1/3) for the final exam preparation, which confirms the survey result. Question Q2 gives more insight into the motivation of the student's study organisation. From all alternatives offered in the survey, being "always available" and "self-paced learning" are the key advantages that students see in the system, in other words, these were ranked first (mean=1.63) and second (mean=2.22), respectively. Less than 4 percent of the students actually gave non-listed answers in the open-ended part.

Acceptance

Question S1 in Table 2 shows an overall acceptance of tool-mediated active learning as the pedagogical approach, which becomes even stronger when referring to the course with a strong practical element (database programming) in particular (Question S2). A positive attitude towards the approach usu-

Table 1. Student survey: Motivation

(Q1)	For what purpose have you been using the interactive tutorial ?						
	Answer	Count	Rank	(1st	2nd)		Mean
	preparation for coursework	98	1	71%	21%		1.13
	preparation for the exam	94	2	29%	67%		1.63
(Q2)	What were the main values of the interactive tutorial for you ?						
	Answer	Count	Rank	(1st	2nd	3rd	4th)	Mean
	always available	92	1	62%	22%	10%	6%	1.63
	self-paced learning	86	2	30%	41%	9%	20%	2.22
	easy Web access	82	3	12%	28%	38%	22%	2.70
	integrated with lectures	84	4	8%	14%	39%	39%	3.11

Table 2. Student survey: Acceptance (S1,S2) and usage (S3)

STATEMENT	Count	strongly agree	agree	undecided	disagree	strongly disagree
(S1) virtual courses are in general suitable for undergraduate courses	102	44%	25%	10%	13%	8%
(S2) having a virtual course as part of your degree programme is a good idea	102	53%	25%	9%	9%	4%
(S3) mixing the use of lectures and interactive tutorial is a good idea	101	49%	33%	10%	7%	1%

Table 3. Student survey: Acceptance (comparison)

QUESTION	Count	traditional	rathertraditional	undecided	rather virtual	virtual
(Q6) in combination with virtual lectures, would you prefer tradit. or virtual tutorials?	100	30%	13%	13%	18%	25%
(Q7) Do you think your exam results would be better in a traditional or a virtual course?	102	21%	17%	29%	15%	18%

ally goes hand in hand with frequent and regular usage—a correlation between these two variables confirms this. Comparing traditional and virtual tutorials—see Table 3—gives a more differentiated view on acceptance. Answers to question Q3 show no favourite—which demonstrates that students accept virtual tutorials as equally suitable and effective as traditional tutorials. We have asked the students about their preference of delivery mode with respect to performance in exams (Q4). The opinion is split. Nonetheless, this result shows the acceptance of virtual tutorials—virtual tutorials are at least as good as traditional ones—as a means to support one of the students' major objectives—good coursework and exam performance. The answers to questions Q3 and Q4 demonstrate that, given an adequate online tool, virtual tutorials are feasible and they are accepted by learners as equally suitable and effective as traditional tutorials.

Another indicator for the acceptance of self-directed active learning is reflected by frequent and regular usage, in particular when alternatives are available. According to Web statistics students have worked in 19 sessions on average. About eight percent of students have used the system twice or less. While Web mining shows that the tutorial system has not been used frequently and regularly over the whole term, it has, however, been used intensively in certain periods to fulfil a particular purpose.

Organisation

The organisation is reflected by the frequency and regularity of the usage. The access times in the

Web log show high usage during later afternoon and early evening hours. The distribution over the week shows high usage in the middle of the week with 66 percent on Tuesdays and Wednesdays (weekly discussion meetings were held on these days), but also significant usage at weekends with close to 20 percent. The study organisation overall—the self-paced learning aspect expressed through Q2 and Web mining results concerning frequency and regularity—shows a just-in-time learning approach with high usage immediately before coursework deadlines during the semester and before examinations.

Education-specific Web usage mining (Zaiane & Luo, 2001; Pahl, 2004) can give us a clearer picture about the organisation than Web-usage statistics. Session classification allows us to determine the purpose of sessions, for instance attending virtual lectures or practising in virtual tutorials, and to compare the session purposes of different periods. Table 4 shows percentages for the whole course system for two periods—the lecturing period during term and the exam preparation period following the lectures. A session can serve multiple purposes; thus, cumulative percentages can exceed 100 percent. The purpose "Organisational" includes downloading course notes and other material and look-up of course schedule and coursework results. "Exploring" refers to an explorative behaviour, typical for the first sessions of a user. Surprising is the high number of organisational visits—even though these visits tend to be much shorter than lecture or tutorial sessions. Time series of session classifications allow us monitor the changing focus over time. We found dramatic changes in the classifications over time. Interactive services are heavily used during term, but less so for the exam preparation. Another change is the transition from a novice user with substantial explorative behaviour to an experienced one with more targeted behaviour that can be observed over time.

Usage

Besides the long-term study organisation, analysing learning activities within a study session is crucial to understand how students learn. An abstract picture of the purpose(s) of each session is provided by session classifications, but we also need to look at how students interact with the system, whether they repeat units, or whether they combine interactive elements with lectures. A pattern analysis can answer these questions.

Tut_1 ; [$LookUp_1$ | $ExecQuery_1$]* ;
Tut_2 ; [$LookUp_2$ | $ExecQuery_2$]* ; ... ;
Tut_{12} ; [$LookUp_{12}$ | $ExecQuery_{12}$]* ;

This is a behavioural pattern describing a usage pattern for the interactive tutorial. The tutorial consists of 12 units—Tut_1, ... , Tut_{12} –to be worked on sequentially, which is indicated through a semicolon (;). Within each unit students can iteratively (*) either look up background lecture resources (*LookUp*) or can execute an SQL query (*ExecQuery*)—options are separated by a vertical bar (|). A behavioural pattern analysis can extract such a pattern, and can, given a pattern, determine the overall support of the pattern by the class. For instance, 84 percent of all student sessions actually follow this pattern of mixing active tutorial learning and lecture look up—most of those sessions who do not are either very short or use a different order.

Question S3 (Table 2) shows that students recognise the potential of virtual courses to use tutorials and lectures at the same time, overcoming time and space constraints that apply to a traditional delivery. Question S3 gives an indication of the appreciation of this new style of learning mixing tutorials and lectures. This is confirmed by pattern analysis results, which show the proportion of students using the tutorial on its own or in combination with lecture resources. Lectures are usually used on their own. Interactive tutorials,

Table 4. Web usage mining: Session classification, in percent

Activity Period	Lectures	Tutorials	Organisational	Exploring	Unclassified
Semester	56%	39%	56%	12%	17%
Exam Preparation	43%	12%	41%	1%	4%

however, are used to a large extent—Web mining confirms 77 percent—in combination with lecture resources. The students have looked up background material to solve specific problems interactively. While nearly all students avail of this feature in their first sessions, we observed a decrease of lecture usage during tutorials over time, indicating the increased knowledge, skills, and self-reliance of students.

DISCUSSION

Looking at the four behaviour aspects we found common expectations about motivation, essentially coursework and exam preparation, confirmed by the survey. Tool-mediated active learning is accepted as an equally effective means for learning — a positive result. The organisation analysis shows expected, but more undesirable results. In order to overcome the just-in-time approach to study organisation, the instructional design and course organisation would need to encourage a more regular use. The usage analysis provides again encouraging results. It shows that active and multi-modal learning are accepted and adopted. Overall, we have seen changing patterns, indicating both changes in the short-term focus of learning, but also in the long-term strategies that are used.

This behaviour evaluation shows the feasibility of tool-mediated active learning. Active learning can be supported by Web-based technologies. Students accept and use the system as a proper alternative to traditional forms of learning in particular for practical course elements. The analysis confirms steps we have already taken to facilitate tool-mediated active learning in a Web environment. The analysis shows that scaffolding techniques providing feedback, self-assessment functionality, and links to other services and background material including lectures, are as important as the learning activity itself within the interactive tutorial. One reason is that Web-based tutorials are used in a self-paced and self-reliant way. For most of the students, our course was the first substantial exposure to tool-mediated learning. However, a substantial group had used similar systems before and we found evidence of a correlation between previous experience and high usage. This is an indicator for the change and refinement of learning strategies among students. Initially, students have used scaffolding and navigation support substantially, but over time a change towards self-reliance with respect to the content and also the usage of the tool was observed.

Important instructional design issues emerging from the analysis concern multi-modal learning, feedback, and the learning organisation. Using the tutorial integrated with lecture resources can result in more problem-oriented multi-modal learning, which organises different aspects such as theory and practice around a realistic problem. A wide awareness of this potential exists; however, a better implementation of this learning behaviour seems possible. Student responses, erratic behaviour, and some examples of repeated behavioural patterns show that individual feedback and scaffolding features are prerequisites to enable efficient and satisfactory usage of the system. Just-in-time learning resulting is usage peaks is seen as undesirable with respect to knowledge retention. Weekly discussions that were introduced have helped to flatten these peaks and to encourage a more regular attendance.

CONCLUSION

The detection, analysis, and understanding of student learning processes in new forms of learning environments—such as tool-mediated active learning—is a prerequisite for the development of effective instructional design. The central problem is the adaptation of support for learning processes using new technologies. We have demonstrated the benefits of a learning behaviour analysis method based on combined survey and Web mining techniques that addresses the behaviour aspects motivation, acceptance, organisation, and usage. Although Web mining has limitations related to caching and other technical Web features, usage mining techniques enable constant, non-intrusive monitoring of student behaviour and the detection of behaviour changes, which supports the adoption of Web-based instructional techniques in an incremental process, (Coates & Humphreys, 2001). This technique can be deployed for interactive Web-based tutorials where access logs reflecting learning activities are automatically generated.

Ravenscroft, Tait, and Hughes (1998) stress the importance of the appropriate level of student interaction with learning or knowledge media, referring to their experience with text-based and editable material for online lectures. Often, a distinction is made between educational content aimed at developing conceptual knowledge, problem solving, and analytical skills on one hand, and skills development, recognition, and memorisation on the other, (Weston & Barker, 2001). The students' motivation in our case is the acquisition of skills, rather than knowledge, and good performance in practical coursework and examinations. Consequently, the form of interaction with course material supporting active learning of skills is different from knowledge-based learning. Other aspects such as a realistic setting for the interaction, for example in relation to project work or exam settings, become more important. Active learning provides this necessary type of interaction. Our conclusion—that the right level of interaction has to be designed and supported—is the same. The support of active learning through interactive tools needs to facilitate skills development in a realistic setting.

We found, based on our behaviour evaluations, that active self-controlled learning is an effective approach for practical, skills- rather than knowledge-oriented subjects. Interactive tools in a realistic setting that engage the students allow students to interact with the course content through its tool-based delivery medium in an adequate way. According to the students' opinion, tool-mediated active learning effectively replaces the instructor as a means for coursework and exam preparation to a large degree, in particular when direct contact with the instructor is not possible. The integration with other forms of learning provides an additional, beneficial context. However, using this technology, students are required to change their learning strategies. A constant analysis of student behaviour can help the instructor to support learning strategies and to accommodate changes in these strategies.

While we have analysed a computing course, Web technologies enable a wider range of subjects to be supported through active and dynamic Web pages; user-controlled animations; or submission, execution, and feedback systems. These subjects need to aim at skills based on activities that involve some form of text processing or manipulation that is supported by Web technologies.

REFERENCES

Coates, D., & Humphreys, B. (2001). Evaluation of computer-assisted instruction in principles of economics. *Educational Technology & Society, 4*(2).

Kinshuk, P., & Russell, D. (2000). A multi-institutional evaluation of intelligent tutoring tools in numeric disciplines. *Educational Technology & Science, 3*(4).

Northrup, P. (2001). A framework for designing interactivity into Web-based instruction. *Educational Technology, 41*(2), 31-39.

Pahl, C. (2004). Data mining technology for the evaluation of learning content interaction. *International Journal on E-Learning IJEL, 3*(4), 48-59. AACE.

Pahl, C., Barrett, R., & Kenny, C. (2004). Supporting active database learning and training through interactive multimedia. In *Proceedings International Conference on Innovation and Technology in Computer Science Education* ITiCSE'04 (pp. 58-62). ACM Press.

Ravenscroft, A., Tait, K. & Hughes, I. (1998). Beyond the media: Knowledge level interaction and guided integration for CBL Systems. *Computers and Education, 30*(1,2), 49-56.

Rose, E. (1999). Deconstructing interactivity in educational computing. *Educational Technology, 39*(1), 43-49.

Weston, T., & Barker, L. (2001). Designing, implementing, and evaluating Web-based learning modules for university students. *Educational Technology, 41*(4), 15-22.

Zaiane, O., & Luo, J. (2001). Towards evaluating learners' behaviour in a Web-based distance learning environment. In *Proceedings IEEE International Conference on Advanced Learning Technologies ICALT'01* (pp. 357-360). IEEE Computer Society.

This work was previously published in the International Journal of Web-Based Learning and Teaching Technologies, Vol. 1, Issue 3, edited by L. Esnault, pp. 45-55, copyright 2006 by Idea Group Publishing (an imprint of IGI Global).

Chapter 7.20
Gender Differences In Education and Training in the IT Workforce

Pascale Carayon
University of Wisconsin-Madison, USA

Peter Hoonakker
University of Wisconsin-Madison, USA

Jen Schoepke
University of Wisconsin-Madison, USA

INTRODUCTION

Historically, women have had lower levels of educational attainment (Freeman, 2004; NCES, 1999), which in turn could negatively affect their opportunities in the labor market. However, in the past decade, this has changed dramatically. In general, more women have completed college, and more women have received bachelor's and master's degrees than men. Only in the highest level of education (PhD), men hold more degrees than women (NCES, 1999, 2002). In a recent study by the National Center for Education Statistics (NCES), Freeman (2004) presents an overview of the latest developments with regard to gender differences in educational attainment. Historically, females have tended to account for the majority of bachelor's degrees in fields that often lead to lower paying occupations, such as education and health professions, while males have typically predominated in higher paying fields, such as computer science and engineering. While some of these disparities persist, many changes have occurred since the 1970s. Certain fields in which men received the majority of degrees in the 1970s, such as social sciences, history, psychology, biological sciences/life sciences, and business management and administrative services, attained relative gender parity or were disproportionately female by 2001. While other fields, such as computer and information sciences, physical sciences and science technologies, and engineering, continue to have a larger proportion of males, the percentages of females majoring in those fields is increasing (Freeman, 2004). Between 1970 and 2001, the percentages of master's, doctoral and first-professional degrees earned by females increased substantially in many fields. However, advanced degrees conferred still tend to follow traditional patterns, with women accounting for the majority of master's and doctor's degree recipients in education and health, and men accounting for the majority of recipients in computer and information sciences and engineering. Higher levels of educational attainment are associated with certain labor market outcomes, such as higher labor force participation rates, higher rates of employment, and higher earnings (Freeman, 2004). A study by Igbaria, Parasuraman and Greenhaus (1997) looked at gender differences in the *information technology (IT) work force* with regard to education and experience, career history and attainments and career orientation. The results showed significant differences in educational attainment. A larger percentage of female IT employees in the study ended their formal education after attaining a bachelor's degree.

BACKGROUND

IT companies face many dilemmas when hiring new employees. Among these dilemmas are which recruits are qualified hires and how to ensure that their current IT employees amass critical skills needed for the company to stay competitive (Schwarzkopf, Mejias, Jasperson, Saunder, & Gruenwald, 2004). To answer this dilemma, many companies look for employees that have a formal post-secondary educational background in a technical field. The U.S. Department of Commerce (2003) has found that a four-year technical degree helps IT professionals get their foot in the door and get promoted. This is further emphasized in the projection from the Bureau of Labor Statistics (BLS) that between 2000 and

2010 almost 75% of the job openings within the IT professional level will require a minimum of a bachelor's degree (U.S. Department of Commerce, 2003). The IT training landscape is diverse, complete with traditional four-year university degree programs in computer science to newer training models, such as IT vendor-related training and certification programs and online learning (U.S. Department of Commerce, 2003). This vast array of education and training options provides a multitude of knowledge and skill sets that IT employees may enjoy. However, with such a diversity of IT training and educational pathways, it quickly becomes apparent that there is no "one-size fits all" approach to training for companies to take (U.S. Department of Commerce, 2003). Thus, IT companies are faced with the following challenge: how to keep pace with technological changes that have short life cycles (U.S. Department of Commerce, 2003). This challenge to keep pace with the ever-changing technology is felt by IT employees within companies as well. Stress has been recognized as a key factor affecting IT productivity and turnover, and can increase the costs that companies endure (Sethi, King, & Quick, 2004). It is essential for IT employers to examine the factors that contribute to their employees' stress. There are two stressors associated with training: one involves the need for appropriate training and the other involves the development of skills to complete tasks (Sethi, King, & Quick, 2004).

MAIN THRUST OF THE CHAPTER

The data analyzed in this chapter comes from the database of the project on "*Paths to Retention and Turnover in the IT Workforce: Understanding the Relationships between Gender, Minority Status, Job and Organizational Factors*" (http//cqpi2. engr.wisc.edu/itwf/index.html). Participants within the selected companies were identified based on two characteristics: (1) their job was within the information technology workforce, and (2) they have worked in their current job for two months or more. The data collection tool used is a 139-item Web-based questionnaire (Carayon, Schoepke, Hoonakker, Haims, & Brunette, 2005, in press).

Sample

The sample consists of five companies of varying size. Company 1 is a medium-sized Midwestern IT firm with 190 professionals. Company 2 is an eastern health care provider network with 895 IT professionals. Company 3 is a small western IT firm with 11 IT professionals. Companies 4 and 5 are both small eastern IT firms with 9 and 11 IT professionals respectively. Since the large company is not an IT company per se, the sample exemplifies the literature that 92% of IT professionals work in non-IT companies (ITAA, 2002). The total sample size is 624 with 46% women and 54% men (27 respondents did not report their gender). The average age is 40 years, with women being significantly older than men (t-test; $p<0.05$). Marital status (e.g., living with someone vs. not) is significantly different between women and men ($\chi 2$ test; $p<0.05$): 65% of the women live with a spouse/partner, compared with 73% of the men. Parental status is not significantly different between women and men.

Measures

The items on training received were adapted from Lehto & Sutela (1999). To measure satisfaction with training opportunities, we developed our own scale, based on in-depth interviews we conducted in the pilot study (Carayon, Brunette, Schwarz, Hoonakker, & Haims, 2003). Respondents are asked whether they strongly disagree, disagree, agree, or strongly agree with statements such as: "I receive ongoing training which enables me to do my job better" (see Figure 1). Cronbach's alpha for the scale of satisfaction with training

opportunities is 0.93. To measure quality of working life (QWL) we used existing scales that were found to be valid and reliable in previous research. Our own analysis has confirmed the validity and reliability of the scales used (Carayon et al, 2005, in press). All scales we used in the questionnaire were converted to scores from 0 (lowest) to 100 (highest). The following QWL factors were measured: job satisfaction (Quinn et al., 1971; $\alpha = 0.78$); organizational involvement (Cook & Wall, 1980; $\alpha = 0.72$) and stress or burnout (Leiter & Schaufeli, 1996; Maslach & Jackson, 1986, $\alpha = 0.91$). Turnover intention was measured using a single item: "How likely is it that you will actively look for a new job next year?" on a seven point scale ranging from 1: not at all likely-2-3: somewhat likely-4-5: quite likely –6- to 7:extremely likely (mean = 2.87, sd = 1.83).

Procedure

We used a Web-based survey to collect the data. For a detailed description of the Web-based survey system, see Barrios (2003). As described above, five IT companies participated in the study. The participating company sent out an e-mail to notify their employees of the survey and two days later, we sent employees an e-mail, describing the study, asking for their participation and providing them with a link to the Web-based survey. An informed consent procedure is integrated in the Web-based survey management system. The total response rate was 56%.

Results

Education and Training

First, we looked at gender differences in the pathways to an IT career. Tables 1, 2, and 3 show the results of this analysis.

Results show that the majority of participants in our study are highly educated: more than three-quarters of the respondents have at least a

Table 1. Highest level of education by gender

	Men	Women	Total
High school or GED	5%	4%	5%
Some college	21%	13%	17%
Bachelor's degree	39%	37%	38%
Some graduate or professional study	12%	15%	14%
Graduate or professional degree	23%	30%	26%
Total	100%	100%	100%

Table 2. Formal IT education by gender

	Men	Women	Total
No IT/computer-related formal schooling	18%	27%	22%
Some high school computer-related courses	2%	2%%	2%
High school computer-related degree/certificate	2%	0.4%	1.3%
Some technical college computer-related courses	10%	11%	10%
Technical college computer-related degree	8%	3%	6%
Some university computer-related courses	28%	33%	30%
University computer-related bachelor's degree	24%	13%	19%
University computer-related Graduate (MS, PhD) degree	8%	11%	9%
Total	100%	100%	100%

Table 3. Informal IT-education by gender

	Men	Women	Total
Training for certification	19%	6%	13%
Company provided training courses/seminars (besides certification training)	25%	40%	32%
Other training sources/seminars (not provided by a company; not including Web-based training)	7%	4%	6%
Self taught (without formal courses or training, but including Web-based training)	33%	17%	26%
Former work experience	12%	23%	17%
Other informal training	3%	9%	6%
Total	100%	100%	100%

bachelor's degree. The differences between men and women in level of education are statistically significant. Female employees in our sample are higher educated than men.

Results show some remarkable differences between men and women with regard to formal IT schooling. Women (27%) in our sample have more often no IT-related formal schooling than men (18%). Men are more likely to have a university computer-related bachelor's degree as compared to women.

Women (40%) received company-provided training more often than men (25%) and had their training through former work experience (23%) more often than men (12%). Men were more likely to receive training for certification and to be self-taught as compared to women.

Decision to Seek an IT Career

In the questionnaire, we asked respondents a question developed by Leventhal et al.: "At what point in your life did you first decide to seek an IT career?" Table 4 shows the results.

Results show that most respondents (38%) choose to seek an IT career during non-IT-employment: this result is similar for both men and women. Few of the respondents (3%) already knew prior to high school that they would seek an IT career. During their formal education, most respondents (21%) choose to seek an IT career during their undergraduate degree program. Analysis of the category "other" shows that many respondents choose to seek an IT-career after being laid off at a previous job.

Training

In the questionnaire, several questions were asked about training received in the past 12 months.

Different scenarios were possible. Table 5 shows the results.

Results show that most respondents receive company sponsored training on company time (CSCT). There are significant gender differences in CSCT: women (74%) receive significantly more often CSCT than men (65%). The average number of days of CSCT for women and men is 5.7 days per year. Men were more likely to get self-sponsored training than women, both on company time (SSCT) as well as in their own time (SSOT). The differences between men and women in number of days of training received over the past 12 months are not statistically significant.

Table 4. Decision to seek an IT career

	Men	Women	Total
Prior to high school	3%	2%	3%
During high school	13%	6%	10%
During undergraduate degree program	24%	17%	21%
Prior to entering graduate degree program	4%	8%	6%
After earning highest academic degree	11%	16%	14%
During non-IT employment	37%	38%	38%
Other	7%	12%	10%
Total	100%	100%	100%

Table 5. Training scenarios over the past 12 months

	Percentage of employees that received training			If yes, how many days (in full days)?		
	Men	Women	All	Men	Women	All
Company sponsored/On company time (definition: company pays for training and pays you while you are attending training)	65%**	74%**	70%	6	5.5	5.7
Company sponsored/On your own time (definition: company pays for training, but you are not paid by your company for the time you spend training)	9%	8%	9%	14.5	4.6	9.8
Self sponsored/On company time (definition: you pay for the training, but your company pays for your time while you are attending the training)	9%***	3%***	3%	5.6	1.8	4.9
Self sponsored/On your own time (definition: you pay for the training and you are not paid by your company for the time you spend on the training)	21%**	12%**	17%	15.6	8.8	13.6

Satisfaction with Training Opportunities

Figure 1 depicts the results of the analysis on gender differences with regard to satisfaction with training opportunities. Results shows that half to two-thirds of the respondents are satisfied with training opportunities to improve their skills. There are no gender differences in satisfaction with training opportunities.

Impact of Training on QWL and Turnover

Although more than half to two-thirds of the respondents are satisfied with different aspects of training opportunities, one-third to nearly half of the respondents are not satisfied with several aspects of training. We divided the sample into two groups. The first group is not satisfied with the different training opportunities (a score between 0 and 50 on the satisfaction-with-training scale) and a second group that is satisfied with the training opportunities (a score between 50 and 100 on the satisfaction-with-training scale). Figure 2 shows the effect of satisfaction with training opportunities on QWL, (i.e., job satisfaction, organizational involvement, and stress).

Obviously, satisfaction with training opportunities has an impact on QWL. Employees who are satisfied with training opportunities report significantly more job satisfaction and greater organizational involvement and suffer less from job stress than employees who are not satisfied with the training opportunities.

Furthermore, employees who are not satisfied with the training opportunities their organization offers are significantly more likely to look for another job (mean=3.07) than employees who are satisfied with the training opportunities (mean=2.13) (t-test, p<.001).

Figure 1. Satisfaction with training opportunities

Note: Percentage of employees that agree or strongly agree with the statements

Figure 2. Effects of satisfaction with training opportunities on QWL

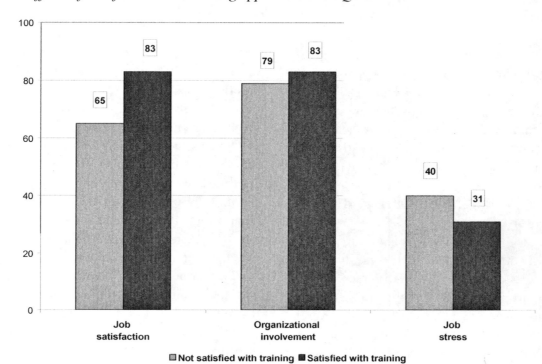

FUTURE TRENDS

Historically, women have had lower levels of educational attainment than men. However, in the past decade, this has changed dramatically. In general, more women have completed college, and more women have received bachelor's and master's degrees than men. Women still lag behind in science and engineering, although the percentage of women in those fields is also increasing. Higher levels of educational attainment are associated with certain labor market outcomes, such as higher labor force participation rates, higher rates of employment, and higher earnings.

CONCLUSION

Results of this study show significant gender differences in the pathways to IT employment. Women have a higher general education as com-

pared to men. Women have more often completed some graduate or professional study and/or have a graduate or professional degree. These results are different from the study by Igbaria et al (1997) who found that a larger percentage of female IT employees ended their formal education after attaining a bachelor's degree. The results are aligned with changes observed in the past decade, as reported by Freeman (2004), who found that more women have completed college, and more women have received bachelors' and masters' degrees than men. However, men completed *specific IT education* more often than women. Significantly more men in our sample have a university computer-related bachelor's degree as compared to women (24% vs. 13%). With regard to informal IT education, women (40%) received more often than men (25%) company-provided training and more often than men had their training through former work experience. Men, more often than women, had training for certification and were self-taught. With regard to on-going-training, results show that most employees (70%) receive company-sponsored training on company time. Female employees receive more company-sponsored training on company time than men. The average number of days of training is the same for men and women. Men have had more self-sponsored training than women, both on company time as well as on their own time. The majority of men and women are satisfied with opportunities for training. Half to two-thirds of the employees are satisfied with various aspects of training opportunities. They are most satisfied with training opportunities that were received *in the past* and that enabled them to do their job better. They are relatively less satisfied with training opportunities offered to them to keep up with latest technologies and training that enables them to do their job better *in the future*. On-going training and education are important in the IT work force. Results of this study highlight the variety of educational and training pathways and opportunities offered and used by IT workers. A few differences between men and women were found regarding some of the educational and training pathways and activities. But, overall, both groups of male and female IT workers report relatively high levels of satisfaction with training opportunities. Employees who are satisfied with training opportunities report higher levels of job satisfaction and organizational commitment, and lower levels of stress than employees who are not satisfied with training opportunities. Those satisfied with training opportunities are less likely to look for another job. Those results underscore the importance of the diversity of educational and training pathways for working in the IT. Once in the IT workforce, men and women report relatively high levels of satisfaction with training opportunities. Companies employing IT workers should pay attention to the variety of training opportunities offered to their IT workers, and ensure that those workers are satisfied with those opportunities. Dissatisfaction with training opportunities is related to a higher likelihood of looking for another job.

NOTE

Funding for this research is provided by the NSF Information Technology Workforce Program (Project #EIA-0120092, PI: P. Carayon).

REFERENCES

Barrios, E. (2003). *Web Survey Mailer System (WSMS 1.1)*.CQPI Technical Report No. 186. Madison, WI: Center for Quality and Productivity Improvement.

Carayon, P., Brunette, M. J., Schwarz, J., Hoonakker, P., & Haims, M. C. (2003). *Paths to retention and turnover in the IT workforce: Understanding the relationships between gender, minority status, job, and organizational factors.* An NSF information technology workforce study: Interim report: Phase 1 pilot study. Madison, WI: University of Wisconsin- Madison.

Carayon, P., Schoepke, J., Hoonakker, P., Haims, M., & Brunette, M. (2005). *Evaluating the causes and consequences of turnover intention among IT users: The development of a questionnaire survey.* Accepted for publication by Behaviour and Information Technology.

Cook, J., & Wall, T. D. (1980). New work attitudes measures of trust, organizational commitment, and personal need non-fulfillment. *Journal of Organizational Psychology, 53,* 39-52.

Freeman, C. E. (2004). *Trends in Educational Equity of Girls & Women: 2004* (No. NCES 2005-016). Washington, DC: U.S. Government Printing Office: U.S. Department of Education, National Center for Education Statistics.

Igbaria, M., Parasuraman, S., & Greenhaus, J. H. (1997). Status report on women and men in the IT workplace. *Information Systems Management, 14,* 44-53.

ITAA. (2002). *Bouncing back: Jobs, skills, and the continuing demand for IT workers.* Information Technology Association of America.

Lehto, A. M., & Sutela, H. (1999). *Efficient, more efficient, exhausted: Findings of Finnish quality of work life surveys 1977-1997.* Helsinki, Finland: Statistics Finland.

Leiter, M. P., & Schaufeli, W. B. (1996). Consistency of the burnout construct across occupations. *Anxiety, Stress, and Coping, 9,* 229-253.

Leventhal. (2005). Personal communication.

Maslach, C., & Jackson, S. (1986). *Maslach Burnout Inventory Manual.* Palo Alto, CA: Consulting Psychologists Press.

National Center for Education Statistics: Digest of Education Statistics (1999). Retrieved January 18, 2005, from http://nces.ed.gov/edstats/

National Center for Education Statistics (NCES): Digest of Education Statistics (2002). Retrieved January 18, 2005, from http://nces.ed.gov/edstats/

Quinn, R., Seashore, S., Kahn, R., Mangion, T., Cambell, D., Staines, G., & McCullough, M. (1971). *Survey of working conditions: Final report on univariate and bivariate tables.* Washington, DC: U.S. Government Printing Office, Document No. 2916-0001.

Schwarzkopf, A. B., Mejias, R. J., Jasperson, J. S., Saunder, C. S., & Gruenwald, H. (2004). Effective practices for IT skills staffing. *Communications of the ACM, 47*(1), 83-88.

Sethi, V., King, R. C., & Quick, J. C. (2004). What causes stress in information systems professionals? *Communications of the ACM, 47*(3), 99-102.

U.S. Department of Commerce. (2003). *Education and training for the information technology workforce.* Retrieved October 12, 2004, from http://www.technology.gov/reports/ITWorkForce/ITWF2003.pdf

KEY TERMS

Education: Refers to the formal courses of study that one takes to obtain a formal degree.

Information Technology Work Force (ITWF): Information technology work force (ITWF) refers to the people who work in the information technology domain.

Job Satisfaction: Describes how content an individual is with their job. There are a variety of factors that can influence a person's level of job satisfaction; these factors include the level of pay and benefits, the perceived fairness of the promotion system within a company, the quality of the working conditions, leadership and social relationships, and the job itself (the variety of tasks involved, the interest and challenge the job generates, and the clarity of the job description/requirements).

Job Stress: Job stress is measured as burnout or the exhaustion of emotional and/or physical strength as a result of prolonged stress and/or frustration.

Organizational Involvement: The extent of an individual's involvement in an organization.

Quality of Working Life (QWL): Represents the quality of the relationship between employees and their total working environment, with human dimensions added to the usual technical and economic considerations.

Training: Training refers to the skills and knowledge taught to individuals within the information technology work force to make them proficient in their area of expertise.

Turnover Intention: The intention of an individual to move in or out of employment with a particular firm or organization.

Web-Based Survey: In the broad sense of the notion of a survey ("looking over or upon with a purpose of reporting the results"), any hypertext markup language (HTML) form that solicits input from respondents can be considered a survey. In our definition, a Web-based survey is a well-defined questionnaire, that has been proven to be reliable and valid in research and that, with the use of HTML, is put on the Web and solicits responses from specifically sampled respondents.

This work was previously published in the Encyclopedia of Gender and Information Technology, edited by E. Trauth, pp. 535-542, copyright 2006 by Idea Group Reference (an imprint of IGI Global).

Chapter 7.21
Preparing African Higher Education Faculty in Technology

Wanjira Kinuthia
Georgia State University, USA

INTRODUCTION

One of the most difficult challenges facing African higher education institutions (HEIs) is the successful resolution of the inherent tension that underlines efficient and effective utilization of existing resources on one hand and intensified demand for more and better education on the other (Okuni, 2001; Sawyerr, 2004). Although the potential of information and communication technologies (ICTs) to enhance participation in African HEIs has been widely recognized, its transformational capacity has barely been reached because of limited infrastructure, technological capacity, funding and sustainability of resources, and human resources and expertise. Poor infrastructure and weak regulatory policies and frameworks have resulted in inadequate access to affordable telephones, broadcasting, computers, and the Internet (Johnson, 2002).

For many countries, the uneven use of ICT presents the equity dilemma, where the gap between the information-rich and information-poor further marginalizes disadvantaged groups, inadvertently widening the digital divide (Dunne & Sayed, 2002). Statistics by African Internet Connectivity (2002), for example, indicates that although all African countries have Internet connectivity, there are only about four million Internet users. A report by the Association of African Universities (2002) indicates that ICT in African HEIs is limited and varied. Consequently, the benefits of ICT are not being fully realized due to factors such as struggling economies and rising enrollment.

Despite policy pronouncements, the status of ICT shows that the continent is at a growing disadvantage with respect to the global information and technological revolution (Association of African Universities, 2000). Meanwhile, educators are expected to be at the forefront, helping to plan and develop national and international systems that facilitate rapid dissemination of information while simultaneously keeping current with the literature in their various academic disciplines.

BACKGROUND

Two features characterize higher education in Africa. First, until the 1960s, HEIs consumed few public resources, because they were not central to the economic needs of the society. As countries achieved independence, higher education expanded as a symbol of autonomy and autarchic development. The second feature was the major response of higher education to social and economic change has been curricular change. As noted by Dunne and Sayed (2002) and Sawyerr (2004), the 1970s witnessed growing cynicism and skepticism that replaced the initial optimism about higher education development. By the 1980s, it was clear that while substantial progress was being made, it was evident that higher education was in need of change. The proposed changes harnessed the potential of ICT and faculty development to improve access, quality, and efficiency of higher education.

The massification of higher education is now associated with increased access for those who have been previously excluded (Dunne & Sayed, 2002). HEIs have invariably been cast in the role of producing skilled human capital, coupled with the responsibility of acting as catalyst in the search for quality and relevance in terms of teaching, research, and service (Seddoh, 2003). Paradoxically, during the transition, higher education has been characterized by increased competition and decreased funding, and slow rates of economic development have contributed to the perception that HEIs are not making significant economic and social contributions. Faced by financial constraints, African governments question investing in higher education, and donor agencies primarily focus on primary education. The following is an overview of the state of ICT in African HEIs in relation to challenges and opportunities. The importance of decision making in the selection and implementation of ICT in higher education is discussed, followed by a discussion of techniques

that are useful in preparing faculty to use ICT. Recommendations and the future of ICT are also presented.

TECHNOLOGY INTEGRATION IN HIGHER EDUCATION

The goal of higher education is to expand educational opportunities, seek pedagogical alternatives, and accommodate new theoretical assumptions that potentially enhance teaching and learning (Minishi-Majanja, 2003). While not everyone agrees with the assumption that technology-enhanced instruction is a viable method of delivery, experimenting with new modes of delivery has been one of the means of accommodating enrollment pressures. When ICT are well-implemented and utilized, they can add new resources to existing course content in the learning environment and introduce unique options for teaching and learning.

Access to the Internet offers users the possibility of interaction that transcends the boundaries of time and space, enhances the range of information available to learners, and expands the opportunities for international communication (Donat, 2001; Minishi-Majanja, 2003). Distance education, open learning, and e-learning have all made considerable use of various resources, such as Web-based and Web-enhanced learning. Many of these include satellite links, computers, telephone conferencing, fax, and interactive video. Donat (2001) and Okuni (2000) present the case of the African Virtual University based in Kenya as an example of a first attempt to use on a large-scale various ICT to meet the growing demand for access to quality higher education throughout the continent. Bhalalusesa (1999) and Minishi-Majanja (2003) also present examples of the Open University of Tanzania and the University of South Africa as HEIs that have also been providing quicker and more effective access to higher education in the continent. The University of Ghana also has increased use of the preexisting External Degree Centers (Sawyerr, 2004).

One goal of technology integration is to reach new levels of productivity, but several barriers can obstruct the process. These barriers include the characteristics of the adopter and the organization, the innovation itself, communication, institutional culture, barriers in organizational structures, access to the innovation, and autonomy to implement the innovation (Rogers, 1995; Surry & Land, 2000). Hence, administrators seeking to institute rapid change must consider three issues: First, a percentage of the faculty may resist change. Second, a variety of strategies are needed to address the needs of individuals with differing rates of innovativeness. Third, the strategies should be ongoing over the life cycle of the innovation (Bennett & Bennett, 2003; Surry & Land, 2000).

Ely (1990, 1999) identified eight conditions that should be present when implementing, planning, and monitoring educational technologies: (1) dissatisfaction exists with the status quo; (2) knowledge and skills exist for the implementers; (3) resources are available and accessible; (4) time to learn and integrate technology is available; (5) rewards or incentives are available; (6) participation is expected and encouraged; (7) commitment exists at all levels; and (8) good leadership is evident. The important questions to ask are, "How many of the conditions currently exist?" and "Which conditions require improvement to help in our situation?"

Several setbacks occur when HEIs invest in technology for political and commercial purposes, or without adequate funding to maintain efficiency of operations. First, technology policies are not coordinated with availability of resources, supporting infrastructures, and training. Second, ICT are introduced without adequate understanding of the organizational culture and context, or the political, physical, economic, social, and technological environment. Third, ICT are introduced

hastily and arbitrarily in a top-down manner. To maximize the benefits of ICT, they should be viewed as a set of tools for solving specific problems, and not as a universal remedy for all educational challenges. In these circumstances, technology plans should establish explicit connections between the proposed physical infrastructure of the ICT and instructional and professional development strategies needed in addition to evaluation processes that monitor progress. Above all, decisions should be made about whether adoption of ICT is a priority of the HEI and what resources they are willing to invest in.

The importance of faculty support and training to the success of technology in instruction has been widely acknowledged in higher education (Bennett & Bennett, 2003; Erasmus, 2002; Wanzare & Ward, 2000). While many studies have primarily focused on program design, evaluation, and logistics, there is still a need to investigate the effectiveness of these activities. However, no study has yet fully documented the expertise among educators in African HEIs despite ICT being demanded as a means of ensuring that the few resources available are used optimally (Minishi-Majanja, 2003).

Many faculty have received little or no explicit training in how to teach, or in the theories and processes of teaching, yet they are responsible for teaching learners who are expected to master a large knowledge base or to perform a broad variety of procedures (Fourie & Alt, 2000; Wanzare & Ward, 2000). Erasmus discussed the importance of performing needs assessment before initiating a faculty development program. The needs assessment should be followed up with frequent reviews to highlight the progress made and areas to be addressed. It should also focus on proficiency in ICT applications, the ability to adapt to new teaching styles and ability to adapt to instructional environments that may be more demanding and time consuming.

While one aspect of developing technological expertise involves mastering the skills required to use various software and programs, an equally time-intensive task includes translating those skills to specific curriculum content (Fourie & Alt, 2000). Instructors who use technology in instruction are often chosen based not on their personality, motivation, and teaching style, but on the subject matter of the course. Research also indicates that faculty are motivated, interested, and have positive attitudes toward the use of instructional consultation on a personal basis but not by coercion (Bennett & Bennett, 2003; Rups, 2003). To increase motivation, Minishi-Majanja (2003) recommends academic rewards and entrepreneurial benefits, especially in respect to intellectual property ownership for their course content.

The competencies necessary for the development and training of effective educators and found that issues include the introduction of new curriculum with little training and development afforded to educators and the promotion of personnel who lack experience and managerial skills. One challenge of faculty development is to break the barrier of unfamiliarity with and phobia of computers. Another challenge is to help faculty grasp instructional techniques that are specific to the technology-enhanced environment, while at the same time respecting and building upon their experience in the traditional classroom.

Faculty development is more effective when participants set their own goals, plan their own learning, use experiential learning, and evaluate outcomes in terms of achievement of their own goals. Thus, it is important to provide participants with opportunities to improve their skills and make the personal rewards of teaching more compelling (Bennett & Bennett, 2003; King & Fawley, 2003; Minishi-Majanja, 2003; Rups, 2003). Faculty development programs not only hold the possibility of helping educators learn to use technology but also provide forums for them to share their questions and solutions and to discover alternatives together (King & Lawler, 2003).

HEIs are faced with the task of incorporating and increasing ICT usage so as to be relevant, visible, and competitive. At a time of crisis and scarce resources, HEIs must be effective, efficient, and enhanced by administration that focuses on faculty development, the infrastructure, and equipment. Thus, shared vision, attitudes, clearly defined and communicated policies, goals, and objectives are important in ICT implementation, development, and utilization (Seddoh, 2003).

FUTURE TRENDS

Rapid changes in ICT are inescapable, and the globalization imperative requires potential role players not only to possess requisite ICT skills but also to have regular access to corresponding facilities. Technology is crossing borders and continents, offering opportunities to pool knowledge resources for sustainable global development. While the digital divide in higher education can be bridged, it requires tolerance, patience, and adequate, well-targeted resources for capacity building, connectivity, and access to broaden participation and awareness (Johnson, 2002). Notwithstanding the potential benefits of ICT, access to higher education, computing, and communication technology resources remains one of the great challenges that makes it increasingly difficult for educators and learners to keep abreast of current developments in their fields.

There are several guiding questions that should be addressed when planning ICT programs. The question posed for the administrator is: How can I assist faculty to integrate new technologies in instruction? The questions posed for faculty intending to adopt new technologies are: Which instructional applications of technology are easy enough to use and helpful to myself and my learners? How much will I have to change the way I teach? There, ICT adoption and faculty development programs should specify objectives and how they relate to professional activities, because programs that are mandatory or inflexible are likely to be unsuccessful. The programs should also be task centered, with an immediacy of application (King & Lawler, 2003).

Many institutions are at various stages of planning and infrastructural development, and case studies are underway to document and analyze the experiences of selected HEIs so as to draw out lessons and best practices, as well as to identify potential pitfalls (Sawyerr, 2004). Continued collaboration is recommended especially for HEIs that have limited financial resources or are in the initial stages of technology adoption. Collaborative initiatives are particularly significant, because networking, which is the backbone of effective harnessing of ICT, often transcends departmental, institutional, and national jurisdictions. Some of the areas of collaboration include finance and budgeting, selecting hardware and software, instructional development, implementation of programs, administration of ICT projects, and faculty development (Minishi-Majanja, 2003).

While administrators cannot force improvement, it is important for them to provide opportunities and environments that free faculty to attend training programs. Providing release time for training, for instance, demonstrates the value that the administration places on providing meaningful incentives, long-term achievement, and opportunities to improve the quality of instruction. To attain success, HEIs must find connections between curriculum, pedagogy, technology, and administration. They should also address how, when, and what types of technology would most benefit the instructional situation.

CONCLUSION

Because little is known about the effectiveness of different types of pedagogical approaches in African higher education, it would be useful to investigate the pedagogical and technological

issues that are unique to higher education, with an aim of building models and programs for improving instruction. Although most public higher education institutions have often been constrained by lack of finance and human resources, if well implemented and maintained, it is likely that ICT will continue to be used to strengthen higher education. Continued support in these areas of research is important in order to investigate technological needs from both an intra- and an interinstitutional perspective.

The proliferation of ICT forces users to rethink how they use technology in their work, whether in instructional development, planning, gathering information, or communicating. Technology certainly facilitates communication, but it is also a cause for reflection and dialogue to encourage and empower educators to capture its greatest potential. In a knowledge age, successful societies will be those that effectively and efficiently integrate ICT with teaching and learning. Technology should be seen as a tool for reform, where a new vision of technical cooperation can be used to create a new development paradigm and provide the potential tools for building sustainable capacity.

REFERENCES

African Internet Connectivity. (2002). *The African Internet: A status report*. Retrieved July 19, 2004, from http://www3.sn.apc.org/africa/afstat.htm

Association of African Universities. (2000, September). *Technical experts meeting on the use and application of information and communication technologies in higher education institutions in Africa*. Retrieved July 19, 2004, from http://www.aau.org/english/documents/aau-ictreport-toc.htm

Bennett, J., & Bennett, L. (2003). A review of factors that influence the diffusion of innovation when structuring a faculty training program. *Internet and Higher Education, 6*(1), 53-63.

Bhalalusesa, E. (1999). The distance mode of learning in higher education: The Tanzanian experience. *Open Learning, 14*(2), 14–23.

Donat, B, N. P. (2001). International initiatives of the virtual university and other forms of distance learning: The case of the African virtual university. *Higher Education in Europe, 26*(4). 577-588.

Dunne, M., & Sayed, Y. (2002). Transformation and equity: Women and higher education in sub-Saharan Africa. *International Studies in Educational Administration, 30*(1), 50-65.

Ely, D. P. (1990). Conditions that facilitate the implementation of educational technology innovations. *Journal of Research on Computing in Education, 23*, 298-305.

Ely, D. P. (1999). *New Perspectives on the Implementation of Educational Technology Innovations.* (ERIC Document Reproduction Services No. ED427775).

Erasmus, A. (2002). Pitfalls, challenges, and triumphs: Issues in an international capacity development project. *Higher Education in Europe, 27*(3), 273-282.

Fourie, M., & Alt, H. (2000). Challenges to sustaining and enhancing quality of teaching and learning in South African universities. *Quality Higher Education, 6*(2), 115-124.

Johnson, P. (2002). New technology tools for human development? Towards policy and practice for knowledge societies in Southern Africa. *Compare, 32*(3), 381-389.

King, K. P., & Lawler, P. A. (2003). Trends and issues in the professional development of teachers of adults. *New Directions for Adult and Continuing Education, 98*, 5-13.

Minishi-Majanja, M. K. (2003). Mapping and auditing information and communication technologies in library and information science education in Africa: A review of the literature. *Education for Information, 21*, 159-179.

Okuni, A. (2000). Higher education through the Internet expectations, reality and challenges of the African virtual university. *D+C Development and Cooperation, 2*, 23-25. Retrieved June 18, 2002, from http://www.dse.de/zeitschr/de200-4.htm

Rogers, E. M. (1995). *Diffusion of innovations* (4th ed.). New York: The Freeman Press.

Rups, P. (1999). Training instructors in new technologies. *T.H.E. Journal, 26*(8), 66-99.

Sawyerr, A. (2004). Challenges facing African universities. Selected Issues. *Association of African Universities.* Retrieved July 19, 2004, from http://www.aau.org/english/documents/asa-challengesfigs.pdf

Seddoh, K. F. (2003). The development of higher education in Africa. *Higher Education in Europe, 28*(1), 33–39.

Surry, D., & Land, S. (2000). Strategies for motivating higher education faculty to use technology. *Innovations in Education and Training, 37*(2), 145–153.

Wanzare, Z., & Ward, K. L. (2000). Rethinking staff development in Kenya: Agenda for the twenty-first century. *The International Journal for Educational Management, 14*(6), 265–275.

KEY TERMS

Digital Divide: The disparity in the access to and effective use of computers, the Internet, and technology. The term is defined in terms of ethnic groups, household income and composition, age, gender, and geographic location.

Distance Education: Print-based and electronic learning resources used to connect learners, resources, and educators, where the learning group is separated by time or geographical distance.

E-Learning: Instruction that covers a wide set of applications and processes such as Web-based learning, computer-based learning, virtual classrooms, and digital collaboration. It includes the delivery of instructional content via Internet, intranet/extranet, audio- and videotape, satellite, and CD-ROM. However, many organizations only consider it as a network-enabled transfer of skills and knowledge.

Educational Technology: Systematic identification, development, organization, or utilization of educational resources and the management of these processes. The term is occasionally used in a more limited sense to describe the use of multimedia technologies or audiovisual aids as tools to enhance the teaching and learning process.

Faculty Development: A purposeful, institutionalized approach to doing that which helps faculty do their work better as individuals within an institution and within the collective enterprise of higher education.

Instructional Development: The development of learner, instructor, and management materials (both print and nonprint) that incorporates specifications for an effective, efficient, and relevant learner environment. Instructional development includes formative and summative evaluation of the instructional product.

Needs Assessment: Problem identification process that looks at the difference between "what is" and "what should be" for a particular situation. It is an analysis that studies the needs of a specific group, such as employees or learners, and presents the results detailing those needs, for example, training needs and resources needs. Needs assessment also identifies the actions required to fulfill these needs, for the purpose of program development and implementation.

Open Education or Open Learning: Any scheme of education or training that seeks systematically removes one or more barriers to learning. It also means learning in your own time at your own pace and at your own base, using higher education settings for academic assistance and as a base for facilities and equipment. Formats include distance learning and online learning.

Web-Assisted Instruction: A course that uses the Internet to provide a significant amount of course content on the Web to learners outside of class time. Materials may be supplemental in nature or provide content in an alternative form that may be viewed at the learner's convenience. Learners interact with other learners outside of class through e-mail, message boards, or chat.

Web-Based Instruction or Online Instruction: Courses or programs mediated by computers where instructional material is delivered via the World Wide Web. Instruction that is solely delivered through the Internet using message boards, electronic drop boxes, or e-mail attachments to the instructor. Interaction occurs via e-mail, message boards, chats, or conferencing programs. WBI often utilizes the multimedia capacity of the technology.

This work was previously published in the Encyclopedia of Developing Regional Communities with Information and Communication Technology, edited by S. Marshall, W. Taylor, and X. Yu, pp. 576-580, copyright 2006 by Idea Group Reference (an imprint of IGI Global).

Chapter 7.22
Disability, Chronic Illness, and Distance Education

Christopher Newell
University of Tasmania, Australia

Margaret Debenham
Consultant, UK

INTRODUCTION

Distance education may be seen as both enabling and disabling in its application to, and relationship with, people with disability and chronic illness. Cutting-edge work suggests that it can provide a suitable route to support the studies of students with disabilities and those with long-term health problems. However, it is important that this should be regarded in terms of providing choice to students rather than requiring those who are identified as having impairment/chronic illness to undertake studies at a distance. Unless well designed and evaluated, as with any technology, DE can also become disabling in its impact (Goggin & Newell, 2003; Newell & Walker, 1992).

Defining Disability and Chronic Illness in the Context of Models of Disability

Within the Western world some 20% of the population has some degree of disability. Yet there are a wide variety of impairments and diverse ways of understanding disability. Taken together, these have significant implications for curriculum design and pedagogy, as well as research and development of educational technology. Considerable differences in life orientation may be found between those who are born with impairment (and for whom such a condition is "normal") and those who acquire them. Those who have visible disability and those whose conditions are hidden can have markedly different experiences, including whether or not their situation is seen as warranting disability support by institutions.

The UK and U.S. disability studies literature highlights a marked shift in recent years from the so-called "medical" model of disability to a "social model" (Albrecht, Seelman, & Bury, 2001). As Fulcher (1989) observes, medical and charitable discourse still dominates everyday understanding of disability. The *medical model* sees disability as a "personal tragedy" located within a deviant individual, to be overcome by providing aids on an individual basis. On the other hand, those proposing a *social model* argue that it is society that creates disability, and that barriers to participation need to be addressed systemically.

The literature also highlights the importance attached by disabled people[1] to the maintenance of personal control over decision making relating to their needs (e.g., Hunt, 1966; Finkelstein, 1991).

One example that serves to illustrate the marked differences that can occur even within one broad category of disability is the experience of hearing-impaired people. A distinction may be drawn between deaf people (who are usually post-lingually deafened) and Deaf people (who are part of the Deaf community). The Deaf culture consists of people who are born or become deaf, use sign language as their first language, and identify themselves as being Deaf. A very real issue is whether education and training: (a) require Deaf people to conform to the dominant approaches to disability as deficit and be educated with English (or other oral language) as the main language; or (b) will regard Deaf people as being a socio-linguistic minority, delivering training using the appropriate sign language as the dominant form of language (Padden & Humphries, 1988). This is inherently an issue of pedagogy. For example, in some instances a Deaf person may benefit from material delivered via videoconferencing with sign language more than written text, depending upon competence. However, in a comparative study of deaf and non-disabled students undertaken in a distance learning environment, Richardson (2001) concludes that, in terms of both their persistence and performance, students with a hearing loss are similar to students with no reported disability. Berry (1999) also highlights the diversity occurring amongst members of the blind and partially sighted population in relation to the issue of access to the World Wide Web.

Debenham (2001) identifies differences between the needs of those with disabilities that are stable (or have stabilised) and those with long-term health problems. Her research explores the experience of distance learners with chronic illnesses in tertiary education, terming these "long-term health problems." In particular the impact of such conditions can be variable, and because they are often hidden, they may not be well understood in terms of "disability" with its stereotypes such as the wheelchair. A further illustration of this is provided by Roulstone (1994), who wryly describes his own experience of disability in terms of being regarded as "a fit person fallen from grace" rather than finding an acceptance of his limitations and ways he needs to work in light of his condition.

Anti-Discrimination Law

In most Western countries there is specific legislation prohibiting discrimination on the grounds of disability. Such legislation provides exemptions for unreasonable hardship imposed by complying with such legislation, as well as specific exemptions. In countries such as the U.S., there have also been moves to revise the broad protections offered by such law, reclaiming narrow and stereotypical definitions of, and approaches to, disability (Johnson, 2003). While people with long-term health conditions may be covered by provisions of such anti-discrimination legislation, they may not be aware of their rights or identify as having disability. Indeed, many learners with functional impairment may choose not to reveal their disabilities to institutions.

Studying at a Distance

There are particular advantages associated with studying via distance, flexible, and open learning. These include overcoming the inability to attend (or difficulty in attending) traditional educational establishments and the flexibility of study hours, which permit the individual to work when feeling freshest. The latter addresses problems encountered with severe fatigue and pain that can fluctuate from day to day.

Disadvantages can include: (a) a lack of social engagement, (b) capital and running costs, and (c) the use of distance education to avoid the issue of making campuses accessible. In the first case, access to a "virtual campus" environment of a distance learning institution can provide one possible way to address isolation for those studying at a distance (Debenham, 2001; Jennison, 1997). In the second case, capital and running costs can present a formidable barrier to participation for those on a low income as a result of their disability (for further discussion in this area, see also Moisey & Moore, 2002; Ommerborn, 1998; Paist, 1995).

Accordingly a very real issue is whether distance education (particularly via computer-mediated communication) promotes autonomy? Certainly, there is a danger that it might reinforce control on decision making on "special needs" by professionals (the old medical model). Yet at its best, educational options available to learners with disability/long-term health needs are enhanced (Debenham, 2001; Newell & Walker, 1991; Walker, 1989, 1994).

Methods of Course Delivery: Multimedia

Traditionally distance education utilised print media as its dominant form of presentation. From the 1970s onward this was variously supplemented, in particular with audiotapes, videotapes, and TV in an effort to present material in as approachable a form as possible. Such an approach recognises that learners absorb, process, and operate in different ways. In addition, a variety of institutions around the world have sought to supplement packaged material with face-to-face opportunities, especially in terms of "on-campus schools" geared to the needs of distance learners. While DE has been operational for some time, it is only comparatively recently that issues of disability have been more formally considered in the mainstream literature.

With the growth of the Internet in recent years, the emphasis has shifted to e-mail, Web access (library, course materials, and other resources), computer conferencing (including the "virtual campus environment"), resource material provided on CD-ROM, and video and/or audio conferencing via the Internet. While the media may change, there are continuing issues for people with disability and long-term health needs. This leads to a consideration of the need to ensure adequate and equitable access, particularly when mainstream technology has often historically served to disable people (Goggin & Newell, 2003).

Access Issues

There are a variety of issues of access for students who need to study/be connected via various types of service. A vital issue is *the cost of disability*, and especially capital and running costs. In the UK these have to some extent been addressed by the government-funded Disabled Student Allowance (DSA) introduced over recent years (see Department of Education and Skills, 2003).

Major differences in approach to access as a construct may be discerned. These include the difference between *mainstream* and *adaptive/special solutions*. Mainstream approaches seek to move people with disabilities from specialist, often isolated, educational settings to the mainstream, seeking to integrate them in such regular services (Fulcher, 1989). At all levels of education in the Western world, this has tended to raise important issues of ensuring equity, not just in terms of opportunity, but making sure that there is adequate support and an understanding of the needs and aspirations of learners with disabilities.

The incorporation of needs into the *mainstream* can include a variety of specific educational programs. A general rule of thumb is that programs designed with disability access in mind will usually be more accessible for the general population. In the IT area there have been a variety of positive initiatives that have made computers and the World Wide Web more accessible, thus removing some need for special solutions. (See for example the WAI Web Content Accessibility Guidelines: *www.w3.org/wai*). As Foley and Regan (2002) suggest: "An essential part of Web design today is designing for individuals with disabilities" (p.1).

Regardless of how effective mainstreaming is, *adaptive/special solutions* to support individual requirements may also be needed. An example of this is found in the Open University (UK) offering to arrange assessments in which areas of need can be explored with individual students. Such assessments may take place either at an Access Centre or exceptionally (in the case of severely disabled students who are unable to travel) in the student's home. For eligible students, adaptive equipment and running costs may be paid for from an individual's DSA, within the limits of the yearly allowance (Open University, 2002).

Universal Design

There is a further body of literature which concerns people with disabilities and the benefits of a universal design approach (e.g., Burgstahler, 2000a, 2000b; French, 2002). Cutting-edge work with regard to DE and open and flexible learning has revolved around the concept of universal design, which may be defined as:

The design of products and environments to be usable by all people, to the greatest extent possible, without the need for adaptation or specialised design. (Connell et al., 1997)

This includes the work of the "Trace Center," a U.S. federally funded centre focussed on developing more accessible products and services for people with disabilities with a particular focus on technology. An online design tool to assist in the development of more usable products has recently been launched (Trace Center, 2004).

In 1997 a working group of architects, product designers, engineers, and environmental design researchers collaborated to establish the following "Principles of Universal Design":

1. **Equitable use:** The design is useful and marketable to people with diverse abilities.
2. **Flexibility in use:** The design accommodates a wide range of individual preferences and abilities.
3. **Simple and intuitive use:** Use of the design is easy to understand, regardless of the user's experience, knowledge, language skills, or current concentration level.
4. **Perceptible information:** The design communicates necessary information effectively to the user, regardless of ambient conditions or the user's sensory abilities.
5. **Tolerance for error:** The design minimises hazards and the adverse consequences of accidental or unintended actions.
6. **Low physical effort:** The design can be used efficiently and comfortably, and with a minimum of fatigue.
7. **Size and space for approach and use:** Appropriate size and space is provided for approach, reach, manipulation, and use regardless of the user's body size, posture, or mobility (Connell et al, 1997).

Such work has a variety of parallels, including the work of the UK "Disability and Elderly Advisory Group" (http://www.acts.org.uk/diel). A further example is the development of BOBBY, a software tool established to ensure that text-based pages meet the criteria of accessibility (BOBBY, 2002).

Examples of Successful Implementation

There are many examples of institutions and projects that have had an enabling impact. These include:

The Open University UK

The Open University UK is a distance learning institution that supports a large number of students studying at the tertiary level within the UK, in mainland Europe, and beyond. Having admitted its first students in 1971, it is the UK's largest university, with over 200,000 students and customers. Predominantly, these are part-time distance learners. However, around 280 full-time postgraduate research students study on campus. In the year 2000 the student population is reported to have included 7,696 students who declared a disability, or learning or mental health difficulty, from a student body of 181,197 (Open University, 2002b). A variety of special arrangements are available to support the needs of such students, with the aim of equalising access.

Historically the university has employed a multimedia approach to support student study. In the early days these included specially written course units (which contain the core teaching material), set books, TV and radio programmes and cassette tapes, in addition to a number of face-to-face tutorials with a course tutor and week-long residential schools. More recently, the advent of CMC has led to dramatic developments in the use of e-learning. Media such as computer conferencing, e-mail, CD-ROMs, DVDs, and the Internet are integral to the study process. Access to a "virtual campus" (using dedicated computer conferencing software such as FirstCLass™ is a key feature of the learning environment.

DO-IT Project (University of Washington)

This award-winning project began in 1992. It aims to increase participation of students with disabilities in academic programs and careers in challenging fields, including science, technology, engineering, and mathematics. DO-IT supports hundreds of high school and college students with disabilities year-round in an online mentoring community, work-based learning experiences, and other activities. Each year around 22 high school students with a wide range of disabilities are selected to be DO-IT Scholars to prepare for college, careers, and leadership roles. Scholars are provided with computers, assistive technology, and Internet connections in their homes. They communicate year-round with mentors, peers, and DO-IT staff, and participate in multiple summer programs and internships as they prepare for college (see also Burgstahler, 2003, 2002c, n.d.).

EASI

Based at the Rochester Institute of Technology, EASI arose from the work of Professor Norman Coombs, a blind professor who first became enthused by the potential of text-based computer conferencing to support the studies of blind students in the 1980s (Coombs, 1989). The project provides access to resource information and guidance on access-to-information technologies for those with disabilities. Online training courses are also offered on adaptive technology to promote understanding of ways in which institutions can address barriers of access for individuals with disabilities in the field of computer and information technology systems.

JOB Project (Bournville College, England)

Created in 1998, this project delivers pre-vocational guidance and training to adults with disability, mental illness, and those who feel themselves to be disadvantaged in returning to the labour market. Training is delivered via CMC, with 24-hour access to vocational guidance. Information is also provided in text format in a "resource room." This is an example of a project where people with disability/chronic illness can also teach. For example, in the case of one former student who became a staff member, working online meant that there was minimal aggravation of physical symptoms. It eliminated the need for travel, and she could work at times to suit herself. This had the effect of removing any anxiety about teaching in a traditional face-to-face environment (Dilloway, Ball, & Sutherland, 2002).

Students Becoming Experts and Educators

One of the real advantages of DE is the way in which learners with disability/chronic illness can become educators and professionals through their experience and use of DE technology. Both the authors of this piece are examples of this (see also Evans & Newell, 1993). Tobin (2002) also documents how vision-impaired instructors can participate in the online world.

Lance (2001), who identifies herself as having Cerebral Palsy, documents how she has been enabled to fulfil her ambition to become a teacher by working as an instructor on a Web-based course delivered online. She identifies a number of advantages of working in this way. Firstly, use of CMC enabled her to take time to answer questions. Preparation of text off-line allows for slow typing, and text can be entered into the conference area as a coherent whole. Secondly, without the hindrance of impaired speech and mobility, her communication partners need not know about her disability unless she chooses to disclose it. She explores the complex issues of whether or not she should reveal her disability to students.

Critical Perspectives

Disabled American professor Art Blaser, writing in the influential disability journal *Ragged Edge Online,* argues that "disabled people, more often than not, simply fall through the 'net'" and that disability is an important variable in terms of the "digital divide" (Blaser, 2001, p. 2). He goes on to show the way in which universities need to address the culture and norms that operate, not only in university environments, but also indeed in the technologies themselves. Similarly, Goggin and Newell (2003) propose that disablism is built into many online technologies, and that the online world does not meet all of the needs of people with disabilities. Such a critique builds on the work of writers such as Oliver (1996), who proposes that disablism operates in structures and taken-for-granted social norms. Communities such as people with intellectual disability, and those with complex communication needs, have largely been forgotten, and this can result in a ghetto effect for blind users and those who rely upon screen readers, given the widespread use of graphical interfaces. Newell and Walker (1992) also express concern that so-called "open" approaches to learning have the potential to decrease access and increase control of people with disabilities if not well done.

CONCLUSION: TOWARDS ENABLING DE

Distance education can provide a useful and helpful route to support the studies of students with disabilities, provided that it takes place in an environment that is well designed and sensitive to their needs. It should not however be regarded as a universal panacea. The maintenance of individual choice is vital. Increasingly learning environments utilise not only the discourse of customers, but also that of diversity. Yet this is a discourse that has largely failed to connect with disability (Campbell, 2002), although other demographic variables such as race and sex have figured in fostering teaching and learning environments that embrace diversity. A very real challenge for society, educational institutions, and educators is to ensure that disability is included in DE learning environments that explicitly embrace and evaluate their programs against diversity guidelines. Accordingly, it is argued that the experience of disability and chronic illness constitutes a crucial dimension for the evaluation and benchmarking of DE in the future.

REFERENCES

Albrecht, G.L., Seelman, H.L., & Bury, M. (Eds.). (2001). *Handbook of disability studies*. Thousand Oaks, CA: Sage Publications.

Berry, J. (1999). Apart or a part? Access to the Internet by visually impaired and blind people, with particular emphasis on assistive enabling technology and user perceptions. *Information Technology and Disabilities Journal, 6*(3). Retrieved 22 March 2004, from http://www.rit.edu/~easi/itd/itdv06n3/article2.htm

Blaser, A. (2001). Distance learning—boon or bane? *Ragged Edge Online*, 5(September). Retrieved 29 August 2003, from http://www.raggededg emagazine.com/0901/0901ft1.htm

BOBBY. (2002). Retrieved 29 March 2004, from http://bobby.watchfire.com/bobby/html/en/about.jsp

Burgstahler, S. (2002a). Distance learning: Universal design, universal access. *Educational Technology Review, 10*(1). Retrieved 29 August 2003, from www.aace.org/dl/index.cfm/fudeaction/ViewPaper/id/11562/searchvars/

authors%3Dburgstahler%26publications%3D AACEJ%26start%5Frow%3D1

Burgstahler, S. (2002b). Universal design of distance learning. *Information Technology and Disabilities Journal, 8*(1). Retrieved 29 August 2003, from www.rit.edu/~easi/itd/itdvo8.htm

Burgstahler, S. (2002c). Working together: People with disabilities and computer technology. Retrieved 22 March 2004, from http://www.washington.edu/doit/Brochures/Technology/wt-comp.html

Burgstahler, S. (2003) DO-IT: Helping students with disabilities transition to college and careers. In *Research practice brief: Improving student education and transition services through research*. National Center on Secondary Educational Transition, University of Minnesota, USA. Retrieved 22 March 2004, from http://www.ncset.org/publications/viewdesc.asp?id=1168

Burgstahler, S. (n.d.). *Use of the Internet in DO-IT*. Retrieved 22 March 2004, from http://www.washington.edu/doit/Brochures/Technology/internet.html

Campbell, J. (2002). Valuing diversity: The disability agenda—we've only just begun. *Disability & Society, 17*(4), 471-478.

Connell, B.R., Jones, M., Mace, R., Mueller, J., Mullick, A., Ostroff, E., Sanford, J., Steinfeld, E., Story, M., & Vanderheiden, G. (1997). *The principles of universal design, version 2.0 dated 4/1/97*, NC State University, The Center for Universal Design. Retrieved 25 March 2004, from http://www.design.ncsu.edu:8120/cud/univ_design/principles/udprinciples.htm

Coombs, N. (1989). Using CMC to overcome physical disabilities. In R. Mason & A.R. Kaye (Eds.), *Mindweave: Communication, computers and distance education*. Oxford: Pergamon Press. Retrieved 25 March 2004, from http://www.icdl.open.ac.uk/lit2k/external.ihtml?loc=http://icdl.

open.ac.uk/literaturestore/mindweave/mind-weave.html

Debenham, M. (2001). *Computer mediated communication and disability support: Addressing barriers to study for undergraduate distance learners with long-term health problems.* Unpublished PhD Thesis, The Open University, UK.

Department for Education and Skills. (2003). *Bridging the gap: A guide to the disabled students allowances (DSAs) in higher education in 2003/2004.* Retrieved 28 March 2004, from http://216.239.59.104/search?q=cache:XdIq o5B-skr0J: www.dfes.gov.uk/studentsupport/uploads/Bridging-the-Gap-2003-Web-version.doc+Disabl ed+Student+Allowances+UK&hl=en&ie=UTF-8

Dilloway, M., Ball, A., & Sutherland, A. (2002). Computer mediated conferencing case study: A JOB project student. In L. Phipps, A. Sutherland, & J, Seale (Eds.), *Access all areas York: JISC TechDis Service and ALT* (pp. 70-72). Retrieved 28 March 2004, from www.techdis.ac.uk/accessallareas/AAA.pdf

Evans, T. & Newell, C. (1993). Computer mediated communication for postgraduate research: Future dialogue. In T. Nunan (Ed.), *Distance education futures* (pp. 81-91). University of South Australia.

Finkelstein, V. (1990). Services for clients or clients for services; annual course: 'working together'. Northern Regional Association for the Blind. Centre for Disability Studies, University of Leeds, UK. Retrieved 29 August 2003, from http://www.leeds.ac.uk/disability-studies/publish.htm

Finkelstein, V. (1991). Disability: An administrative challenge (the health and welfare heritage). In M. Oliver (Ed.), *Social work—disabling people and disabling environments.* London: Jessica Kingsley.

Finkelstein, V. (1996, May). Modelling disability. *Proceedings of the Breaking the Moulds Conference,* Dunfermline, Scotland. Retrieved 29 August 2003, from http://www/leeds.ac.uk/disability-studies/archiveuk/finkelstein/models/models.htm

Foley, A. & Regan, B. (2002). Web design for accessibility: Policies and practice. *Educational Technology Review, 10*(1). Retrieved 29 August 2003, from http://www.aace.org/pubs/etr/issue2/foley.cfm

Fulcher, G. (1989). *Disabling policies?* London: Falmer Press.

Goggin, G. & Newell, C. (2003). *Digital disability: The social construction of disability in new media.* Boulder, CO: Rowman & Littlefield.

Hunt, P. (1966). A critical condition. In *Stigma: The experience of disability.* London: Geoffrey Chapman. Retrieved 22 August 2003, from http://www.leeds.ac.uk/disability-studies/archiveuk/archframe.htm

Jennison, K. (1997). Mutual support on the virtual campus. *Proceedings of the New Learning Environment; A Global Perspective; ICDE '97* (CD ROM), Penn State, USA.

Johnson, M. (2003). *Make them go away.* Louisville, KY: The Advacado Press.

J.T.E. Richardson. (2001) The representation and attainment of students with a hearing loss at the Open University. *Studies in Higher Education, 26*(3), 299-312.

Lance, G.D. (2001). Distance learning and disability: A view from the instructor's side of the virtual lectern. *Information Technology and Disabilities Journal, 8*(1). Retrieved 22 March 2004, from http://www.rit.edu/~easi/itd/itdv08n1/lance.htm

Moisey, S.D. & Moore, B. (2002). Students with disabilities: Their experience and success at

Athabasca University. *ICDE/CADE Calgary 2002 Conference Papers,* CADE, Canada. Retrieved 22 March 2004, from http://www.cade-aced. ca/icdepapers/moiseymoore.htm

Newell, C. & Walker, J. (1991). Disability and distance education in Australia. In T. Evans & B. King (Eds.), *Beyond the text: Contemporary writing on distance education* (pp. 27-55). Geelong, Australia: Deakin University Press.

Newell, C. & Walker, J. (1992). "Openness" in distance and higher education as the social control of people with disabilities: An Australian policy analysis. In T. Evans & P. Juler (Eds.), *Research in distance education 2* (pp. 68-80). Victoria, Australia: Institute of Distance Education, Deakin University.

Oliver, M. (1996). *Understanding disability from theory to practice.* Houndmills: MacMillan.

Open University. (2002a) Learners guide—services for disabled students. What services are available? *Disabled Students' Allowances.* Retrieved 22 March 2004, from http://www3.open. ac.uk/learners-guide/disability/services_available/factsheet_dsa.htm

Open University. (2002b) *Learners guide—services for disabled students.* Numbers of disabled students receiving services in 1999/2000. Retrieved 25 March 2004 from http://www3.open. ac.uk/learners-guide/disability/services_available/chart.htm

Padden, C. & Humphries, T. (1988). *Deaf in America: Voices from a culture.* Cambridge, MA: Harvard University Press.

Paist, E.H. (1995). Serving students with disabilities in distance education programs. *The American Journal of Distance Education, 9*(1), 61-70.

Ommerborn, R. (1998). *Distance study for the disabled: National and international experience and perspectives.* Hagen, Germany: Fern Universitat.

Roulstone, A. (1994). *New technology and the employment experience of disabled people: A barriers approach.* Unpublished PhD thesis, The Open University, UK.

Tobin, T.J. (2002). Issues in preparing visually disabled instructions to teach online: A case study. *Information Technology and Disabilities, 8*(1). Retrieved 29 August 2003, from www.rit. edu/~easi/itd/itdvo8n1/tobin.htm

Trace Center. (2004). Retrieved 28 March 2004, from http://www.trace.wisc.edu/world/tool_nav. html

Walker, J. (1989). Mark's story: A disabled student's case study in distance education. *Distance Education, 10*(2), 289-297.

Walker, J. (1994). Open learning: The answer to the government's equity problems? A report on a study of the potential impact of the Open Learning initiative on people with disabilities. *Distance Education, 15*(1), 94-111.

KEY TERMS

Chronic Illness: A long-term health condition, usually persisting for more than one year. People may or may not identify as having disability, but it will often impact upon major life functions such as learning.

Computer-Mediated Communication: Includes a wide range of applications, for example, Web-based applications such as library access, e-mail, computer conferencing, videoconferencing, audio conferencing, shared whiteboards, material on CD ROM, and software permitting online 'virtual' scientific experiments.

Disability: There are two basic approaches. The *medical model* sees disability as a 'personal tragedy' or 'deficit' located within an individual. The *social model* argues that it is society that creates disability, with barriers to participation needing to be addressed systemically.

Disablism: Similar to sexism and racism as a concept. Writers such as Oliver (1996) point to the structural and ideological way in which disablism occurs in everyday society.

Deaf People: There are two major types: deaf (or hard of hearing) people who are usually post-lingually deafened, and Deaf people who are part of the Deaf community. The Deaf culture consists of people who use sign language as their first language and for participating in activities within this community. Socio-linguistically they are a cultural minority.

Universal Design: An approach to the design of products and environments fostering usability by as many people as possible, without adaptation.

Virtual Campus: An institution-based interactive online learning environment that may be supported either by dedicated computer conferencing software or on the Internet.

ENDNOTE

[1] It is worth noting that in the UK, the preferred language is "disabled people" whereas in the U.S. and countries like Australia, disability advocates favour the language "people with disabilities." In deference to authors within these countries, we have sought to reflect the language used by the writers.

This work was previously published in the Encyclopedia of Distance Learning, Vol. 2, edited by C. Howard, J. Boettcher, L. Justice, K. Schenk, P.L. Rogers, and G.A. Berg, pp. 591-598, copyright 2005 by Idea Group Reference (an imprint of IGI Global).

Chapter 7.23
A Socio–Technical Analysis of Factors Affecting the Integration of ICT in Primary and Secondary Education

Charoula Angeli
University of Cyprus, Cyprus

Nicos Valanides
University of Cyprus, Cyprus

ABSTRACT

We live in a world that is constantly impacted by information and communication technology (ICT). ICT is considered an important catalyst and tool for inducing educational reforms and progressively extending and modifying the concept of literacy. With the extensive use of ICT in schools and everyday life, the term computer literate has already been established. Schools are open systems that interact with their environment, and the effective use and integration of technology is directly associated with the role of various socio-technical factors that may impact the integration of ICT in schools. In this chapter, we report on an exploratory study undertaken in Cyprus schools to examine the status of using ICT from the perspective of socio-technical systems. Specifically, teachers' knowledge of ICT, frequency of using ICT for personal purposes, frequency of using ICT for instructional purposes in different subject matters, attitudes toward ICT, self-confidence in using ICT in teaching and learning, and school climate were examined. The findings provide useful guidance to policymakers for planning, implementing, managing, and evaluating the integration of ICT in schools. Implications for the concept of computer literacy are discussed.

INTRODUCTION

Due to rapid technological advancements, we live in a world that is constantly impacted by information and communication technologies (UNESCO, 1999). Some key-markers that characterize differences between 19th-century societies (i.e., industrial-age societies) and 20th-century[1] societies (i.e., information age societies) are: (a) standardization vs. customization, (b) bureaucratic organizations vs. team-based organizations, (c) adversarial relationships vs. cooperative relationships, (d) parts oriented vs. process oriented, (e) compliance vs. initiative, and (f) conformity vs. diversity (Reigeluth & Garfinkle, 1994). By virtue of these differences, we are obliged to evaluate once more the worth of our existing educational systems. Are our current educational systems, with their emphasis on content coverage and teacher-centered classroom practices, conducive to preparing students to survive in a changing world that is steadily shaped by developments in information technologies? How do we prepare our future citizens to become computer or technology literate? Do new computer technologies herald the beginning of an era of broader literacy, and if we are educating children to be active citizens in an information society, what forms of literacy are required? What does it mean to be literate, an active reader, a writer, and a communicator of meaning in the information society?

Countries in North America, South America, Europe, Asia, and Africa have all identified a significant role for information and communication technology (ICT[2]) in improving education and reforming curricula for the purpose of preparing future citizens to be productive and actively involved in an information society (Kozma & Anderson, 2002; Pelgrum, 2001). ICT is considered by many not only to be the "backbone of the Information Society, but also to be an important catalyst and tool for inducing educational reforms that change our students into productive knowledge workers" (Pelgrum, 2001, p. 165). For these reasons, schools have made major investments and continue to invest heavily in increasing the number of computers in schools and the networking of classrooms.

ICT is thus steadily becoming part of classroom life, and it progressively changes the concept of literacy (Brindley, 2000; Watt, 1980). The traditional concept of literacy as the ability to read and write (Crystal, 1987x) is changing, and ICT opens up a further definition of literacy—one that goes beyond the acquisition of basic skills. Brindley (2000) argues that "schooled literacy, which traditionally sees the acquisition of the ability to construct and interpret text as largely an individual activity, bounded by the concept of text as linear and fixed, is no longer adequate" (p. 13). With the enduring introduction of computers in schools and the extensive use of ICT in our everyday life, the term computer literate has been established and flourished.

For many, being computer literate simply means acquiring technical expertise to be able to competently use computer software and hardware. In this chapter, we consider a much more complex and exciting concept of computer literacy—one that is directly associated with the affordances of ICT and the concept of visual literacy. "Visual literacy refers to the use of visuals for the purposes of communication, thinking, learning, constructing meaning, creative expression, [and] aesthetic enjoyment" (Baca, 1990, p. 65). Thus, the extensive use of multimedia in schools and everyday life opens up the way to an extended concept of literacy. For example, ICT reinvents the text and leads us to a new form of literacy, which encompasses a range of media by which students learn and communicate, such as graphics, video, and sound (Papert, 1993). Similarly, McFarlane (2000) argues that multimedia allow students to record and present their own meaning using multiple media. Thus, the technology of multimedia does not restrict reading and writing to the mere coding and decoding of text. Using a computer, children can represent their creativity with text, graphics, speech, video, animation, and more. Technology offers us new forms of representation and expression that extend the traditional and limited concept of literacy. In the information society one has to be able to "read the text and the image and the moving image and the ability to secure understanding through reading the pictures as well as text in a rich and organic way" (Kempster, 2000, p. 25). For these reasons, Papert (1993) argues for an emerging model of literacy, which encompasses a range of media by which students learn to express themselves and communicate. In the emerging model of literacy, the ability to read and write are enhanced and sometimes replaced by images, graphics, and sound. These new forms of expression and communication can be used by students to construct meaning and represent their understandings after selecting the most appropriate form from multiple alternatives.

McFarlane (1997) also argues that with computers, literacy extends beyond simple manual encoding and decoding of text. "It involves the habit of viewing writing as a way of developing and communicating a child's thoughts. The use of word processors, for example, helps to present text as something to be experimented with, redrafted and developed as ideas develop, or the demands of purpose or audience change. It liberates the writer from the heavy burden of manual editing and presentation" (McFarlane, 1997, p. 119). Word processors are not only great productivity tools to write faster, or to make fewer mistakes, but also tools that fundamentally change the authoring process children are engaged in (Heppell, 2000).

Moreover, a literate person in the 21st century makes judgments about the quality and value of information. The Internet presents new challenges in the way text is presented, and requires a new set of reading skills that go beyond linear book print to screen print. The inclusion of graphics, hyperlinks, and bookmarks also requires an understanding of how text in a Web page is structured

in a non-linear way. Thus, a computer-literate person should be able to access this non-linear text with speed and accuracy and construct understanding. Similarly, the National Grid for Learning in the United Kingdom (DfEE, 1998) refers to the new form of literacy as information or network literacy that is defined as the capacity to use electronic networks to access resources, create resources, and communicate with others. The use of e-mail, for example, to electronically communicate with others dramatically extends and alters students' written language in many ways. E-mail can rapidly expand the audience that children write for, and secondly, it can engage children in a different type of writing than the traditional one (Easingwood, 2000). It can also be used to support students' collaborative work with other students in different schools. The notion of children working collaboratively using the new technologies enables them to develop skills that are needed in the workplace and modern society. Thus, it is clear that computers have the potential to make new things possible in new ways, creating new forms of literacy that are critical to our present and future society.

It therefore becomes important to examine whether schools offer the education needed so that students develop the new forms of literacy that are so important for surviving in the information age. Several researchers (Eraut & Hoyles, 1989; Snyder, 1994) argue that teachers need to be trained in order to utilize ICT appropriately in teaching and learning, and that computers by themselves cannot develop the new form of literacy. Pelgrum (2001) also suggests that ICT in education is an area that is in turmoil and in which many participants play a role. For example, forces operating in schools and in classrooms may be influential in bringing about changes or inhibiting them. Hence, it is important to regularly monitor the status of ICT in education in order to not only account for the financial investments, but also to inform policymakers regarding the content and direction of future policies.

Apple (1986) states that technology in the schools has usually been seen as an autonomous process. "It is set apart and viewed as if it had a life of its own, independent of social intentions, power, and privilege" (p. 105). Along the same line of reasoning, Street (1987) also argues that computer literacy erroneously rests on the assumption that ICT is a neutral tool that can be detached from other specific and social contexts. According to Kling (2000), the integration of ICT in the school system should be examined within a socio-technical framework, where people need to sufficiently interact with the technological tools within the system for the change to be effective. The term "socio-technical systems" was coined in the 1950s (e.g., Trist & Bamforth, 1951; Trist, 1982) to capture the interdependencies between the social and technical aspects of a system. Put simply, a socio-technical system is a mixture of people and technology, hence the concept of "socio-technical system" was established to stress the reciprocal interrelationship between humans and machines. More importantly, the socio-technical systems approach provides us with a comprehensive and systemic methodology for holistically examining factors that may impact the integration of ICT in elementary and secondary education.

In this chapter, we present an exploratory study undertaken in Cyprus schools to examine the current status of using ICT in primary and secondary education, and discuss implications for the development of the new form of literacy given the current status of ICT in Cyprus schools.

THEORETICAL FRAMEWORK

Schools, like other organizations, are open systems that continuously interact with their environment (Getzels & Cuba, 1957; Hanna, 1997; Hoy & Miskel, 2001). The interaction of a system with its external environment is vitally important, because as a result of this interaction, the system receives feedback and appropriately adapts to new demands and circumstances. A system that does not adapt to the needs of its external environment will gradually become extinct (Hoy & Miskel, 2001).

Thus, the introduction of ICT in the school system has created a need for extending the theory of social systems into a theory of socio-technical systems such that the interaction between teachers and technology can be examined and understood. Social informatics is an area of research, which systematically examines the design, uses, and consequences of technology, taking into consideration the context of the organization, the people who work within the organization, and the interactions between people and technology (Denning, 2001; Friedman, 1998; Kling, 2000). One key idea of social informatics is that ICT, in practice, is socially shaped and the uses of technology in an organization are contingent upon several social and technical dependencies. The concept of socio-technical networks or systems is used to describe the interdependencies between technology and people, and to explain that the culture of an organization and people's beliefs, attitudes, and feelings play an important role in shaping the organization's mood and determining the effectiveness of the integration of technology in the organization (Kling & Lamb, 2000; Kling, 2000; Markus & Benjamin, 1987). Thus, the effective use of technology in different organizational settings is directly associated with the intertwining of technical and social elements (Friedman, 1998; Heracleous & Barrett, 2001; Kling, 2000).

Trach and Woodman (1994) support that the socio-technical model constitutes a flexible model for successfully implementing systemic changes in an organization. Thus, technology should be viewed as a catalyst for pursuing systemic as opposed to piecemeal changes across the different subsystems of an educational system (Angeli, 2003; Valanides & Angeli, 2002). Teachers, for example, constitute an important subsystem of every educational system. According to Fullan (1991), every reform effort should take into consideration the knowledge, skills, beliefs, and attitudes of the people who will implement the changes. In general, a precondition for a successful implementation of any change effort is adaptation, which includes all adjustments an organization makes in order to realize the changes (Hoy & Ferguson, 1985). Adaptation is a broad term and may include multiple criteria. In this chapter, six areas of adaptation are examined: (1) teachers' knowledge of ICT, (2) teachers' frequency of using ICT for personal purposes, (3) teachers' frequency of using ICT for instructional purposes in different subject matters, (4) teachers' attitudes toward ICT, (5) teachers' self-confidence in using ICT in teaching and learning, and (6) school climate.

According to Huberman and Miles (1984), many educational reform efforts died out because teachers were not supported in their change efforts and thus never accepted or understood the changes they had to implement. Therefore, the adoption of changes requires educating teachers to understand and accept the nature of the restructuring effort, and develop the knowledge, skills, and attitudes that are required for bringing about the change in their classrooms (Fullan, 1991; Louis & Miles, 1990). As Barth (1990) characteristically stated, nothing influences students more than their teachers' own professional and personal development. The process of integrating ICT in teaching and learning is demanding and requires teachers' continuous professional development (Picciano, 2002). In addition, Picciano (2002) believes that school principals and inspectors also need to participate in ICT training so that they understand how ICT integration affects the classroom (micro level) and the school (macro level). Fullan (1991) explains that educational change efforts often create feelings of uncertainty that are not only related to lack of knowledge and skills, but also confidence, as teachers often feel inadequate and uncertain about their new roles (Fullan & Hargreaves, 1992). For these reasons, teachers'

ICT professional development is important not only for the development of ICT knowledge and skills, but also for the development of positive attitudes and confidence in using ICT in teaching and learning. Hoy and Miskel (2001) point out that the style of leadership in an organization is also an important factor. For example, a principal who encourages the use of ICT in teaching and allows teachers to create collaborations within the school and between schools for the exchange of ideas will play an important role in successfully institutionalizing the change effort. Thus, as Hoy and Miskel (2001) state, what is needed is transformational leadership. Transformational leaders are those who foresee the need for change, create new visions for education, and encourage teachers to take responsibility for their professional and personal development in order to successfully fulfill their new obligations and roles.

METHODOLOGY

The Context of the Study

The public educational system in Cyprus consists of the primary and secondary levels, while just recently new attempts have been made to also include pre-primary (3-5½-year-old children) education. Grades 1 to 6 constitute the primary level, and grades 7 to 12 the secondary level. Education is free for all grade levels and mandatory until grade 9 or the age of 16, but an overwhelming 95% of students complete all grade levels. The majority of students attend public schools, and there are only a small number of private schools mainly at the secondary level. Fifty teachers were randomly selected from each one of the 12-grade levels in primary and secondary public education. During the spring semester of 2004, a questionnaire and a pre-stamped self-addressed envelope were delivered to each individual teacher with the help of research assistants. Each teacher was asked to individually complete the questionnaire and return it to the researchers. The majority of the teachers (520) returned their completed questionnaires within a week, and only 22 teachers did not return their questionnaires at all, even after a second reminder by telephone two weeks after the questionnaires were delivered to them. Thus, the data from 578 questionnaires were used in the study.

Data Collection

The questionnaire consisted of seven parts. The first part collected demographic data related to teachers' age, number of computer labs in each school, number of computers in each lab and teachers' classrooms, teachers' ownership of a personal computer, and teachers' participation in an ICT professional development training program. The other six parts collected data related to: (1) teachers' knowledge of computer software, (2) teachers' frequency of software use for personal purposes, (3) teachers' attitudes towards integrating ICT in teaching and learning, (4) teachers' self-confidence in integrating ICT, (5) teachers' frequency of using ICT for instructional purposes in the classroom, and (6) school climate and support.

Specifically, the second part of the questionnaire used a Likert-type scale from 1 to 5 (I do not know how to use it, I somewhat know how to use it, I know how to use it satisfactorily, I know how to use it well, I know how to use it very well) to measure teachers' knowledge of various software, and the third part used a Likert-type scale from 1 to 5 (never, once or twice every three months, once or twice a month, once or twice a week, almost every day) to measure frequency of software use for personal purposes. Similarly, the fourth part measured teachers' attitudes with a Likert-type scale from 1 to 5 (absolutely disagree, disagree, neither disagree nor agree, agree, absolutely agree), and the fifth part used the same Likert-type scale that was used in the fourth part to measure teachers' self-

confidence. For the sixth part, which measured teachers' frequency of using various computer programs in classroom practices, teachers had to write how many times a week they were using different software in their teaching. Finally, the seventh part of the questionnaire measured school climate and support. A Likert-type scale from 1 to 5 (absolutely disagree, disagree, neither disagree nor agree, agree, absolutely agree) was also used. Thus, the questionnaire collected demographic data and data related to the six areas of adaptation from the perspective of the socio-technical systems perspective.

RESULTS

Demographic Data

Among the 578 teachers, 446 (77.14%) of them were females and 132 (22.86%) were males. The average age of the participating teachers was 31.98 ($SD = 8.107$), but the average age was significantly smaller ($t = -5.501$, $p = .000$) for female teachers (*Mean* = 30.94 years) than for male teachers (*Mean* = 35.24 years). All teachers owned their own personal home computer (*Mean* = 1.05, *SD* = 1.05), and in some cases, there were teachers who owned two computers at home. Also, 67.5% of teachers recently participated in an ICT teacher professional training program where they learned how to use several computer programs, such as Word, Excel, PowerPoint, and the Internet. Teachers taught in schools where there was a computer lab with an average of 5.55 computers in each lab and at most one computer in each classroom. The computer-student ratio differed for each school, but in all schools computer access was not prolific.

These data signify the tremendous effort that has been undertaken for successfully integrating ICT in the teaching-learning environment, considering the fact that not too long ago there were no computers in the classrooms and no computer labs in most of the schools in Cyprus. It is however more important to investigate how and to what extent ICT is used in Cyprus schools in relation to teachers' ICT knowledge, frequency of software use for personal purposes, attitudes, self-confidence, frequency of using ICT for instructional purposes in the classroom, and various factors pertaining to the socio-technical character of the school system.

Teachers' Knowledge of Computer Software and Frequency of Use for Personal Purposes

Descriptive statistics related to teachers' knowledge of computer software and teachers' frequency of using software for personal purposes are presented in Tables 1 and 2, respectively.

The results in Table 1 indicate that teachers' knowledge of software varied according to the type of software. Specifically, teachers appeared to be more familiar with Word than with any of the other software such as PowerPoint, the Internet, e-mail, Excel, graphics, authoring software, and databases. More analytically, the results in Table 1 indicate that teachers knew how to use well only Word, while their knowledge about PowerPoint, the Internet, and e-mail was rated just above the level of satisfactory use. Teachers' knowledge regarding the rest of the software was rather poor and below the level of satisfactory

use. It is also important to mention that these data refer to teachers' self-reported estimates of their knowledge and do not necessarily represent their actual knowledge.

The results in Table 2 indicate that teachers used Word most frequently and rarely used databases. The Internet, e-mail, educational CD-ROMs, and PowerPoint were less frequently used than Word, but more frequently used than graphics, spreadsheets, authoring software, and databases. For example, teachers used Word once or twice a week, while they almost never used databases or authoring software. Thus, the collective results from Tables 1 and 2 clearly indicate that teachers' knowledge of computer software was dependent upon the type of software, and that their frequency of software use followed almost the same order, in terms of magnitude, as their knowledge of computer software. The means of the variables related to knowledge of software and frequency of software use were found to be highly and significantly correlated ($r = .905$, $p = .01$). Correlation, of course, does not mean a direct causal relation, and it cannot explain whether better knowledge of a computer program causes its frequent use, or whether the need to frequently use a computer program causes better knowledge of it. The only valid conclusion is that the participants reported better knowledge for some kinds of software and higher frequency of use for the same kinds of software.

Table 1. Descriptive statistics of teachers' knowledge of software

Software	M	SD	n
Word Processing (i.e., Word)	4.17	1.11	577
Databases (i.e., Access)	2.01	1.16	569
Spreadsheets (i.e., Excel)	2.76	1.33	572
Graphics (i.e., Paint)	2.68	1.34	570
Presentation (i.e., PowerPoint)	3.34	1.43	575
Authoring Software (i.e., Hyperstudio)	2.17	1.35	576
Internet	3.28	1.53	570
E-Mail	3.19	1.60	575
Knowledge of Software	**2.95**	**1.03**	**560**

Table 2. Descriptive statistics of teachers' frequency of software use for personal purposes

Software	M	SD	n
Word Processing (i.e., Word)	4.11	1.16	576
Databases (i.e., Access)	1.03	.44	569
Spreadsheets (i.e., Excel)	1.92	1.08	576
Graphics (i.e., Paint)	2.07	1.06	576
Presentation (i.e., PowerPoint)	2.15	1.06	576
Authoring Software (i.e., Hyperstudio)	1.28	.64	576
Internet	3.70	1.38	574
E-Mail	3.20	1.53	573
Educational CD-ROMs	2.55	1.24	572
Frequency of PCU	**2.95**	**1.03**	**560**

Teachers' Attitudes

Dealing effectively with ICT relates not only to knowledge of ICT tools, but also to individuals' attitudes and perceptions regarding ICT tools. Attitudes and perceptions act as a filter through which all learning occurs (Marzano, 1992), and are considered as a constituent part of learners' "self-esteem" that oversees all other systems (Markus & Ruvulo, 1992). Thus, learners continually filter their behaviors through their self-belief system to the extent that they even attempt to modify the "outside world" and make it more consistent with the "inside world" (Glaser, 1981). The limited teachers' knowledge or skills about the use of several software and computer applications seem to have an impact on their attitudes and concerns.

For these reasons, teachers' attitudes towards ICT were also examined in this study. Table 3 shows descriptive statistics related to the 19 items measuring teachers' attitudes towards the use of ICT in education. The results in Table 3 indicate that the majority of teachers expressed rather positive attitudes towards the use of ICT tools in education. Teachers felt rather comfortable in using ICT for instructional purposes, and expressed positive attitudes towards applying ICT in teaching and learning, because ICT could make learning easier, meaningful, and useful. However,

there were a lot of teachers who expressed skepticism or even fear, because they felt incompetent to resolve potential technical problems with the computer. In general, teachers expressed a somewhat overall positive attitude towards the use of computers in education, although some of them also expressed concerns pertaining to technical problems that might hinder their work and students' learning. Of course, even though ICT-related attitudes seem to play an important role in how ICT is used in teaching and learning (Levine & Donitsa-Schmidt, 1998), research indicates that positive attitudes alone are not always good indicators of teachers' eventual use of ICT in the classroom (Wild, 1996). This is due to the fact that teachers often times have positive attitudes about ICT integration without realizing how difficult the task is, or how much effort they need to invest to successfully complete the task. Thus, despite teachers' rather positive disposition towards ICT integration, they may still find the task of integrating computers in the classroom difficult, once they realize what it really entails. This also seems to be a reasonable conclusion from the current results, taking into consideration that the participants of the present study had limited knowledge of the full range of affordances of several ICT tools and their applications in the teaching-learning environment.

Table 3. Descriptive statistics of teachers' attitudes

Item	M	SD	n
I feel comfortable with the computer as a tool in teaching and learning.	3.78	1.07	577
The use of computers makes me stressful.	3.84	1.06	575
If something goes wrong with the computer, I will not know what to do.	2.63	1.12	576
The use of computers in teaching and learning makes me skeptical.	2.70	1.13	576
The use of computers in teaching and learning makes me enthused.	3.78	.83	574
The use of computers in teaching and learning interests me.	3.99	.93	575
The use of computers in teaching and learning scares me.	2.28	1.14	572
I believe the computer is a useful tool for my profession.	4.33	.81	575
The use of computers in teaching and learning will mean more work for me.	3.61	1.00	574
Computers will change the way I teach.	3.77	.83	575
Computers will change the way my students learn.	3.74	.88	571
Whatever the computer can do, I can do it equally well in another way.	2.50	.90	572
Computers make learning harder because they are not easy in their use.	2.16	1.71	575
Computers make learning harder because often times there are technical problems associated with them that students cannot resolve.	2.67	1.07	574
The computer supports and enhances student learning.	4.05	.72	574
The computer makes learning more meaningful.	3.86	.75	576
The computer helps students represent their thinking better.	3.85	.74	577
The computer is a meaningful tool for the teacher because it can help him/her teach a topic more effectively.	3.92	.79	575
The computer hinders teaching because of the technical problems it may cause.	3.62	.92	576
ICT Attitudes	**3.36**	**.26**	**545**

Teachers' Confidence

Table 4 shows descriptive statistics related to teachers' confidence. The results indicate that teachers felt somewhat confident in selecting appropriate software to be used in their teaching, and felt about the same with designing and implementing classroom activities with ICT tools. Several factors seem to play an important role in affecting how individuals use ICT (Fullan, 1991). These factors include not only ICT knowledge and the amount and nature of prior ICT experience, but also ICT-related attitudes and learners' beliefs in their ability to work successfully with ICT tools (self-confidence or self-efficacy) (Levine & Donitsa-Schmidt, 1998; Liaw, 2002; Murphy, Coover, & Owen, 1989). Attitudes and beliefs are considered as predictors of behaviors and behavioral intentions that are linked to self-confidence. Beliefs about an object usually lead to attitudes towards it, and in turn, attitudes lead to behavioral intentions regarding the object, which affect actual behaviors towards the object. Finally, there is a feedback loop where behavioral experience modifies preexisting beliefs about the object. In terms of ICT use, attitudes toward ICT affect users' intentions or desire to use ICT. Intentions in turn affect actual ICT usage or experience, which modifies beliefs and consequent behaviors or behavioral intentions (future desire), and self-confidence or self-efficacy in employing ICT in learning. Thus, teachers' actual ICT usage in the classroom is directly associated with their knowledge, attitudes, and self-confidence, although at-

Table 4. Descriptive statistics of teachers' confidence

Item	M	SD	n
I feel confident in selecting appropriate software to use in my teaching.	3.51	1.12	572
I feel confident in preparing classroom activities with ICT for my students.	3.30	1.20	569
Confidence	**3.40**	**1.04**	**568**

titudes and confidence are directly dependent on knowledge and improve with success and frequent use. It seems that teachers' self-confidence was delimited by teachers' knowledge of software and frequency of software use.

Teachers' Frequency of Using ICT in the Classroom for Instructional Uses

Table 5 shows descriptive statistics of teachers' frequency of using ICT for instructional purposes in the classroom. The results in Table 5 draw a rather pessimistic picture in terms of actual instructional use of ICT in the classroom. First, none of the teachers reported any use of electronic communication (i.e., E-mail), authoring software (i.e., Hyperstudio), or graphics (i.e., Paint) in their teaching. Second, computer applications such as spreadsheets, databases, and PowerPoint were minimally used, and teachers' reported mean frequencies of use for these software were .309, .031, and .711, respectively. Third, only the mean frequencies of use for the Internet, Word, and

educational CD-ROMs had values higher than one indicating that they were used infrequently and only by very few teachers. Finally, in comparison with the results in Table 2, teachers were using the same software much less frequently for instructional purposes than for personal purposes. This seems to suggest that teachers' knowledge, attitudes, and self-confidence had probably less impact on teachers' instructional use of the software or that there were some other reasons inhibiting the instructional uses of ICT in teaching and learning. For example, the existing socio-technical character of a school could substantially constitute a significant factor in supporting or inhibiting both use and frequency of use of certain software, despite teachers' technical expertise, attitudes, and self-confidence in employing ICT in teaching and learning.

Socio-Technical Environment

Table 6 shows descriptive statistics of various factors related to the socio-technical character

Table 5. Descriptive statistics of teachers' frequency of using ICT in the classroom

Software	M	SD	n
Internet	1.233	2.46	578
Word Processing (i.e., Word)	1.606	2.78	578
Spreadsheets (i.e., Excel)	.309	.97	578
Databases (i.e., Access)	.031	.21	578
Presentation (i.e., PowerPoint)	.711	1.45	578
Educational CD-ROMs (e.g., drill and practice, tutorials, etc.)	1.524	2.75	578
Frequency of Instructional Use	**5.413**	**8.22**	**576**

of a school. Based on the results, there were participants who felt that their superiors, the computer coordinator, and other colleagues tended to encourage them to use ICT in the classroom, but there were also other participants who did not share the same point of view. Also, teachers in general neither agreed nor disagreed about the availability of technical or instructional support in their school, or whether there was adequate computer equipment or software available. Lastly, teachers expressed mixed views on whether the subject of ICT integration was sufficiently discussed in faculty meetings. The only valid conclusion from the results in Table 6 is that at least teachers did not feel discouraged for their attempts to integrate ICT tools in their classrooms. They perceived a rather neutral socio-technical environment in their schools. It is also possible that teachers did not have a clear understanding of the situation in their school, and thus were unsure about it. Obviously, there was not a strong momentum, nor systematic plan of action for effectively integrating ICT in the participants' schools. Teachers' somewhat positive attitudes and perceived self-confidence were rather compatible with the existing socio-technical environment, which, as the results indicated, was not overwhelmingly supportive, but rather neutral.

Personal Computer Use and Instructional Computer Use

The fact that teachers reported rather infrequent instructional use or no instructional use for most of the software is contradictory to teachers' frequency of using computer software for personal purposes, as well as contradictory to their subjective self-confidence, attitudes, reported knowledge of several software, and the socio-technical environment.

In order to better examine the existing discrepancy between the frequency of ICT use for personal purposes and frequency of ICT use for instructional purposes, five composite variables were created, namely, teachers' knowledge (KNOW), frequency of personal computer use (PCU), attitudes (ATT), self-confidence (SCF), and frequency of instructional computer use (ICU). These five variables represented the mean value of the single items in Tables 1, 2, 3, and 4, respectively, with the exception of ICU, which represented the sum of the individual items in Table 5. Two regression analyses were consequently conducted with the frequencies of PCU and ICU as the dependent variables for the first and the second analyses, respectively. The independent variables for both analyses were the

Table 6. Descriptive statistics of socio-technical factors

Item	M	SD	n
There are teachers in my school who help me integrate ICT in my teaching.	3.12	1.120	564
The computer coordinator encourages me to use ICT in my classroom.	3.38	1.171	565
The principal encourages me to use ICT in my classroom.	3.15	1.083	566
The inspector encourages me to use ICT in my classroom.	3.43	1.087	565
We often talk about ICT integration during our faculty meetings.	2.92	1.109	567
There are many software available in my school.	3.04	1.072	566
There is technical support readily available in my school.	3.09	1.086	565
There is ICT instructional support readily available in my school.	2.89	1.080	566
There is adequate computer equipment in my school.	3.04	1.223	566

other three composite variables (KNOW, ATT, and SCF), teachers' age, participation in an ICT professional development training program, and the nine individual items shown in Table 6 measuring aspects of the socio-technical environment (STE1 to STE9). Other variables from the first part of the questionnaire were excluded from the analyses, because they were found not to be discriminating nor redundant, as they were highly and significantly correlated with other variables.

For example, *years of teaching experience* was considered to be a redundant variable, because it was highly and significantly correlated with age ($r = .960$, $p = .01$). *Ownership of a personal computer* was also not a discriminating variable, because all teachers owned a personal computer, with the exception of some of them who reported that they owned two personal computers. Table 7 shows the correlations between all possible variables (dependent and independent) that were

Table 7. Correlations between all possible pairs of criterion, predictor, and selected demographic variables

Variable	1	2	3	4	5	6	7	8	9
Age (1)	1.00								
ICT Inservice (2)	-.190**	1.00							
KNOW (3)	-.401**	-.075	1.00						
PCU (4)	-.274**	-.077	.772**	1.00					
ICU (5)	-.046	-.087*	.283**	.354**	1.00				
SCF (6)	-.201**	-.134**	.597**	.592**	.303**	1.00			
ATT (7)	-.185**	-.056	.537**	.527**	.309**	.592**	1.00		
STE$_1$ (8)	-.015	-.046	.085*	.128**	.202**	.220**	.136**	1.00	
STE$_2$ (9)	-.003	-.069	.166*	.195**	.202**	.321**	.232**	.548**	1.00
STE$_3$ (10)	-.019	-.086*	.191*	.242**	.211**	.325**	.273**	.517**	.575**
STE$_4$ (11)	.005	-.046	.160*	.195**	.191**	.285**	.238**	.420**	.452**
STE$_5$ (12)	.031	-.094*	.036	.044	.048	.168**	.036	.294**	.282**
STE$_6$ (13)	-.079	-.104*	.094*	.068	.110**	.185**	.119*	.275**	.353**
STE$_7$ (14)	.003	-061	-.008	-.023	.074	.149**	.052	.273**	.375**
STE$_8$ (15)	.016	-.104*	.042	.050	211**	.188**	.059	.343**	.436**
STE$_9$ (16)	.069	-.111*	.077	.035	.057	.173**	.108*	.236**	.310**
Gender (17)	.225**	-.034	.132**	.207**	.118**	.082	.171**	-.008	.108*
TE (18)	.960**	-.200**	-.407**	-.290**	-.057	-.217**	-.205**	-.020	-.024

Variable	10	11	12	13	14	15	16	17
Age (1)								
ICT Inservice (2)								
KNOW (3)								
PCU (4)								
ICU (5)								
SCF (6)								
ATT (7)								
STE$_1$ (8)								
STE$_2$ (9)								
STE$_3$ (10)	1.00							
STE$_4$ (11)	.656**	1.00						
STE$_5$ (12)	.440**	.400**	1.00					
STE$_6$ (13)	.361**	.267**	.302**	1.00				
STE$_7$ (14)	.267**	.252**	.278**	.438**	1.00			
STE$_8$ (15)	.375**	.316**	.354**	.363**	.560**	1.00		
STE$_9$ (16)	.321**	.210**	.248**	.437**	.448**	.384**	1.00	
Gender (17)	.044	.039	-.0119	.003	.003	.044	.102*	1.00
TE (18)	-.002	.002	.036	-.078	-.001	.022	.047	.146*

Note: SCF = teachers' self-confidence, KNOW = teachers' knowledge, ATT = teachers' attitudes, STE1 to STE9 = individual items of the socio-technical environment corresponding to the items in Table 6, TE = years of teaching experience, ICT inservice = participation in ICT professional training, Gender = 1 for females and 2 for males.

used in the two regression analyses and some additional variables from the first part of the questionnaire.

Regarding the items in the first part of the questionnaire (demographic information), it was considered more appropriate to use them as individual variables, since they could not be considered dimensions of the same construct, as it was, for example, the case with teachers' self-confidence or attitudes. Regarding the items in the last part of the questionnaire, although they were measuring aspects of the socio-technical environment, we used them as individual variables, because we were interested in identifying which dimensions of the socio-technical environment seemed to play an important role in ICT integration. In most cases, the guiding principle for including or excluding a variable in the regression analyses was their overlapping meaning and high significant (positive or negative) correlation with other variables. Table 7 shows the correlations between all possible pairs of dependent and independent variables, as well as some additional variables that were considered important to further clarify and interpret the results of the regression analyses.

Table 8 displays the results of the first stepwise multiple regression analysis with the frequency of PCU as the dependent variable and the independent variables that were determined as significant predictors of the dependent variable. The independent variables that contributed significantly to the prediction of frequency of PCU were teach-

ers' knowledge, self-confidence, and gender, and from the socio-technical factors those related to inspector support and computer equipment in the school. Teachers' knowledge was found to be the best predictor of teachers' frequency of PCU and alone explained 59% of the variance.

Teachers' self-confidence was found to be the second best predictor that contributed to a significant increment in *R2,* from .590 to .617. There were three other variables, namely gender, inspector support, and computer equipment that also contributed significantly to the prediction of teachers' frequency of PCU, but the amount of variance in frequency of PCU attributable to each one of them was much smaller, namely, .9%, .4%, and .6%, respectively. Although the correlations between teachers' frequency of PCU and some other variables (i.e., age, ICT inservice, and the remaining socio-technical factors) were significant, these variables were not found to be significant predictors of PCU.

For the purpose of further clarifying and interpreting the results of the first regression analysis, a careful examination of the pair-wise relationships between teachers' age, gender, knowledge, and ICT inservice training in Table 7 indicates that these variables were highly and significantly correlated, but only teachers' knowledge and gender proved to be significant predictors of the frequency of PCU. This can be explained by the fact that the younger teachers tended to have more knowledge than the older ones ($r = -.274$, $p = .01$), female teachers were in

Table 8. Multiple regression analysis of factors predicting teachers' frequency of personal computer use

Model	Variables	R	R^2	Adjusted R^2	Adjusted ΔR^2	F Change	Significance
1	Knowledge	.768	.590	.590	.590	693.205	.000
2	Confidence	.786	.617	.616	.026	33.594	.000
3	Gender	.792	.628	.625	.009	13.349	.000
4	Inspector Support	.795	.632	.629	.004	5.568	.019
5	Computer Equipment	.799	.635	.635	.006	9.361	.002

general younger than male teachers ($r = .225$, $p = .01$), and younger teachers were more inclined to participate in inservice ICT training ($r = -.190$, $p = .01$). Similarly, as shown in Table 7, all items corresponding to the socio-technical factors were highly and significantly correlated among each other, but not all of them were found to be significant predictors, as the unique contribution of many of them was not found to be significant. In multiple regression:

It is possible for a variable to appear unimportant in the solution when it actually is highly correlated with the dependent variable. If the area of that correlation is whittled away by other independent variables, the unique contribution of the independent variable is often very small despite a substantial correlation with the dependent variable. (Tabachnick & Fidell, 1989, p. 143)

Thus, from the list of the socio-technical factors, only inspector support and availability of computer equipment in the school were found to be significant predictors of the frequency of PCU. Interestingly, computer equipment was found to be an important predictor of PCU, although there was not a significant positive correlation between the two variables. This outcome is really difficult to explain, because computer equipment in the schools does not seem to be directly related to teachers' frequency of PCU outside the classroom. One possible explanation is that teachers began to use computers for personal purposes after the

introduction of computers in the schools, which possibly served as the impetus for teachers to learn how to use computers.

Table 9 displays the results of the second stepwise multiple regression analysis between frequency of ICU as the dependent variable and the significant independent variables. The independent variables that contributed significantly to the prediction of the frequency of ICU were teachers' self-confidence, instructional support in the school, teachers' attitudes, support from colleagues, and knowledge. Teachers' self-confidence was the best predictor of the frequency of ICU, but it could explain only 9.1% of the variance. The variables of instructional support, teachers' attitudes, support from other colleagues, and teachers' knowledge also contributed to a significant increment in $R2$. These variables could explain 2.4%, 2.5%, 0.8%, and 0.8% of the variance in the frequency of ICU, respectively. The total amount of variance attributable to these variables was only 16.8%.

Interestingly enough, the significant predictor of self-confidence was removed from the regression equation in step six, as its unique contribution was no longer significant. When, in step 7, age was introduced, there was a significant increase of $\Delta R2$ that made the total amount of variance, after partialling out the contribution of teachers' self-confidence, significantly higher. In stepwise regression, "independent variables are added one at a time if they meet statistical criteria, but they also may be deleted at any step where they

Table 9. Multiple regression analysis of factors predicting teachers' frequency of instructional use

Model	Variables	R	R²	Adjusted R²	ΔR²	F Change	Significance
1	Self-Confidence	.301	.091	.089	.089	48.778	.000
2	Instructional Support	.342	.111	.113	.024	14.358	.000
3	Attitudes	.378	.143	.138	.025	14.785	.000
4	Colleagues	.391	.152	.146	.008	5.448	.020
5	Knowledge	.404	.163	.154	.008	6.125	.014
6	**Self-Confidence (removed)**	**.400**	**.160**	**.153**	**-.003**	**1.761**	**.185**
7	Age	.410	.168	.160	.008	4.743	.030

no longer contribute significantly to prediction" (Tabachnick & Fidell, 1989, p. 147). Thus, the total amount of variance attributable to instructional support, teachers' attitudes, and support from colleagues, knowledge, and age, was found to be 16.8%. Teachers' self-confidence that was initially found to be the best predictor of ICU was in the end excluded from the list of significant predictors for two reasons. First, teachers' self-confidence, after the inclusion of four other variables (instructional support, attitudes, support from colleagues, and knowledge), did not contribute significantly to the prediction of the frequency of ICU and could be excluded without any significant decrease in *R2*. Second, the combination of teachers' knowledge and age was a better predictor of the frequency of ICU than teachers' self-confidence.

DISCUSSION AND IMPLICATIONS

The socio-technical systems model provides us with a framework to systematically identify factors that could possibly affect the integration of ICT in education. From this perspective, a questionnaire was used in this study to collect demographic data and information related to teachers' knowledge of ICT, frequency of using ICT for personal and instructional purposes, attitudes toward ICT, self-confidence in using ICT in teaching and learning, and school climate. The findings tend to support that female teachers were more inclined to participate in ICT inservice training; had better knowledge, attitudes and self-confidence related to ICT; and used ICT tools more frequently both for personal and instructional uses than male teachers. Teachers' knowledge of computer software and frequency of use for personal and instructional purposes were dependent on the type of software. For example, teachers' knowledge and frequency of use for personal purposes was mainly restricted to word processing, and to a much smaller extent to the Internet, e-mail, and educational CD-ROMs, while other software were almost unknown

and rarely used. Teachers expressed somewhat positive attitudes towards the use of computers in education, but they also expressed concerns pertaining to technical computer problems that might hinder their work and students' learning. They also felt, to some extent, self-confident in selecting appropriate software to be used in their teaching, and somewhat confident in designing and implementing classroom activities with ICT tools. Teachers' attitudes and self-confidence seem to be delimited by their restricted knowledge of software and frequency of software use for personal and instructional purposes.

Teachers also reported infrequent instructional use or no instructional use even for software that they frequently used for personal purposes outside the classroom. This discrepancy seems to be attributable to the rather neutral socio-technical environment that existed in their schools, but also to other factors. Specifically, a stepwise regression analysis indicated that 63.5% of the variance in teachers' frequency of personal computer use could be predicted by teachers' knowledge, self-confidence, gender, inspector support, and computer equipment. However, according to a second stepwise regression analysis, only 16.8% of teachers' frequency of instructional computer use could be predicted by instructional support from officials, teachers' attitudes, support from the other teachers in the school, knowledge, and age.

The findings indicate that teachers in Cyprus are not illiterate in terms of having ICT skills and in terms of using ICT for personal purposes. What the findings clearly show is that teachers do not feel empowered to actively use ICT in authentic teaching and learning activities. Teachers need to develop confidence in their own professional activities and realize that what they are doing is right and important for their students' education. As Gable and Easingwood (2000) state, it will take time to train teachers to fully appreciate the power of ICT, but it is crucial to invest in such efforts, so that teachers fully appreciate the philosophical aspects of what they are doing, rather than just learning how to use the computer. These results indicate that policymakers in Cyprus have to seriously consider the lack of learning opportunities for the development of new literacy in Cyprus schools, and make coordinated efforts for providing a different and better kind of training to teachers. This training should pay attention not only to teachers' technical expertise, but also their attitudes, self-confidence, and in-depth understanding of ICT's affordances and added value in teaching and learning targeting an extended concept of literacy. Teacher professional development about the instructional uses of ICT in the classroom and about computers as learning tools for providing us with new forms of media that can enrich learner communication and expression is absolutely in great need.

Along the same line of reasoning, teacher education departments must also consider the quality of their curricula and adapt them appropriately, so that they adequately prepare teachers to integrate ICT in teaching and learning. We argue that teacher preparation, inservice or preservice, should focus on new interactive computer-based technologies, such as electronic communication systems, visualization and dynamic systems modeling tools, simulations, and networked multimedia environments, for scaffolding and amplifying students' thinking (Bransford, Brown, & Coccking, 2001). These tools are known as cognitive tools or mindtools (Jonassen, 2000), because they engage learners in meaningful thinking to analyze, critically think about the content they are studying, and organize and represent what they know. Jonassen, Carr, and Yueh (1998) state that "learning with mindtools depends on the mindful engagement of learners in the tasks afforded by these tools and that there is the possibility of qualitatively upgrading the performance of the joint system of learner plus technology" (p. 40). Therefore, mindtools require learners to think harder about the content being studied, and engage them in thinking that would be impossible without the tools. Finally, the tools we use and the way we use them shape our experiences and our thinking (Vygotsky, 1978) and impact our literacy. Thus, "if technology is to be viewed as an add-on in the learning environment that is pursued for the sake of technology alone, then it will not change education" (Valanides, 2003, p. 45), because technology, in and of itself, cannot influence learning, no matter how powerful it might be. On the other hand, if technology is utilized as a cognitive tool that has added value in certain instructional situations, then it will become a driving force for systemic educational change to help teachers and students to experience deep learning and acquire an extended concept of literacy that is compatible with the needs of our society.

The overall findings of the study indicate that ICT is not systematically integrated in Cyprus schools and is not an important part of everyday classroom practices. This seems to be related to several reasons, such as teachers' limited knowledge of a variety of software, limited instructional support provided to teachers by the Ministry of Education, teachers' somewhat positive attitudes, lack of a true community of practice in the schools where teachers help each other to integrate ICT in teaching and learning, and teachers' age. It seems that a supportive school environment could play an important role in effectively and successfully integrating ICT in teaching and learning. Teacher support can be provided in each school by the more experienced teachers in the school or even by more experienced teachers in different schools, by inspectors who visit the school in order to assist teachers in their ICT integration efforts, or by an Instructional Support Service in the Ministry of Education that is responsible for providing instructional guidance to practicing teachers. Moreover, a supportive school environment can eliminate teachers' feelings of isolation in the school, and can encourage effective communication and collaboration among teachers for achieving common goals and literacy in education.

In addition, when ICT is integrated into the classroom environment, the learning environment becomes more learner centered than before, and new assessment strategies are needed in order to capture the essence of learning that takes place in these environments. Traditionally, assessment has been used to sort out students, as well as distinguish the good students from the weak students and, as the end-point of instruction, to assess students' understandings after the instruction ended (Graue, 1993). Hence, the focus of evaluating student learning has been on the products or outcomes of learning, such as facts and information, and not the processes of learning. In ICT-enhanced classrooms, learning objectives vary from achieving deep understanding of concepts to developing critical thinking, decision making, and problem-solving skills, to cultivating positive attitudes towards learning. Therefore, as Shepard argues (2000), the form and content of assessment must change to "capture important learning goals and processes and to more directly connect assessment to ongoing instruction" (p. 5). If the focus of assessment does not change and if new assessment strategies are not developed and accepted as valid methods for assessing student performance, then, as we strongly believe, teachers will hesitate to generously use ICT in their teaching.

Another factor that we consider important, even though it was not found to be a significant predictor in this study, is the lack of adequate computer access in Cyprus schools. For example, given the current situation in Cyprus schools, a teacher who wants to use ICT in a lesson must first make special arrangements to reserve the computer lab in the school in order to be allowed to use it. It seems, however, that because at this point teachers do not use ICT regularly in their classroom practices, they feel that the one computer lab in the school provides them with sufficient computer access. McFarlane (2000) also argues that computer access is a key factor in inhibiting teaching with ICT and states that "until children come to school with a powerful portable computer of their own, access will remain a key brake on the use of digital media in school" (p. 22).

In conclusion, the schools in Cyprus do not seem to be adequately preparing students to develop the new forms of literacy skills that are needed in the information society, and have not been affected to a great extent by new modes of communication, new tools for expression, and new ways of the representation of knowledge. Given the current situation, it is hard to see how new forms of literacy can be satisfactorily developed in Cyprus schools. These findings have implications for Cyprus' international competitiveness. If the educational system in Cyprus will not invest in learning with ICT, then the students in Cyprus will not develop the competencies and the literacy skills that are needed to fairly compete with the students of other countries, which have a better status of ICT in education. Technological illiteracy "could lead to becoming a member of an underclass with a similar status to those who, in previous generations, could not read and write" (Easingwood, 2000, p. 97).

The development of an extended concept of literacy is not an easy matter and many factors seem to affect its development. The implications of this study for the development of an extended concept of literacy are important and need to be seriously taken into consideration by policymakers. It seems that policymakers and government officials have to systemically approach the issue of ICT integration in primary and secondary education, so that ICT is infused in a system that is ready to accept the new educational change. The results imply that a systemic effort for the development of an extended concept of literacy should include a focus on creating a supportive school environment, and a revised focus on teaching and teacher training.

Another implication for the development of an extended concept of literacy is that plans of action have to be developed to identify areas in the curriculum that can be enhanced with the use of ICT. Currently, the curriculum in Cyprus does not include a focus on ICT, and does not appear to have a direction and urgency in systematically integrating ICT. Specifically, the official curriculum in primary and secondary education does not currently include the use of ICT in the teaching of the subject domains despite the fact that ICT integration has been proclaimed as a top priority in the agenda of policymakers. Thus, in the present system, the teacher has to decide how and when to integrate ICT in teaching and learning. Curriculum restructuring efforts need to be undertaken so that teachers receive better guidance about how ICT can be integrated in different subject matters and how ICT can extend the traditional concept of literacy.

CONCLUSION

The purpose of this chapter was to examine factors that may affect teaching with ICT in primary and secondary education, and thus ultimately hinder or delay the development of new literacy skills that are important for citizens to survive in a rapidly changing world. Based on the findings of the study, the development of an extended concept of literacy is not easy and many factors seem to affect its growth, such as teachers' knowledge of ICT, attitudes, self-confidence, age, and instructional support from colleagues and superiors. We argued in this chapter that policymakers need to carefully plan the development of the new forms of literacy in Cyprus schools by systemically integrating ICT in the schools so teachers and students together can develop an extended concept of literacy that is critical for surviving in the information society.

REFERENCES

Angeli, C. (2003). A systemic model of technology integration. Paper presented at the *American Educational Research Association Conference*, Chicago.

Apple, M.W. (1986). *Teachers and texts: A political economy of class and gender relations in Education.* London: Routledge and Kegan Paul.

Baca, J.C. (1990). *Identification by consensus of the critical constructs of visual literacy: A Delphi study.* Unpublished doctoral dissertation, East Texas State University, USA.

Barth, R. (1990). *Improving schools from within: Teachers, parents and principals can make the difference.* San Francisco: Jossey-Bass.

Bransford, J. D., Brown, A. L., & Cocking, R. R. (Eds.). (2001). *How people learn: Brain, mind, experience, and school.* Washington, DC: National Academy Press.

Brindly, S. (2000). ICT and literacy. In N. Gamble & N. Easingwood (Eds.), *ICT and literacy: Information and communications technology, media, reading and writing* (pp. 11-18). London: Continuum.

Crystal, D. (1987). *The Cambridge encyclopedia of language.* Cambridge: Cambridge University Press.

Denning, P. J. (2001). The IT schools movement. *CACM, 44*(8), 19-22.

DfEE. (1998). *The national literacy strategy: Framework for teaching.* London: HMSO.

Easingwood, N. (2000). Electronic communication in the twenty-first-century classroom. In N. Gamble & N. Easingwood (Eds.), *ICT and literacy: Information and communications technology, media, reading and writing* (pp. 45-57). London: Continuum.

Eraut, M., & Hoyles, C. (1989). Group work with computers. *Journal of Computer Assisted Learning, 5,* 12-24.

Friedman, B. (Ed.). (1998). *Human values and the design of computer technology.* Cambridge: Cambridge University Press.

Fullan, M. (1991). *The new meaning of educational change.* New York: Teachers College Press.

Fullan, M., & Hargreaves, A. (1992). *What's worth fighting for in your school?* Buckingham: Open University Press.

Gamble, N., & Easingwood, N. (Eds.). (2000). *ICT and literacy: Information and communications technology, media, reading and writing.* London: Continuum.

Getzels, J. W., & Cuba, E. G. (1957). Social behavior and the administrative process. *Social Review, 65,* 423-441.

Glaser, W. (1981). *Stations of the mind.* New York: Harper & Row.

Graue, M. E. (1993). Integrating theory and practice through instructional assessment. *Educational Assessment, 1,* 293-309.

Hanna, D. (1997). The organization as an open system. In A. Harris, N. Bennett, & M. Preedy (Eds.), *Organizational effectiveness and improvement in education* (pp. 13-20). Philadelphia: Open University Press.

Heppell, S. (2000). Foreword. In N. Gamble & N. Easingwood (Eds.), *ICT and literacy: Information and communications technology, media, reading and writing* (pp. xi-xv). London: Continuum.

Heracleous, L., & Barrett, M. (2001). Organizational change as discourse: Communicative actions and deep structures in the context of informational technology implementation. *Academy of Management Journal, 44*(4), 755-778.

Hoy, W. K., & Ferguson, J. (1985). A theoretical framework and exploration of organizational effectiveness in schools. *Educational Administration Quarterly, 21,* 117-134.

Hoy, W. K., & Miskel, G. C. (2001). *Educational administration: Theory, research, and practice* (6th ed.). New York: McGraw Hill.

Huberman, M., & Miles, M. B. (1984). *Innovation up close.* New York: Plenum.

Kemster, G. (2000). Skills for life: New meanings and values for literacies. In N. Gamble & N. Easingwood (Eds.), *ICT and literacy: Information and communications technology, media, reading and writing* (pp. 25-30). London: Continuum.

Jonassen, D. H. (2000). *Computers as mindtools for schools: Engaging critical thinking* (2nd ed.). Upper Saddle River, NJ: Prentice-Hall.

Jonassen, D. H., Carr, C., & Yueh, H. -P. (1998). Computers as mindtools for engaging learners in critical thinking. *TechTrends, 34*(2), 24-32.

Kling, R. (2000). Learning about information technologies and social change: The contribution of social informatics. *The Information Society, 16*(3), 217-232.

Kling, R., & Lamb, R. (2000). IT and organizational change in digital economies: A socio-technical approach. In B. Kahin & E. Brynjolfsson (Eds.), *Understanding the digital economy: Data, tools, and research.* Boston: MIT Press.

Kozma, R., & Anderson. R. E. (2002). Qualitative case studies of innovative pedagogical practices using ICT. *Journal of Computer Assisted Learning, 18,* 387-394.

Lenine, T., & Donitsa-Schmidt, S. (1998). Computer use, confidence, attitudes, and knowledge: A causal analysis. *Computers in Human Behavior, 14*(1), 125-146.

Liaw, S. -S. (2002). Understanding user perceptions of World Wide Web environments. *Journal of Computer Assisted Learning, 18,* 137-148.

Louis, K. S., & Miles, M. M. (1991). *Improving the urban high school: What works and why.* London: Cassell.

Markus, H., & Ruvulo, A. (1992). "Possible selves." Personalized representation of goals. In L. Pervin (Ed.), *Goal concepts in psychology.* Hillsdale, NJ: Lawrence Erlbaum.

Markus, M. L., & Benjamin, R. I. (1987). The magic bullet theory in IT-enabled transformation. *Sloan Management Review, 38*(2), 55-68.

Marzano, R. J. (1992). *A different kind of classroom: Teaching with dimensions of learning.* Alexandria, VA: ASCD.

McFarlane, A. (2000). Communicating meaning—Reading and writing in a multimedia world. In N. Gamble & N. Easingwood (Eds.), *ICT and literacy: Information and communications technology, media, reading and writing* (pp. 19-24). London: Continuum.

McFarlane, A. (Ed.). (1997). *Information technology and authentic learning: Realizing the potential of computers in the primary classroom.* London: Routledge.

Murphy, C. A., Coover, D., & Owen, S. V. (1989). Development and validity of the computer self-efficacy scale. *Educational and Psychological Measurement, 49,* 893-899.

Papert, S. (1993). *The children's machine: Rethinking school in the age of the computer.* New York: Basic Books.

Pelgrum, W. (2001). Obstacles to the integration of ICT in education: Results from a worldwide educational assessment. *Computers and Education, 37,* 163-178.

Picciano, A. G. (2002). *Educational leadership and planning for technology* (3rd ed.). Upper Saddle River, NJ: Prentice-Hall.

Reigeluth, C. M., & Garfinkle, R. J. (Eds.). (1994). *Systemic change in education.* Englewood Cliffs, NJ: Educational Technology Publications.

Salomon, G., Perkins, D. N., & Globerson, T. (1991). Partners in cognition: Extending human intelligence with intelligent technologies. *Educational Researcher, 20*(3), 2-9.

Shepard, L. (2000). The role of assessment in a learning culture. *Educational Researcher, 29*(7), 1-14.

Snyder, I. A. (1994). Writing with word processors: A research overview. *Journal of Curriculum Studies, 26,* 43-62.

Street, B. V. (1987). Models of computer literacy. In R. Finnegan (Ed.), *Information technology social issues.* London: Sevenoaks, Hodder and Stoughton.

Tabachnick, B. G., & Fidell, L. S. (1989). *Using multivariate statistics* (2nd ed.). New York: Harper & Row.

Trach, L., & Woodman, R. (1994). Organizational change and information technology: Managing on the edge of cyberspace. *Organizational Dynamics, 23,* 30-46.

Trist, E. L. (1982). The development of socio-technical systems as a conceptual framework and as an action research program. In A. H. Van de Ven & W. F. Joyce (Eds.), *Perspectives on organizational change and behavior* (pp. 19-75). New York: John Wiley & Sons.

Trist, E. L., & Bamforth, K. W. (1951). Some social and psychological consequences of the Longwall method of goal setting. *Human Relations, 4,* 3-38.

UNESCO. (1999). *The science agenda—Framework for action.* Paris: UNESCO.

Valanides, N. (2003). Learning, computers, and science education. *Science Education International, 14*(1), 42-47.

Valanides, N., & Angeli, C. (2002). Challenges in achieving scientific and technological literacy: Research directions for the future. *Science Education International, 13*(1), 2-7.

Vygotsky, L. S. (1978). *Mind in society.* Cambridge, MA: Harvard University Press.

Watt, D. H. (1980). Computer literacy: What should schools be doing about it? *Classroom Computer News, 1*(2), 1-26.

Wild, M. (1996). Technology refusal: Rationalizing the future of student and beginning teachers to use computers. *British Journal of Educational Technology, 27*(2), 134-143.

ENDNOTES

[1] We consider the last part of the 20th century to be the beginning of immense developments in information, communication, and network technologies.

[2] The term *ICT* is used in this study interchangeably with *computer applications*, and includes the Internet, the World Wide Web, and all types of computer software.

This work was previously published in the Handbook of Research on Literacy in Technology at the K-12 Level, edited by L. Tan and R. Subramaniam, pp. 604-625, copyright 2006 by Idea Group Reference (an imprint of IGI Global).

Chapter 7.24
Critical Success Factors for Distance Education Programs

Wm. Benjamin Martz, Jr.
University of Colorado at Colorado Springs, USA

Venkateshwar K. Reddy
University of Colorado at Colorado Springs, USA

INTRODUCTION

Distance education is playing an ever-growing role in the education industry. As such, it is prudent to explore and understand driving conditions that underlie this growth. Understanding these drivers and their corresponding concerns (Table 1) can help educators in the distance education field better prepare for the industry.

BACKGROUND

Distance education's primary driver is that it is the major growth segment in the education industry. In 1999, nearly 80% of the public, four-year institutions and over 60% of the public, two-year institutions offered distance education courses. Over 1.6 million students are enrolled in distance courses today. Over 90% of all colleges are expected to offer some online courses by 2004 (Institute of Higher Education Policy, 2000). Corporations envision online training warehouses saving large amounts of training dollars. Combined, the virtual education market and its sister market, corporate learning, are predicted to grow to over $21 billion by the end of 2003 (Svetcov, 2000).

A second major driver is employer expectations. Fundamental job market expectations are changing. Today, employees are not expected to stay in the same job for long periods of time; 20-plus year careers are not expected. The current modes of careers include multiple careers, combinations of part-time work in multiple jobs, telecommuting, leaving and re-entering into the full-time work force, switching jobs, and so forth, and today's employee easily accepts the need to maintain a level of knowledge current with the career demands (Boyatzis & Kram, 1999). To complement these changes in employer expectations, employees have begun to accept the need for life-long learning.

Table 1. Influences on the distance education industry

Drivers	Concerns
Growth segment in education industry	Retention
Job market expectations	Fading Back
Life-long learning as an education paradigm	Less social learning
Profit center for educational institutions	Trust & isolation
Possible strategic competence	Impact of technology

A third driver is the profit potential. Cost savings may be obtained and if significant enough may drive up demand and costs may be lowered. For example, elective classes that do not have enough students enrolled in them on-campus may pick up enough distance students to make teaching the course more feasible (Creahan & Hoge, 1998). A final driver is the institution's mission. Most educational institutions serve a geographical region, either by charter or mission, and a distance-learning program may be a practical method to help satisfy this strategic mission (Creahan & Hoge, 1998).

However, the "commercialization" of education raises its own concerns about the basic process of learning (Noble, 1999). For example, are there any problems fundamental to the distance environment because of limited social interaction?

Retention may be one such problem. Carr (2000) reports a 50% drop-out rate for online courses. Tinto (1975) compared the learning retention of distance groups with traditional groups and found that the social integration was a key factor in successful retention of traditional groups. Haythornthwaite et al. (2000) think they found another one. They looked at how social cues such as text without voice, voice without body language, class attendance without seating arrangements, and students signing in without attending Internet class impacted students "fading back." They found that the likelihood of students "fading back" is greater in distance-learning classes than in face-to-face classes. From the United Kingdom, Hogan and Kwiatkowski (1998) argue that the emotional aspects of this teaching

method have been ignored. Similar concerns are raised from Australia, where technology has been supporting distance- teaching for many years, as Hearn and Scott (1998) suggest that before adopting technology for distance teaching, education must acknowledge the social context of learning. Finally, two other factors, trust and isolation, have been researched by Kirkman et al. (2002), whereby communication helped improve the measures of trust in students using the virtual environment.

By definition, the paradigm of distance education changes the traditional education environment by expanding it to cover geographically dispersed learning. In turn, this means that students will probably respond differently to this environment than they do to the traditional classroom. In addition, academic researchers have always been interested in explaining how people react to the introduction of technology. This body of work can be useful to the distance education environment.

Poole and DeSanctis (1990) suggested a model called adaptive structuration theory (AST). The fundamental premise of the model is that the technology under study is the limiting factor or the constraint for communication. It further proposes that the users of the technology, the senders and the receivers, figure out alternative ways to send information over the channel (technology). A good example here is how a sender of e-mail may use combinations of keyboard characters or emoticons (i.e., :) – sarcastic smile, ;) – wink, :o – exclamation of surprise) to communicate more about their emotion on a subject to the receiver.

Ultimately, the key to realizing the potential of distance education is trading off the benefits and the concerns to produce a quality product. In the new Malcolm Baldridge evaluation criteria, companies are asked to better show a program's effectiveness through customer satisfaction. In turn, Gustafsson et al. (2000) show customer satisfaction linked significantly to quality at Volvo Car Corporation. Finally, in their more broad analysis of well-run companies, Peters and Waterman (1982) deemed customer satisfaction as a key factor contributing to the companies' performance.

With these perspectives in mind, we suggest that these areas interact to identify satisfaction as one important measure of quality for distance education programs. Therefore, one of the key factors to a program's success will be the satisfaction of one of its key stakeholders – its students. If one can identify what helps satisfies students in a distance education environment, one has a better chance to develop a successful program.

THE RESEARCH STUDY

The distance program used in this study is one of the largest, online, AACSB-accredited MBA programs in the world (US News and World Report, 2001). The methodology used a questionnaire with a battery of 49 questions to gather the data. The questions were developed using the concepts and ideas from literature discussed earlier as a guide.

Once the subject identified his or her reference course, that subject's grade was obtained from administrative records and recorded. In addition, four other demographic questions gathered information on gender, number of courses taken, student status, amount of time expected to spend in the reference course, and the amount of time actually spent in the reference course (Martz et al., 2004).

Two sets of questions were used. The first set asked about the student's use of different technologies (i.e., chat, e-mail, streaming video, etc.) in the class and if used, how effective (five-point Likert: 1 = LO …. 5 = HIGH) did they believe the technology to be in helping them with the class. We created a new variable, LOHITECH, for analysis purposes. Using LOHITECH, respondents can be placed in one of two groups: one group that reported using three or less technologies, while the second group reported using four or more technologies in their reference class. The second set of questions asked students to rate (five-point Likert: 1 = Strongly Agree …. 5 = Strongly Disagree) their experience with the reference distance course against statements concerning potential influences for satisfaction. These questions associated a five-point rating scale to statements about the issues identified earlier. The order of the questions was randomly determined and the questionnaire was reviewed for biased or misleading questions by non-authors.

The questionnaire was sent to 341 students enrolled in the distance MBA program. In Fall 2002, the program served 206 students from 39 states and 12 countries. The majority of these students are employed full-time. The program used in this study has been running since Fall 1996 and has over 179 graduates. It offers an AACSB accredited MBA and its curriculum parallels the on-campus curriculum. Close to 33% of the enrolled students are female. The oldest student enrolled is 60 years old and the youngest is 22. The average age of all students enrolled is 35. Over 25 PhD qualified instructors participate in developing and delivering the distance program annually. Recently, the news magazine *US News and World Report* (2001) classified the program as one of the top 26 distance education programs.

There were 131 useable questionnaires returned. The students' final grade for their reference course was obtained and added to the questionnaire record as a variable. These were separated into two groups: 30 that had not yet taken a course and 101 that had completed at least one course. This second group, those students who had completed at least one course, provided the focus for this study.

RESEARCH RESULTS

Question 24, "Overall, I was satisfied with the course," was used as the subject's level of general satisfaction. The data set was loaded into SPSS for analysis. Table 2 shows that 23 variables, including LOHITECH, proved significantly correlated to satisfaction (Q24).

The large number of significant variables leads to the need for a more detailed analysis on how to group them (StatSoft, 2002). Kerlinger (1986, p. 590) suggests the use of factor analysis in this case "to explore variable areas in order to identify the factors presumably underlying the variables". An SPSS factor analysis was performed with a Varimax Extraction on those questions that had proven significantly correlated to satisfaction. All reliability coefficients (Cronbach Alpha) are above .7000 and all Eigenvalues are above 1.00, indicating an acceptable level for a viable factor (Kline, 1993; Nunnally, 1978). Finally, the five components explain 66.932% of the variance.

In summary, 22 variables from the questionnaire proved significantly correlated to satisfaction. A factor analysis of those 22 variables extracted five possible constructs. These constructs were labeled: Interaction with the Professor; Fairness; Content of the Course; Classroom Interaction; and Value, Technology & Learning, based upon the key characteristics of the underlying questions. Table 3 shows the results of combining the ratings for the questions in each construct and correlating each of them to satisfaction. As can be seen

Table 2. Questions that correlate significantly to satisfaction

ID	Question Statement	Correlation	
		Coef.	Sign.
16	I was satisfied with the content of the course	.605	.000
17	The tests were fair assessments of my knowledge	.473	.000
18	I would take another distance course with this professor	.755	.000
19	I would take another distance course	.398	.000
20	The course workload was fair	.467	.000
21	The amount of interaction with the professor and other students was what I expected.	.710	.000
22	The course used groups to help with learning	.495	.000
23	I would like to have had more interaction with the professor.	-.508	.000
26	The course content was valuable to me personally	.439	.000
28	Grading was fair	.735	.000
30	Often I felt "lost" in the distance class	-.394	.000
31	The class instructions were explicit	.452	.000
33	Feedback from the instructor was timely	.592	.000
34	I received personalized feedback from the instructor	.499	.000
36	I would have learned more if I had taken this class on-campus (as opposed to online)	-.400	.000
37	This course made me think critically about the issues covered.	.423	.000
38	I think technology (email, web, discussion forums) was utilized effectively in this class	.559	.000
39	I felt that I could customize my learning more in the distance format	.254	.001
42	The course content was valuable to me professionally	.442	.000
43	I missed the interaction of a "live," traditional classroom	-.341	.002
46	Overall, the program is a good value (quality/cost)	.258(1)	.017
LOHITECH	Aggregate of Yes votes in Q6 through Q15	.270(1)	.012
(1) While significant, the low correlation coefficient below .300 should be noted			

from the table, the constructs hold up well as five indicators of satisfaction.

FUTURE TRENDS

As mentioned earlier, the organization, the school in this case, is a key stakeholder in the success of a distance education program. The future success of distance programs depends largely on satisfying these critical success factors. Distance education courses and programs are not only used for providing an alternative delivery method for students but also to generate revenues for the offering unit/college/university. As the number of distance courses and programs increase at an exponential rate, the necessity to enhance quality and revenues also takes prominence. We conclude with a set of operational recommendations that can impact online program success (Table 4).

The data in this study indicate that a timely and personalized feedback by professors results in a higher level of satisfaction by students. The administrators therefore have to work closely with their faculty and offer them ways to enrich the teacher-student relationships. Paradoxically, a faculty member needs to use technologies to add a personal touch to the virtual classroom. For example, faculty should be encouraged to increase the usage of discussion forums, respond to e-mail within 24 to 48 hours, and keep students up-to-date with the latest happenings related to the course.

The data also indicate that good course content and explicit instructions increase student satisfaction in the virtual classroom. It may well be that this basically sets and manages the expectations for the distance student. This result suggests that faculty should have complete Web sites with syllabi and detailed instructions. In turn, this

Table 3. Correlation of final constructs to satisfaction

Construct (Component: Loading)	Correlation	Significance
Professor Interaction (Q18: .576, Q21: .643, Q33: .794, Q34: .849)	.771	.000
Fairness (Q17: .722, Q20: .738, Q28: .626, Q31: .512)	.695	.000
Course Content (Q16: .596, Q26: .850, Q39: .689, Q42: .825)	.588	.000
Classroom Interaction (Q23: -.354, Q30: -.514, Q36: -.809, Q43: -.770)	-.515	.000
Technology Use & Value (LOHITECH: .508, Q19: .596, Q22: .542, Q37: .494, Q38: .478, Q46: .700)	.624	.000

Table 4. Recommendations to increase online program success

1	Have instructors use a 24-48-hour turnaround for e-mail.
2	Have instructors use a 1-week turnaround for graded assignments.
3	Provide weekly "keeping in touch" communications.
4	Provide clear expectation of workload.
5	Provide explicit grading policies.
6	Explicitly separate technical and pedagogical issues.
7	Have policies in place that deal effectively with technical problems.
8	Provide detailed unambiguous instructions for coursework submission.
9	Provide faculty with instructional design support.
10	Do not force student interaction without good pedagogical rationale.
11	Do not force technological interaction without good pedagogical purpose.
12	Collect regular student and faculty feedback for continuous improvement.

suggests that distance education administrators should focus their attention on providing faculty with support such as good Web site design, instructional designer support, test design, user interaction techniques, and so forth, appropriate for distance learning.

Since distance students' notion of value intertwines learning and technology, it is imperative that distance administrators offer, and faculty use, the available technology in the distance program. Technology in this case not only refers to the actual software and hardware features of the platform but also how well technology is adapted to the best practices of teaching. The results imply that if technology is available but not used, it lowers satisfaction. So, technology options that are not being used in a course should not appear available. For the program administrator, this would suggest adoption of distance platforms that are customizable at the course level with respect to displaying technological options.

CONCLUSION

This study attempts to identify potential indicators for satisfaction with distance education. A body of possible indicators was derived from the literature surrounding the traditional versus virtual classroom debate. A 49-question questionnaire was developed from the indicators and was administered to MBA students in an established distance education program. One hundred and one questionnaires from students with one or more distance classes were analyzed with the result that 22 variables correlated significantly to satisfaction. A factor analysis of the questionnaire data extracted five basic constructs: Professor Interaction, Fairness, Course Content, Classroom Interaction and Technology Use & Value. Several recommendations for implementing and manag-

ing a distance program were extracted from these constructs and discussed.

REFERENCES

Boyatzis, R.E., & Kram, K.E. (1999, Autumn). Reconstructing management education as lifelong learning. *Selections, 16*(1), 17-27.

Carr, S. (2000, February 11). As distance education comes of age the challenge is keeping students. *Chronicle of Higher Education.*

Creahan, T.A., & Hoge, B. (1998, September). *Distance learning: Paradigm shift of pedagogical drift?* Presentation at Fifth EDINEB Conference, Cleveland, OH.

Gustafsson, A., Ekdahl, F., Falk, K., & Johnson, M. (2000, January). Linking customer satisfaction to product design: A key to success for Volvo. *Quality Management Journal, 7*(1), 27-38.

Haythornthwaite, C., Kazmer, M.M., Robins, J., & Showmaker, S. (2000, September). Community development among distance learners. *Journal of Computer-Mediated Communication, 6*(1).

Hearn, G., & Scott, D. (1998, September). Students staying home. *Futures, 30*(7), 731-737.

Hogan, D., & Kwiatkowksi, R. (1998, November). Emotional aspects of large group teaching. *Human Relations, 51*(11), 1403-1417.

Institute for Higher Education Policy. (2000). *Quality on the line: Benchmarks for success in Internet distance education.* Washington, D.C.

Kerlinger, F.N. (1986). *Foundations of behavioral research* (3rd ed.). Holt, Rinehart & Winston.

Kirkman, B.L., Rosen, B., Gibson, C.B., Etsluk, P.E., & McPherson, S. (2002, August). Five challenges to virtual team success: Lessons from Sabre, Inc. *The Academy of Management Executive, 16*(3).

Kline, P. (1993). *The handbook of psychological testing.* London: Routledge.

Martz, W.B, Reddy, V., & Sangermano, K. (2004). Assessing the impact of Internet testing: Lower perceived performance. In C. Howard, K. Schenk & R. Discenza (Eds.), *Distance learning and university effectiveness: Changing educational paradigms for online learning.* Hershey, PA: Idea Group Publishing.

Noble, D.F. (1999). *Digital diplomas mills.* Retrieved November 28, 2002, from http://www.first-monday.dk/issues/issue3_1/noble/index.html

Nunnally, J. (1978). *Psychometric theory.* New York: McGraw-Hill.

Peters, T.J., & Waterman, R.H., Jr. (1982). *In search of excellence.* New York: Harper and Row.

Poole, M.S., & DeSanctis, G. (1990). Understanding the use of group decision support systems: The theory of adaptive structuration. In J. Fulk & C. Steinfeld (Eds.), *Organizations and communication technology* (pp. 173-193). Newbury Park, CA: Sage Publications.

Rockart, J.F. (1979, March-April). Chief executives define their own data needs. *Harvard Business Review.*

Statsoft. (2002). Retrieved November 30, 2002, from *http://www.statsoftinc.com/textbook/stfacan.html*

Svetcov, D. (2000). The virtual classroom vs. the real one. *Forbes,* 50-52.

Tinto, V. (1975). *Leaving college.* University of Chicago Press.

US News and World Report. (2001, October). *Best online graduate programs.*

KEY TERMS

Classroom Interaction: The interaction that can only be achieved face-to-face in a classroom. For example, the real-time feedback of facial expressions is not (yet) available in a distance course and so would be considered "classroom interaction".

Concerns of "Commercialization": The negative factors that the implantation and use of distance education may create.

Course Content: The main themes covered in a course.

Critical Success Factors: The few key areas in which activities must "go right" so that a project of program succeeds (Rockart, 1979).

Exploratory Factor Analysis: A process used to identify statistically significant constructs underlying a set of data.

Fairness: A subjective term defining the level to which a student feels he or she was treated fairly by the professor with respect to the class, including but not limited to test questions, grading, schedule flexibility, and so forth.

Market Drivers for Distance Education: The key elements that seem to be driving the diffusion and usage of distance education in the marketplace.

Professor Interaction: The amount of communication (e-mail, phone calls, video, chat rooms, etc.) that occurs between a student and the professor.

Satisfaction Constructs for Distance Education: Five constructs identified that seem to help identify satisfaction in distance education programs.

Technology Use: The usage of a technology whether it be e-mail, chat rooms, automated tests, software, and so forth.

Technology Value: The user's benefits (perceived and actual) over the costs (perceived and actual) created by the use of technology.

This work was previously published in the Encyclopedia of information Science and Technology, Vol. 1, edited by M. Khosrow-Pour, pp. 622-627, copyright 2005 by Idea Group Reference (an imprint of IGI Global).

Chapter 7.25
Understanding Cognitive Processes in Educational Hypermedia

Patricia M. Boechler
University of Alberta, Canada

INTRODUCTION

Cognitive load theory (CLT) is currently the most prominent cognitive theory pertaining to instructional design and is referred to in numerous empirical articles in the educational literature (for example, Brünken, Plass, & Leutner, 2003; Chandler & Sweller, 1991; Paas, Tuovinen, Tabbers, & Van Gerven, 2003; Sweller, van Merri‚nboer, & Paas, 1998). CLT was developed to assist educators in designing optimal presentations of information to encourage learning. CLT has also been extended and applied to the design of educational hypermedia and multimedia (Mayer & Moreno, 2003). The theory is built around the idea that the human cognitive architecture has inherent limitations related to capacity, in particular, the limitations of human working memory. As Sweller et al. (pp. 252-253) state:

The implications of working memory limitations on instructional design cannot be overstated. All conscious cognitive activity learners engage in occurs in a structure whose limitations seem to preclude all but the most basic processes. Anything beyond the simplest cognitive activities appear to overwhelm working memory. Prima facie, any instructional design that flouts or merely ignores working memory limitations inevitably is deficient. It is this factor that provides a central claim to cognitive load theory.

In order to understand the full implications of cognitive load theory, an overview of the human memory system is necessary.

BACKGROUND

The Human Memory System: The Modal Model of Memory

It has long been accepted that the human memory system is made up of two storage units: long-term memory and working memory. There is an abundance of behavioral (for example, Deese & Kaufman, 1957; Postmand & Phillips, 1965) and neurological evidence (Milner, Corkin, & Tueber, 1968; Warrington & Shallice, 1969) to support this theory. Long-term memory is a repository for information and knowledge that we have been exposed to repetitively or that has sufficient meaning to us. Long-term memory is a memory store that has an indefinable duration but is not conscious; that is, any information in long-term memory must first be retrieved into working memory for us to be aware of it. Hence, any conscious manipulation of information or intentional thinking can only occur when this information is available to working memory. The depth and duration of processing in working memory determines whether information is passed on to long-term memory. Once knowledge is stored in long-term memory, we can say that enduring learning has occurred.

Working Memory Limitations

Unfortunately, working memory has some very definite limitations. First, there is a limit of volume. Baddeley, Thomson, and Buchanan (1975) reported that the size of working memory is equal to the amount of information that can be verbally rehearsed in approximately 2 seconds. A second limitation of working memory concerns time. When information is attended to and enters working memory, if it is not consciously processed, it will decay in approximately 20 seconds.

CLT AND EDUCATIONAL HYPERMEDIA

The modal model of human memory, specifically these limitations of working memory, is the basis for CLT. A version of CLT, Mayer and Moreno's (2003) selecting-organizing-integrating theory of active learning, is specifically targeted to learn-

ing in hypermedia environments. The theory is built upon three core assumptions from the modal model of memory: the dual channel assumption, the limited capacity assumption, and the active processing assumption. The dual channel assumption is based on the notion that working memory has two sensory channels, each responsible for processing different types of input. The auditory or verbal channel processes written and spoken language. The visual channel processes images. The limited capacity assumption applies to these two channels; that is, each of these channels has a limit as to the amount of information that can be processed at one time. The active processing assumption is derived from Wittrock's (1989) generative learning theory and asserts that substantial intentional processing is required for meaningful learning. With these assumptions as a foundation, Mayer and Moreno have focused on three key mental activities that can place demands on available cognitive resources: attention, mental organization, and integration.

Improving Working Memory Capacity Directly

How does CLT advocate improving working memory limitations? To date, the solution for reducing cognitive load has focused on directly reducing the demands on working memory. Mayer and Moreno (2003) outline a number of methods for reducing cognitive load in hypermedia: (a) Resting on the dual channel assumption, cognitive load on one channel can be relieved by spreading information across both modalities, that is, by providing information in both a visual and auditory format, (b) presenting material in segments and providing pretraining on some material can reduce overload, (c) the redundancy of information can be eliminated, and (d) visual and auditory information can be synchronized.

Mayer and Moreno (2003) also refer to "incidental processing" as "cognitive processes that are not required for making sense of the presented material but are primed by the design of the learning task" (p. 45). Incidental processing is considered undesirable as it relates to the cognitive resources that are needed to process extraneous, irrelevant material that may be included on the presentation. Mayer and Moreno advocate weeding out this extraneous material to reduce cognitive load.

Measuring Cognitive Load

If the premise of cognitive load theory is correct, then certainly a primary activity in designing instructional materials must be the meaningful measurement of cognitive load. This is not a simple task as the method of measurement is dependent on the constructs that different researchers use to describe cognitive load. For example, Paas et al. (2003) propose that three constructs define cognitive load: mental load, which reflects the interaction between task and subject characteristics; mental effort, which reflects the actual cognitive reserves that are expended on the task; and performance, which can be defined as the learner's achievements. Previous research in cognitive load measurement has relied on three types of measures to assess the cognitive load of the user: (a) physiological measures such as heart rate and pupillary responses, (b) performance data on primary and secondary tasks, and (c) self-reported ratings (Paas et al.). These tasks have been used in various configurations to measure overall cognitive load (Brünken et al., 2003; Chandler & Sweller, 1996; Gimino, 2002; Paas, 1992). To date, most efforts to measure cognitive load have focused on self-reported ratings (see Paas et al.).

FUTURE TRENDS

Our ability to reduce cognitive load in educational hypermedia rests on our thorough definition of the underlying constructs of cognitive load as well as the design of test mechanisms that allow us to measure cognitive load and detect situations where cognitive resources are overtaxed. Future research directed at these two issues will contribute to the explanatory power of the theory and allow us to apply these theoretical principles to educational settings that make use of hypermedia materials.

CONCLUSION

The cognitive load theory for educational hypermedia has emerged as a prominent theory for guiding instructional designers in the creation of educational hypermedia. It is based on the modal model of human memory, which posits that there are limits to the working memory store that impact the amount of cognitive effort that can be expended on a given task. When available cognitive resources are surpassed, performance on memory and learning tasks is degraded, a condition referred to as cognitive overload. CLT for educational hypermedia advocates that educational materials must be designed that take into account these limitations. In order to do this, two obstacles to using CLT to its full advantage must be resolved: (a) the diversity of the descriptions of its underlying constructs and (b) the lack of valid and reliable methods for the measurement of cognitive load.

REFERENCES

Baddeley, A. (1992). Working memory. *Science, 255*, 556-559.

Baddeley, A., Thomson, N., & Buchanan, M. (1975). Word length and the structure of short-term memory. *Journal of Verbal Learning & Verbal Behavior, 14*, 575-589.

Brünken, R., Plass, J. L., & Leutner, D. (2003). Direct measurement of cognitive load in multi-dimensional learning. *Educational Psychologist, 38*(1), 53-61.

Chandler, P., & Sweller, J. (1991). Cognitive load theory and the format of instruction. *Cognition and Instruction, 8*, 293-332.

Chandler, P., & Sweller, J. (1996). Cognitive load while learning to use a computer program. *Applied Cognitive Psychology, 10*, 151-170.

Deese, J., & Kaufman, R. A. (1957). Serial effects in recall of unorganized and sequentially organized verbal material. *Journal of Experimental Psychology, 54*, 180-187.

Frensch, P. A., & Miner, C. S. (1994). Effects of presentation rate and individual differences in short-term memory capacity on an indirect measure of serial learning. *Memory and Cognition, 22*, 95-110.

Gimino, A. (2002). *Students' investment of mental effort.* Paper presented at the annual meeting of the American Educational Research Association, New Orleans, LA.

Mayer, R., & Moreno, R. (2003). Nine ways to reduce cognitive load in multimedia learning. *Educational Psychologist, 38*(1), 43-52.

Milner, B. S., Corkin, S., & Tueber, H. L. (1968). Further analysis of the hippocampal amnesic syndrome: 14 year follow-up study of H. M. *Neuropsychologica, 6*, 215-234.

Paas, F. G. (1992). Training strategies for attaining transfer of problem-solving skill in statistics: A cognitive approach. *Journal of Educational Psychology, 84*, 429-434.

Paas, F. G., Tuovinen, J. E., Tabbers, H., & Van Gerven, P. W. M. (2003). Cognitive load measurement as a means to advance cognitive theory. *Educational Psychologist, 38*(1), 63-71.

Postmand, L., & Phillips, L. W. (1965). Short-term temporal changes in free recall. *Quarterly Journal of Experimental Psychology, 17*, 132-138.

Reber, A. S. (1993). *Implicit learning and tacit knowledge: An essay on the cognitive unconscious* (Oxford Psychology Series No. 19). New York: Oxford University Press.

Shanks, D. R., & St. John, M. F. (1994). Characteristics of dissociable learning systems. *Behavioral & Brain Sciences, 17*, 367-395.

Squire, L. R. (1992). Memory and the hippocampus: A synthesis from findings with rats, monkeys and humans. *Psychological Review, 99*, 195-231.

Sweller, J., van Merri͵nboer, J. J., & Paas, F. G. (1998). Cognitive architecture and instructional design. *Educational Psychology Review, 10*(3), 251-296.

Tulving, E. (2000). Concepts of memory. In E. Tulving & F. I. M. Craik (Eds.), *The Oxford handbook of memory* (pp. 33-43). New York: Oxford University Press.

Tulving, E., & Schacter, D. L. (1990). Primary and human memory systems. *Science, 247*, 301-306.

Warrington, E. K., & Shallice, T. (1969). The elective impairments of auditory verbal short-term memory. *Brain, 92*, 885-896.

Wittrock, M.C. (1989). Generative processes of comprehension. *Educational Psychologist, 24*, 345-376.

KEY TERMS

Active Processing Assumption: The active processing assumption asserts that intentional and significant mental processing of information must occur for enduring and meaningful learning to take place.

Cognitive Load Theory: Cognitive Load Theory asserts that the capacities and limitations of the human memory system must be taken into account during the process of instructional design in order to produce optimal learning materials and environments.

Dual Channel Assumption: The dual channel assumption is based on the notion that working memory has two sensory channels, each responsible for processing different types of input. The auditory or verbal channel processes written and spoken language. The visual channel processes images.

Limited Capacity Assumption: The limited capacity assumption applies to the dual channels of verbal and auditory processing. The assumption is that each of these channels has a limit as to the amount of information that can be processed at one time.

Long-Term Memory: Long-term memory is a repository for information and knowledge that we have been exposed to repetitively or that has sufficient meaning to us. Long-term memory is a memory store that has an indefinable duration but is not conscious; that is, any information in long-term memory must first be retrieved into working memory for us to be aware of it.

Mental Effort: A second construct related to measuring cognitive load, "mental effort is the aspect of cognitive capacity that is actually allocated to accommodate the demands imposed by the task" (Paas et al., 2003, pp. 64).

Mental Load: One of three constructs devised by Paas et al. (2003) to assist in the measurement of cognitive load. Mental load reflects the interaction between task and subject characteristics. According to Paas et al. (2003), " it provides an indication of the expected cognitive capacity demands and can be considered an a priori estimate of cognitive load"(pp. 64).

Performance: Performance is the third construct in Paas et al.'s (2003) definition of cognitive load and is reflected in the learner's measured achievement. Aspects of performance are speed of completing a task, number of correct answers and number of errors.

Short-Term or Working Memory: Short-Term or Working Memory refers to a type of memory store where conscious mental processing occurs, that is, thinking. Short-term memory has a limited capacity and can be overwhelmed by too much information.

Chapter 7.26
The Online Discussion and Student Success in Web–Based Education

Erik Benrud
American University, USA

INTRODUCTION

This chapter examines the performance of students in a Web-based corporate finance course and how the technologies associated with communication on the Internet can enhance student learning. The chapter provides statistical evidence that documents that the online discussion board in a Web-based course can significantly enhance the learning process even in a quantitative course such as corporate finance. The results show that ex ante predictors of student performance that had been found useful in predicting student success in face-to-face classes also had significant predictive power for exam performance in the online course. However, these predictors did not have predictive power for participation in the online discussion. Yet, online participation and exam performance were highly correlated. This suggests that the use of the online discussion board technology by the students enhanced the performance of students who otherwise would not have performed as well without the discussion.

The online discussion in a Web-based course promotes active learning, and active learning improves student performance. Educators have long recognized the importance of an active learning environment; see Dewey (1938) and Lewin (1951). It is no surprise, therefore, that later research such as Dumant (1996) recognized the online discussion as one of the strengths of Web-based learning. Some researchers, such as Moore and Kearsley (1995) and Cecez-Kecmanovic and Webb (2000) have gone on to propose that the online discussion may even challenge the limits of the face-to-face (F2F) environment.

To explore the effect of the discussion on students' grades, we must first measure the amount of variation in the grades explained by ex ante measures that previous studies have used. The Graduate Management Aptitude Test[1] (GMAT) score, gender, and age were used. A variable that indicated whether the student considered himself or herself someone who took most courses on the Web, that is, a "Web student," was also included, and these four ex ante predictors of student performance explained over 35 percent of the variation of the final course grades in a sample of 53 students. This level of explanatory power using these predictors was similar to that of previous studies concerning F2F finance classes; see Simpson and Sumrall (1979) and Borde, Byrd, and Modani (1998). In this study, with the exception of the condition "Web student," these determinants were poor predictors of online discussion participation; however, there was a significant relationship between online discussion participation and performance on the exams. These results provide evidence that multimedia technologies that promote student interaction can aid the learning process in a course that is largely quantitative in nature.

THE ROLE OF THE ONLINE DISCUSSION

The Internet is ideally suited for a learning tool such as a discussion board where the students can interact and discover answers for themselves. The overall effect of this combination of computer and teaching technology appeared to stimulate student interest and enhanced the learning process. The data gathered in this study indicates that the students appreciated the use of the technology and that each student tended to benefit to a degree that was commensurate with his or her level of participation.

The online discussion consisted of a Socratic dialogue that was led by the instructor. This is an ancient technique that recognizes that student activity aids the learning process. As applied here, it is a learning technique that begins with a single question and then requires participants to continually answer a series of questions that are generated from answers to previous questions with the goal of learning about a topic. The Socratic dialogue is widely used in F2F classes around the world, see Ross, (1993). Using the interactive technology of the discussion board over the Internet seemed

especially beneficial. Having the discussion over a week's time on the Internet allowed students time to think and reflect both before and after their contribution. The students were motivated to participate because the discussion made up 25 percent of their final grade, which was equal to the weight of each of the two exams. The remaining 25 percent was earned from small assignments and one project.

The students earned a portion of the discussion grade each week. At the beginning of each week, a question would be posed such as: "Corporations must pay institutions like Moody's and S&P to have their debt rated. What is the advantage to the corporation of having its debt rated?" The students would post answers and, with the guidance of the instructor, would explore a number of related issues. The students earned credit by "adding value" to the dialogue each week. Students were invited to contribute reasoned guesses, personal anecdotes, and examples from the Internet. One well-thought-out and thorough contribution would earn a student a perfect score for the week. Several small contributions would earn a perfect score as well. The grades earned from discussion participation were generally good. The average discussion grade earned, as a percentage of total points, was 92.81 with a standard deviation of 8.75. The results were highly skewed in that nine of the 53 students earned 100 percent of the online discussion grade. The corresponding percent of total points earned for the course without the discussion had an average equal to 86.21 and a standard deviation equal to 7.26 for all students.

The students generally reacted favorably to the online discussion. All 53 students took a confidential survey that asked them questions about their perceptions of the online discussion. The results reveal that 60 percent felt that this course used the online discussion *more* than the average Web-course they had taken; 76 percent rated the quality of the discussion *higher* than the average they had experienced in other Web-classes; and 55 percent said that the online discussion *significantly aided* their understanding of corporate finance.

STATISTICAL ANALYSIS

To begin the analysis, this study used the variables gender, age, GMAT score, and whether a student was a Web-MBA student to explain performance in the course. Table 1 lists the correlations of various components of these ex ante characteristics with the grades and discussion-participation data. The variables are defined in the list below. The letter "N" appears at the end of a definition if the data for that variable has a bell-shaped or normal distribution, which means the test results for those variables are more reliable.[2]

- **AGE:** The age of the student at the beginning of the class; the range was 21 to 55 with a mean of 31.47, N.
- **DE:** Number of discussion entries, a simple count of the number of times a student made an entry of any kind in the discussion, N.
- **DISC:** Grade for student participation in the online discussion.
- **FAVG:** Final average grade for the course, N.
- **FINEX:** Final exam grade, N.
- **GEN:** Gender, this is a dummy variable where GEN=1 represents male and GEN=0 represents female, the mean was 0.540.
- **GMAT:** Graduate Management Aptitude Test score.
- **GWD:** Grade for the course without discussion, to get this the discussion grade was removed from the final average and that result was inflated to represent a score out of 100 percent, N.
- **MT:** Midterm exam grade, N.
- **PROJ:** Grade on a project that required the creation of a spreadsheet.
- **WC:** Word count; the total number of words the student wrote in the discussion over the

entire course, the range was 391 to 5524 with a mean equal to 2164, N.

- **WMBA:** Whether the student considered him/herself a Web-MBA student as opposed to student who takes most courses in a F2F environment, WMBA=1 for Web-MBA students, else 0; the mean was 0.684.

For each pair of variables, Table 1 lists both the correlation coefficient and the probability value associated with a hypothesis that the correlation is zero. In those cases where an assumption of normality could not be rejected for both variables, the correlation and p-value on the table are in bold font. The correlation coefficient is a measure of the strength of the linear relationship between the variables.

Table 1 displays several interesting phenomena. AGE was positively correlated, but not at a significant level, with most measures of performance. The GMAT score served as a good predictor of test scores and the project score (symbols: MT, FINEX, and PROJ). The correlation of the GMAT score with the three measures of student participation in the online discussion was much weaker. Those three measures of student participation in the online discussion were the number of discussion entries, the word count, and the discussion grade for each student (symbols: DE, WC and DISC).

Some interesting observations concern the discrete binomial, or "zero/one variables," GEN and WMBA, and they are included on Table 1 for descriptive purposes. As found in previous studies, males had a higher level of success on exams. Students who consider themselves Web students, that is, WMBA=1, had a superior performance in all categories too.

Analysis of variance tests (ANOVA) allow us to determine if the effects of WMBA and GEN were statistically significant. Consistent with the requirements of ANOVA, Table 2 reports the results for the normally distributed measures of performance: FINEX, GWD, FAVG, DE, and WC.

Table 1. Correlation matrix of grades, discussion data, and student characteristics

Correlation coef. with p-value underneath, e.g., corr(Disc,MT)=0.087 and p-value=0.460.
Cells in BOLD indicate both variables pass tests for normality.

	DISC	MT	FINEX	PROJ	FAVG	GWD	DE	WC	GEN	AGE	WMBA
MT	0.087										
	0.460										
FINEX	0.297	**0.673**									
	0.010	**0.000**									
PROJ	0.143	0.366	0.41								
	0.222	0.001	0.000								
FAVG	0.573	**0.755**	**0.88**	0.528							
	0.000	**0.000**	**0.000**	0.000							
GWD	0.281	**0.85**	**0.913**	0.564	0.948						
	0.015	**0.000**	**0.000**	0.000	**0.000**						
DE	0.513	**0.329**	**0.322**	0.104	**0.471**	**0.351**					
	0	**0.004**	**0.005**	0.374	**0**	**0.002**					
WC	0.515	**0.362**	**0.428**	0.19	**0.57**	**0.466**	0.755				
	0.000	**0.001**	**0.000**	0.102	**0.000**	**0.000**	0.000				
GEN	0.032	0.346	0.451	0.248	0.369	0.419	0.053	0.136			
	0.787	0.002	0.000	0.032	0.001	0.000	0.649	0.245			
AGE	0.099	**-0.013**	**0.093**	0.053	**0.124**	**0.107**	**0.13**	**0.169**	0.034		
	0.398	**0.914**	**0.427**	0.650	**0.288**	**0.362**	**0.265**	**0.147**	0.773		
WMBA	0.246	0.288	0.274	0.125	0.316	0.271	0.293	0.211	0.122	-0.052	
	0.034	0.012	0.017	0.285	0.006	0.019	0.011	0.069	0.298	0.657	
GMAT	0.178	**0.301**	**0.368**	0.404	**0.421**	**0.414**	0.063	**0.245**	0.509	**-0.171**	0.273
	0.203	**0.029**	**0.007**	0.003	**0.002**	**0.002**	0.652	**0.077**	0.000	**0.221**	0.048

The condition WMBA=1 had a positive effect in all categories, and the effect was significant at the 10 percent level in all five cases. The condition WMBA=1 was the one ex ante predictor that had a significant relationship with DE and WC, and it probably indicated those students who had more experience with Web-based activities. This points to how a student's familiarity with the learning technologies employed in a course will affect that student's performance.

The ANOVA results show that males had significantly higher scores for the final exam,

the grades without the discussion grade, and the course grade (FINEX, GWD, FAVG). This is congruent with previous research. The reason for the lower level of significance of GEN with respect to FAVG is explained by the fact that there was not a significant difference in the student participation in the online discussion for males and females, and that discussion grade is included in FAVG. For the raw measures DE and WC, there was not a significant difference in the participation rates of males and females. We should also note that for the non-normally distributed variable DISC,

Table 2. ANOVA results for dummy variables GEN and WMBA

F-statistic and probability value are reported in each cell.		FINEX	GWD	FAVG	DE	WC
GEN	F-stat.=	18.66	15.55	11.53	0.210	1.38
	p-value=	0.000	0.000	0.001	0.649	0.245
WMBA	F-stat.=	5.94	5.77	8.10	6.84	3.41
	p-value=	0.017	0.019	0.006	0.011	0.069

Table 3. Regression of student performance on ex ante variables

Results in each cell in the explanatory variables columns are the coefficient, (t-statistic), probability value.
For example, for the first equation for FAVG, the intercept coefficient is 78.307, the t-statistic is 18.5, and the probability value is 0.000.

Dependant Variable		explanatory variables					R^2 adj.R^2	F-stat P-value
		Constant	GEN	AGE	WMBA	GMAT		
FAVG	coef=	78.307	4.701		3.129	0.0105	0.347	8.690
	t-stat=	(18.5)	(2.92)		(1.81)	(1.20)	0.307	0.000
	p-val=	0.0000	0.005		0.077	0.234		
DISC		90.550	3.175			0.004	0.0740	2.000
		(24.0)	(1.58)			(0.57)	0.0370	0.146
		0.000	0.121			0.571		
GWD		74.412	5.163		3.968	0.013	0.331	8.096
		(14.5)	(2.72)		(1.86)	(1.21)	0.290	0.000
		0.000	0.009		0.069	0.231		
PROJ		86.983	1.670			0.018	0.183	5.600
		(21.2)	(1.05)			(2.34)	0.150	0.006
		0.000	0.300			0.023		
WC		-376.5		32.300	375.31	2.482	0.130	2.400
		(-0.29)		(1.41)	(1.42)	(1.39)	0.077	0.075
		0.774		0.165	0.161	0.169		

the discussion grade, males only slightly outperformed females. The average grades for males and females were 93.1 and 92.5 respectively with an overall standard deviation of 8.75.

Ordinary least squares (OLS) regressions can measure predictive power of the ex ante variables. Table C lists the results of regressions of the final grades (FAVG), the discussion grades (DISC), the grades without the discussion (GWD), the project grade (PROJ), and the word count (WC) on the indicated variables.

The equations for FAVG and GWD had the highest explanatory power, and the equation for PROJ had significant explanatory power too. The explanatory power of the equations for WC and DISC were much lower; although the equation for WC was significant at the ten-percent level, the results for WC and DISC were not significant at the 5 percent level.

As we would expect from past research, GEN had very significant coefficients in the equations for FAVG and GWD, which means that the condition "male" was associated with higher final averages and grades without discussion. GMAT was only marginally significant in most cases, but this could be the result of the high correlation of GMAT with GEN.

We can use OLS regressions to demonstrate how online discussion performance, as measured by DISC and WC, affected FINEX because the discussion occurred before FINEX was determined. In a regression of FINEX on GEN, MT, and DISC, the coefficient for DISC had a t-statistic greater than that for GEN. These results are on Table 4a.

Since WC was normally distributed and was a raw measure of effort, a second specification on Table 4a replaces DISC with WC. The t-statistic for the discussion variable decreased slightly, as did the coefficient of determination symbolized by R2. Both t-statistics were significant, however, and both R2 values exceeded 50 percent. The coefficient of determination is a measure of variation explained, which means that in this case the variables in each equation explained over half of the differences in the grades on the final exams.

Table 4b gives the result of a second set of equations, which used two-stage least squares (TSLS) to estimate the effect of WC on GWD and then GWD on WC. TSLS was required here because GWD and WC developed simultaneously during the course. The results show that each had a significant relationship with the other.

The purpose of this section has been to report the statistical results and point out the interesting relationships. Many of those relationships are congruent with earlier work. For example, male students and those who had higher GMAT scores had higher exam grades. The most interesting results concern the grades for the online discussion, which had a low correlation with the ex ante student characteristics GEN and AGE but were highly correlated with exam grades. The next section discusses some of the implications of these results.

DISCUSSION OF EMPIRICAL RESULTS

Consistent with previous research concerning F2F finance classes, the following were significant ex

Table 4a. OLS regression of final exam on ex ante performance in the course

Results in each cell in the explanatory variables columns are the coefficient, (t-statistic), probability value.							
Dependant Variable	**explanatory variables**					R^2 adj.R^2	**F-stat**
	Constant	**GEN**	**MT**	**DISC**	**WC**		
FINEX	-21.70	6.010	0.766	0.334		0.564	30.653
	(-2.42)	(3.11)	(7.93)	(3.54)		0.546	P=0.000
	0.018	0.003	0.000	0.001			
FINEX	9.669	5.961	0.692		0.0026	0.545	28.386
	(1.05)	(3.01)	(5.84)		(2.48)	0.526	P=0.000
	0.300	0.004	0.000		0.0015		

Table 4b. Two-stage least squares estimation

Instrument list: C AGE AGE^2 AGE^{-1} WMBA GEN							
Results in each cell in the explanatory variables columns are the coefficient, (t-statistic), probability value.							
Dependant Variable	**explanatory variables**					R^2 adj.R^2	**F-stat. p-value**
	Constant	**Gen**	**GWD**	**WC**	**Age**		
WC	-8493		120.15		10.212	0.033	3.781
	(1.98)		(2.33)		(0.59)	0.006	0.027
	0.0518		0.023		0.555		
GWD	70.923	4.534		0.006		0.216	11.038
	(14.08)	(2.77)		(2.44)		0.194	0.000
	0.000	0.007		0.017			

ant predictors of performance: gender, GMAT score, and age. As we might expect for a Web-based course, students who considered themselves web students or Web-MBA students, performed significantly higher for most of the grade variables. The most interesting point is that the traditional ex ante characteristics did not predict performance in the online discussion very well, yet there was a strong relationship between the online discussion and exam grades.

Gender displayed a very weak relationship with measures pertaining to the online discussion. The word count was normally distributed and did not have a significant relationship with gender in either an ANOVA or OLS regression. Word count was an unrefined measure, but this was an advantage in that it served as a direct measure of effort, and it was unaffected by the subjective opinions of the instructor. Word count was significantly correlated with each of the student's class scores. In summary, gender, age, GMAT score, and whether a Web-MBA student explained success on exams. With the exception of whether a Web-MBA student, these predictors were not significantly correlated with performance in the online discussion. Although these variables had low explanatory power for the online discussion measures, there was a high correlation between performance in the online discussion and exam grades.

Using the discussion score or word count in an equation with the gender variable and the midterm exam grades explained over 50 percent of the variation of the final exam grades. In fact, the discussion grade's coefficient had a larger t-statistic than the gender variable. The coefficient for word count in the equation for the final exam grade has a significant coefficient equal to 0.0026. This means that for every 385 words of writing in the online discussion, on average, a student's final exam grade was about one point higher: $385*0.0026 \approx 1$. The TSLS results on table D2 indicate the effect of word count on the grades

without the discussion grade. The coefficient was significant and estimated to be 0.006. This means that for every 167 words, on average, there was an associated increase in the grade without the discussion of one point: $167*0.006=1$.

The results of this study show that a student's use of a technology such as an Internet discussion board can enhance that student's performance in other areas of a course. The ex ante measures of gender, age, and GMAT were not useful in predicting who would participate in and thus benefit from the discussion board. Use of the discussion board technology by the students had a significant and positive effect on the grades earned on the exams. Furthermore, the fact that students who considered themselves Web-MBA students had superior performance means that training and experience in the use of multimedia technologies is important in order to allow students to benefit from such technologies to a greater degree.

REFERENCES

Borde, S., Byrd, A., & Modani, N. (1998). Determinants of student performance in introductory corporate finance courses. *Journal of Financial Education, Fall*, 23-30.

Cecez-Kecmanovic, D. & Webb, C. (2000). A critical inquiry into Web-mediated collaborative learning. In A. Aggarwal (Ed.), *Web-based learning and teaching technologies: Opportunities and challenges*. Hershey, PA: Idea Group Publishing.

Dewey, J. (1938). *Experience in education*. New York: Macmillan.

Dumant, R. (1996). Teaching and learning in cyberspace. *IEEE Transactions on Professional Communication, 39*, 192-204.

Lewin, K. (1951). *Field theory in social sciences*, New York: Harper and Row Publishers.

Moore, M. & Kearsley, G. (1995). *Distance education: A systems view*. Belmont, CA: Wadsworth Publishing.

Ross, G.M. (1993). The origins and development of socratic thinking. *Aspects of Education 49*, 9–22.

Simpson, W., Sumrall, G, & Sumrall, P. (1979). The determinants of objective test scores by finance students. *Journal of Financial Education*, 58-62.

KEY TERMS

Active Learning: Learning where students perform tasks, that is, post notes on a discussion board, to help in the learning process.

Coefficient of Determination: A statistical measure of how well predictive variables did indeed predict the variable of interest.

Correlation Coefficient: A statistical method of measuring the strength of a linear relationship between two variables.

Ex Ante Predictors of Student Performance: Student characteristics prior to beginning a class which have been determined to help forecast the relative performance of the students.

Online Discussion Board: Often called a "forum," it is a technology which allows students to interact by posting messages to one another at a particular URL on the Internet.

Probability Value: A statistical measure that attempts to assess whether an outcome is due to chance or whether it actually reflects a true difference; a value less than 5 percent means that a relationship very likely exists and the result probably did not occur by chance.

Regression: A statistical method of estimating the exact amount a variable of interest will change in reaction to another variable.

Socratic Dialogue: A learning technology which requires the participants to answer a series of questions to discover an answer or truth concerning a certain topic; the questions are typically generated spontaneously from previous answers.

Student Participation in the Online Discussion: The level to which a student contributed in an online discussion as measured, for example, by the number of words posted or the number of entries or the grade issued by an instructor.

Web Student: A student that plans to take the majority, if not all, of his or her classes in a particular program of study over the Internet.

ENDNOTES

[1]   The GMAT is a standardized entry exam that is administered on certain dates around the world. Most graduate business schools require students to take this test as part of the application process.

[2]   More specifically, the "N" signifies that we cannot reject a null hypothesis of the variable being distributed normally distributed based on a Kolmogorov-Smirnov test using a 5 percent level of significance. If "N" does not appear, then the null hypothesis of normality was rejected.

This work was previously published in the Encyclopedia of Multimedia Technology and Networking, edited by M. Pagani, pp. 778-784, copyright 2005 by Idea Group Reference (an imprint of IGI Global).

Chapter 7.27
The Influences and Responses of Women in IT Education

Kathryn J. Maser
Booz Allen Hamilton, USA

INTRODUCTION

This chapter highlights findings from an empirical study that explores the nature of female underrepresentation in information technology. Specifically, this research focuses on (a) identifying key sociocultural factors that can facilitate the pursuit of IT at the undergraduate level, and (b) testing Trauth's (2002) Individual Differences Theory of Gender and IT through a comparison of female responses to the social construction of IT. To answer the author's research questions, interviews were conducted with 10 female seniors in an IT department at an American university in the mid-Atlantic region (MAU).[1]

Although experiences with social factors vary, comparing the stories of women who have successfully navigated their way into and through an IT undergraduate degree program reveals common influences and motivations. In addition, though some common factors may facilitate female entry into the field, the Individual Differences Theory of Gender and IT explains that women will react differently to the social constructions of gender and IT. By gaining a better understanding of the gender imbalance, applying appropriate theories to explain the problem, and uncovering the challenges that women of our society face in their entry to the field of IT, collegiate programs can more effectively implement strategies that will improve the recruitment and retention of female students.

BACKGROUND

IT-related undergraduate degree programs such as computer science (CS), management information systems (MIS), and information-science and -technology programs are important gateways to the IT industry, providing valuable exposure and experience to students interested in pursuing IT careers. Research suggests that women are entering undergraduate IT programs in smaller numbers (e.g., Camp, 1997; Freeman & Aspray, 1999) and may be doing so with less formal and informal IT experience (e.g., Craig & Stein, 2000; Fisher, Margolis, & Miller, 1997; Margolis &

Fisher, 2002; Teague, 1997). Thus, education at the undergraduate level is critical in the foundation of their skills, their interests in IT, and their pursuit of work in the field. Moreover, actively recruiting and retaining females in IT-related undergraduate degree programs can have a significant impact on the diversification of the IT workforce. As Margolis and Fisher (2002, p. 3) explain, "women must be part of the design teams who are reshaping the world, if the reshaped world is to fit women as well as men."

This study first focuses on identifying sociocultural factors influential in women's decisions to pursue IT at the undergraduate level. The social-construction perspective of gender and IT explains that, reflective of the social norm in America, cultural expectations and influences often convey the message that women are unsuitable for the IT world (e.g., Trauth, 2002; von Hellens, Nielsen, & Trauth, 2001). By the time young women reach college, there is evidence of the effects of these social norms and expectations. For example, in the years prior to college, certain studies have revealed that, in comparison with males, females exhibit lower levels of self-efficacy in computing, are less likely to explore computing independently through informal channels (e.g., within peer groups, computer camps, and clubs), and elect to take advanced computing courses less frequently; in addition, some women have misconceptions about the IT workforce and IT work (e.g., Beise & Myers, 2000; Craig & Stein, 2000; Fisher et al., 1997; Margolis & Fisher, 1997, 2002; Nielsen, von Hellens, & Wong, 2000; Symonds, 2000; Teague, 1997; von Hellens, Nielsen, Doyle, & Greenhill, 1999; Woodfield, 2000).

An examination of the factors that enable women to confront and circumvent these social barriers is an important part of understanding the gender imbalance; however, it should not be assumed that all women have the same reactions to these barriers. The Individual Differences Theory of Gender and IT embraces the notion that gender is a fluid continuum rather than a dichotomy. This theory focuses on women as individuals, having distinct personalities, experiencing a range of sociocultural influences, and therefore exhibiting a range of responses to the construction of the IT field (Trauth, 2002). Comparing and contrasting females' responses to the social construction of IT tests the individual-difference theory of gender and IT.

RESEARCH APPROACH

This research focuses on women at a critical point of IT entry: the undergraduate level of education. In examining the trends of female underrepresentation discussed in the literature and the theoretical perspectives used to explain the problem, the following research questions emerged.

1. What significant sociocultural factors in the lives of women are influential in their pursuit of IT at the college level?
2. How similar are female responses to the social construction of gender and IT?

To investigate these questions, in-depth interviews were conducted with a sample of 10 female seniors in MAU's IT department in the spring of 2003. The IT department at this university was chosen because of its proactive stance with respect to the recruitment and retention of women students. The department also has a diversity committee and a student organization, Women in Information Technology (WIT), that was established to provide support and mentoring for female students in the program. At the time of these interviews, the student enrollment in the department was 21% female. Interviews were open ended and lasted approximately 40 minutes in duration. The qualitative format was selected as it was most appropriate for capturing in detail the participants' broad range of influences and experiences. Interview questions were derived from the themes of family background, educational history, personal traits and interests, discovery and selection of the IT program at MAU, experiences in the program, and future plans.

FINDINGS

Comparing Sociocultural Influences

In comparing participant experiences, the women study reported modest levels of formal education and informal experimentation in IT; these experiences made little impact on their decisions to pursue the IT degree. Participants consistently described their education in high-school computer classes as basic. The two women who elected to exceed the minimum computing requirements and complete C++ classes felt they lacked a clear understanding of the extent to which the language could be applied in real-world scenarios. On the whole, these high-school computer classes served the purpose of familiarizing these women with computers, but did little more. Although a few of the participants were aware of certain IT careers, the majority did not have a clear and complete understanding of the IT field prior to college. In terms of computing exposure and use in the home, experiences were quite consistent and corresponded strongly with the literature (e.g., Margolis & Fisher, 2002). The primary functions of home computers were education and communication: word processing for homework, and e-mail and instant messaging for chatting with friends.

Family influence and encouragement was a key social factor identified as impacting the participants' decisions to pursue the IT program at MAU. Despite differences in family environments, common to each of the women's experiences was a high level of parental academic support, encouragement, and expectation. The participants had mothers, fathers, and siblings that were, to varying degrees, actively involved in their academic careers. Many of the participants were pushed for academic achievement, and many

were also specifically encouraged to choose the IT program at MAU. Other participants reported less direct academic involvement, though expectations and encouragement remained strong. This encouragement, with little exception, helped the women create high personal expectations for their future careers.

In developing an understanding of gender roles, the participants grew up in homes where mothers and fathers assumed diverse roles and responsibilities. Regardless of their family environments (e.g., dual income, sole breadwinner, single parent), the children in these households, male and female, were treated equally. The participants were raised believing they were able to achieve whatever they wished, and that gender was not a factor that should steer them in one direction or deter them from another. This confidence facilitated their selection of MAU's IT program. This confidence was also revealed in the way the women dealt with male domination in the IT department; half of the participants reported being largely unaware of the gender imbalance, and the remaining half, though initially intimidated by the experience gap, learned quickly that they were equally capable of achieving success in the program.

The presence of role models was another significant factor that influenced participants' decisions to pursue IT. Lacking a significant amount of formal and informal experience in IT, the women's IT understanding was strongly correlated with the presence of a role model. Consistent with the literature (e.g., Beise & Myers, 2000; Craig & Stein, 2000; Symonds, 2000; Teague, 1997), IT role models affected how the women perceived and related to the field, exposed them to opportunities in the field, and helped them develop an interest in IT work. The majority of females with IT role models entered college with a general understanding of the field they wanted to pursue.

Specific characteristics of MAU's IT program were also influential in the participants' decisions to pursue IT. In particular, the program was perceived by the women as new, exciting, cutting-edge, and as offering a wide range of both technical and nontechnical business-related educational opportunities. A number of the participants also described positive experiences with IT program advisors. Finally, the fact that the IT program was formed with close ties to industry and emphasized postgraduation employment opportunities was a principal factor in the participants' decisions.

Exploring Individual Responses

Findings regarding personal future expectations and outlooks provide clear support for the Individual Differences Theory of Gender and IT. Individual differences were most clearly revealed through the participants' outlooks on female participation in IT and expectations for their own futures.

Although the participants arrived at the same source of IT education, differences in their interests, values, and priorities will cause some women to maintain their participation in the IT industry, and others to reevaluate and change careers. Half of the women in this study strongly believed IT was the field they wanted to pursue, did not envision themselves switching careers, and were committed to balancing their work and family lives. The remaining half of the participants expressed uncertainty over whether or not they would remain in the IT field. Some of these women were unsure they would enjoy the IT industry and indicated a desire to explore other fields, while others anticipated an incompatibility in balancing an IT career with expectations for family commitments in the long term.

Differences in formative experiences have also led these women to hold a variety of opinions about female participation in IT: differences in their explanations of the gender imbalance, and differences in their opinions on how and whether or not the issue should be addressed. In offering explanations for the underrepresentation of fe-

males in IT, opinions were split between those who believed women were simply less interested in the technical subject matter, and those who believed the imbalance persists due to a lack of female exposure to the field. Additionally, some of the women felt that the gender imbalance in IT was a significant issue and that, in certain situations, intervention was necessary to provide support for females. However, the perspective most frequently described by this group of participants was the belief that hypersensitivity about gender issues can be problematic and can place unnecessary emphasis on the division between the sexes.

CONCLUSION

The analysis of females in the IT program at MAU reveals high levels of encouragement from family, exposure to IT through role models, and balanced perceptions of gender roles and expectations to be the primary social factors facilitating participant decisions to pursue this IT-related undergraduate degree program. Specific characteristics of MAU's IT program were also influential in attracting the participants to the department. In particular, the program was perceived by the women as a cutting-edge program with a comprehensive curriculum, and as capable of providing access to great career opportunities. Finally, in investigating participant outlooks on female participation in IT and expectations for their own futures, this research found strong support for the Individual Differences Theory of Gender and IT.

The participants' stories also revealed that their high schools were an untapped source of potential influence, exposure, and encouragement. Findings suggest that improved IT education in high school could provide females with exposure to the field they might not receive elsewhere, which would greatly support the development of interest in IT. Findings also indicate that high-school IT courses should be developed with an emphasis on both technical knowledge and real-world business applications of the subject matter. Additionally, women should be actively encouraged to participate in these courses and explore computing beyond minimum requirements. Finally, high school would be an appropriate place to provide young women with exposure to female role models in the industry, a factor that is particularly important when role models are not present within the home.

Finally, in discussing the women's expectations for the future, implications for female retention in the IT industry emerged. As exemplified by this participant group, there are both unpredictable success stories and unexpected stumbling blocks. One participant, for example, entered MAU with great uncertainty and little career orientation. Having been strongly encouraged by her father to achieve academically and pursue the IT program, she secured a summer internship where she discovered a specific technical focus within IT she enjoyed and became intent on pursuing. In contrast, another woman described growing up as a tomboy and was the only participant to exemplify the "boy-wonder syndrome"[2] (e.g., Margolis & Fisher, 1997). Her conditions for IT selection were very favorable: a strong interest in technology, exposure to IT, strong role models, encouragement, and confidence. Yet, she felt her participation in the industry was temporary because of her desire to pursue other interests she viewed as more compatible with the kind of family environment she wants to create. To increase female participation in IT, we should first strive to make conditions for the selection of IT education most favorable so that with exposure and experience in the field, women may find their niche among the many lines of work that IT has to offer. Retention strategies at the undergraduate and industry levels are the next step; however, they are beyond the scope of this research.

REFERENCES

Beise, C., & Myers, M. (2000). A model for examination of underrepresented groups in the IT workforce. In *Proceedings of the 2002 ACM SIGCPR Computer Personnel Research Conference,* Kristiansand, Norway (pp. 106-110).

Camp, T. (1997). The incredible shrinking pipeline. *Communications of the ACM, 40*(10), 103-110.

Craig, A., & Stein, A. (2000). Where are they at with IT? In E. Balka & R. Smith (Eds.), *Women, work and computerization: Charting a course to the future* (pp. 86-93). Boston: Kluwer Academic Publishers.

Fisher, A., Margolis, J., & Miller, F. (1997). Undergraduate women in computer science: Experience, motivation and culture. In *Proceedings of the 1997 ACM SIGCSE Technical Symposium,* San Jose, CA (pp. 106-110).

Freeman, P., & Aspray, W. (1999). *The supply of information technology workers in the United States.* Washington, DC: Computing Research Association.

Margolis, J., & Fisher, A. (1997). Geek mythology and attracting undergraduate women to computer science. *Impacting Change through Collaboration: Proceedings of the Joint National Conference of the Women in Engineering Program Advocates Network and the National Association of Minority Engineering Program Administrators.* Retrieved from http://www.cs.cmu.edu/~gendergap/working.html

Margolis, J., & Fisher, A. (2002). *Unlocking the clubhouse: Women in computing.* Cambridge, MA: The MIT Press.

Nielsen, S., von Hellens, L., & Wong, S. (2000). The game of social constructs: We're going to WinIT! In *Panel Presentation: Proceedings of the International Conference on Information Systems*, Brisbane, Australia.

Symonds, J. (2000). Why IT doesn't appeal to young women. In E. Balka & R. Smith (Eds.), *Women, work and computerization: Charting a course to the future* (pp. 70-77). Boston: Kluwer Academic Publishers.

Teague, J. (1997). A structured review of reasons for the underrepresentation of women in computing. In *Proceedings of the Second Australasian Conference on Computer Science Education*, Melbourne, Australia (pp. 91-98).

Trauth, E. M. (2002). Odd girl out: An individual differences perspective on women in the IT profession. *Information Technology & People, 15*(2), 98-118.

Trauth, E. M., Quesenberry, J. L., & Morgan, A. J. (2004). Understanding the under representation of women in IT: Toward a theory of individual differences. *Proceedings of the ACM SIGMIS Computer Personnel Research Conference*, Tucson, AZ (pp. 114-119).

Von Hellens, L., Nielsen, S., Doyle, R., & Greenhill, A. (1999). Bridging the IT skills gap: A strategy to improve the recruitment and success of IT students. In *Proceedings of the 10th Australasian Conference on Information Systems*, San Diego, CA (pp. 116-120).

Von Hellens, L., Nielsen, S., & Trauth, E. M. (2001). Breaking and entering the male domain: Women in the IT industry. *Proceedings of the ACM SIGCPR Computer Personnel Research Conference*.

Woodfield, R. (2000). *Women, work and computing*. Cambridge, MA: Cambridge University Press.

KEY TERMS

Boy-Wonder Theory: The belief that true scientific talent, interest, and achievement must be exhibited early in one's lifetime.

Individual Differences Theory of Gender and IT: The perspective of gender and IT that argues that "women as individuals experience a range of different socio-cultural influences which shape their inclinations to participate in the IT profession in a variety of individual ways" (Trauth, 2002, p. 103).

Social-Construction Perspective of Gender and IT: The perspective of gender and IT that attributes female underrepresentation to societies' incompatible constructions of femininity and the IT field.

Tomboy: A female considered boyish or masculine in behavior or manner.

ENDNOTE

[1] The name of this university has been changed.
[2] As Sheila Tobias explains, "one of the characteristics of the ideology of science is that ... both scientific talent and interest come early in life—'the boy wonder syndrome'. If you don't ask for a chemistry set and master it by the time you're five, you won't be a good scientist" (Margolis & Fisher, 1997).

This work was previously published in the Encyclopedia of Gender and Information Technology, edited by E. Trauth, pp. 808-812, copyright 2006 by Idea Group Reference (an imprint of IGI Global).

Chapter 7.28
Factors Affecting the Adoption of Education Technology

Graeme Salter
University of Western Sydney, Australia

INTRODUCTION

There are many in education who appear to think that it is sufficient to purchase and install technology for it to be successfully used (Boddy, 1997). Another common belief is that teachers will "automatically seek to learn about new technology and instructional methods" (Dooley, 1999, p. 38). However, while the investment in technology is there, surveys have consistently found that very few teachers integrate technology into either the K-12 (Newhouse, 1999) or the university classroom (Spotts, 1999). One research study found that even when the technology is readily available and staff accept the functionality of it, they "might not anticipate their personal use of it" (Mitra, Hazen, LaFrance, & Rogan, 1999).

Even with intensive staff development, results may be disappointing. Staff developers with Apple Computer tried a range of staff-development approaches with teachers involved in the Apple Classrooms of Tomorrow (ACOT) project. They held workshops after school over a day and even over a week during vacation. As one said, they used the "spray and pray" approach. The most successful of these was a week-long workshop introducing constructivist learning strategies. They hoped that after returning to their classrooms, the teachers would modify their teaching practices. However, on follow-up visits to their classrooms, they "did not see that teaching strategies had changed much or that teachers were implementing the units they had designed during the workshop" (Apple Computer Inc, 2000).

BACKGROUND

There is a significant body of knowledge concerned with the diffusion or adoption of innovations that can provide a theoretical base. An increasing number of instructional technologists are turning to these theories after realizing that innovative products and practices are underutilized (Surry, 1997).

There is no unified theory of diffusion. Among the most widely cited theories are those of Rogers from his book *Diffusion of Innovations*, originally published in 1960 and now in its fourth edition (Surry, 1997). These theories include the rate-of-adoption theory, which "states that innovations are diffused over time in a pattern that resembles an *s*-shaped curve" (Surry & Farquhar, 1997)(see Figure 1).

The rate of adoption rises slowly at first. When around 20% of the population has joined, the adoption "takes off." The rate increases to a maximum when adoption reaches about 50% of the population. After this period of rapid growth, the rate of adoption gradually stabilizes and may even decline. This theory is related to the individual-innovativeness theory, which states that "individuals who are predisposed to being innovative will adopt an innovation earlier than those who are less predisposed" (Surry & Farquhar, 1997). Individuals can be placed into adopter categories based on specific characteristics in relation to a proposed innovation. These categories are innovators, early adopters, early majority, late majority, and laggards. The *s*-shaped curve relates to the timing of adoption by the various categories. These are, of course, "ideal" types, and in reality there are no pronounced breaks between the categories. Nevertheless, they are useful for guiding research efforts, planning professional development strategies, and anticipating reactions to change (Dooley, Metcalf, & Martinez, 1999; Edmonds, 1999; Rogers, 1983). These theories highlight that change is a process and that characteristics of individuals will affect when or if they will adopt a change during this process.

As well as the general diffusion theory, there are theories specifically related to the diffusion of instructional technology. These can be divided into two categories based on their underlying philosophies regarding technological change: technological determinism and technological instrumentalism (Surry & Farquhar, 1997).

TECHNOLOGICAL DETERMINISM

Must society be shaped by the available technology, or may society shape technology? (Jones, 1982, p. 211)

Many theories of diffusion are based on a deterministic view of technology. Technology is seen as an inevitable, autonomous force. Utopian determinists, such as Alvin Toffler, feel that it will lead to prosperity and be the salvation of humanity. On the other hand, dystopian determinists, such as George Orwell, view technology as morally corrupt and that it will eventually lead to the destruction of humanity (Surry & Farquhar, 1997).

Determinist or developer-based models of diffusion focus on the technical characteristics in order to promote change. They assume that technological superiority is all that is required to bring about the adoption of innovative products and practices (Hansen, Deshpande, & Murugesan, 1999). However, successful adoption entails continued use. There are classic examples, such as the results of the contests between Beta and VHS video or the Dvorak and QWERTY keyboard, which demonstrate that technical superiority alone is not sufficient to ensure change. Clearly, other factors influence change.

Instrumentalist or adopter-based theories of diffusion emphasize the importance of the social context of change and the need to address the

Figure 1. Rate of adoption (Source: Surry, 1997)

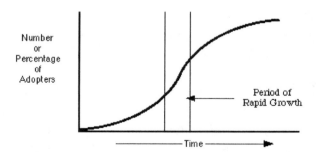

3300

knowledge, beliefs, feelings, and concerns of the users (Crawford, Chamblee, & Rowlett, 1998). Successful change is seen in relation to meeting the real and perceived needs of the bulk of the users, not just the innovators. Technology is viewed as being under human control and its use can lead to beneficial or disastrous consequences. Jones (1982) notes that technologically determined decisions do not just happen. They are "contrived, pushed and promoted by conscious human agencies—specialists in particular fields, many bureaucrats, advertising agencies, manufacturers, newspapers and television—people who argue a position and, in default of any effective alternative view being put, win the debate" (p. 210).

However, care needs to be taken, as a totally instrumentalist approach that turns out "technically inferior and pedagogically weak products that people want to use is not the answer" (Surry & Farquhar, 1997). Nevertheless, the history of the adoption of innovations is littered with failure due to the lack of attention to the concerns of the people ultimately responsible for the change: the end users. For example, in the 1960s a number of "teacher proof" curricula were developed in the United States, partly in response to the fear of being left behind raised by the launch of Sputnik. A study of a number of high-school teachers using one of the supposedly teacher-proof curricula found that in the classroom, teachers still had strikingly different patterns of practice (Hall & Hord, 1987).

Many authors comment on the importance of addressing the concerns of the teachers who will actually implement the change in the classroom (Bondaryk, 1998; Dooley, 1999; Harasim, Hiltz, Teles, & Turoff, 1998; Marx, Blumenfield, Krajcik, & Soloway, 1998). Concerns theory and research reveals "that concerns change over time in a fairly predictable, developmental manner" (Dooley et al., 1999, p. 109). Individuals go through stages during the change process with differing needs through these stages. Change strategies that meet these needs are more likely to be effective

(Crawford et al., 1998; Schiller & Mitchell, 1993). In order to address these concerns, it is necessary to identify them.

FACTORS AFFECTING ADOPTION

There are a number of potential barriers to online teaching and learning. If the barriers are seen to outweigh benefits, adoption is unlikely. The potential user must see a clear advantage in order to change his or her way of teaching (Collis, 1998; Mitra et al., 1999).

Perceptions

Staff perceptions of the difficulties in using a technological innovation may be quite different from the reality. Nevertheless, it is the perceptions rather than the reality that will influence their decision to adopt. This is one of the reasons why a concerns-based model is particularly useful. Staff with positive perceptions about the use of technology tend to use it more (Lu, Zhu, & Stokes, 2000; Mitra et al., 1999), particularly those who perceive a valuable benefit (Collis, 1998; Spotts, 1999). Higher level users tend to be more forgiving of frustrations encountered (Spotts). While some staff find technology intriguing, others find it intimidating and frightening (Bondaryk, 1998; Tennant, 1999).

Of course, negative perceptions may indeed match the reality. Online teaching can represent a significant departure from traditional teaching styles (Bondaryk, 1998). Academics who are already doing a good job using traditional delivery may ask, "Why change?" especially given the real possibility that they may not do as well online (Hagner, 2000). There may also be healthy skepticism of the motives behind the change. Is it change for change's sake? Is it change because of a technology push? Is the development of a "virtual university" simply a way for politicians to reduce funding? (Collis, 1998) Will this change result in educated students or an "entire class of victims" (Hagner, p.31)? Of particular concern to some staff is whether technology may eventually replace them (Boddy, 1997; Bondaryk, 1998). According to Cooley, "Any teacher who *can* be replaced by a computer *deserves* to be" (as cited in McInerney & McInerney, 1994, p. 279).

The perceptions of administrators can also affect the teaching staff. For example, some administrators may consider that lower level staff can "play back" material after initial development (Harasim et al., 1998). Some may even be "lured into the fantasy" that online delivery can supplant the need for teachers (Schneiderman, Borkowski, Alavi, & Norman, 1998). These views appear to be predicated on the role of the teacher as information transmitter, which is "a limited conception of what it means to be a teacher" (Freeman, 1997). Educators need to be aware that some managers hold this view and that they may need to promote a better understanding of what it means to be a good teacher.

It is important to recognize that information technology is not just a way of accessing a vast, impersonal data bank, but that it represents a human information network, developed and maintained by humans (Fowler & Dickie, 1997). Some people mistakenly speak of "computer communication" when they really mean "computer-mediated communication": the object of communication is another person, not a computer.

In particular, the significance of relationship in the process of learning cannot be underestimated (Pettit, 1998; Schneiderman et al., 1998). Even if it is interactive, computer-based subject matter cannot guarantee learning any more than the presence of a library on campus (Reeves, 1993). Content cannot carry itself. Noddings (1992) goes so far as to claim that, apart from some rare exceptions, relation precedes engagement with subject matter.

It is unlikely in the near future that this aspect of teaching will be successfully simulated. Incorporating human elements into courses run entirely online will be one of the main challenges for distance educators. While it is doubtful that technology will replace teachers, it has the potential to significantly influence approaches to teaching, and it may become increasingly important for teachers to be technology literate (Barritt, Ashhurst, & Pearson, 1995; Porter & Foster, 1998). According to Barrit et al. (p. 33), "Computers will never replace teachers, but teachers who can use computers will replace teachers who can't." Whether the threat of being made obsolete is real or perceived, it should be remembered that it will affect teacher attitudes and the rate of adoption of technology in education (Bondaryk, 1998).

Experience

Experience with technology can influence use. Teachers who have had positive experiences with technology are more likely to use technology in their own teaching (Freeman, 1997). A study into patterns of use of e-mail at Wake Forest University found that users differed significantly from nonusers on a number of demographic characteristics (Mitra et al., 1999). Users tended to be relatively new to the university, held junior ranks, reported the first use of computers earlier in their academic career, and had higher self-evaluation of their computer skills. Nonusers of e-mail tended to be nonusers of other computer applications as well. Gender or discipline did not show any significant differences.

In a study where all participants had above-average computer skills, Schoenfeld-Tacher and Persichitte (2000) found that teachers had a wide variation of knowledge about distance education theories. This in turn influenced their satisfaction with teaching online.

Increased Workload

Staff workload is a critical issue in the successful adoption of technology in education (Bondaryk, 1998; Leigh, 2000; McNaught, Kenny, Kennedy, &

Lord, 1999). Innovators often put in large numbers of unpaid hours (Mason & Bacsich, 1998). This is clearly unsustainable. While good teaching is important, it should not be overwhelming (Collis, 1998). Teachers need to carefully manage their online commitment (Harasim et al., 1998). Mainstream users are much less likely than innovators to devote a large amount of extra time and will resist change if there is even a perception of increased workload without corresponding benefits.

Developing online material and learning how to use online teaching tools is time consuming. Most staff guard their time and are reluctant to give it up, particularly when the amount of extra time required or the steepness of learning curves are largely unknown (Hagner, 2000; Spotts, 1999). There are also legitimate concerns that materials and skills developed may have only a short shelf life given the rapid changes in technology (Slough, 1999).

Once an online subject is under way, extra time may be needed to respond to large numbers of student messages and postings. A student question that might be answered in 30 seconds in the office may require a much longer response in electronic form when visual cues are not present. Encouraging electronic communication with students can also result in increased administration (Littlejohn & Sclater, 1998). Time and date stamping on messages creates greater accountability to provide prompt feedback. In some cases, students tend to expect almost instant responses to questions (Freeman, 1997).

For some teachers, the move to online teaching may also correspond with a move to a more facilitative form of teaching, which, whether face to face or online, requires an inordinate amount of time for it to be successful (Pettit, 1998).

Institutional Barriers

Even where staff are fully trained and eager to proceed, attempts to integrate technology in teaching may still fail due to a range of institutional barriers (Edmonds, 1999). For example, some institutions seem to operate on the assumption that time in front of the class is the only valid use of a teacher's time (Grant, 1996). A number of researchers highlight the importance of providing release time to allow for activities such as planning, collaborating with colleagues, developing new skills, developing online materials, and evaluating new practices (Edmonds; Grant; McKinnon & Nolan, 1989). On the other hand, some question the value of release time, particularly for individuals, given that most academic staff do not have the technical skills "required to develop educationally stimulating, digital learning materials that involve students actively in the learning process, and accommodate diverse learner needs" (Brahler, Peterson, & Johnson, 1999, p. 46). However, even if the teaching role is kept distinct from that of resource development, there are still many demands on time relating to changing teaching practices.

Apart from the difficulty in obtaining release time, there may actually be disincentives to change given that reward structures often give little, if any, weight to this work (Hagner, 2000). Many authors argue that there is little reason for academics to invest time in teaching innovations when this is not recognized in tenure or promotion decisions (Carr, 1999; Donovan & Macklin, 1999; Edmonds, 1999; Spotts, 1999). Similarly, there may be some reluctance if there are questions regarding intellectual property (Donovan & Macklin).

The success of online teaching is strongly correlated with the "timely provision of equipment and support" (Mason & Bacsich, 1998, p.256). Without such support, technology can actually interfere with learning (Cifuentes, Murphy, Segur, & Kodali, 1997; Hara & Kling, 1999). Both students and staff may have problems with connection to the Internet, lack of training, incompatibility problems, software and hardware difficulties, and network outages. With attempts to mainstream online teaching, problems may magnify as the use of technology increases if adequate attention has not been paid to problems of scaling up (Mason & Bacsich).

While innovators are prepared to be relatively understanding of technical problems, the bulk of users are not likely to be as forgiving (Freeman, 1997). There have been many failures in attempting to introduce technology in education because of the "lack of equipment, time and training" (Hara & Kling, 1999).

As well as the level of technological support, lack of access to technology is a major barrier (Apple Computer Inc., 2000; Boddy, 1997; Dooley et al., 1999; Leigh, 2000). Even where technology is available, having to transport it for use in the classroom is often considered too much effort (Spotts, 1999).

CONCLUSION

Good teaching is never easy. Changing one's way of teaching is never easy. Handling new technologies is never easy. Finding the time to do new things is never easy. Doing things differently than the traditional ways of the university is never easy. Being rewarded and appropriately supported (in terms of human and financial support) for putting effort into one's teaching in higher education is unfortunately often not easy (Collis, 1998, p. 391).

Concerns about workload and lack of incentives are major issues that can affect adoption (Carr, 1999; Grant, 1996). This requires a commitment to staff development. At the University of Florida, they were able to successfully integrate technologies in teaching. Rather than simply making sure that everyone had access to technology on their desk, they took preemptive steps to ensure that staff development was strong and continued to have value (Goral, 2001). For "e-learning to succeed, supervisors need to schedule some time in the workflow" (LGuide, 1999). There was a time when teachers were only considered

to be working when they were face to face with students in the classroom (Cook, 1997; Grant). However, with modern workload agreements, there is perhaps more scope to schedule time for training or development.

Incentives do not always have to take the form of funds or time release. Staff appreciate "pleasant and well equipped facilities, materials, access to course credit for professional development, and technological and pedagogical assistance" (Grant, 1996). In many instances, simple encouragement or acknowledgement in front of peers may be all that is needed to provide sufficient motivation (Hagner, 2000).

A concerns-based approach takes into account the concerns of the users: those who ultimately have to work with the innovation on a regular basis. Techniques for consulting users throughout the design process include "peer review, walk-through of a rapid prototype, observation of target group using the software, user-tracking, and interviews individually or in focus groups" (Kennedy, 1998, p. 383). As well as consulting the early adopters, it is probably even more important to co-opt potential resistors so that their concerns and needs can be met (Harasim et al., 1998).

REFERENCES

Apple Computer Inc. (2000). *Teacher-centered staff development*. Retrieved February 11, 2000, from http://www.apple.com/education/kl%202/staffdev/tchrcenterstaff.html

Barritt, M., Ashhurst, C. J. S., & Pearson, M. (1995). *Integrating information technology into teaching.* Canberra, Australia: CELTS (University of Canberra) & CEDAM (Australian National University).

Boddy, G. (1997). Tertiary educators' perceptions of and attitudes toward emerging educational technologies. *Higher Education Research & Development, 16*(3), 343-356.

Bondaryk, L. (1998). *Publishing new media in higher education: Overcoming the adoption hurdle.* Retrieved March 21, 2003, from http://www-jime.open.ac.uk/98/3/bondaryk-98-3.pdf

Brahler, C., Peterson, N., & Johnson, E. (1999). Developing on-line learning materials for higher education: An overview of current issues. *Educational Technology & Society, 2*(2), 42-54.

Carr, V. (1999). Technology adoption and diffusion. In E. Ullmer (Ed.), *An online education sourcebook.* The Collaboratory for High Performance Computing and Communication. Retrieved from http://collab.nlm.nih.gov

Cifuentes, L., Murphy, K., Segur, R., & Kodali, S. (1997). Design considerations for computer conferences. *Journal of Research on Computing in Education, 30*(2), 177-201.

Collis, B. (1998). New didactics for university instruction: Why and how? *Computers & Education, 31,* 373-393.

Cook, C. (1997). *Critical issue: Finding time for professional development.* Retrieved July 12, 2002, from http://www.ncrel.org/sdrs/areas/issues/educatrs/profdevl/pd300.htm

Crawford, A., Chamblee, G., & Rowlett, R. (1998). Assessing concerns of algebra teachers during a curriculum reform: A constructivist approach. *Journal of In-service Education, 24*(2), 317-327.

Donovan, M., & Macklin, S. (1999). The catalyst project: Supporting faculty uses of the Web...with the Web. *Cause/Effect, 22*(3). Retrieved from http://www.educause.edu./ir/library/html/cem/cem99/cem9934.html

Dooley, K. (1999). Towards a holistic model for the diffusion of educational technologies: An integrative review of educational innovation studies. *Educational Technology & Society, 2*(4), 35-45.

Dooley, L., Metcalf, T., & Martinez, A. (1999). A study of the adoption of computer technology by teachers. *Educational Technology & Society, 2*(4), 107-115.

Edmonds, G. (1999). *Making change happen: Planning for success.* Retrieved April 6, 2001, from http://ts.mivu.org/default.asp?show=article&id=40

Fowler, S., & Dickie, B. (1997). *Making a difference: Equipping teachers for curriculum change* (No. 2). Canberra, Australia: Commonwealth Department of Employment, Education, Training and Youth Affairs.

Freeman, M. (1997). Flexibility in access, interaction and assessment: The case for Web-based teaching programs. *Australian Journal of Educational Technology, 13*(1), 23-39.

Goral, T. (2001). Professional development in a high-tech world. *Curriculum Administrator, 37*(2), 48-52.

Grant, C. (1996). *Professional development in a technological age: New definitions, old challenges, new resources.* Retrieved June 20, 2001, from http://ra.terc.edu/publications/terc_pubs/tech-infusion/prof_dev/prof_dev_frame.html

Hagner, P. (2000). Faculty engagement and support in the new learning environment. *Educause Review, 35*(5), 27-37.

Hall, G., & Hord, S. (1987). *Change in schools: Facilitating the process.* Albany, NY: State University of New York Press.

Hansen, S., Deshpande, Y., & Murugesan, S. (1999). *Adoption of Web delivery by staff in educational institutions.* Paper presented at AusWeb99.

Hara, N., & Kling, R. (1999). *Students' frustrations with a Web-based distance education course.* Retrieved December 13, 1999, from http://firstmonday.org/issues/issue4_12/hara/index.html

Harasim, L., Hiltz, S., Teles, L., & Turoff, M. (1998). *Learning networks: A field guide to teaching and learning online.* Cambridge, MA: The MIT Press.

Jones, B. (1982). *Sleepers, wake: Technology & the future of work.* Melbourne: Oxford University Press.

Kennedy, D. (1998). *Software development teams in higher education: An educator's view.* Paper presented at the 15th Annual Conference of the Australasian Society for Computers in Learning in Tertiary Education (ASCILITE).

Leigh, G. (2000). Key markers in Victoria's information technology journey into the knowledge age. *Australian Educational Computing, 15*(1), 7-12.

LGuide. (1999). *E-learning: Beyond the hype.* Retrieved March 13, 2001, from http://www.lguide.com/index.cfm?item_id=4108

Littlejohn, A., & Sclater, N. (1998). *The virtual university as a conceptual model for staff development.* Retrieved April 23, 1999, from http://cvu.strath.ac.uk/admin/cvudocs/webnet98/concepts.html

Lu, A., Zhu, J., & Stokes, M. (2000). The use and effects of Web-based instruction: Evidence from a single-source study. *Journal of Interactive Learning Research, 11*(2), 197-218.

Marx, R., Blumenfield, P., Krajcik, J., & Soloway, E. (1998). New technologies for teacher professional development. *Teaching and Teacher Education, 14*(1), 33-52.

Mason, R., & Bacsich, P. (1998). Embedding computer conferencing into university teaching. *Computers & Education, 30*(3/4), 249-258.

McInerney, D., & McInerney, V. (1994). *Educational psychology: Constructing learning.* Sydney, Australia: Prentice Hall.

McKinnon, D., & Nolan, P. (1989). Using computers in education: A concerns-based approach to professional development for teachers. *Australian Journal of Educational Technology, 5*(2), 113-131.

McNaught, C., Kenny, J., Kennedy, P., & Lord, R. (1999). Developing and evaluating a university-wide online distributed learning system: The experience at RMIT University. *Educational Technology & Society, 2*(4), 70-81.

Mitra, A., Hazen, M., LaFrance, B., & Rogan, R. (1999). *Faculty use and non-use of electronic mail: Attitudes, expectations and profiles.* Retrieved March 3, 2000, from http://www.ascusc.org/jcmc/vol4/issue3/mitra.html

Newhouse, P. (1999). Examining how teachers adjust to the availability of portable computers. *Australian Journal of Educational Technology, 15*(2), 148-166.

Noddings, N. (1992). *The challenge to care in schools.* New York: Teachers College Press.

Pettit, A. (1998). Teaching first, technology second: The possibilities for computer mediated communication. In J. Cameron (Ed.), *Online teaching* (pp. 18-35). NTU, Centre for Teaching and Learning in Diverse Educational Contexts. Northern Territory, Australia: Darwin.

Porter, T., & Foster, S. (1998). From a distance: Training teachers with technology. *Technological Horizons in Education Journal, 26*(2), 69-72.

Reeves, T. (1993). *Conducting science and pseudoscience in computer-based learning.* Paper presented at the 10th Annual Conference of the Australasian Society for Computers in Learning in Tertiary Education (ASCILITE).

Rogers, E. (1983). *Diffusion of innovations.* New York: The Free Press.

Schiller, J., & Mitchell, J. (1993). Interacting at a distance: Staff and student perceptions of teaching and learning via video conferencing. *Australian Journal of Educational Technology, 9*(1), 41-58.

Schneiderman, B., Borkowski, E., Alavi, M., & Norman, K. (1998). Emergent patterns of teaching/learning in electronic classrooms. *Educational Technology, Research and Development, 46*(4), 23-42.

Schoenfeld-Tacher, R., & Persichitte, K. (2000). *Differential skills and competencies required of faculty teaching distance education courses.* Retrieved June 30, 2004, from http://www.ao.uiuc.edu/ijet/v2n1/schoenfeld-tacher/index.html

Slough, S. (1999). *Some concerns about the concerns-based adoption model (CBAM) and technology.* Paper presented at SITE 99.

Spotts, T. (1999). Discriminating factors in faculty use of instructional technology in higher education. *Educational Technology & Society, 4*(2), 92-99.

Surry, D. (1997). *Diffusion theory and instructional technology.* Paper presented at the Annual Conference of the Association for Educational Communications and Technology (AECT), Albuquerque, NM. Retrieved March 9, 2000, from http://www.gsu.edu/~wwwitr/docs/diffusion/

Surry, D., & Farquhar, J. (1997). *Diffusion theory and instructional technology.* Retrieved May 11, 1999, from http://www.usq.edu.au/electpub/e-jist/docs/old/vol2no1/article2.htm

Tennant, J. (1999). Teleteaching with large groups: A case study from the Monash experience. *Australian Journal of Educational Technology, 15*(1), 80-94.

KEY TERMS

Computer Communication: A term often mistakenly used in relation to online education. Typically, the object of communication is another person, not a computer, and the term computer-mediated communication would be more accurate.

Concerns Theory: Concerns theory suggests "that concerns change over time in a fairly predictable, developmental manner" (Dooley et al., 1999, p. 109). Individuals go through stages during the change process, with differing needs through these stages. Change strategies that meet these needs are more likely to be effective (Crawford et al., 1998; Schiller & Mitchell, 1993).

Early Adopters: Those who embrace change and are the most likely to adopt technological innovations quickly.

Rate-of-Adoption Theory: This "states that innovations are diffused over time in a pattern that resembles an *s*-shaped curve" (Surry & Farquhar, 1997). The rate of adoption rises slowly at first. When around 20% of the population has joined, the adoption "takes off." The rate increases to a maximum when adoption reaches about 50% of the population. After this period of rapid growth, the rate of adoption gradually stabilizes and may even decline.

Resistors: Those who are hesitant about change, often with good cause, and are the least likely to adopt technological innovations quickly. It may include the categories termed late majority and laggards.

Technological Determinism: Determinist or developer-based models of diffusion focus on the technical characteristics in order to promote change. They assume that technological superiority is all that is required to bring about the adoption of innovative products and practices.

Technological Instrumentalism: Instrumentalist or adopter-based theories of diffusion emphasize the importance of the social context of change and the need to address the knowledge, beliefs, feelings, and concerns of the users (Crawford et al., 1998). Successful change is seen in relation to meeting the real and perceived needs of the bulk of the users, not just the innovators.

Chapter 7.29
EBS E-Learning and Social Integrity

Byung-Ro Lim
Kyung Hee University, South Korea

ABSTRACT

In an effort to support public education, the Korean government utilized e-learning and established EBS e-learning system. Educational Broadcasting System (EBS) e-learning is developed especially for high school students who are preparing for the Korean college entrance examination (KSAT). This case study is to introduce EBS e-learning and analyze its merits and weaknesses in order to investigate some meaningful implications for future use of a similar learning system. First, this case study analyzes characteristics of EBS e-learning; secondly, this study identifies its outcomes and issues. Lastly, some implications are presented for further development and more effective use of e-learning system. The outcomes of the case study showed that EBS e-learning system has brought (1) cost reduction on private education on a short-term basis, (2) positive effect on social integrity, and (3) a possibility of supporting public education. Some issues identified in this case study are in the following areas: (1) contents, (2) school use, (3) learning management and operation strategies, (4) faculty members, and (5) user interface. Lastly, some implications and suggestions are made to ensure the provision of quality programs in the future: (1) it is necessary to reinforce interactive ways of learning; (2) customized education and level-specific learning programs should be developed and provided; (3) it should utilize effective learning management systems; and (4) it needs to provide qualified, specialized learning contents to the learners.

INTRODUCTION

Today, we witness rapid growth of knowledge production and distribution more than any other age in human history. Also, knowledge does not belong to particular groups or dominant classes anymore, and it has become public goods to share with grassroots. In the past, social oppressions have been operated by exclusive accessibility to knowledge and information. Currently, however, the general public is able to access knowledge at their convenience and can change the quality of their lives through utilizing this accessible information and knowledge. Moreover, both accessibility and utility of knowledge and information allow even people who have limited educational opportunity to receive quality education.

However, it is obvious that there is a need to bring new communication tools in order to achieve meaningful educational experiences. People can exchange information and knowledge beyond the limit of time and space using the Internet, collaborate with other people for study at any time and any place, and take e-learning courses, which are developed in other countries for their individual achievement. E-learning, a revolutionary method of learning in this global age using the Internet, seems to have a special value in realizing meaningful learning and achieve equal educational opportunities to everyone regardless of their ethnic, local differences, and income gap. It is clear that people can be no longer competitive if they depend only upon off-line, formal education from the traditional educational institutions because of the extensive growth of information and knowledge. In order to learn newly produced knowledge along with creating new knowledge, learners need to take life-long learning and self-directed learning. E-learning can be considered as one of the best alternatives. E-learning enables people not only to get beyond the limitation of time and space, to have a good education anytime and anyplace, but also to build learning communities with other people including peers, teachers, and professionals. Building learning communities through e-learning is a realistic alternative for people to learn in authentic learning environments.

By far, e-learning has been developed in various arenas. But, the benefits of e-learning were mainly enjoyed by a few who belong to urban, middle class until now. So it is necessary to make e-learning affordable and accessible to anyone who wants to learn. In this purpose, the Korean government made a grand project in 2004 and initiated the EBS e-learning program especially for high school students who are preparing KSAT (i.e., college entrance examination in Korea). The predecessor of the Educational Broadcasting System (EBS), the Broadcasting Department of the Korean Educational Development Institute (KEDI), produced and broadcast radio programs for the Air and Correspondence High School in 1974. EBS was established as an affiliate of the KEDI with two channels (EBS TV and EBS FM) in 1990. It was separated from KEDI in 1997 and started e-learning services from the year 2004. EBS, headquartered in Seoul, Korea, is the only broadcasting corporation dedicated to public education in Korea. EBS has grown into a major broadcaster for entertaining and educational programs. Accordingly, this case study is to introduce the EBS e-learning and analyze its merits and weaknesses in order to gain some educational insights from them, especially in terms of social integrity.

EBS E-LEARNING PROGRAM

Background

The Korean government has energetically carried out a policy of utilizing information technology in public education since 1996, which results in connecting every school across the nation with the high-speed network, EDUNET, which provides educational portal service in national scale. Another development is EBS e-learning. From the year 2000, not only ISST (ICT Skill Standard for Teachers), and ISSS (ICT Skill Standard for Students), but also contents such as teaching guides and multimedia materials were developed and disseminated widely. In the year 2004, the EBS e-learning program and National Center for Teaching and Learning opened. Figure 1 shows e-learning history in public education from 1966 to 2004.

Figure 1. History of e-learning in Korea

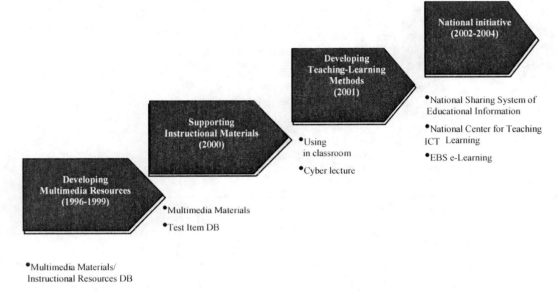

Traditionally, Koreans have considered the college diploma as a symbol of high status and a way of being successful in society. Therefore, high stake tests such as KSAT, which is held once a year, are highly valued and getting a good score in the test ensures a successful future to some degree. Many Korean students are willing to do their best when they come to high school.

Thus, the Ministry of Education and Human Resources (MEHR) initiated EBS e-learning service to help high school students to study KSAT preparation courses in more convenient and cheaper ways. It is one of the important projects, which were planned to support public education and maintain normal schooling practices. In 2004, MEHR announced a plan to reduce cost in the private educational sector in order to normalize public education and established learner-support educational systems through e-learning. EBS was delegated to perform this plan, and it began to provide e-learning service from April 1, 2004. The purposes of EBS e-learning are as follows:

- To help students' KSAT preparation study
- To establish learner-support programs

- To support public education
- To normalize public schooling
- To reduce the cost of extra-curriculum study
- To solve the problem of social inequality

At first, EBS e-learning was considered to be a short-term measure to reduce costs in the private educational sector, which caused economic crisis and intensified social conflicts between the rich and the poor. Table 1 shows the overall cost reduction plan by MEHR.

EBS E-LEARNING MENUS AND PROGRAMS

EBS e-learning focuses on the e-learning courses and programs for KSAT. The EBS e-learning system consists of online lectures (VOD), bulletin boards (announcement board, Q&A, FAQs, free board), diagnosis test service and personalized area ("my page"). The lectures cover almost every area of KSAT (refer to Table 2). EBS e-learning is unrivaled in the range of school courses it covers

Table 1. Plan for cost reduction in private educational sector to normalize public education

> 1. Short-term plan: Public education taking over private tutoring demand
>   • Substituting for extracurricular studies: Establishing e-Learning systems
>   • Absorbing formal, curricular studies: Level-specific supplementary learning program
>   • Giving satisfaction to the gifted learning or English study programs: Using extracurricular program
>   • Nursery programs: Running after-school programs to 1~3 grade students
> 2. Mid-term project: Substantiating public schooling
>   • Rethinking reliability on public education: Hiring highly qualified faculties
>   • Recovering functions of public education: Improving teaching and assessment methods
>   • Supplementing educational standard for high school students: Level-specific education and expanding learner's choices
>   • Normalizing public education: Improving college entrance examination and enhancing instruction for student's life goal
>   • Guaranteeing the grounding knowledge of citizens: Enhancing responsibility-taking for basic knowledge-level education
> 3. Long-term project: Improving socio-cultural climate
>   • Reforming projects for social systems and consciousness

*Note: * Source: Ministry of Education and Human Resources (2004). Plan for Cost Reduction in Private Educational Sector to Normalize Public Education.*

with regards to the college entrance exam. It offers a wide selection of subjects to accommodate the various needs of students, including non-regular courses for those interested in vocational colleges. In addition, EBS courses are available in three levels: beginning, intermediate, and advanced for more demanding students.

Other special lectures are also provided: Theme-based lecture, today's new lecture, this month's new lecture, special lecture, and EBSi special. In addition to the lectures, EBS e-learning provides subscribers with various menus such as news, Q&A, KSAT information, and "my page" to facilitate interaction with its users (Table 3).

Table 2. EBS e-learning lecture list

Lecture list
Korean, English, Mathematics, Social Studies, Science, Occupation education
Linguistic domain, Foreign language domain, Mathematical domain
Social Studies domain, Science domain, Occupation education domain, Oral interview/
Analytical writing, Entrance exam special, Listening

Table 3. Functional menus of EBS e-learning

	Content
My Page	Pre-chosen lecture, My lecture, My Q&A, My information management Personal schedule management, Personal note, My friend, My bookmark
Entrance	EBS news, Entrance examination/College-oriented info examination
Info	GPA calculation/a trial application
My teacher	EBS e-Learning teacher, Q&A teacher, Learning guide, Consulting with teacher, FAQ
EBS Empathy	New info, Thank you EBS, Excellent EBSi, Revisiting lecture, Let's get together, Let's share with, Let's recommend, Photo gallery, E-column

EBS e-learning ensures all course materials are put together by the best, qualified writers and authors available, and takes an active part in the production process, from planning to editing. However, EBS e-learning contents are developed in VOD method, that is, the media specialist shoots the lectures and uploads them on the Internet. The instructional design is more teacher-centered and lecture-based. As shown in Figure 2, the VOD window has two parts: VOD viewer window (left half of the window) and index of the VOD (right half of the figure). It has speed control system, which enables users to control the VOD speed at their convenience, and electronic note with which users can take a note during the lecture.

Characteristics of EBS E-Learning

EBS e-learning system has the following characteristics:

1. It maximizes benefits of e-learning.

    *   It frees learners from constraints of time and location.
    *   It utilizes a wide variety of EBS educational contents.

    *   It offers two-way communication.
    *   It has group discussion sessions.

2. It customizes students' learning.

    *   It allows students to select video-on-demand and audio-on-demand services when they choose EBS educational contents.
    *   Students can plan their own study schedule according to EBS broadcast programs.
    *   Students can create their own homepage containing their history of taking EBS courses.
    *   It enables learners to choose lectures suitable to their own levels.
    *   It provides a caption service and Internet sites for the deaf/blind students.

3. It has other beneficial services.

    *   Students can download lecture video clips around the clock, thanks to a server powerful enough to handle up to 100,000 simultaneous connections.
    *   It hires cyber tutors who consult learners.
    *   It provides reserved download services for online lectures.
    *   It provides updated new lectures every day.

Figure 2. Interface of the VOD window

Efforts on Social Integrity

EBS provides support for students from local areas and low-income families as follows:

- Every cable TV company should provide EBS Plus 1 channel (broadcasting service of EBS e-learning).
- About 10,000 high school students from local areas were provided with a satellite broadcasting receiver, and some of them were paid for the Internet service.
- Four thousand and five hundred students who could not afford the Internet cost were provided with a satellite-broadcasting receiver and were paid for the Internet service.
- Students from low-income families (about 28,000 persons) got the textbooks for EBS e-learning for free.
- Seventy thousand students from low-income families were provided with personal computers and were paid for the Internet services.

EBS also provides support for the handicapped as follows:

- The EBS e-learning site was developed for the blind to use a screen reader so that they can read the contents.
- Textbooks for EBS e-learning were provided to the National Special Education Bureau, which in turn lets schools for the handicapped use.
- Textbooks in Braille were published for the blind.
- VOD services for about 500 lectures to the people who have hearing difficulties.

ACADEMIC AND ADMINISTRATIVE ISSUES

Content Related Issue

Currently, the main type of content is test-oriented VOD lectures. Just as profit-making private institutions put an emphasis on teaching for test and test-oriented learning, EBS e-learning programs focus on teaching learners how to get good test scores.

Secondly, it has a large number of faculty members from profit-making private institutions. Some instructors seem to use their popularity to advertise their own private academy. In this situation, it is hard to expect sound contents in an educational sense.

Lastly, the lecture-oriented teaching method is another issue. Since the main type of content is VOD, it is hard to use various instructional methods such as debate, small group activities, problem-based learning, and so on. The content quality is just dependent on the instructor's personal teaching ability. Relying too much on the individual competency may cause a problem in content quality, especially when the instructor is not willing to continue his/her teaching any more.

Issues Regarding School Use

The primary goal of EBS e-learning is to provide high quality of teaching and learning materials and to normalize the public education. Accordingly, it is important to effectively use these materials within the public school system. Reviews on school use of EBS e-learning programs do not guarantee that the primary goal of EBS e-learning has been achieved (KERIS, 2005). Kim (2004) worried that EBS e-learning would weaken the functions of the public education because of its extensive use in the regular class and self-study time. Current

EBS e-learning experiments may have a negative effect on the public education since EBS e-learning makes teachers and students concentrate more on the EBS programs and college entrance examination instead of normal schooling.

Learning Management and Operation Strategies Issues

For the learning management strategies, there are several issues to consider:

- Lectures only depend upon teacher's individual teaching ability.
- LMS is not fully supportive, so there is insufficient learning support for the learner.
- Contents are generally unsatisfactory since they are not based on robust instructional design.
- There are insufficient learning assessment tools to evaluate learner's achievement level.

Issue on Recruiting Faculty Member

Current status of faculty members consists of middle and high school teachers (about 70%) and profit-making private institution teachers (about 30%). Moreover, the ratio of private institution teachers in faculty members is much higher than that of public school teachers in the beginner and advanced levels. From this statistics, market concerns and commercialism are likely to affect the program. This fact may cause some problems in improving the quality of EBS e-learning.

User-Interface Related Issue

User interface is not bad, but too much information on the main page can interrupt learners' choice. Also, there is a lack of clear understanding in menu structure and expression in the main page.

PROGRAM EVALUATION

Statistics

The number of users and members of EBS e-learning service has rapidly increased since its providing KSAT courses on the Internet in April 2004. Access numbers tended to decrease after the KSAT in November 2004. In January 2005, an average number of signing up members per day was 3,925 and an average number of visitors per day was 1,010,168. For the membership number, total number of visitors from the beginning, average number of VOD access per day, number of download per day, and number of contents, refer to Table 4.

Figure 3 shows the changes of average number of VOD access per day during August 2004 through January 2005.

Effects of Cost Reduction on Profit-Making Private Institutions

Even though there are some differences in the amount of reduced money after taking EBS e-learning courses, researchers agreed with the cost reduction effects of EBS e-learning:

Table 4. Statistics of membership, access number, and number of download and contents (January 23, 2005)

	N. of membership	Total N. of visitors	Average N. of VOD access per day	N. of download per day	N. of contents
Number	1,433,330	127,435,643	91,467	23,050	5,526

Figure 3. Average number of VOD access per day during August 2004 through January 2005 (Source: Report of EBSi weekly statistics, 2005)

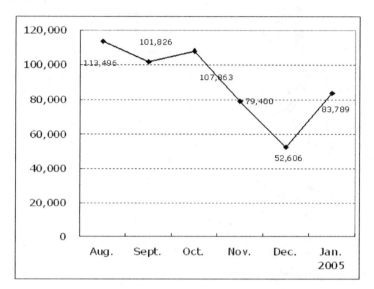

- According to the survey by "Research & Research Institute" (May 17, 2004), there happened extensive cost reduction from 237,000 won (about $230) to 194,700 won (about $190) per month after EBS e-learning programs was operated, only saved $40.
- According to the survey by "MB Zone & CNC" (November 15, 2004), research participants said that they saved 153,000 won (about $150) per month after taking EBS e-learning courses.
- Research by EBS (2005) show that after taking EBS e-learning programs, an average cost for private education reduced from 346,000 won ($340) to 326,000 won (about $320) per month/household, only saved 20,000 won (about $20).

Effect on the Social Integrity in Terms of Solving Inequity of Local and Class Differences

EBS e-learning seems to contribute to solving educational inequity from local and class differences. Two halves of all senior high school students used EBS e-learning or at least accessed the program. On average, students from middle class and local areas took much more EBS e-learning than others (EBS, 2005).

- Almost 80% of high school teachers and 63% of parents believed that EBS e-learning programs contributed to reduce regional differences in education.
- Seventy-seven percent of students in small and medium cities took EBS e-learning programs, while 70% of students in large cities did.

- In terms of parents' income, 32% of students who came from high-income families took private tutoring and used private institutions after school. But only 23% of students whose parents were middle class and 16% of students whose parents were low class took private tutoring and private institutions. But, more students from the low class (24.5%) and the middle class (16.8%) than others from the upper class (11.8%) took EBS e-learning courses instead of private tutoring.

Possibility of Supporting Public Education

It seems that social and collective inquiries are not encouraged in the EBS e-learning system. The main type of learning is isolated learning, that is, listening to the lectures, memorizing, practicing, solving problems in his/her own house without any collaboration with others. When teachers introduce EBS e-learning in their classroom, students usually learn on their own, and there is little discussion, collaboration, and collective inquiry. This means that EBS e-learning is not appropriate when the learning objective is to develop higher-order thinking skills. Also, students and teachers can enjoy meaningful learning experiences when interactive methods are used appropriately. However, EBS e-learning is limited in facilitating interaction among students. It has limitation to provide students with community of learners. Currently, EBS e-learning has some limitations in realizing the ideal of public education.

IMPLICATIONS AND CONCLUSION

Based on the analysis of the benefits and weaknesses of EBS e-learning systems, some implications are made to improve EBS e-learning (Lim, 2005). They include:

- It is necessary to reinforce interactive ways of learning.
- Customized education and level-specific learning should be taken place.
- It should utilize effective learning management systems.
- It needs to provide qualified, specialized learning contents to the learners.

It is necessary to reinforce interactive ways of learning. EBS e-learning has provided one-way VOD services, which largely depend on teacher's individual teaching ability. It does not actively use interactive strategies between learners, learners and content knowledge, learners and teachers, but rather it focuses on providing learners with content knowledge in a teacher-controlled environment. Also, it does not facilitate learners to learn in interactive ways even though it tries to provide prompt feedback to the students who have questions to ask during their studies. It is necessary to develop contents in consideration of diverse interactions and to overcome teacher-directed instruction.

Customized education and level-specific learning should be taken place. Currently, EBS e-learning provides contents on three difficulty levels: beginner, intermediate, and advanced. But it is difficult for students to perceive their learning levels in order to select appropriate contents. The reason is that EBS e-learning does not provide users with any diagnostic testing and evaluation tool. EBS e-learning needs to develop a diagnostic testing to help learners acknowledge their own level. It is also necessary to develop level-specific learning materials fitted into students' needs.

It should utilize effective learning management system (LMS). Major e-learning institutions usually have an effective LMS so that they can provide learners with excellent services. Effective LMS should be developed so as to manage a personal learning history and help individual learners to study on their own learning patterns and learning styles, and with their own learning strategies.

It needs to provide qualified, specialized learning contents to the learners. EBS e-learning needs to provide qualified, specialized learning contents to the learners. Specialized content, design, and method will ensure quality learning. Since the main type of content of EBS e-learning is VOD, the instructor, lecture method, and lecture organization are very important. Based on the needs analysis, it should select specialized contents, provide the best instructor, let him/her choose the appropriate method, and organize the contents in more favorable ways. Also, in order to provide quality learning, it is necessary not only to provide VOD type lectures, but also to organize lectures with various multimedia technologies.

In conclusion, EBS e-learning should go beyond simple test-oriented lectures. If it is used well enough, it has a potential to revolutionize the current educational system. In order to do that, EBS e-learning has to give up the traditional, teacher-centered paradigm of education and accept a new learner-centered paradigm. If EBS e-learning is continuously test-oriented and maintains the "teaching for test" strategy, normalization of the public education will not be realized. In order to design and develop EBS e-learning better, macro-level and long-term perspective is needed along with paradigm shifts.

LESSONS LEARNED

1. One of the successful factors to e-learning is consistent, steady efforts by the government. Especially in the early stage, when e-learning vendors are premature and the market is not active, governmental investment can provide a momentum for e-learning development and control the direction of e-learning. Korean government initiated e-learning on a national scale already in the late 1990s. Even though the governmental efforts result in e-learning development to some degree, it does not seem to facilitate competition among vendors for the creative instructional design (remember that the content type is mainly "lecture video"). Now it is time to recognize the limitation of government's effort and to find out a way of quality assurance and creative design.

2. E-learning is not just a new type of teaching method, but also a tool for satisfying social needs. People want quality education in cheaper and more convenient way. However, in Korea too much money is wasted for extracurricular activities and private lessons, which makes Korean economy weaker and unsound. Through providing cheaper and good education programs, EBS e-learning can satisfy social needs and change the direction of money or investment to the more productive industry.

3. E-learning can contribute to social integrity. EBS e-learning experience shows that e-learning can be an effective social mechanism to lessen the differences among classes and localities. It lessens inequity between the rich and the poor and reduces the gap between the rural and the urban in terms of equal access to educational opportunities. For example, poor students who cannot afford private lessons can take KSAT courses offered by EBS e-learning instead. This is important in the Korean situation where the more educated can have a better chance to earn the money and get the power.

4. E-learning needs to be practiced with educational purposes. When we consider introducing e-learning, we need to ask what education is for. E-learning is one type of educational activity, so its purpose and methods should comply with educational purposes. Education is not simply "teaching for test," but is inspiration and should culti-

vate human mind. In this sense, educators criticize EBS e-learning because its main purpose is to make learners ready for the test, and its methods consist of lectures and explanation of test items. So, EBS e-learning is limited and targets only specific users who want to prepare the KSAT test. If it continues to do current practices and is not based on the sound educational principles, it can disappear after short-term, temporary glory.

5. E-learning is only a supplementary tool for normal class, and it should strengthen teachers' status in his/her classroom. In the current K-12 school system, it is difficult for teachers to get used to use technology. Many technologies have failed in the last 50 years. The main reason for the failure is mainly because of teachers' unconcern or unwillingness. If e-learning is to be used widely in schools, it should get teachers involved in the decision-making process and let them acknowledge its convenience and merits. If it is used in a way of alienating teachers, however, it will weaken public education and normal schooling in the end. Current EBS e-learning programs are good resources to teachers. The programs should not replace teachers, but supplement classroom activities.

6. E-learning can cover diverse topics and subjects beyond formal curriculum so that learners select appropriate ones and widen their knowledge. EBS e-learning has many beneficial programs that can draw students' attention. E-learning should provide students with interesting programs so that students attend (watch) the lectures when they want and where they feel comfortable.

7. E-learning can provide a different point of view or colorful explanations on a certain subject, and facilitate multiple perspectives. E-learning materials are based on multi-media, and it can represent information in many different ways. For example, historical simulation can show the dynamic situation of the historical event, which is not possible in oral or written expression. Representation of information in many different forms helps students learn the subject in a more authentic and deeper way.

8. Facilitating interactive ways of learning using Q&A, prompt feedback, and discussion ensures success of e-learning. EBS e-learning provides many ways to communicate between tutors and learners and among learners. For example, when students are puzzled, they post a question in the Q&A bulletin board. EBS tutors are supposed to answer the question within 12 hours.

9. The community of learner approach makes e-learning more dynamic and interesting. EBS e-learning uses several techniques to facilitate sense of community. For example, it provides bulletin boards such as "Buzz session', 'Postscript," "Impression after class," and "Recommending best teacher." Students use these bulletin boards to express their feelings, share stories they experienced during their study, exchange their own know-how about how to study, and express their gratitude for their tutors.

10. E-learning success depends on not only quality content, but also management for individualized learning. In the EBS e-learning system, a same content has three levels—beginning, intermediate, and advanced—and the system allows learners to choose one of them based on their competencies.

Best Practices

1. In the early stage, a consistent and comprehensive effort by the government is a very important factor to the development of e-learning.

2. E-learning is not just a method of teaching and learning, but also a way to strengthen social integrity through access to quality contents regardless of social status and regional differences.

3. Effectiveness of e-learning depends on orchestrating many factors such as level-specific contents, tutoring, community of learners, management, as well as technical infrastructure (efficient LMS, server stability, etc.).

REFERENCES

EBS. (2005). *A study on strategies for activation of EBS KSAT preparation courses*. Report 2005-1.

Ministry of Education and Human Resources. (2004). *A study on mid- and long-term development strategies of EBS KSAT preparation courses project for growth of e-learning*. Report on policy study 2004 dedicated.

KERIS. (2005). *Report of Local Educational Bureau on EBS e-learning*. Unpublished report.

Kim, H. B. (2004). Educational value of EBS e-learning and analysis of its effectiveness. In *Proceedings of KAEMS Conference* (2004.12.4).

Lim, B. R. (2005). EBS e-learning: Current status, issues, and improvement strategies. In *Proceedings of KAEMS Conference* (2005.9.10)

Chapter 7.30
Educational Geostimulation

Vasco Furtado
University of Fortaleza, Brazil

Eurico Vasconcelos
Integrated Colleges of Ceará (FIC), Brazil

ABSTRACT

In this work we will describe EGA (educational geosimulation architecture), an architecture for the development of pedagogical tools for training in urban activities based on MABS (multi-agent based simulation), GIS (geographic information systems), and ITS (intelligent tutoring systems). EGA came as a proposal for the lack of appropriate tools for the training of urban activities with high risk and/or high cost. As a case study, EGA was used for the development of a training tool for the area of public safety, the ExpertCop system. ExpertCop is a geosimulator of criminal dynamics in urban environments that aims to train police officers in the activity of preventive policing allocation. ExpertCop intends to induce students to reflect about their actions regarding resources allocation and to understand the relationship between preventive policing and crime.

INTRODUCTION

Simulation aims to represent one phenomenon via another. It is useful to measure, demonstrate, test, evaluate, foresee, and decrease risks and costs. Computational simulation can be considered as experimentation based on a computer model that provides a safe experimental environment for the inquiry of system properties. In educational terms, simulation is important because it allows learning through the possibility of doing (Piaget, 1976). Simulation has proven to be a good teaching tool, especially for complex situations, with high cost and risk. Practical application can be seen in various areas, such as in the aeronautical industry, nuclear industry, space exploration, petrochemical industry, and military research (Roger, 1994).

Multi-Agent paradigm has been widely adopted in the development of complex systems. In particular, if there are heterogeneous entities or organizations with different (possibly conflicting) goals and proprietary information, then a multi-agent system (MAS) is useful to handle their interactions. A MAS is also appropriate whenever there is a need to represent each entity of the modeled domain individually or if these entities have an intelligent behavior to be modeled.

Social or urban environments are dynamic, non linear, and made of a great number of interacting entities, characterizing a complex system. The use of MAS to simulate social environments has become broadly used (Billari & Prskawetz, 2003; Gilbert & Conte, 1995; Khuwaja, Desmarais, & Cheng, 1996). Aggregating a GIS (geographical information system) to an MAS in the simulation of social or urban environments characterizes geosimulation (Benenson & Torrens, 2004). With the computational development of GIS, bringing precision and realism to simulation (Wu, 2002), multi-agent based simulations (MABS) benefited from them in terms of geographical representation of the areas to be simulated.

Analyzing the existing proposals and tools, and in accordance with Gibbons (Gibbons, Lawless, Anderson, & Duffin, 2001), there are few or even no adequate tools for developing educational computer systems where intelligent agents support the interaction between the simulation model and the user. Despite recent proposals on new models and implementations of instructional layers in simulators (Gibbons et al., 2001; Mann & Batten, 2002), few tools have been created specifically for urban activities, none of them with adequate support to the education process.

This chapter describes the educational geosimulation architecture (EGA), an architecture for training in urban activities based in the synergy among MABS, GIS, and ITS (intelligent tutoring systems) that we consider an optimal and complementary set of technologies for building educational geosimulation. We also describe the ExpertCop system, a training tool developed for the area of public safety based on EGA. ExpertCop is a geosimulator of criminal dynamics in urban environments, which aims to train police officers in the activity of preventive policing allocation. ExpertCop intends to induce students to reflect about their actions regarding resources allocation. Assisting the user, the pedagogical agent aims to define interaction strategies between the student and the geosimulator in order to make simulated phenomena better understood.

This software, based on a police resource allocation plan made by the user, produces simulations of how crime behaves in a certain period of time based on the defined allocation. The goal is to allow a critical analysis by students (police officers) who use the system, allowing them to understand the cause-and-effect relation of their decisions.

With the aim of helping the user to understand the causes and effects of his/her process of allocation, ExpertCop uses an intelligent tutorial agent provided by the architecture endowed with strategies and pedagogical tools that seek to aid the user in understanding the results obtained in the simulation. The agent offers the student a chronological, spatial, and statistical analysis of the results obtained in the simulation. Using a machine learning concept formation algorithm, the agent tries to identify patterns on simulation data, to create concepts representing these patterns, and to elaborate hints to the student about the learned concepts. Moreover, it explores the reasoning process of the domain agents by providing explanations, which help the student to understand simulation events.

ExpertCop was applied in a set of training classes, making it possible to analyze its effectiveness quantitatively as an educational tool.

BACKGROUND KNOWLEDGE

Geosimulation and Intelligent Tutoring Systems

The MABS is a live simulation that differs from other types of computational simulations because simulated entities are individually modeled with the use of agents. There is a consensus about the adoption of multi-agent approach (bottom-up) to the study of social and urban systems (Benenson & Torrens, 2004; Billari & Prskawetz, 2003; Gilbert & Conte, 1995; Gimblett, 2002). Social or urban environments are dynamic, nonlinear, and made up of a great number of variables. MAS are also appropriate when the environments are composed of a great amount of entities whose individual behaviors are relevant in the general context of the simulation.

A particular kind of simulation, called geosimulation, treats an urban phenomena simulation model with a multi-agent approach to simulate discrete, dynamic, and event-oriented systems (Benenson & Torrens, 2004). In geosimulated models, simulated urban phenomena are considered a result of the collective dynamic interaction among animate and inanimate entities that compose the environment. Geosimulation has been applied in the study of a variety of urban phenomena (Torrens, 2002). The GIS is responsible to provide the "data ware" in geosimulations.

Simulation is widely used as an educational tool because computerized simulation of the studied activity allows the user to learn by doing (Piaget, 1976) and to understand the cause-and-effect relationship of his/her actions.

The simulation *per se* is not a sufficient tool for education. It lacks the conceptual ability of the student to understand the simulation model. Therefore, some works (Siemer, 1995; Taylor & Siemer, 1996) have tried to integrate the notions of intelligent tutoring system (ITS) and simulation in order to better guide learning and to improve understanding of the simulation process. The idea of an ITS is the integration of artificial intelligence in computer learning systems. It aims at emulating the work of a human teacher that has knowledge of the content to be taught, and how and to whom it should be taught. To achieve this, we need to represent I) the domain of study, II) the pedagogical strategies, and III) the student to whom the teaching is provided. A fourth component may also be considered (Kaplan & Rock, 1995; Woolf & Hall, 1995) the interface with the user. Figure 1 shows the most common architecture for an ITS.

The user interface determines how the interaction with the system will be. Through the interaction of these components, the ITS adapts pedagogic strategies on a domain at the level of the student for his/her individual needs.

Figure 1. Components of an ITS architecture

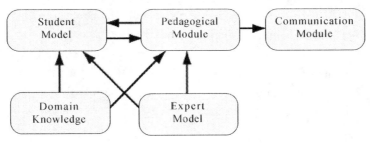

EDUCATIONAL GEOSIMULATION ARCHITECTURE

Motivation and Proposal

A major part of the world's population today lives in large urban centers, dynamic environments composed of an infinity of heterogeneous entities that are interrelated in a series of activities.

Good management of these urban centers depends on an understanding of the dynamics of the urban social relationships and activities occurring among the entities that make up these environments as well as of the environment itself.

Public safety, urban traffic control, migration control, population growth, and urbanization are a few of the complex examples of urban activities or processes that are part of urban relationships. The management of such activities demands an understanding of the factors, entities, and processes involved directly or indirectly in these activities. In addition to the various factors involved and the natural complexity of the relationships between these factors, such activities involve, for the most part, very high risks and costs since they deal with values of public domain and mainly because they involve human lives. This complexity inherent to urban activities makes it difficult, if not impossible, to study and manage such activities.

One can observe a lack of support to the study of these areas, means, or tools that allow a broader base for actions undertaken in such areas.

It is fitting in this context to construct computational tools that make it possible to represent the complexity of the entities and relationships involved in urban activities, in a way as to provide subsidy to studies, training, and management by those persons responsible directly or indirectly for the administration or enforcement of such activities.

Focusing on the complexity of studying urban social activities and the lack of tools to support the study of such activities, a bibliographical search was performed in the area of the educational computational tools, seeking proposals, architectures, or even tools that could aid in the understanding and training of urban activities.

By means of this research, it was possible to verify there are no specific pedagogical tools, nor even approaches, geared toward the presented problematic. We thus confirmed Gibbons' affirmation that there are no, or very few, educative tools that give adequate support to the interaction between the student and studied environment (Gibbons et al., 2001).

However, it was possible to identify approaches geared toward parts of our problem. We observed, for example, that when the object of study is dynamic and the practical study of which is impracticable by involving high risks and costs, the computational simulation of this object proves to be a good strategy. Simulating allows for the representation of the dynamics of the studied object and prevents or minimizes the risks and costs of onsite study, and according to Piaget, simulating offers the pedagogical advantage of "learning by doing." The urban activities we propose to study are made up of a great number of different interrelating entities. In order to simulate the activity of these entities, it is necessary to represent them appropriately. Various authors (Gilbert & Conte, 1995; Khuwaja et al., 1996) describe the approach of multi-agent systems as appropriate for the computational representation of independent entities that interrelate within the same environment. In this way, the multi-agent based simulation seems to encompass two interesting characteristics for our study—process and entities. A third important characteristic in the process of study of urban activities is the representation of the environment. For such, the geographic information systems allow for the appropriate representation and study of geographic environments. The aggregation of the MABS with the GIS in representing the urban environment was named geosimulation by Benenson and Torrens (2004). Finally, another practice noted in the research is the adoption of Intelligent Tutorial Systems as support to computerized learning.

Based on the observations made in our research, we elaborated an architecture that brings together what we consider ideal for the development of urban activities training systems—the architecture for educational geosimulation.

The Architecture

Educational geosimulation architecture is composed of four main components: a user interface, an MAS, a GIS, and a database as is shown in Figure 2. Each part of EGA architecture will be discussed in detail in the following items.

User Interface

The interface is the means of interaction between the user and the rest of the system. It contains the map of the region to be studied, generated by the GIS, whereby the user will interact with the agents of the system in the simulation process. It is also the means by which the pedagogical agent will apply its pedagogical techniques, aimed at assisting the user. The interface is constructed in a logical and organized way and in accordance with ergonomic standards.

Geographical Information System

The GIS is responsible for generating, manipulating, and updating a map on a small scale of the studied region. The map contains a set of layers, representing the geographic, social, and urban characteristics of the area as quadrants, streets, avenues, buildings, parks, slums, and so forth. The GIS offers a set of controls that allows the user and the other parts of architecture to manipulate the map, by amplifying, minimizing, or demarcating areas or by creating, manipulating, or moving objects on the map, among other functions.

System Database

The system database contains (a) information about each user and about his/her simulations, (b) configuration data, (c) statistical data about the studied domain, (d) the domain ontology. Such an ontology is a definition of basic concepts used in the domain and was produced by an expert of the domain.

MABS Platform

The structure, communication, administration, and distribution of the agents, the simulation process, and flux are provided by the MABS platform.

Figure 2. Educational geosimulation architecture

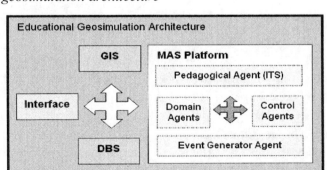

The multi-Aagent based simulation platform is made up of four groups of agents: control agents, domain agents, the event generator agent, and the pedagogic agent.

Control Agents

The control agents are responsible for the control, communication, and flow in the system. The control agents are:

- **GIS agent:** Which is responsible for answering requests from the graphical interface, domains, and control agents. It is responsible for updating the map with the generated simulation data. The GIS agent represents the environment.
- **Manager agents:** Which manage the types of agents of the studied domain. They are responsible for the coordination and interaction with agents of their type or class. This agent is a middleware among agents of its own group and the rest of the system. They control pre-programmed activities of their agents as activation and deactivation and controlled events.
- **Log agent:** Which is responsible for recording all interactions among system agents in order to contain all data about the simulation and the configuration.
- **Graphical agent:** Which compiles pertinent data to the domain, exchanging information with the log agent, and dynamically generating statistical graphics to the graphical interface.

Domain Agent

Domain agents are the actors of the domain, acting actively in the simulation process. The number, kind, behavior, and relations of these agents depend on the studied problem. The idea is that these agents are modeled with the most representative characteristics and actions of the behavior of the entities being proposed for representation.

Event Generator Agent

The event generator agent is responsible for generating the events that will motivate the simulation process. If the studied domain were that of traffic of urban vehicles, it would generate, for example, events such as traffic accidents. If the domain were that of the dynamics of the fire department, it would generate fires or other occurrences to be responded to. If the domain were that of natural disaster control, the generator could be responsible for generating floods, landslides, or earthquakes. The type of event generated by the agent depends on the domain to be studied, being up to the developer to define what he/she wishes to be generated by the agent. This definition is provided by filling in statistical tables in which the agent bases itself on data about the events in question. These tables associate the type of event with its geographical or temporal pattern of occurrence. Each type of event is characterized by sets of geographic and temporal factors, each factor with a set of possible values and each value with weights that define its probability in relation to the event.

The parameters assessed by the agent for the process of events generation are: the period to be simulated, the area selected, and the student's level of experience. The agent simulates the occurrence of events for the entire period to be simulated, making the geographic and temporal distribution of the events according to statistical data from the system database. This process occurs within of a set of successive steps:

1. **Input of parameters:** Initially the agent receives, as input parameters, the period, the area, and the user of the simulation.
2. **Total events calculation:** Based on the parameters received, the agent searches through specific tables of the database,

statistical information referring to the occurrences of the events.

Data of the occurrence of events in similar and previous periods are joined with those that the user desires to simulate, taking into account the seasonableness of the occurrence of such events.

Based on these statistical data, the agent makes a projection (prediction) of the tendency of the total number of events per type for the selected period. For such, we used one of the analysis methods of time series represented by formula $Y = TCSI$, where "Y" represents the value of time series to be estimated, "T" represents the trend, "C" represents the cyclical movements, "S" represents the seasonal movements, and "I" represents the irregular movements. The method in question is the method of moving averages that allows us to ignore the cyclical, seasonal, and irregular variations, analyzing only the tendency. To analyze the tendency, the agent will consider the tendency of the last periods and the tendency of periods similar to the desired one as in Figure 3.

3. **Temporal and geographical distribution of events:** Once the numbers of events per type are obtained for the period, the second step is to distribute them geographically and temporally within the area to be simulated. For temporal distribution, the generator verifies the statistical data of the selected area in relation to the number of events of each type per day of the week and per shift. These statistics allow the agent to make the distribution of the total of events per type (calculated in the first step), on each day to be simulated.

For the geographical distribution (latitude and longitude) of the events, the simulator uses the map's geographical and social information seeking to associate the type of event to its geographic pattern of occurrence. A geoprocessed map is composed of various layers of information and characteristics. For this reason it is possible to subdivide the map by sets of characteristics. These characteristics are mapped in a table (by the developer) with the association among areas with specific characteristics and types of events. In Figure 4 we can see four layers of a map (A, B, C, D), each layer showing areas demarcated by a specific type of characteristic. In this way we can create a table relating, on a percentile basis, the types of events to the characteristics of the environment in which they occur.

4. **Adequacy to the student's level:** The final step is to consider the student's level of experience. The system can characterize the student as beginner, intermediary, and experienced. Each level functions as a type of filter for steps 2 and 3 described. For example, at the inexperienced level, the simu-

Figure 3. Selection of data for calculation of the totals of an event for any given period

	PERIODS	Events	
Seasonableness is observed by the fact that only periods similar to the one selected were studied (month of January).	January 1998	20	The projection of **tendency** (temporal series) observes the behavior of the values in a temporal way; in this case in particular there is a tendency of growth over the years, especially for the months of January.
	January 1999	21	
	January 2000	24	
	January 2001	24	
	January 2002	26	
	January 2003	27	
	January 2004	30	
	January 2005	?	

Figure 4. Association between area characteristics and event types

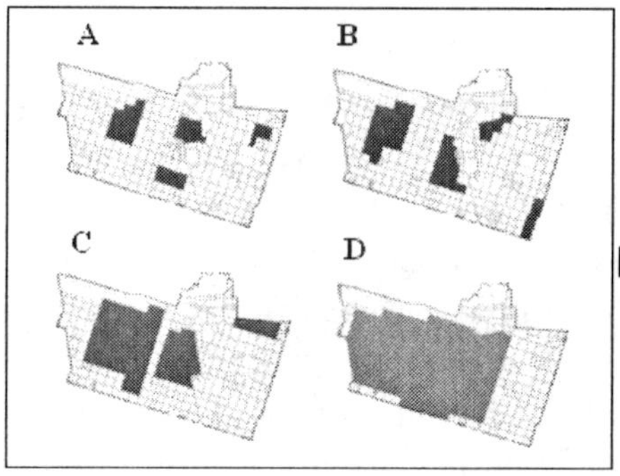

Characteristics / Events	Event X	Event Y	Event Z
Characteristic A	10 %	20%	30%
Characteristic B	20 %	5%	10%
Characteristic C	0%	25%	25%
Characteristic D	70%	50%	35%

lator—after calculating the total number of events per type—applies a percentile reduction to the total, and when geographically distributing this total of crimes on the map for the given period, the simulator will seek to encase the event into a smaller and more accurate number of geographic, social, and temporal characteristics, so as to facilitate the student's analysis and identification of patterns. These characteristics are defined previously in a table for each level of student. At the professional level, instead of having a percentage of reduction, there is a percentage of increase. When registering with the system, the student is automatically placed at the beginner level. This presumption will be validated and readjusted according to the results obtained by the student during his or her simulation. Since the results of the simulations are kept individually per student in the system's database, the student's growth in using the system can be accompanied and adapted according to the student's performance. The system does not delimit the number of simulation that one user can perform nor the area and period that the student wishes to simulate; the accompaniment is done according to the results.

The agent also permits the simulation to be focused on a specific type of event, for example, if the student/user is showing results of a certain type of event with discrepancies in relation to the others, the simulator increases the percentage of events of the type with which the user has the most difficulty, reducing the percentage of the type that the user shows better performance.

The level of difficulty rises with the increase of factors that characterize each type of event and with each increase in the total number of events.

Pedagogical Agent

The pedagogical agent will be discussed in detail in the pedagogical proposal of the EGA section.

Pedagogical Proposal of EGA

The pedagogical model of EGA is based on the concept of intelligent tutoring simulation system where in addition to the simulation itself, an agent provides adaptive explanations to the student at different levels.

The Simulation as a Pedagogical Tool

In EGA we consider simulation as one part of a pedagogical tool. The student can learn by doing. He/she initially interacts with the system by allocating the police. It is a moment to expose his/her beliefs on the allocation of the resources. A simulation of the interaction of the agents is then done, and the students' beliefs can be validated by means of a phase of result analysis. This cycle can be repeated as many times as the student finds necessary.

The Learning Process

According to Kolb (1984), learning is favored when the learning process occurs within successive steps as shown in Figure 5. These steps are:

- **Concrete experience:** Obtained through the activity itself or its simulation in a virtual environment.
- **Reflexive observation:** The experience is followed by the reflection phase. It is recreated internally in the user's mind under different perspectives.
- **Abstract conceptualization:** In this stage, the experience is compared, and its patterns, processes, and meanings are analyzed. In this context, abstract concepts and new knowledge are created. The knowledge is

generated in two moments of the cycle, in this step and in that of solid experience. The knowledge generated in the solid experience phase comes only from the simple observation of the external event, while the knowledge generated in the abstract conceptualization phase emerges as a consequence of an internal cognitive process of the student.

The Pedagogical Agent as a User Support

The pedagogical agent gives support to the user during the reflexive observation and abstract conceptualization steps of the simulation process. It uses two distinct strategies to help the user in the comprehension of the simulation results. The explanation is provided at a micro level and at a macro level.

THE EXPERTCOP SYSTEM

Motivation

Police resource allocation in urban areas in order to perform preventive policing is one of the most important tactical management activities and is usually decentralized by sub sectors in police departments of the area. What it is intended by those tactical managers is that they analyze the

Figure 5. Steps that enhance learning

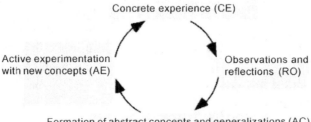

disposition of crime in their region and that they perform the allocation of the police force based on this analysis. We agree with the principle that by knowing where the crime is happening and the reasons associated with this crime, it is possible to make an optimized allocation of resources and consequently, to decrease the crime rate.

The volume of information that police departments have to analyze is one of the main factors to provide society with efficient answers. Tactical managers who perform police allocations, for instance, have a lack of ability related to information analysis and decision-making based on this analysis. In reality, understanding criminal mapping activities, even using GIS, is a nontrivial task. In addition to that, experiments in this domain cannot be performed without high risks because they result in loss of human lives. In this context, simulation systems for teaching and decision support are a fundamental tool.

Objectives

The ExpertCop system aims to support education through the induction of reflection on simulated phenomena of crime rates in an urban area. The system receives as input a police resource allocation plan, and it makes simulations of how the crime rate would behave in a certain period of time. The goal is to lead the student to understand the consequences of his/her allocation as well as understanding the cause-and-effect relations.

In the ExpertCop system, the simulations occur in a learning environment and along with graphical visualizations that help the student's learning. The system allows the student to manipulate parameters dynamically and analyze the results.

ExpertCop Architecture

ExpertCop Architecture is based totally on EGA. As the generic parts of the architecture were previously described, we will deal only with the specification of the domain agents and how the pedagogical agent implemented the pedagogical strategies.

MABS Platform in ExpertCop

There is a great diversity of existing frameworks for the development and maintenance of an MABS platform. We adopted the JADE (Java Agent Develop Framework) (TILab, 2003) due to the fact that it is based on FIPA specifications, to offer tools for creation, maintenance, and control of the agents in addition to being open source, freeware, offering a good documentation, and being based on a robust freeware and simple programming language, JAVA.

ExpertCop Domain Agents

Domain agents are the actors of the domain, performing actively in the simulation process:

- **Police teams:** The mission of the police teams is to patrol the areas selected by the user during the work period and work shifts scheduled for the team. An agent represents each team and has a group of characteristics defined by the user, such as means of locomotion, type of service, and work shift that will influence his patrol. The team works based on its work period and work shift. The work period determines the beginning and end of work, and the work shift determines the work and rest periods. The patrol areas are composed of one or more connected points. The patrol areas are given to the police team as a mission. These areas are associated with intervals of time so as to fill out the work period of the team.
- **Criminals:** The criminal manager creates each criminal agent in the simulation, with the mission of committing a specific crime. After the selection of the area and simulation period by the user, the criminal manager

loads, from the system database, all the crimes pertaining to the area and selected period and places the crimes in chronological order. When beginning the simulation, observing the chronological order of the events, it creates a criminal agent for each crime. The criminal's task is to evaluate the viability of committing the crime. The evaluation is based on risk, benefit, and personality factors, defined on the basis of a set of interviews with specialists in crime of the public safety secretariat and on research in the area of criminal psychology.

The values of the variables regarding crime (type of crime, type of victim, geographical location of crime, date, and time) are sent to the criminal by the criminal manager. But to obtain the data on the environment (geographical factors), the criminal exchanges messages with the GIS agent who furnishes the geographical location, date, and time of the crime.

Having collected all the necessary information for the decision support process of the crime to be executed, the agent uses a PSM of evaluation (abstract and match) (Pinheiro & Furtado, 2004) that will evaluate the viability of committing the crime. The PSM is made up by sets of inference rules containing the structure of the decision support process and an inference machine, in our case JEOPS (Figueira & Ramalho, 2000), that sweeps these rules associating them with the data collected on the crime. This process results in the decision of committing the crime or not. In the sequence below, we demonstrate an example of rules contained in the PSM:

IF **distance_police** = *close* AND **type_crime** = *robbery* AND **type_victim** = *bank* THEN **risk** = *high*

IF **type_victim** = *bank* THEN **benefit** = *high*

IF **benefit** = *high* AND **risk** = *high* AND **personality** = *bold* THEN **decision** = *commit_crime*

We observed the logical structure of the rule with capital letters, the variables that make up the agent's internal state in boldface type, and the values of the variables coming from the data of the crime and the exchange of messages with the GIS Agent in italic type.

After deciding, the criminal sends a message to the GIS Agent informing his decision so that the GIS Agent may mark on the map exhibited by the user, the decision made—in red if the crime is committed and in green if it is not.

- **Notable points:** Notable points are buildings of a region relevant to the objective of our simulation, such as shopping centers, banks, parks, and drugstores. They are located on the simulation map having the same characteristics of the buildings they represent.
- **Pedagogical agent (PA):** PA represents the tutorial module of an ITS proposal. Endowed with pedagogical strategies, this agent aims to help the user in the understanding of the simulation process and results. PA will be discussed in detail in the pedagogical proposal section of this work.

ExpertCop Pedagogical Proposal: Adaptive Explanations

The pedagogic agent uses two distinct forms to explain the events of the system, the explanation at a micro level and at a macro level.

Individual Explanation of the Events of the System (Micro-Level)

To explain the simulation events (crimes), the system uses a tree of proofs describing the steps of reasoning of the criminal agent responsible for the event. This tree is generated from the process of the agent's decision making. The agent's evaluation of a crime is represented by a set of production rules explored by an inference engine called JEOPS (Figueira & Ramalho, 2000). The student can obtain the information on the crime and the process that led the agent to commit it or not, by just clicking with the mouse on the point that represents the crime on the map. Each crime represented by mean of a point at the screen is associated with a proof tree. An example of this can be viewed in Figure 6.

Explanation of the Emerging Behavior or Starting at the Agents' Interactions (Macro-Level)

In ExpertCop, we understand as emerging behavior, the effects of individual events in crime, its increase or reduction, criminal tendencies, and seasonableness. For the explanation of the emerging behavior of the system, the pedagogical agent tries to identify patterns of behavior from the database generated in the simulation. The pedagogical agent (PA) does the KDD process automatically, illustrated in Figure 7.

First, the agent takes (requesting the LOG agent) the simulation data (events generated for the interaction of the agents as crimes (date, hour,

motive, type) and patrols (start time, final time, stretch), and preprocesses it, adding geographic information such as escape routes, notable place coordinates, distance between events, agents and notable places, and social and economical data associated with geographic areas. After preprocessing, in the mining phase, PA identifies patterns by means of a probabilistic concept formation algorithm. This algorithm generates a hierarchy of probabilistic concepts. The generated concepts are characterized according to their attribute/value conditional probabilities. That is to say, a conceptual description is made of attribute/values with high probability. Having the probabilistic concept formation hierarchy constructed, the agent identifies and filters the adequate concepts for being transformed in questions to the student. The heuristics used to filter which concepts will generate questions to the student and which features will compose these questions follow the next steps:

- The root of the hierarchy is ignored (not appraised), because it aggregates all the concepts being too generic.

Figure 6. Individual events explanation

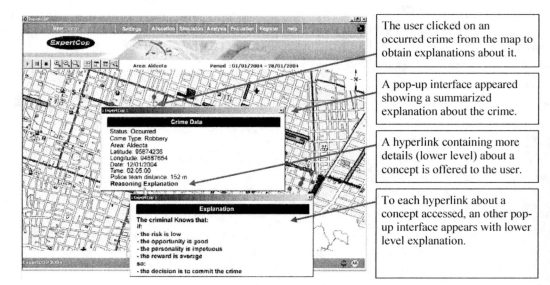

Figure 7. Macro-level explanation process

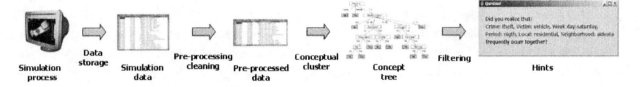

Simulation process — Data storage — Simulation data — Pre-processing cleaning — Pre-processed data — Conceptual cluster — Concept tree — Filtering — Hints

- The hierarchy is read in a bottom-up fashion from the most specific to the most generic concepts.
- The criteria used in the analysis of the concepts for selection are:
    - A concept must cover at least 10% of the total of examples. We believe that less than 10% of the examples would make the concept little representative.
    - An attribute value is only exhibited in the question when it is present in at least 70% of the total of the observations covered by an example.
    - A question must contain at least three attributes.
- When going through a branch of the tree considering the previous items, in case a concept is evaluated and selected, the nodes superior to this concept (parent, grandparent...) will no longer be appraised in order to avoid redundant information. This does not exclude the nodes in the same level of the hierarchy of this node that may be appraised in the future.

An example of COBWEB result displayed to the user in a simulation process is the following hint: "Did you realize that: *crime*: theft, *victim*: vehicle, *weekday*: Saturday, *period*: night, *local*: residential street, *neighborhood*: Aldeota frequently occur together?" Having this kind of information, the user/student can reflect on changes in the allocation, aiming to avoid this situation.

System Functioning

Initially, the student must register with the system and configure the simulation parameters using a specific interface. After that he/she determines the number of police teams to be allocated and the characteristics of these teams. Based on the geographical and statistical data available on the map about the area and his/her knowledge about police patrol, the student determines the areas to be patrolled and allocates the police teams available on the geoprocessed map. To perform the allocation process, the student selects the patrol areas on the map for each team. After that he/she defines the period of time that the police team will be in each patrol area. The sum of each period of time must be equal to the team's workload.

Agents representing the police teams monitor the patrol areas defined by the user following the programmed schedule. The patrol function is to inhibit possible crimes that could happen in the neighborhood. We presume that the police presence is able to inhibit crimes in a certain area scope. The goal of the student is to provide a good allocation, which prevents to the greatest extent the occurrence of crimes.

After the configuration and allocation process, the user can follow the simulation process in the simulation interface. At the end of the simulation process, the user accesses the pedagogical tools of the system.

In addition to the visualization functionalities, the student can access the explanation capabilities. A micro-level explanation can be obtained from the click of the mouse on any red or green point on the screen that indicates crimes that have occurred or have been avoided, respectively. Figure 6 is a screen-shot of the screen at the moment of a micro-level explanation. The student can request a macro-level explanation pushing the hint button represented on the screen. Figure 7 also shows how the concepts discovered by the probabilistic concept formation algorithm are presented on the screen. A set of questions is shown to the student in order to make him/her reflect about possible patterns of crimes.

Upon each new performed allocation, the system will comparatively evaluate the simulated moments, showing the student whether the modification brought a better effect to the crime rate or not. PA also makes comparisons among results obtained in each simulation tour for evaluating learning improvements done by the student. The student can also evaluate the results among a series of simulation at the evaluation screen. On this screen, the results of all simulations made by the student are shown in a bar graphic.

EGA and ExpertCop Evaluation

ExpertCop was used to support a course at the Ministry of Justice and the National Secretariat of Public Safety—SENASP. The course had the objective of emphasizing the importance of information technologies in public safety. ExpertCop was intended to help police officers reflect on the forms of treatment and analysis of information and how these influence the understanding of crime. The audience was made up of three groups of 30 professionals in the area of public safety: civil police officers, chiefs of police, and military police (which are the majority). This use of the system allows us to validate the tool itself and the proposed architecture.

The use of the system was introduced in a training context, which allows us not only to teach the students how to evaluate the tool regarding its usability and effectiveness for the learning

proposals. Regarding the architecture, the tool was completely implemented according to the proposal, showing itself to be appropriate for representing and simulating the studied domain. According to the analysis of the obtained results, we observed a significant statistical growth (test T-Test mean-pair with 1% of error) in students' performance, which we consider as learning. We used a dispersion graph to evaluate the relation between the results obtained and the use of the system's support tools. According to the graph, a positive linear relation indicates that the use of the pedagogical tools made it possible for the students to gain from their performance. We also observed that the tool showed itself to be ergonomic regarding its usability and aroused interest in the students, who showed themselves to be motivated to use it even after the course was over. More specific data on the methodology of application for the course and the methods of attainment and analysis of the results can be obtained in Furtado and Vasconcelos (2005).

RELATED WORK

MAS simulation in education (Gibbons, Fairweather, & Anderson, 1997; Khuwaja et al., 1996; Querrec et al., 2003), ITS (Johnson, Rickel, & Lester, 2000; Ryder, Scolaro, & Stokes, 2001), social simulation to support decision-making, and GIS works (Gimblett, 2002) strongly influenced this research work. These propose an intersection among these areas. There are a great number of works that describe solutions with parts of our proposal:

- Virtual environments for training as SECUREVI proposed by Querrec (Querrec et al., 2003). The system is based on MASCARET model that uses multi-agents systems to simulate realistic, collaborative, and adaptive environments for training simulation.

- Intelligent GIS as the proposed system by Djordjevic (Djordjevic-Kajan, Mitrovic, Mitrovic, Stoimenov, & Stanic, 1995) that intends to provide computer support in fire rescue. The system has a "Fire Trainer," an intelligent agent that covers the activities connected to education.
- Multi-agents with GIS in urban simulation (geosimulation) as a computer model in the approach of Wu (2002) and Benenson (2004).
- Intelligent tutoring systems as those proposed by Wisher (Wisher, MacPherson, Abramson, Thorndon, & Dees, 2001) that describe an intelligent tutoring for field artillery training or Sherlock system by Lesgold (Lesgold, Lajoie, Bunzo, & Eggan, 1992), which provides advice when impasses appear while using a simulated system.
- Phoenix system (Cohen, Michael, David, & Adele, 1989), a discrete event simulator based on an agent architecture. The system is a real-time, adaptive planner that simulates the problem of forest fires.
- The architecture proposed by Atolagbe (Atolagbe & Hlupic, 1996) and Draman (1991) for educational simulation also has similar points with this work although they do not emphasize the power of simulation in GIS with the use of KDD to improve student learning.
- Pedagogical agents in virtual environments as proposed by Jondahl and Morch (Jondahl & Morch, 2002).
- Several works in games and entertainment (Galvão, Martins, & Gomes, 2000; Leemkuil, Jong, Hoog, & Christoph, 2003) use simulation with an educational propose. Even though they present some similarities with our approach, game simulators have a different pedagogical strategy. They focus on the results of the simulation while we believe that most important is the process itself. Another differential is that few games

are adapted to the student level. In order to diminish this, some have proposed to insert ITS features in games (Angelides & Siemer, 1995).

CONCLUSION AND FUTURE WORK

This work described the educational geosimulation architecture, an architecture for the development of tools for training in urban activities based on the interaction among MABS, GIS, and ITS. As a case study, we develop a system based on EGA for the area of public safety, the ExpertCop system. ExpertCop is a pedagogical geosimulator of crime in urban areas. The ExpertCop architecture is based on the existence of MAS with a GIS to perform geosimulations and of a pedagogical agent that follows the simulation process and can define learning strategies as well as use a conceptual clustering algorithm to search relations in the facts generated in the simulation. ExpertCop is focused on education for police officers, in relation to resources allocation.

Initial training courses with police officers interacting with the system were performed aiming to evaluate learning by using this tool. As a complement to the use of the system, a course was made where ExpertCop was used as a tool for analysis and reflection of practical situations. The methodology adopted to analyze the learning of students in ExpertCop has shown a significant improvement in the students' data analysis abilities, in the process of resource allocation with ExpertCop, and in the identification of factors that influence crime.

We intend to continue this research on the ExpertCop system, enhancing its functionalities, and increasing the training support, aiming to make it not only an educational tool but a decision-making support tool as well. The next steps are to implement systems based on EGA in other domains of urban activities.

REFERENCES

Angelides, M. C., & Siemer, J. (1995). Evaluating intelligent tutoring with gaming-simulations. In C. Alexopoulos & K. Kang (Eds.), *Proceedings of 1995 Winter Simulation Conference* (pp. 1376-1383). Arlington, VA: ACM.

Atolagbe, T., & Hlupic, V. (1996). A generic architecture for intelligent instruction for simulation modelling. In J. M. Charnes, D. J. Morrice, & D. T. Brunner (Eds.), *Proceedings of the 1996 Winter Simulation Conference* (pp. 856-863). San Diego, CA.

Benenson, I., & Torrens, P. M. (2004). Geosimulation: Object-based modeling of urban phenomena. *Computers, Environment and Urban Systems, 28*(1-2), 1-8.

Billari, C. F., & Prskawetz, A. (2003). *Agent-based computational demography: Using simulation to improve our understanding of demographic behaviour.* Germany: Phisica-Verlag.

Cohen, P. R., Michael, L. G., David, M. H., & Adele, E. H. (1989). Trial by fire: Understanding the design requirements for agents in complex environments. *AI Magazine, 10*(3), 32-48.

Djordjevic-Kajan, S., Mitrovic, D., Mitrovic, A., Stoimenov, L., & Stanic, Z. (1995). Intelligent GIS for fire department services. In S. Folving, A. Burrill, & J. Meyer-Roux (Eds.), *Proceedings of Eurocarto XIII* (pp. 185-196). Ispra, Italy: Joint Research Centre, European Commission.

Draman, M. (1991). A generic architecture for intelligent simulation training systems. In *Proceedings of the 24th Annual Symposium on Simulation*, (pp. 30-38). New Orleans, LA: IEEE Computer Society.

Fayyad, U. M., Piatetsky, G., Smyth, P., & Uthurusamy, R. (1996). From data mining to knowledge discovery: An overview. In U. Fayyad, G. Pi-atesky-Shapior, P. Smith, & R. Uthurusamy (Eds.), *Advances in knowledge discovery and data mining* (pp. 1-34). Menlo Park, CA: AAAI Press.

Fensel, D., Motta, E., Benjamins, V. R., Decker, S., Gaspari, M., Groenboom, et al. (2003). The unified problem-solving method development language UPML. *Knowledge and Information Systems, An International Journal, 5*(1), 83-127.

Figueira, F. C., & Ramalho, G. (2000). JEOPS—The Java Embedded Object Production System. In M. Monard & J. Sichman (Eds.), *Advances in artificial intelligence* (pp. 52-61). London: Springer-Verlag.

Fisher, D. (1987). Knowledge acquisition via incremental conceptual clustering. *Machine Learning, 2*(2), 139-172.

Furtado, V., & Vasconcelos, E. (2005). A pedagogical agent on mining, adaptation and explanation of geosimulated data. In M. M. Veloso & S. Kambhampati (Eds.), *Proceedings of the 18th International IAAI* (pp. 1521-1528). Pittsburgh, PA: AAAI Press.

Galvão, J. R., Martins, P. G., & Gomes, M. R. (2000). Modeling reality with simulation games for a cooperative learning. In P. A. Fishwick (Ed.). *Proceedings of the 2000 Winter Simulation Conference* (pp. 1692-1698). Orlando, FL: Society for Computer Simulation International.

Gibbons, A. S., Fairweather, P. G., & Anderson, T. A. (1997). Simulation and computer-based instruction: A future view. In C. R. Dills & A. J. Romizowski (Eds.), *Instructional development: State of the art* (pp. 772-783). Englewood Cliffs, NJ: Educational Technology Publications.

Gibbons, A. S., Lawless, K. A., Anderson, T. A., & Duffin, J. (2001). The Web and model-centered instruction. In B. H. Khan (Ed.), *Web-based training* (pp. 137-146). Englewood Cliffs, NJ: Educational Technology Publications.

Gilbert, N., & Conte, R. (Eds.). (1995). *Artificial societies: The computer simulation of social life.* London: UCL Press.

Gimblett, H. R. (Ed.). (2002). *Integrating geographic information systems and agent-based modeling techniques for simulating social and ecological processes.* University of Arizona & Santa Fe Institute: Oxford University Press.

Johnson, W. L., Rickel, J. W., & Lester, J. C. (2000). Animated pedagogical agents: Face-to-face interaction in interactive learning environments. *International Journal of AI in Education, 11*(1), 47-78.

Jondahl, S., & Morch, A. (2002). Simulating pedagogical agents in a virtual learning environment. In G. Stahl (Ed.), *Proceedings of Computer Support for Collaborative Learning* (pp. 531-532). Boulder, CO: Lawrence Erlbaum.

Kaplan, R., & Rock, D. (1995). New directions for intelligent tutoring systems. *AI Expert, 10*(1), 30-40.

Khuwaja, R., Desmarais, M., & Cheng, R. (1996). Intelligent guide: Combining user knowledge assessment with pedagogical guidance. In C. Frasson, G. Gauthier, & A. Lesgold (Eds.). *Proceedings of International Conference on Intelligent Tutoring Systems* (pp. 225-233). Berlin: Springer Verlag.

Kolb, D. A. (1984). *Experiential learning: Experience as the source of learning and development.* Englewood Cliffs, NJ: Prentice Hall.

Leemkuil, H. H., Jong, T. de, Hoog, R. de, & Christoph, N. (2003). KM quest: A collaborative internet-based simulation game. *Simulation & Gaming, 34*(1), 89-111.

Lesgold, S., Lajoie, M., Bunzo, G., Eggan. (1992). SHERLOCK: A coached practice environment for an electronics troubleshooting job. In J. Larkin & R. Chabay (Eds.), *Computer assisted instruction and intelligent tutoring systems* (pp. 201-238).

Hillsdale, NJ: Lawrence Erlbaum Associates.

Piaget, J. (1976). *Le comportement, moteur de l'évolution.* Paris: Ed. Gallimard.

Pinheiro, V., & Furtado, V. (2004). Developing interaction capabilities in knowledge-based systems via design patterns. In A. Bazzan & S. Labidi (Eds.), *Proceedings of Brazilian Symposium of Artificial Intelligence-* (pp. 174-183). São Luis, Brazil: ACM.

Querrec, R., Buche, C., Maffre, E., & Chevaillier, P. (2003). SecuReVi: Virtual environments for fire fighting training. In S. Richir & B. Taravel (Eds.), *Proceedings of Conférence Internationale sur la Réalité Virtuelle* (pp. 169-175). Laval, France.

Roger, B. (1994). *The swarm multi-agent simulation system* (Tech. Rep.). In Object-Oriented Programming Systems, Languages, and Applications (OOPSLA) Workshop on "The Object Engine." (Request report number and publisher/publication location)

Ryder, J. M., Scolaro, J. A., & Stokes, J. M. (2001). An instructional agent for UAV controller training. In *Proceedings of UAVs—16th International Conference* (pp. 3.1-3.11). Bristol, UK: University of Bristol.

TILab S.p.A. Java Agent Development Framework—JADE. (2003). Retrieved January 20, 2005, from ttp://sharon.cselt.it/projects/jade

Torrens, P. M. (2002). Cellular automata and multi-agent systems as planning support tools. In S. Geertman & J. Stillwell (Eds.), *Planning support systems in practice* (pp. 205-222). London: Springer-Verlag.

Wisher, R. A., MacPherson, D. H., Abramson, L. J., Thorndon, D. M., & Dees, J. J. (2001). *The virtual sand table: Intelligent tutoring for field artillery training* (ARI Research Report 1768). Alexandria, VA: U.S. Army Research Institute for the Behavioral and Social Sciences.

Woolf, B., & Hall, W. (1995). Multimedia pedagogues—Interactive systems for teaching and learning. *IEEE Computer, 28*(5), 74-80.

Wu, F. (2002). Complexity and urban simulation: Towards a computational laboratory. *Geography Research Forum, 22*(1), 22-40.

Section 8
Emerging Trends in Online and Distance Learning

This concluding section highlights research potential within the field of online and distance learning while exploring uncharted areas of study for the advancement of the discipline. The introductory chapters set the stage for future research directions and topical suggestions for continued debate. Providing a fresh, alternative view of distance education, colleagues from universities all over the world explore the adaptive traits necessary as disseminators of knowledge within this evolving platform of education; a reminder that not only is the role of the learner rapidly evolving, but so too is the role of the facilitator. Educational programs throughout the world have witnessed fundamental changes during the past two decades—changes that are emphasized in the 25 rigorously researched chapters included in this section. With continued technological innovations in information and communication technology and with on-going discovery and research into newer and more innovative techniques and applications, the online and distance learning discipline will continue to witness an explosion of knowledge within this rapidly evolving field.

Chapter 8.1
Trends and Perspectives in Online Education

Bruce Rollier
University of Baltimore, USA

Fred Niederman
Saint Louis University, USA

INTRODUCTION

Although the Internet has been in existence since 1969, it was not widely used for educational purposes in its first two decades. Few students had access to e-mail, and few educators could visualize its value as a teaching tool. Programs to serve students from remote locations, often called "distance education," became popular; these were generally delivered synchronously through television broadcasts and did not involve the Internet. When the World Wide Web was created in the early 1990s (Berners-Lee, 1999) and the first browsers became available (Waldrop, 2001), the enormous potential for education began to be recognized. New global users came online at a fantastic pace, and the value of all this connectivity was increasing even more rapidly in accordance with Metcalf's Law (Gilder, 1996). Nearly all students used e-mail regularly, and college professors were putting syllabi and course assignments online and creating Web pages with increasing sophistication. Soon entire programs were offered completely via the Internet, with students from all over the globe taking courses together.

According to a major survey conducted by Allen and Seaman (2003), there were more than 1.6 million students who took at least one online course during the Fall 2002 semester and 11% of U.S. higher education students took at least one such course. These numbers were projected to increase rapidly, and most institutions considered online education as a critical long-term strategy.

BACKGROUND

At first, online courses tended to be offered sporadically, often by a few technically savvy faculty members exploring how best to use the Web. Developing the courses to fit this new medium was difficult and time-consuming, and these professors began to demand some recognition for their efforts. Soon, university administrators noticed that the online courses were very popular, that they could attract students from distant locations, and that programs in other institutions were proliferating. The need for uniform policies was recognized. What is a reasonable class size? Should development of a new Web course count toward promotion and tenure? Should Web courses be taught the same way as face-to-face classes?

It soon became clear that there were major differences between face-to-face classes and online classes in providing high quality instruction. Online, at least with the currently available technologies, must be almost completely asynchronous, whereas face-to-face is primarily synchronous. The term "virtual university" (Morrissey, 2002; Schank, 2002; Stallings, 2001) has recently come into vogue, but with varying meanings. The term is often applied to an institution with no physical campus and that is completely online; Jones International University, Cardean University, and University of Phoenix Online are examples (Mason, 2000). These are usually for-profit companies and may be spin-offs from private universities. The term can also signify an institution that has a large physical campus but that has a coordinated approach to online education and devotes significant resources to its online programs. Some of the largest of these include Penn State's World Campus and the University of Maryland's University College (Stallings, 2001). There are also a large number of Virtual University Consortiums (VUCs) that have been formed to offer online programs in multi-institutional partnerships. A recent study identified 63 of such consortia, primarily state-based (Twigg, 2003).

There are many reasons for the rapid growth of VUCs. In many countries, including the U.S., governmental funding of education has decreased in recent years, forcing universities to look for new markets and new sources of revenue (Mason,

2000). Lifelong learning has become much more critical for the workforce as skills and careers become obsolete at a rapid clip. There is high demand from adult learners for retraining for new careers or for upgrading of skills, and online programs are ideal for this.

Twigg (2003) lists several factors motivating the growth of VUCs and other online programs: increased demand for adult education and training; the educational needs of underserved communities; coping with the increasing frequency of interuniversity transfers; streamlined access to the state's institutions via a portal; providing increased variety of degrees; lowering costs, and "overcoming the possibility that the state's institutions will be left behind in the new, highly competitive online environment" (p. 5). She comments that most states have decided that "stand-alone virtual university initiatives are too expensive to initiate and sustain both fiscally and politically" (p. 7). Most VUCs do not offer degrees because that would put them in direct competition with the other state institutions. Instead, they have adopted the collaborative model, which can mean anything from a passive posting of available courses or library sharing to a fairly aggressive stance as an alternative degree path.

Twigg notes that the most successful VUCs have adopted the following policies (2003, p. 10): (1) focus on increasing access for new students that might not otherwise attend; (2) find out what students need and create a viable response rather than merely aggregating what the member institutions can offer; (3) do not get involved in irrelevant higher education policy issues; (4) create a business plan for long-term viability without reliance on state aid, and (5) use a cost-effective development and delivery model.

Morrissey (2002, pp. 460-461) cites the following conflicts which may inhibit achieving the virtual university's full potential: compensation and ownership issues; lack of recognition for course development; poor course support; push for large classes, which may result in fewer faculty positions, and possible adverse impact on research.

ISSUES OF QUALITY

The societal benefits of a first class online education program are obvious, in providing quality instruction for those in remote locations or the physically handicapped or those whose daily responsibilities do not permit them to attend on-campus classes. It is not clear that the programs are being established for such altruistic objectives. Schank (2002) is skeptical, and suspects that Web courses are developed primarily for revenue and academic prestige. He also believes that courses are being modeled to fit existing programs and are not taking advantage of the unique characteristics of the Web.

Are online courses equivalent in quality to courses in face-to-face classrooms? Quality is of course dependent on many factors, and each medium has both advantages and disadvantages. A major difference is synchrony; physical classes are largely synchronous, and current virtual courses are almost completely asynchronous. This characteristic severely limits student interaction and team activities. Chat rooms are synchronous, but quickly become confusing. On the other hand, asynchrony makes it possible to maintain availability 24 hours a day, seven days a week (Pittinsky, 2002). Schank (2002) argues that Web courses would be superior to face-to-face if the design were less dependent on our current concepts, such as performing tasks rather than listening, and length of course and material covered based on student need.

Spicer (2002) warns university presidents that they must be prepared to expend significant resources over long time periods to maintain high quality and avoid shortcuts. The view of online education as a source of revenue and profit is misguided, and costs are typically underestimated (Niederman & Rollier, 2001). Teaching

Web classes is difficult and labor-intensive, and huge classes are not feasible if quality is to be preserved. The required faculty skills are not the same as in the classroom, and the necessary faculty development is often overlooked (Agee, Holisky & Muir, 2003). Procedures must be established and controls enforced to ensure that instructors are responding quickly to student needs and evaluating student learning effectively (Hall, 2002; Stein, 2001).

FUTURE TRENDS

Market share is much more important on the Web than in traditional businesses. A large firm has a major marketing advantage, especially in advertising. The largest virtual universities of the future will offer programs in almost every discipline, at every level, and will compete for students in every locality and most languages. The competition is global; note that the largest distance education institution in the world is in China (Dunn, 2000). These schools will gradually lose their identification with a particular country. Most universities will adopt a niche strategy, competing in those markets for which they perceive a competitive advantage, or markets thought to be underserved.

With continuing advances in technology it seems likely that Web teaching techniques will change greatly. Online education will continue to be basically asynchronous, but with the availability of synchronous capabilities to provide far more interaction. When a critical mass of online students has access to broadband, the instructor will be able to see the students on screen, and the students to see the instructor. Highly interactive group assignments will be possible. The instructor will be able to demonstrate concepts and skills visually and audibly rather than just describing them. Examinations can be monitored visually. Although keyboards may still be necessary, voice will be the primary means of communication, thus reducing the importance of typing skill. Simulations and virtual reality will be widely employed for user interaction and feedback (Morrison & Aldrich, 2003).

Wireless will be a major communications technology (PriceWaterhouse Coopers, 2001), greatly increasing flexibility of location. Internet2 will make digital video available, bringing large-scale indexed video archives and digital-video conferencing. As Tsichritzis (1999) points out, professors can deliver lectures to more than one university at the same time with little additional overhead. Universities may become increasingly like brokers, providing facilities for buyers and sellers of educational services to meet virtually.

Mergers, joint ventures, and strategic alliances will be increasingly common (Makri, 1999). A growing number of states are centralizing services for distance education (Morrissey, 2002), claiming that such collaborations save money and provide more choices for students. Collaborations between universities in different countries can provide obvious benefits to both. There is an increasing trend for students to take courses at more than one university, and online education seems likely to accelerate that trend. Such alliances will lead to an emphasis on standardization of curricula and on ease of transfer between institutions.

Peter Drucker (2000) has said: "Online Continuing Education is creating a new and distinct educational realm, and it is the future of education" (p. 88). If high quality can be maintained and innovative approaches are developed to take advantage of the wealth of resources available on the Web, the chief beneficiary may be society itself, and particularly the Third World societies. Quality education at reasonable cost can be made available virtually anyplace in the world. As the population ages in the wealthier countries, a steady supply of skilled workers will be needed, but an increasing proportion of younger workers will be from economic classes whose members do not traditionally attend college. At the same time, most of the fastest growing occupations require an increasingly high level of education (BLS, 2004). In Third World countries, there is a critical need for an educated workforce and for high quality instruction and effective educational materials. Online education can potentially provide all of these elements.

CONCLUSION

Online education is here to stay, and seems certain to be an increasingly important component of every university's future. It will also be increasingly competitive, and to survive in this market institutions will have to invest significant resources to achieve high quality and favorable name recognition. The ultimate arbiters of success will be the global employers of college graduates. If the online students prove to be competent employees with knowledge and skills equivalent to the graduates of traditional programs, the students will land good jobs and others will flock into the online programs. If the online graduates are seen as weak, this will have a disastrous effect on program success. Skimping on quality would be a huge mistake.

Surveys such as that of Allen and Seaman (2003) indicate that the perception of quality is high among the institutions who offer the programs, but this is weak evidence; student and employer perceptions are much more significant. Research is critically needed to determine whether the necessary quality is being achieved, whether the students are being prepared for successful careers, whether some students are better candidates for online education than others, and whether college courses are equally adaptable to online delivery.

It should not be assumed that the courses will be taught the same way in the future as they are today. Wide availability of broadband and more sophisticated software will make possible much greater interaction among students and between students and instructors. Successful programs will incorporate the best blend of synchrony and asynchrony. Staying competitive may prove to be very expensive, and less well-endowed institutions may be wise to seek out strategic partnerships for sharing of faculty and other resources.

REFERENCES

Agee, A.S., Holisky, D.A., & Muir, S.A. (2003, September/October 1-6). Faculty development: The hammer in search of a nail. *The Technology Source, Faculty and Staff Development.*

Allen, I.E., & Seaman, J. (2003). *Sizing the opportunity: The quality and extent of online education in the United States, 2002 and 2003.* Sloan Center for Online Education.

Berners-Lee, T. (1999). *Weaving the Web: The original design and ultimate destiny of the World Wide Web by its inventor.* San Francisco: Harper.

BLS. (2004). Employment projections. Bureau of Labor Statistics. *http://www.bls.gov/emp/home.htm*

Drucker, P.F. (2000, May 15). Putting more now into knowledge. *Forbes,* 88.

Dunn, S.L. (2000, Spring). The virtualizing of education. *The Futurist, 34*(2), 34-38.

Gilder, G. (1996). *Telecosm.* New York: Simon & Schuster.

Hall, R. (2002). Aligning learning, teaching and assessment using the Web: An evaluation of pedagogic approaches. *British Journal of Educational Technology, 33*(2), 149-158.

Makri, M. (1999). Exploring the dynamics of learning alliances. *The Academy of Management Executive, 13*(3), 113-114.

Mason, R. (2000, November). Wiring up the ivory towers. *The Unesco Courier,* 31-32.

Morrison, J.L., & Aldrich, C. (2003, September/October). Simulations and the learning revolution: An interview with Clark Aldrich. *The Technology Source,* Vision section, 1-7.

Morrissey, C.A. (2002). Rethinking the virtual university. *Communications of AIS, 9,* 456-466.

Niederman, F., & Rollier, B. (2001). How are you going to keep them in the classroom after they've seen MTV? Online education in a virtual world. In L. Chidambaram & I. Zigurs (Eds.), *Our virtual world: The transformation of work, play, and life via technology* (pp. 56-73). Hershey, PA: Idea Group Publishing.

Pittinsky, M.S. (2002). Transformation through evolution. In M. Pittinsky (Ed.), *The wired tower: Perspectives on the impact of the Internet on higher education.* Upper Saddle River, NJ: Prentice-Hall.

PriceWaterhouseCoopers. (2001). *Technology forecast: 2001-2003.* Menlo Park, CA: Pricewaterhousecoopers Technology Centre.

Schank, R. (2002). The rise of the virtual university. *The Quarterly Review of Distance Education, 3*(1), 75-90.

Spicer, D.V. (2002). Where the rubber meets the road: An on-campus perspective of a CIO. In M. Pittinsky (Ed.), *The wired tower: Perspectives on the impact of the Internet on higher education.* Upper Saddle River, NJ: Prentice-Hall.

Stallings, D. (2001). The virtual university: Organizing to survive in the 21st century. *The Journal of Academic Librarianship, 27*(1), 3-14.

Stein, B. (2001). Benchmarks for virtual learning. *NEA Today, 19*(7), 21-23.

Tscichritzis, D. (1999). Reengineering the university. *Communications of the ACM, 42*(6), 93-100.

Twigg, C.A. (2003). *Expanding access to learning: The role of virtual universities.* Monograph.

Center for Academic Transformation. Rensselaer Polytechnic Institute.

Waldrop, M.M. (2001). *The dream machine*. New York: Viking.

KEY TERMS

Asynchrony: A condition whereby events occur that are not coordinated in time. In online education, asynchrony makes possible performing course tasks at the most convenient time, not tied to a schedule.

Distance Education: A somewhat broader term that includes online courses but also classes that involve broadcasting from a television-like facility, often transmitted via satellite.

Metcalf's Law: The "value" or "power" of a network increases in proportion to the square of the number of nodes on the network.

Online Education: Instruction conducted on a network, usually the Internet, for students separated geographically. Sometimes called *distance education*.

Synchrony: Simultaneous occurrence of two or more events. Synchronous features would make it possible to perform lectures, demonstrations, and group activities in real time.

Virtual: Being such in force and effect, though not actually or expressly such.

Virtual University: A loosely defined term, but usually refers to a large, coordinated, degree program offered by one institution or a consortium in which students can take classes at any time and from any global location.

Virtual University Consortium (VUC): A group of higher education institutions, usually sponsored by the state, who have formed a strategic partnership to jointly offer online programs.

This work was previously published in the Encyclopedia of Information and Science Technology, Vol. 5, edited by M. Khosrow-Pour, pp. 2861-2864, copyright 2005 by Idea Group Reference (an imprint of IGI Global).

Chapter 8.2
E–Learning in Higher Education:
The Need for a New Pedagogy

Dirk Morrison
University of Saskatchewan, Canada

ABSTRACT

This chapter discusses the imperative prerequisite to the effective adoption of e-learning by institutions of higher education, namely, the adoption of new pedagogical perspectives and methods. It examines the purposes and goals of higher education, some grounded in tradition, others born of contemporary demands. By focusing on thinking skills, deep learning, and mature outcomes, the author underscores the need for such pedagogical foci to be integrated into the very fabric of higher education's adoption of e-learning. The hoped for outcome of such a consideration is a transformed institution, enabled to meet the demands of learners and society in the twenty-first century.

INTRODUCTION

Increasingly, valid critiques have pointed to the lack of empirical evidence that technology-enhanced learning initiatives actually improve learning outcomes, enhance the teaching enterprise, and are cost-effective for the institution (Clark, 1994; Twigg, 2001; Zemsky & Massy, 2004). Each of these claims, of course, needs careful analysis. One of the conclusions coming out of such criticisms is that technology, in and of itself, cannot be expected to solve the problems of an inefficient, even archaic, approach to pedagogy employed by the vast majority of our institutions of higher education. What, then, does the successful implementation of e-learning in postsecondary education look like? And, what does any evaluation of the success of e-learning need to include?

A critical measure of success for any institution employing e-learning technologies will be the quality of the outcomes (Weigel, 2002). This chapter aims to expand discussion beyond pragmatic questions regarding how to make the transition from face-to-face teaching to e-learning, to include questions regarding how to fundamentally shift the core guiding pedagogical principles of our institutions of higher education. The basic premise of this chapter is that current strategies used to address gaps in performance (e.g., technology-focused faculty development) will fail to realize the hoped-for outcome of an institution shifting to e-learning technologies. A focus on methods and techniques designed to improve the effective implementation of technological products will only be partially useful; what is also needed is a deep and critical discussion regarding the fundamental purposes of designing and employing such products, and a focus on the hoped-for outcomes of such efforts. Throughout this chapter, e-learning is defined as electronically mediated learning, using any variety of media and hardware/software combinations, and usually including the use of facilitated transactions software (e.g., Blackboard, WebCT) (Zemsky & Massy, 2004, p. 5).

To take full advantage of the potential of e-learning, institutions of higher education not only have to radically change how they are organized to support technology-enhanced learning (infrastructures and organizational models), but also face the challenge of creating a more appropriate pedagogical foundation upon which to build revitalized educational systems necessary to meet the demands of current and future knowledge users and creators. Put another way, I argue that the entire system of tertiary education needs revamping from the bottom-up. Current approaches to teaching and learning are an awkward fit with the new information and communications technology (ICT) tools currently used for teaching and learning (May & Short, 2003). In many ways, these new technologies have forced this pedagogical issue and are inherently changing the system from within. Dziuban, Hartman, and Moskal (2004) pointed to a report by the National Research Council Panel on the Impact of Information Technology on the Future of the Research University, which speculated that "information technology will alter the university's usual constraints of space and time, transforming how institutions of higher education are organized and financed, as well as altering their intellectual activities" (p. 8). While it is important to consider the range and variety of factors necessary to ready institutions of higher education for the adoption of e-learning technologies, it is also critical to examine and critique current pedagogical approaches. In addition, not only will instructors and learners be challenged to learn new skills and new ways of working as a result of the adoption of ICT, but they will also be required to change their ways of thinking about the purposes of higher education, the learning process, what it means to be literate, and how knowledge is created. In other words, both faculty and learners will need to re-examine their beliefs, values, perspectives, and resultant approaches to teaching and learning when adopting e-learning technologies.

HIGHER EDUCATION: WHAT'S IT ALL ABOUT?

Eisner (1997) claimed that knowing how to pursue and capture broad meanings shaped the minds of learners. These minds, in turn, collectively shaped the culture, effected change in democratic societies, and ultimately transformed the global community—no small matter. Bamburg (2002) claimed that the very definition of what it means to be educated has changed. In the past, the educational system concentrated on providing students with the basic skills for working in an industrial economy. Now the system must focus on higher order thinking skills that are needed in our knowledge-based economy.

The implication here is that institutions of higher education have critical responsibilities to provide learning environments conducive to the development of capable and creative minds—minds readied for the challenges of a complex world. They must empower learners to know how to pursue and capture broad and deep meanings and to use holistic thinking as the conduit to deep learning.

Holistic Thinking and Deep Learning

Most educators would willingly promote the idea that, at least within higher education contexts, there is a need to move away from what they would call a surface or "shallow" approach to learning (e.g., emphasis on memorizing, simple recall of facts) to a form of "deep" learning wherein learners construct and integrate complex representations of knowledge into patterns that are personally meaningful (Barell, 1991; Garrison, 1991; Hillfish & Smith, 1961; Paul, 1995; Ruggiero, 1988). The former approach, often characterized as typical of traditional pedagogical methods (e.g., the transmission model of learning), is concomitant with a superficial understanding of the subject matter. Inhibiting the development of thinking skills, this approach prescribes that learners passively accept knowledge as it is presented to them, rather than critically examining and constructing it based on their own experiences and previous knowledge (Burge, 1988; Garrison, 1993; Lauzon, 1992). On the other hand, teaching methods that use active learning participation and interaction are facilitative of deep learning and require both higher-order understanding of content and the active construction of meaning within personal and global contexts (Kember, 1991; Newman, Webb, & Cochran, 1995). Although some course content should be in the form of basic facts to be remembered or skills to be demonstrated (e.g., procedural skills), many would claim that, ideally, most learning opportunities should be presented in ways that encourage and facilitate deep thinking and learning about the subject at hand (Garrison, 1993). Holistic thinking (i.e., critical, creative, and complex thinking) is seen here as a necessary antecedent to deep learning and is implicit in many discussions regarding the transformative, emancipatory, and neo-utilitarian potentials of education (Brookfield, 1987; Gross,

1991; McLaren, 1994; Mezirow, 1990; Paul, 1995; Sternberg, 1996).

Accepting that holistic thinking and deep learning are integrally related and are also important educational outcomes raises two key questions: What does deep learning actually look like? What are the necessary antecedents to realizing such learning?

Holistic Thinking Skills: Necessary Tools for Deep Learning

Morrison (2004) claimed that deep learning is related to the way we see the world, ultimately tied to actions and change, necessarily integrative in nature, and a cumulative process, not a singular event. Many would contend that deep learning generally results in qualitatively changing knowledge constructs; as these constructs grow in complexity, our understanding of the perceived world simultaneously broadens and deepens (Crotty, 1993). According to this view, knowledge gained through deep learning is holistic, and ideas, concepts, principles, perceptions, etc. are not seen as unrelated bits of information to be constructed Lego-set style (Lai & Biggs, 1994), but as a dynamic, fluid, and organic phenomenon in the sense that each knowledge construct generated is related, affects, and is affected by others within the mind of the learner. It is a kind of learning that is integrated, not segmented, and makes a difference in who we are, how we think, and what we do (Draper, 1998). Learning at this level is both personally meaningful and contextually engaging. Further, the learner is, in many ways, inseparable from the learning, and this learning is inseparable from thinking. In other words, deep learning, fuelled by holistic thinking, is learning that does not dissect facts from context, ideas from world-views, and learners from the things to be learned. Deep learning and the holistic thinking associated with it mean being organized around goals of personal knowledge construction rather than simply those of task performance (Bereiter,

1990). Finally, it is important to point out that deep learning and holistic thinking are not just an individual phenomenon. More often than not, these occur within a social context, within a *community* of learners wherein dialogue and exchange of views and thoughts are the norm (Vygotsky, 1978). According to Cust (1996), a deep learning, holistic thinking approach within a social context would have the highest levels of cognitive and affective engagement and would likely be the most meaningful, facilitating, in turn, the production of structurally more complex and affectively satisfying learning outcomes.

So, holistic cognitive processes, among other factors (e.g., context, learning task, individual preferences, and motivations), influence not only the approach to learning but also the end result. Personally meaningful residual knowledge and change, internally or externally manifested as a result of deep learning, are intimately tied not only to *what*, *why*, and *how* we learn but also to the thinking process itself.

Developing Online Environments for Deep Learning

It is critical, then, to move purposefully toward reconfiguring educational goals that include an emphasis on holistic thinking (i.e., critical, creative, complex thinking) for the purposes of facilitating deep learning. However, conceptual frameworks for specifically identifying and evaluating holistic thinking have not been readily available. This latter deficit has been all too often ignored in educational research, although recent efforts have been promising (Garrison, Anderson, & Archer, 2000, 2001). Furthermore, new tools for learning, afforded by the rapid development and expansion of information technologies, have not proven to be a panacea for the development of holistic thinking, let alone deep learning (Weigel, 2002). Many higher education applications of e-learning, for example, have the potential for facilitating holistic thinking and deep learning but may, for a number of reasons, miss the mark

Figure 1. Example of an integrated online learning environment

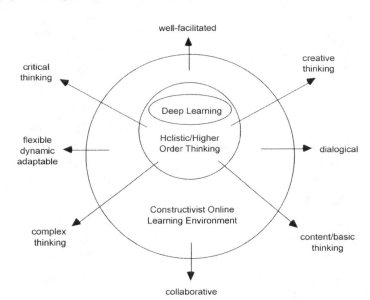

(Gibson, 1995; Weigel, 2002). Within the context of educational applications of e-learning there may be nothing inherently facilitative of holistic thinking, despite the best hopes and intuitions otherwise.

In addition to determining relevant indicators of holistic thinking (Morrison, 2004) and the constellation of factors at play to encourage or discourage thinking (Bullen, 1997), it is equally important to discover, describe, understand, and highlight the critical elements in online learning environments that potentially influence holistic thinking and, by extension, deep learning. The contribution of contextual and process variables, the nature of learning tasks, educational methods utilized, and the "shape" of the technological tools available, among others, are important foci to help illuminate and increase the understanding of the nature of holistic thinking and deep learning in online environments. To support what Weigel (2002) called "depth education" in online learning environments, it would be necessary to include a range of administrative (e.g., faculty training, campus libraries) and technical infrastructures (on/off-campus bandwidth, ICT, educational technologies). Figure 1 provides an example of a conceptual map of interrelated factors important to the construction of an e-learning environment.

While it is useful to focus discussion on the importance of holistic thinking skills as the conduit to the facilitation of deep learning within e-learning environments, it is important to now turn to how this might be translated into a system of learning outcomes for the online classroom.

MATURING OUR OUTCOMES: A SYSTEMS PERSPECTIVE

In their paper "Maturing Outcomes", Costa and Garmston (1998) presented a model of five nested levels of learning outcomes, each level being broader and more encompassing than the level within. What follows is a contextual adaptation of the conceptual framework described in their work.

Outcomes as Activities

The authors characterized this outcome as reflective of "episodic, teacher-centered thinking" with the goal of the online instructor simply being to keep students engaged with the accomplishment of e-learning activities (Costa & Garmston, 1998, para. 5). Success is often measured in terms of whether students made it through the Web resources, completed the online quizzes, and participated in the online discussions. If learners complete all the activities laid out in the online course, the e-learning application is deemed a success.

Outcomes as Content

As instructors gain familiarity with the online learning environment, they are able to ask, What concepts and principles are students learning by completing the embedded activities? The online activities are now employed as vehicles to learn content. The online instructor's focus is on what concepts students will learn, what understandings they will develop, and how that knowledge will be recognized and assessed.

Outcomes as Processes

As online instructors' skills continue to mature, content begins to be selected for its generative qualities (Costa & Garmston, 1998; Perrone & Kallick, 1997). Content becomes both a source and conduit for experiencing, practicing, and applying the cognitive processes needed to think creatively and critically. These processes are basic to lifelong problem solving and include observing and collecting data, forming and testing hypothesizes, drawing conclusions, and posing questions, to name a few.

This shift from a focus on activities and content to cognitive processes is critical. Process outcomes are of central importance because to deeply understand any content, students must know and practice the processes by which that content came into being (Costa & Garmston, 1998; Paul & Elder, 1991; Tishman & Perkins, 1997). At this level, online instructors need to ask, What specific cognitive processes do I want students to practice and develop? How will this online course and the resources I've supplied help them develop those processes? How will I know if they are practicing and developing them?

Outcomes as Dispositions

The realization of this outcome requires the transcendent qualities of systems thinking found in dispositions or habits of the mind, such as enhancing one's capacities to direct and control persistence; managing impulsivity, creativity, and meta-cognition; striving for precision and accuracy; and listening with empathy, risk-taking, and wonderment (Costa, 1991; Costa & Garmston, 1998; Tishman & Perkins, 1997). These universally desirable qualities, exercised within the context of holistic thinking, are valued across disciplines and are a core goal for higher education. Furthermore, a focus on cognitive dispositions assists in developing lifelong capacities and intellectual foundation for continuous learning. Recall that each level of the model presented subsumes the previous. So within the province of outcome as disposition, activities are designed with purpose; content is selected for its generative nature; and critical processes are identified and practiced. These outcomes now build toward a set of superior, more long-range outcomes.

It is important to note that with the three previous outcome levels a single-talented instructor could likely design and implement an online learning environment conducive to the realization of each. At the level of dispositional outcomes, however, it would be desirable to employ the talents of a variety of faculty and course development support staff (e.g., instructional designers, media specialists, etc.). Each instructional team then decides the following: What dispositions do we collectively want online learners to develop and employ? What will we do to assist their development? How can we determine if online learners are developing such dispositions over time? What will we include as evidence of their growth? When an understanding regarding such meta-level outcomes is shared, the entire development team is able to break out of traditional ways of thinking about online learning. As these common goals are achieved collaboratively, they are more likely to be reinforced, transferred, and revisited across the curriculum, the department, and the university.

Outcomes as Mind States

In their model of maturing outcomes, Costa and Garmston (1998) presented five human capacities, or mind states, namely, efficacy, flexibility, craftsmanship, consciousness, and interdependence. These capacities are not trivial and, according to the authors, "act as catalysts or energy sources fueling human thinking, learning and behaviors at the next level of outcomes" and "are the wellsprings nurturing all high performing individuals, groups and organizations and act as beacons toward increasingly authentic, congruent, and ethical behavior" (Costa & Garmston, 1998, para. 15).

At this level, outcomes are drawn not only from the mind states themselves, but also from the ways they interact with the discipline's, department's, or institution's expressed values, culture, and mission. Again, colleagues and instructional development teams need to consider the following questions: In which mind states do we wish students and colleagues to become more resourceful? What will we do to capacitate their development in an online environment? How will we know when they have been amplified?

Costa and Garmston (1998) stated that as a result of a focus on outcomes at this level:

Staff and students learn to draw upon the five mind states to organize and direct their resources as they resolve complex problems, diagnose human frailty in themselves and others, plan for the most productive interventions in groups, and search out the motivations of their own and other's actions. These mind states become desirable meta-outcomes not only for faculty and students, but also for the wider [learning] community as well. (para. 24)

EMPOWERING THE TRANSITION TO E-LEARNING

Attempts to transfer face-to-face courses to an e-learning environment often result in the replication of a limited and inappropriate approach to pedagogy. For example, some authors have pointed out that the adopted outcome expectations fixed at the level of content are reinforced mainly by the ease of measurement afforded by using standard assessment tools (Angelo, 2005; Costa & Garmston, 1998; Cross & Angelo, 1993) and not because of the efficacy of the approach. The focus of most assessment strategies and tools in higher education continues to be on evaluating a learner's demonstration of relatively low levels of knowledge and skills rather than broader, deeper, and more essential outcomes that one would expect. Most of the popular learning management software platforms currently in use in higher education facilitate the continued use of such an approach by building in content-oriented assessment tools. It seems self-evident that transferring such a content-focused approach to learner assessment within an e-learning environment is neither appropriate nor desirable. What needs to be asked is if we are serious about higher order learning outcomes, are we willing to invest the

time, energy, and resources to develop appropriate assessment tools?

Often, examples of failures of e-learning innovations are provided as evidence to support the status quo. In fact, these examples need to be carefully analyzed to determine if the failure was in the technology itself or in the inappropriate application of the technology (Weigel, 2002). If the simple transfer to an electronic environment of an ineffective, low-level approach to face-to-face pedagogy occurred, then one should not be surprised if this translated into poor results. Without a re-configuration of the basic pedagogy, no significant difference should be expected when courses are migrated to an e-learning environment (Clark, 1994). In fact, simply porting a poorly designed course to such an environment might even result in an inferior learning experience (e.g., simply posting lecture notes on a Web site).

A Systems Approach to E-Learning in Higher Education

Dilts (1994), extending work done by Bateson (1972), applied systems thinking to education. A major concept of importance here is that any system of activity is a subsystem embedded within another system. The activities associated with an institution's efforts to make a transition to e-learning are likewise situated within the larger context of that institution. For example, political will, budgets, human resources and skills, extent of entrenched ideals, and resistance to change are all elements within the larger institutional system that will retard or advance progress toward a transformed institution fertile for the growth of e-learning.

The larger educational system within which e-learning is taking place will also influence the type of learning that is facilitated. If an institution of higher education is primarily focused on the measurement of content-based learning outcomes (relying on these to market the institution to employers, funding agencies, etc.), then this focus will dictate the type and range of assessment tools and evaluation methods deemed acceptable. Conversely, if an institution is primarily focused on producing high quality minds and high functioning citizens (focus on dispositional and mind state outcomes), then the tools for assessment and concomitant e-learning strategies will be radically different. Dilts (1994, in Costa & Garmston, 1998) proposed that "learning something on an upper level will change things on lower levels but learning something on a lower level may or may not inform and influence levels above it (para. 40)." The implication here is that while efforts can be made to create innovative e-learning environments at the course or program level, these will not likely result in much change at the institutional level. Only when departments and the wider university undertake a re-engineering of the teaching and learning enterprise, will the transition to an innovative e-learning institution prove to be successful.

BUILDING A NEW FOUNDATION: OUTCOMES, KNOWLEDGE, AND PEDAGOGY

Expectations are high regarding the potential for e-learning to change the face of tertiary education. To help ensure the transition to e-learning results in an improved institution, it is critical that the expected outcomes of higher education are revamped along the lines of those presented by Costa and Garmston (1998). Proponents must inform the university community about the need for higher level outcomes, connecting them to a clear articulation of the generic cognitive skills required (e.g., holistic thinking, states of mind, etc.). Appropriate instructional tools and methods facilitative of the development of such skills must be developed and applied to e-learning contexts. Faculty development opportunities

should be available to enable the necessary shift from a transmittal to a transformative approach to pedagogy. It should be made clear that this emphasis on the development of a wide range and depth of cognitive skills is, in fact, the essential value-addition that higher education can and should offer. I believe, universities should market themselves on the basis of the quality of mental skills acquired and enhanced through their programs—skills that prepare students for a lifetime of work in a knowledge economy—and underscore an increased capacity for critical, creative, and complex thinking, lifelong learning, and citizenry, rather than the many other measures currently used to evaluate the relative quality of one institution against another (e.g., McLean's Guide to Canadian Universities, 2005).

Costa and Liebmann (1997a) point out that our current approach to compartmentalizing knowledge into static disciplines has had utility as a classification system (writing textbooks, hiring faculty, organizing university departments, etc.) but is fundamentally an archaic conception of the disciplines, conveying an obsolete and myopic view of what constitutes knowledge. While an interdisciplinary curriculum may be a difficult sell at universities, more and more it is becoming obvious that areas of innovation and knowledge breakthroughs are a result of cross-fertilization of ideas across the disciplines (Lattuca, 2004). Therefore, in addition to the emphasis on higher-order outcomes as target goals for our instructional efforts, there should be a constant push for interdisciplinary activities. Breaking down the barriers between knowledge areas, challenging the concept of disciplines, and creating opportunities for scholars and students to understand the points of intersection between their disciplines will take us a long way to an expanded view of knowledge.

Finally, it must be made clear to all involved across the institution that any new technologies need to be used wisely. This means adopting and implementing technology within dynamic and adaptive learning environments specifically designed to address and support higher-order learning outcomes, and not just using them as a glossy, high-tech overlay to an outdated and ineffective pedagogy.

Building Capacity for Institutional Change

Currently there is no consensus as to what really effective online education within the context of tertiary institutions should look like. While some have pointed to the eventual emergence of a "dominant design" (Zemsky & Massy, 2004, p. 7) in e-learning, it is important not to search for an ideal model for e-learning, for there is not one model. Instead, what is required is a dynamic, evolving institution that adapts and re-configures appropriate working models for e-learning. This requires openness to new ideas, especially in the arena of pedagogy (King, 1993); the focus should be on building capacity for transforming institutional norms for teaching and learning from within. Faculty adoption of e-learning needs to draw on the traditional strengths of the academy and nurture collaborative individualism. This process includes building connections and inter-dependency between people, organizations, and ideas (Smyre, 2000). A focus on synthesis, using living systems as the metaphor (not the factory), is required to support an adaptive, evolving system of e-learning. Open-minded dialogue and not adversarial debate needs to be the communicative environment within which new ideas for e-learning are be explored. Students must encounter choices (curricular and e-course styles), not rigid standardization in order to use what is known, say, about preferred learning styles (Anderson & Adams, 1992) and multiple intelligences (Gardner, 1993, 2000). Institutions must have a forward thinking orientation, drawing on current research and literature to inform decisions regarding the adoption and implementation of e-learning. And

these activities must incorporate evaluation mechanisms if institutions of higher education are to make the best use of e-learning.

CONCLUSION

One of the core premises of this chapter is that simply overlaying poor pedagogy with the veneer of e-learning as innovation is a sham and is sure to produce results as outlined by Zemsky and Massy (2004) and others. This chapter has identified the need to review and reconfigure this pedagogy to be more in line with contemporary research regarding the nature and purposes of learning within the context of higher education. Current findings in brain research and the cognitive sciences hold much promise for providing guideposts as to how to construct effective learning environments. Underscored was the need to create learning environments that require the development of critical, creative, and complex thinking skills. A brief discussion was presented regarding the need to expand our conception of expected terminal outcomes as a result of singular or collective learning experiences within a university education. The consideration of these suggestions would do well to help ensure an institution's smooth transition to and adoption of e-learning.

All of the previous cannot take place, however, unless the larger context itself is altered. Promotion of change and transformation must occur within both the local and global constellations of activity and innovation in higher education. The context within which universities and by extension e-learning are situated is a rapidly evolving and incredibly dynamic environment. Not known for corporate agility and flexibility, universities may be at risk in terms of quickly adjusting fundamental principles and core values (i.e., ways of doing things) in order to take advantage of opportunities (e.g., e-learning). However, universities are not static, inanimate entities, but rather are the sum total of the people involved. So it is people and their ideas about teaching and learning, including e-learning, that need to be changed if a successful transition to e-learning is to occur. Smyre (2000) asked the correct questions:

If context has emerged as a key concept for education, what do we do to help people learn how to understand how to build capacities for transformation? If the underlying assumptions are changing, how do we coach people to think within a futures context? And possibly the most important question... how do we introduce into educational curricula the need to think about the impact of future trends as well as transforming underlying assumptions? How can schools, community colleges, and universities begin to create a learning environment so that issues are considered within an evolving "futures context?" (p. 7)

These are multiple questions with multiple answers. However, if the university is to undertake a successful transition to e-learning, it must simultaneously undertake a transformation of its approach to pedagogy. Cognitive research from the past decade suggests the instructional strategies that we have been using are no longer appropriate (Bamburg, 2002).

Key decision makers within the higher education sector, as well as those responsible for designing and developing the e-learning opportunities, need to use this knowledge to change the way we do things in higher education. A relevant and vital tertiary educational system is at stake.

REFERENCES

Anderson, J. A., & Adams, M. (1992). Acknowledge the learning styles of diverse student populations: Implications for instructional design. In L. L. B. Chism (Eds.), *Teaching for diversity: New directions in teaching and learning* (No. 42). San Francisco: Jossey Bass.

Angelo, T. (2005). *Doing assessment as if learning matters.* Presentation to the University of Sasktchewan, Saskatoon, SK.

Bamburg, J. D. (2002). Learning, learning organizations, and leadership: Implications for the year 2050. Retrieved May 2005, from http://www.newhorizons.org/trans/bamburg.htm

Barell, J. (1991). *Teaching for thoughtfulness: Classroom strategies to enhance intellectual development.* New York: Longman.

Bateson, G. (1972). *Steps to an ecology of mind.* New York: Chandler.

Bereiter, C. (1990). Aspects of an educational learning theory. *Review of Educational Research, 60*(4), 603-624.

Brookfield, S. (1987). *Developing critical thinkers: Challenging adults to explore alternative ways of thinking and acting.* San Francisco: Jossey-Bass.

Bullen, M. (1997). *A case study of participation and critical thinking in a university-level course delivered by computer conferencing.* Unpublished PhD Thesis: University of British Columbia.

Bullen, M. (1998). Participation and critical thinking in online university distance education. *Journal of Distance Education, 13*(2), 1-32.

Burge, E. J. (1988). Beyond andragogy: Some explorations for distance learning design. *Journal of Distance Education, 3*(1), 5-23.

Clark, R. (1994). Media will never influence learning. *Educational Technology Research and Development, 42*(2), 21-29.

Costa, A. (1991). The search for intelligent life. In A. Costa (Ed.), *The school as a home for the mind.* Palatine, III: Skylight.

Costa, A., & Garmston, J. (1998). *Maturing outcomes.* Retrieved May 2005, from http://www.newhorizons.org/trans/costa_garmston.htm

Costa, A., & Liebmann, R. (1997) Towards renaissance curriculum: An idea whose time has come. In A. Costa & R. Liebmann (Eds.). *Envisioning process as content: Towards renaissance curriculum*. Thousand Oaks, CA: Corwin Press.

Cross, K. P., & Angelo, T. A. (1993). *Classroom assessment techniques* (2nd ed.). San Francisco: Jossey-Bass.

Crotty, T. (1993). *Constructivist theory unites distance education and teacher education*. Ames Teacher Education Center, University of Wisconsin-River Falls, WI. Retrieved November 15, 2003, from http://edie.cprost.sfu.ca/it/constructivist-learning

Cust, J. (1996). A relational view of learning: Implications for nurse education. *Nurse education today, 16*(4), 239-306.

Dilts, R. (1994) Effective presentation skills. In A. Costa & J. Garmston (Eds.), *Maturing outcomes* (p. 3642). Capitola, CA: Meta Publications. Retrieved May 2005, from http://www.newhorizons.org/trans/costa_garmston.htm

Draper, P. (1998). *Understanding student approaches to technology assisted learning*. Queensland Conservatorium Griffith University. Retrieved November 24, 2003, from http://www29.gu.edu.au/staff/draper/tal.html

Eisner, E. (1997). Cognition and representation: A way to pursue the American dream? *Phi Delta Kappan, 78*(5), 348-353.

Gardner, H. (1993). *Multiple intelligences: The theory in practice*. New York: Basic Books.

Gardner, H. (2000). *The disciplined mind: Beyond facts and standardized tests, the K-12 education that every child deserves*. New York: The Penguin Group, Penguin Putnam.

Garrison, D. R. (1991). Critical thinking and adult education: A conceptual model for developing critical thinking in adult learners. *International Journal of Lifelong Education, 10*(4), 287-303.

Garrison, D. R. (1993). A cognitive constructivist view of distance education: An analysis of teaching and learning assumptions. *Distance Education, 14*(2), 199-210.

Garrison, D. R., Anderson, T., & Archer, W. (2000). Critical inquiry in a text-based environment: Computer conferencing in higher education. *The Internet and Higher Education, 2*(2-3), 1-19.

Garrison, D. R., Anderson, T., & Archer, W. (2001). Critical thinking, cognitive presence, and computer conferencing in distance education. *American Journal of Distance Education, 15*(1), 7-23.

Gibson, L. (1995). *Discursive learning spaces: The potential of the Internet as an environment for learning at a distance*. Unpublished MSc Major Research Paper: University of Guelph.

Gross, R. (1991). *Peak learning: A master course in learning how to learn*. Los Angeles, CA: Jeremy P. Tarcher.

Hillfish, G. H., & Smith, P. G. (1961). *Reflective thinking: The method of education*. New York: Dodd, Mead & Company.

Kegan, R. (1994). *In over our heads: The mental demands of modern life*. Cambridge, MA: Harvard University Press.

Kember, D. (1991). Instructional design for meaningful learning. *Instructional Science, 20*(4), 289-310.

King, A. (1993). From sage on the stage to guide on the side. *College Teaching, 31*(1), 30-35.

Kolb, D. A. (1984). *Experiential learning: Experience as the source of learning and development*. NJ: Prentice-Hall.

Lai, P., & Biggs, J. (1994). Who benefits from mastery learning? *Contemporary Educational Psychology, 19*(1), 13-23.

Lattuca, L. (2004). *Creating interdisciplinarity: Interdisciplinary research and teaching among college and university faculty.* Vanderbilt Issues in Higher Education. Nashville, TN: Vanderbilt University Press.

Lauzon, A. C. (1992). Integrating computer-based instruction with computer conferencing: An evaluation of a model for designing online education. *American Journal of Distance Education, 6*(2), 32-46.

MacLean's Guide to Canadian Universities '05. (2005). *MacLean's Magazine.* Toronto, ON: Rogers Media.

May, G. L., & Short, D. (2003). Gardening in cyberspace: A metaphor to enhance online teaching and learning. *Journal of Management Education, 27*(6), 673-693.

McLaren, P. L. (1994). Critical thinking as a political project. In K. S. Walters (Ed.), *Re-thinking reason: New perspectives in critical thinking.* New York: SUNY Press.

Menzies, H. (1996). *Whose brave new world: The information highway and the new economy.* Toronto, ON: Between the Lines.

Mezirow, J. (1990). How critical reflection triggers transformative learning. In J. Mezirow & Associates (Eds.), *Fostering critical reflection in adulthood* (pp. 1-20). San Francisco: Jossey Bass.

Morrison, D. (2004). *A study of holistic thinking in an agricultural leadership development program using asynchronous computer conferencing.* Unpublished doctoral thesis, University of Toronto.

Newman, D. R., Webb, B., & Cochrane, C. (1995). *A content analysis method to measure critical thinking in face-to-face and computer supported groups learning.* Retrieved October, 1998, from listserv@guvm.georgetown.edu

Panel on the Impact of Information Technology on the Future of the Research University. (2002). *Preparing for the revolution: Information technology and the future of the research university.* Washington, DC: The National Academies Press. Retrieved June 2005, from http://www.nap.edu/books/030908640X/html/

Paul, R. (1995). *Critical thinking: How to prepare students for a rapidly changing world.* Sonoma, CA: Foundation for Critical Thinking.

Perrone, V., & Kallick, B. (1997). Generative topics for process curriculum. In A. Costa & R. Liebmann (Eds.), *Supporting the spirit of learning: When process is content.* Thousand Oaks, CA: Corwin Press.

Ruggiero, V. R. (1988). *Teaching thinking across the curriculum.* New York: Harper & Row, Publishers.

Smyre, R. (2000). *Transforming the 20th century mind: The roles of a futures institute.* Retrieved May 2005, from http://www.newhorizons.org/future/smyre.htm

Sternberg, R. J. (1996). *Successful intelligence: How practical and creative intelligence determine success in life.* New York: Simon & Schuster.

Tishman, S., & Perkins, D. (1997, January). The language of thinking. *Phi Delta Kappan, 5*(78), 368-374.

Twigg, C. A. (2001). *Innovations in online learning: Moving beyond no significant difference.* The Pew Symposia in Learning and Technology 2001. Center for Academic Transformation, Rensselaer Polytechnic Institute, Troy, NY.

Vygotsky, L. S. (1978). *Mind in society: The development of higher mental process.* Cambridge, MA: Harvard University Press.

Weigel, V. (2002). *Deep learning for a digital age: Technology's untapped potential to enrich higher education.* San Francisco: Jossey-Bass Publishing.

Zemsky, R., & Massy, W. F. (2004). *Thwarted innovation: What happened to e-learning and why.* A final report for the Weatherstation Project of The Learning Alliance for Higher Education at the University of Pennsylvania.

This work was previously published in Making the Transition to E-Learning: Strategies and Issues, edited by M. Bullen and D. Janes, pp. 104-120, copyright 2007 by Information Science Publishing (an imprint of IGI Global).

Chapter 8.3
Evaluation of an Open Learning Environment[1]

Geraldine Clarebout
University of Leuven, Belgium

Jan Elen
University of Leuven, Belgium

Joost Lowyck
University of Leuven, Belgium

Jef Van den Enden
Institute of Tropical Medicine, Belgium

Erwin Van den Ende[2]
Institute of Tropical Medicine, Belgium

INTRODUCTION

Educational goals have generally shifted from knowing everything in a specific domain to knowing how to deal with complex problems. Reasoning and information processing skills have become more important than the sheer amount of information memorized. In medical education, the same evolution occurred. Diagnostic reasoning processes get more strongly emphasized. Whereas previously knowing all symptoms and diseases was stressed, reasoning skills have become educationally more important. They must enable professionals to distinguish between differential diagnoses and recognize patterns of illnesses (e.g., Myers & Dorsey, 1994).

BACKGROUND

Authentic or realistic tasks have been advocated to foster the acquisition of complex problem-solving processes (Jacobson & Spiro, 1995; Jonassen, 1997). In medical education, this has led to the use of expert systems in education. Such systems were initially developed to assist practitioners in their practice (NEOMYCIN, in Cromie, 1988; PATHMASTER in Frohlich, Miller, & Morrow,

1990; LIED in Console, Molino, Ripa di Meanan, & Torasso, 1992). These systems simulate a real situation and were expected to provoke or develop students' diagnostic reasoning processes. However, the implementation of such expert systems in regular educational settings has not been successful. Instead of developing reasoning processes, these systems assume them to be available. They focus on quickly getting to a solution rather than reflecting on possible alternatives. Consequently, it was concluded that students need more guidance in the development of diagnostic reasoning skills (Console et al., 1992, Cromie, 1988; Friedman, France, & Drossman, 1991); instructional support was lacking.

KABISA is one of the computer programs that, among other things, aims at helping students to develop their diagnostic reasoning skills (Van den Ende, Blot, Kestens, Van Gompel, & Van den Enden, 1997). It is a dedicated computer-based training program for acquiring diagnostic reasoning skills in tropical medicine.

DESCRIPTION OF THE PROGRAM

KABISA confronts the user with cases or "virtual patients". The virtual patient is initially presented by three "characteristics"[3], randomly selected by the computer. After the presentation of the patient (three characteristics), students can ask additional characteristics gathered through anamnesis, physical examination, laboratory and imaging.

If students click on a particular characteristic, such as a physical examination test, they receive feedback. Students are informed about the presence of a certain symptom, or whether a test is positive or negative. If students ask a "non-considered" characteristic, that is, a characteristic that is not relevant or useful in relation to the virtual patient, they are informed about this and asked whether they want to reveal the diagnosis they were thinking about. When they do so, students receive an overview of the characteristics that were explained by their selection and which ones are not. Additionally, they get the place of the selected diagnosis on a list that ranks diagnoses according to their probability given the characteristics at hand. If students do not want to show the diagnosis they were thinking about, they can just continue asking characteristics.

A session is ended with students giving a final diagnosis. KABISA informs them about the correctness. If it is correct, students are congratulated. If the diagnosis is not correct, students may be either informed that it is a very plausible diagnosis but that they do not have enough evidence, or they may get a ranking of their diagnosis and an overview of the disease characteristics that can and cannot be explained by their answer.

Additionally, different non-embedded support devices, that is, tools, are made available to support learners. These tools allow students to look for information about certain symptoms or diseases, to compare different diagnoses, or to see how much a certain characteristic contributes to the certainty for a specific diagnosis. Students decide themselves when and how they use these devices (for a more detailed description, see Clarebout, Elen, Lowyck, Van den Ende, & Van den Enden, 2004).

FUTURE TRENDS

In this section, some critical issues are put forward that raise discussion points for the future design and development of open learning environments.

A Learning Environment vs. a Performance Environment

KABISA is designed as an open learning environment, that is, students are confronted with a realistic and authentic problem; there is a large amount of learner control and tools are provided to learners to guide their learning (Hannafin, Land & Oliver, 1999). However, the performed evaluation study revealed some interesting issues. A first revelation was that students do not follow a criterion path when working on KABISA. Prior to the evaluation, two domain experts in collaboration with three instructional designers constructed a criterion path. This path represented the ideal paths students should go through to optimally benefit from KABISA (following the "normative approach" of Elstein & Rabinowitz, 1993), including when to use a specific tool. Only five out of 44 students followed this path.

A second issue relates to tool use. KABISA offers different tools to support students. These tools can help students in their problem-solving process. Results suggest that students consult some help functions more than others, but overall they do not consult them frequently and if they use them they do not use them adequately. Students also tend to not use the feedback that they can obtain when asking for a "non-considered" characteristic.

Although this environment can be described as an open learning environment, it seems that students do not perceive it as a learning environment, but rather as a performance environment. Thinking aloud protocols reveal that students think they are cheating or failing when consulting a tool. Giving the limited use of these tools, it becomes difficult to gain insight in the effect of tool use on the learning process.

However, in spite of the observation that in only a small number of consultations the criterion path was followed, students do find in 80% of the consultations the right diagnosis. It seems that by trial and error, by not following the criterion path, students can also obtain the right diagnosis.

The results of this evaluation suggest that students do not use KABISA to foster their diagnostic reasoning skills. Rather, KABISA enables them to train readily available skills.

The Use of Design Models for Designing Open Learning Environments

This evaluation shows the importance of an evaluation phase in the design and development of computer-based training programs. It reveals the valuable contribution of (linear) design models, such as the so-called ADDIE-model (Analyse-Design-Development-Implementation-Evaluation). Although it is argued that in open learning environments a linear design process cannot longer be applied, this evaluation shows that it still can contribute to the design. For instance, a more thorough analysis (first phase) of student characteristics could have provided a means to adapt the difficulty level to the level of the students or to identify what guidance students actually need. Apparently, the feedback given to students does not encourage them to adapt their problem-solving process. Being product-rather than process-oriented, feedback may not be adapted to students' actual needs. Or, students' instructional conceptions about computer-based learning environments or their perceptions about KABISA (game versus an educational application) may influence the use of the program. Students' instructional conceptions should be taken into account through the design process of the program. One possible way to influence these conceptions might be the introduction of the program. In the introduction, the aims of the program, the different functionalities and the relationship with the different courses should be clearly defined (see Kennedy, Petrovi, & Keppell, 1998, for the importance of introductory lessons). This relates to the implementation phase.

Given the difficulty of anticipating potential problems and difficulties students might encounter in open learning environments, it might be considered to break the linearity of such design models and to introduce a formative evaluation after each phase. This would enable the redirection of the program while developing it, rather than after the implementation of the program. Rather than only evaluating a final product, the development process should be taken into consideration as well. Rapid prototyping for testing the program at different phases of the development might be indicated. This leads to a more spiral cycle rather than a linear design process.

Amount of Learner Control in Computer-Based Programs

In the design and development of KABISA, a lot of time and effort is spent in the development of tools, similar to other computer-based programs. However, results show that students do not (adequately) use these tools. Other authors have found similar results (see for instance, Crooks, Klein, Jones, & Dwyer, 1996; Land, 2000). This raises questions about the amount of learner control in open learning environments. Should the environment be made less open and provide embedded support devices instead of tools so that students cannot but use these devices? Or should students receive some additional advice towards the use of these tools? In the first case, support might not be adapted to the learners need. This might cause problems, given that either too much or too less support can both be detrimental (Clark, 1991). The second option leaves the environment open. But also here it can be questioned whether this advice should not also be adapted to the learners' needs. A possible solution with respect to this issue might come out of the animated pedagogical agent-research. These agents are animated figures that aim at helping learners in their learning process and adapt their support based on the paths learners followed (Moreno, 2004; Shaw, Johnson, & Ganeshan, 1999).

CONCLUSION

The evaluation of KABISA addressed some general issues important to consider in the design, development and implementation of open learning environments. Although these environments are advocated to foster the acquisition of complex problem-solving skills, there seems still to be gap between the intention of the designers and the use by the learners. This relates to the issue addressed by Winne and Marx (1982) about calibration. In order for an instructional intervention to be effective, calibration is needed between the conceptions of the different people involved. The introduction of a pedagogical agent might help to calibrate the conceptions of students to those of the designers. Moreover these agents might help in encouraging students to adequately use tools without reducing the openness of the learning environment.

REFERENCES

Clarebout, G., Elen, J., Lowyck, J., Van den Ende, J., & Van den Enden, E. (2004). KABISA: Evaluation of an open learning environment. In A. Armstrong (Ed.), *Instructional design in the real world: A view from the trenches* (pp. 119-135). Hershey, PA: Idea Group Publishing.

Clark, R.E. (1991). When teaching kills learning: Research on mathemathantics. In H. Mandl, E. De Corte, N. Bennett, & H.F. Friedrich (Eds.), *European research in an interantional context: Volume 2. Learning and Instruction* (pp. 1-22). Oxford, NY: Pergamon Press.

Console, L., Molino, G., Ripa di Meana, V., & Torasso, P. (1992). LIED-liver: Information, education and diagnosis. *Methods of Information in Medicine, 31*, 284-297.

Cromie, W.J. (1988). Expert systems and medical education. *Educational Researcher, 17*(3), 10-12.

Crooks, S.M., Klein, J.D., Jones, E.E., & Dwyer, H. (1996). Effects of cooperative learning and learner-control modes in computer-based instruction. *Journal of Research in Computing in Education, 29*, 223-244.

Elstein, A.S., & Rabinowitz, M. (1993). Medical cognition: Research and evaluation. In M. Rabinowitz (Ed.), *Cognitive Science Foundation of Instruction* (pp. 189-201). Hillsdale, NJ: Lawrence Erlbaum Associates.

Friedman, C.P., France, C.L., & Drossmann, D.D. (1991). A randomized comparison of alternative formats for clinical simulations. *Medical Decision Making, 11*(4), 265-271.

Frohlich, M.W., Miller, P.L., & Morrow, J.S. (1990). PATHMASTER: Modelling differential diagnosis as "Dynamic Competition" between systematic analysis and disease-directed deduction. *Computers and Biomedical Research, 23*, 499-513.

Hannafin, M.J., Land, S., & Oliver, K. (1999). Open learning environments: Foundations, methods and models. In C.M. Reigeluth (Ed.), *Instructional design theories and models. A new paradigm of Instructional Theory* (Vol. 2, pp. 115-140). Mahwah, NJ: Lawrence Erlbaum Associates.

Jacobson, M.J., & Spiro, R.J. (1995). Hypertext learning environments, cognitive flexibility and the transfer of complex knowledge. *Journal of Educational Computing Research, 12*(4), 301-333.

Jonassen, D.H. (1997). Instructional design models for well-structured and ill-structured problem-solving learning outcomes. *Educational Technology Research and Development, 45*(1), 65-91.

Kennedy, G., Petrovic, T., & Keppell, M. (1998). The development of multimedia evaluation criteria and a program of evaluation for computer aided learning. In R.M. Cordeory, (Ed.), *Proceedings of the 15th Annual Conference of the Australian Society for Computers in Tertiary Education (AS-CILITE)* (pp. 407-415.). Wollongong, Australia: University of Wollongong.

Land, S.M. (2000). Cognitive requirements for learning with open-learning environments. *Educational Technology Research and Development, 48*(3), 61-78.

Moreno, R. (2004, April). Agent-based methods for multimedia learning environments: What works and why? Paper presented at the *annual meeting of the American Educational Research Assocation*, San Diego, CA.

Myers, J.H., & Dorsey, J.K. (1994). Using diagnostic reasoning (DxR) to teach and evaluate clinical reasoning skills. *Academic Medicine, 69*, 429.

Shaw, E., Johnson, W.L., & Ganeshan, R. (1999). Pedagogical agents on the Web. In *Proceedings of the Third Int'l Conf. on Autonomous Agents* (pp. 283-290).

Van den Ende, J., Blot, K., Kestens, L., Van Gompel, A., & Van den Enden, E. (1997). KABISA: An interactive computer-assisted training program for tropical diseases. *Medical Education, 31*, 202-209.

Winne, P.H., & Marx, R.W. (1982). Students' and teachers' view of thinking processes for classroom learning. *The Elementary School Journal, 82*, 493-518.

KEY TERMS

Animated Pedagogical Agents: Animated figures operating in a learning environment and aiming at supporting learners in their learning process and capable of adapting their support to the learners' paths.

Criterion Path: A representation of an "ideal path" to go through a specific learning environment. It specifies for each possible step in the program what the most ideal subsequent steps are.

Embedded Support Devices: Support devices integrated in the learning environment. Learners cannot but use these devices (e.g., structure in a text).

Instructional Conceptions: Conceptions about the functionalities of (elements of) a learning environment. These conceptions can relate to the effectiveness or efficiency of specific features in a learning environment (e.g., tools) or to the environment as a whole (e.g., KABISA as a learning environment).

Non-Embedded Support Devices (synonym: Tools): Support devices that are put to the disposal of learners. Learners decide themselves when and how to use these tools.

Open Ended Learning Environments: A learning environment that aims at fostering complex problem solving skills by confronting learning with a realistic or authentic problem in a learning environment with a large amount of learner control and different tools.

Perceptions: Students' perceptions relate to how they perceive a specific environment (c.q., KABISA), they are the results of an interaction between students' instructional conceptions and a specific learning environment.

ENDNOTES

[1] A more extended version of this manuscript was published in Armstrong, A. (Ed.). (2004). *Instructional design in the real world. A view from the trenches.* Hershey, PA: Idea Group Inc.

[2] The authors express their gratitude to Stefano Laganà who spend a lot of effort in the adaptation of KABISA and in the development of a log file system.

[3] The term 'characteristic' refers to either a symptom or disease characteristics, either a request for results of a physical examination, laboratory test, or imaging. An example of a characteristic can be 'the patient has severe headache' or 'the palpation of the abdomen is negative.'

Chapter 8.4
Faculty Perceptions and Participation in Distance Education

Kim E. Dooley
Texas A&M University, USA

James R. Lindner
Texas A&M University, USA

Chanda Elbert
Texas A&M University, USA

Timothy H. Murphy
Texas A&M University, USA

Theresa P. Murphrey
Texas A&M University, USA

INTRODUCTION

Research in the field of distance education has recognized the need for a change and modification of the faculty role in teaching at a distance (Jones, Lindner, Murphy & Dooley, 2002; Kanuka, Collett & Caswell, 2002; Miller & Pilcher, 2001). While technological advancements are an important part of the distance-learning environment, basic changes in teaching methods, technique, and motivation are needed to make distance education more effective (Purdy & Wright, 1992). Many studies cite faculty resistance to instructional technology as a primary barrier to the continued growth of distance education programs (Jones et al., 2002; McNeil, 1990). McNeil (1990) noted that attitudinal issues related to how faculty perceive and interact with technology are a bigger barrier to adoption and diffusion of distance education than is technology infrastructure.

BACKGROUND

This chapter addresses perceptions of faculty with respect to barriers to adoption, roles and responsibilities, competencies, and rewards. Barriers stem from the lack of perceived institutional support (faculty rewards, incentives, training, etc.) for course conversion to distance education formats (O'Quinn & Corry, 2002; Perreault et al., 2002). As distance education programs continue to proliferate globally, colleges and universities must commit to address the needs of faculty (McKenzie, Mims, Bennett & Waugh, 2000). Despite the fact that much of the literature in distance education discusses the importance of faculty, this group has been largely neglected by the research.

Dooley and Murphy (2000) found that faculty members lacked experience in teaching learners at a distance and that they were much more confident in their technical competence than they were in their methodological ability to use modern technologies in their teaching. These authors further found that faculty perceived training and assistance in the use of instructional technologies to be less available than equipment and facilities. Additionally, faculty members who had not participated in distance education perceived the level of support as lower than those who had taught classes at a distance. The ability of an organization to adapt to these changes is influenced by the following: competence, or the knowledge, skills, and abilities of its staff; value, or the amount of importance the staff places on the role of these technologies to accomplish teaching and learning; information technology support, or the availability of high quality facilities, equipment, technical support, and training (Dooley & Murphy, 2000).

Lindner, Murphy, and Dooley (2002) extended these conclusions by looking at how these factors affect faculty adoption of distance education. Research revealed that faculty members lacked confidence in their ability to use technology in their teaching, perceived technology to be a valuable

addition to the teaching and learning environment, and believed the overall level of support for the use of technology in teaching to be low. Tenure status and academic rank/position for tenure-track faculty were inversely related to overall distance education scores. Non-tenured assistant professors had the highest overall distance education scores and the highest competency scores.

Students learn from competent instructors who have been trained how to communicate effectively through technology. Thomas Cyrs (1997) identified areas of competence important to a distance education environment: course planning and organization, verbal and nonverbal presentation skills, collaborative teamwork, questioning strategies, subject matter expertise, involving students and coordinating their activities at field sites, knowledge of basic learning theory, knowledge of the distance learning field, design of study guides, graphic design and visual thinking (Cyrs, 1997).

Linda Wolcott (1997) conducted an analysis of the institutional context and dynamics of faculty rewards at research universities. She discovered that 1) distance education occupies a marginal status, 2) distance teaching is neither highly valued nor well rewarded as a scholarly activity, 3) distance teaching is not highly related to promotion and tenure decisions, and 4) rewards for distance teaching are dependent on the academic unit's commitment to distance education.

As indicated by Moore (1997), distance education programs with a commitment to faculty support and training result in higher quality programs. As the complexity continues and the desire to integrate distance education programs expands, attention must be given to faculty training and support.

Enhancing Faculty Participation

Overall, faculty members recognize that distance education technologies are—and will be—an important part of the instructional process. However,

Figure 1. Enhancements to increase faculty participation in distance education.

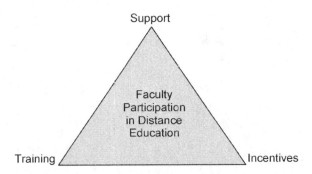

they perceive support and training to be less available than equipment. Enhancing faculty participation requires that resources be directed to provide adequate levels of support and training such that these technologies are used for the benefit of students (Howard, Schenk & Discenza, 2004). It is the integration of incentives, training, and support that promote the adoption of distance education delivery strategies by university faculty.

While faculty recognize the potential, intervention strategies are necessary to alter how people perceive and react to distance education technologies. It is apparent that steps must be taken to increase faculty training and support. Three major areas require consideration: 1) support, 2) training, and 3) incentives. Support extends beyond "verbal" to providing the support/professional staff to assist faculty. Training should not only include technology exposure, but instructional design, pedagogy/andragogy, and "cook-book" strategies and "how-to" manuals. By providing incentives such as release time, mini-grants, continuing education stipends, and recognition in the promotion and tenure process, faculty will have more than verbal encouragement to continue, or begin, using distance education technologies and will have the reason to do so (Dooley & Murphrey, 2000; Murphrey & Dooley, 2000).

Rockwell et al. (1999) found that the primary incentives for faculty participation were *intrinsic* or personal rewards, including the opportunity to provide innovative instruction and apply new teaching techniques. Other incentives included extending educational opportunities beyond the traditional institutional walls, and release time for faculty preparation.

FUTURE TRENDS AND CONCLUSION

Faculty member participation in distance education requires a competence in using technology, an attitude that distance education is important and valuable, and access to quality infrastructure (Hawkes & Coldeway, 2002). Faculty roles and responsibilities must change to accommodate the use of these technologies and it must be recognized that teaching at a distance requires a different set of competencies (Richards, Dooley & Lindner, 2004). Integration of distance education technologies into the teaching and learning process requires a shift of attitude on the part of the faculty members and the removal of barriers created by the lack of institutional support.

REFERENCES

Cyrs, T. (1997). *Teaching at a distance with merging technologies: An instructional systems approach.* Las Cruces, NM: Center for Educational Development, New Mexico State University.

Dooley, K.E., & Murphy, T.H. (2000). College of Agriculture faculty perceptions of electronic technologies in teaching. *Journal of Agricultural Education, 42*(2), 1-10.

Dooley, K.E., & Murphrey, T.P. (2000). How the perspectives of administrators, faculty, and support units impact the rate of distance education adoption. *The Journal of Distance Learning Administration, 3*(4). Retrieved from http://www.westga.edu/~distance/jmain11.html

Hawkes, M., & Coldeway, D.O. (2002). An analysis of team vs. faculty-based online course development: Implications for instructional design. *Quarterly Review of Distance Education, 3*(4), 431-441.

Howard, C., Schenk, K., &Discenza, R. (2004). *Distance learning and university effectiveness: Changing educational paradigms for online learning.* Hershey, PA: Idea Group Publishing.

Jones, E.T., Lindner, J.R., Murphy, T.H., & Dooley, K.E. (2002). Faculty philosophical position towards distance education: Competency, value, and education technology support. *Online Journal of Distance Learning Administration [Electronic Journal], 5*(1). Retrieved from http://www.westga.edu/~distance/jmain11.html

Kanuka, H., Collett, D., & Caswell, C. (2002). University instructor perceptions of the use of asynchronous text-based discussion in distance courses. *American Journal of Distance Education, 16*(3), 151-167.

Knowles, M.S. (1990). *The adult learner: A neglected species.* Houston, TX: Gulf Publishing.

Lindner, J.R., Murphy, T.H., & Dooley, K.E. (2002). Factors affecting faculty perceptions of technology-mediated instruction: Competency, value, and educational technology support. *NACTA Journal, 46*(4), 2-7.

McKenzie, B., Mims, N., Bennett, E., & Waugh, M. (2000). Needs, concerns and practices of online instructors. *Online Journal of Distance Learning Administration [Electronic Journal], 3*(3). Retrieved from http://www.westga.edu/~distance/jmain11.html

McNeil, D.R. (1990). *Wiring the ivory tower: A round table on technology in higher education.* Washington, DC: Academy for Educational Development.

Miller, G., & Pilcher, C.L. (2001). Levels of cognition researched in agricultural distance education courses in comparison to on-campus courses and to faculty perceptions concerning an appropriate level. *Journal of Agricultural Education, 42*(1), 20-27.

Moore, M.G. (1997). Quality in distance education: Four cases. *The American Journal of Distance Education, 11*(3), 1-7.

Murphrey, T.P., & Dooley, K.E. (2000). Perceived strengths, weaknesses, opportunities, and threats impacting the diffusion of distance education technologies for colleges of agriculture in land grant institutions. *Journal of Agricultural Education, 41*(4), 39-50.

O'Quinn, L., & Corry, M. (2002). Factors that deter faculty from participation in distance education. *Online Journal of Distance Learning Administration [Electronic Version], 5*(4). Retrieved from http://www.westga.edu/~distance/jmain11.html

Perreault, H., Waldman, L., Alexander, M. et al. (2002). Overcoming barriers to successful delivery of distance-learning course. *Journal of Education for Business, 77*(6), 313-318.

Purdy, L.N., & Wright, S.J. (1992). Teaching in distance education: A faculty perspective. *The American Journal of Distance Education, 6*(3), 2-4.

Richards, L.J., Dooley, K.E., & Lindner, J.R. (2004). Online course design principles. In C. Howard, K. Schenk & R. Discenza (Eds.), *Distance learning and university effectiveness: Changing education paradigms for online learning.* Hershey, PA: Idea Group Publishing.

Rockwell, S.K., Schauer, J., Fritz, S.M., & Marx, D.B. (1999). Incentives and obstacles influencing higher education faculty and administrators to teach via distance. *Online Journal of Distance Learning Administration, 2*(4). Retrieved from http://www.westga.edu/~distance/rockwell24.html

Rogers, E.M. (2003). *Diffusion of innovations* (5th ed.). New York: Freepress.

Wolcott, L.L. (1997). Tenure, promotion, and distance education: Examining the culture of faculty rewards. *The American Journal of Distance Education, 112*, 3-18.

KEY TERMS

Adoption: A decision to make full use of a new idea as a preferred method (Rogers, 2003).

Andragogy: The art and science of teaching adults (Knowles, 1990).

Competence: A measure of perceived level of ability by faculty in the use of electronic technologies often associated with distance education (Jones, Lindner, Murphy & Dooley, 2002).

Distance Education: Any education received by learners that occurs when the instructor and learner are separated by location and/or time.

Incentive: Intrinsic or extrinsic motivational factors that impact faculty decisions to participate in distance education.

Pedagogy: The art and science of teaching children (Knowles, 1990).

Value: A measure of the importance perceived by faculty of the role that technology will have on distance delivery (Jones, Lindner, Murphy & Dooley, 2002).

This work was previously published in the Encyclopedia of Information Science and Technology, Vol. 2, edited by M. Khosrow-Pour, pp. 1186-1189, copyright 2005 by Idea Group Reference (an imprint of IGI Global).

Chapter 8.5
How Distance Programs Will Affect Students, Courses, Faculty, and Institutional Futures[1]

Murray Turoff
New Jersey Institute of Technology, USA

Richard Discenza
University of Colorado at Colorado Springs, USA

Caroline Howard
Techknowledge-E Systems, USA

ABSTRACT

Designed properly, distance education classes can be at least as effective and, in some ways, even more effective than face-to-face courses. The tools and technologies used for distance education courses facilitate learning opportunities not possible in the face-to-face classroom. Distance programs are accelerating changes that are challenging students, faculty, and the university, itself. Currently, most faculty are rewarded for the quality of instruction, as well as their external funding and their research. Often, university administrators focus more attention on the efficiency of teaching than on its effectiveness. In the future, as the quality of distance learning increases, the primary factor for success will be the faculty's commitment to excellence in teaching. Many institutions will be forced to reevaluate the quality of teaching as the institution becomes more visible to the public, to legislators who support higher education, and to prospective students.

INTRODUCTION

People usually assume that students in distance education programs are at a disadvantage. On the contrary, it is probably not the distance student who is disadvantaged, but rather many face-to-face students. Learning is enhanced by the physical and social technologies typically used in distance education. Students in distance programs have access to tools that allow them to repeat lectures and interact with their fellow students and faculty. Contrast these students with a student sitting in a 500 student lecture. Which student is most at a distance?

In the early 1980s, a research group introduced a computer-mediated system to a regular face-to-face class. The group felt that there was enormous potential for this technology to enhance learning. The system was introduced to students in a number of Computer Science and Information System courses. Due to the amount of material covered in lectures, there was not much time for dialogue and only a few students participated when there was a class discussion. The instructors introduced asynchronous group communication technologies to communicate discussion questions and assigned grade point credits for student participation. One hundred percent of the students participated in these discussions outside of regular classroom hours. The extent and depth of the discussions changed the nature of the classes. Most importantly, student contributions were comprehensive, with more well-thought-out comments, because students had the time to reflect on the ongoing discussion before participating. Also very significant was that students, for whom English was a second language, became equal participants. They could reread the online discussion as many times as needed before replying. The computer-based activity monitoring and transcripts, electronic recordings of the discussions, showed that foreign students spent two to three times more in a reading mode and reread many discussions, far more than the American students.

This ability to monitor activities and review the electronic transcripts gives the instructor insights into how students are learning. By reviewing the transcripts of the online discussions, it becomes obvious what and how students are learning. For courses with high pragmatic content, such as upper level and graduate courses in topics like the design and management of computer applications, students are required to utilize problem-solving approaches to evaluate the trade-offs between conflicting objectives. In a traditional classroom environment, especially in large classes, it is very difficult to detect whether students are accurately incorporating the problem-solving mental models that the instructor is attempting to convey. Reviewing the transcripts of class discussions can provide insight into the approaches students are taking to master the material.

Unfortunately, in the early 1980s no one wanted to hear about a revolution in normal classroom teaching or was willing to expend the effort to dramatically improve classroom education. It was only those interested in distance education who were interested in learning about the educational potential of the technology. As a result, in the mid 1980s the researchers at New Jersey's Institute of Technology (NJIT) obtained research funding to investigate distance education applications of Computer Mediated Communications (CMC). Since NJIT, at the time, had no distance program they created distance sections of regular courses that were used with regular on-campus students taking most other courses face-to-face.

This effort (Hiltz, 1994) utilized quasi-experimental studies that compared a population of students (only familiar with face-to-face classroom education) to a population of students taking the same courses in pure face-to-face sections with pure distance sections using only CMC technology. The students in the matched sections had the same material, the same assignments, the same exams and the same instructor. They found no significant difference in the amount of learning or the rate of student satisfaction. This finding is much more significant than a determination based on a study that included a population of distance learners already familiar with traditional correspondence classes. Two critical underlying variables driving the success of this approach were identified by Starr Roxanne Hiltz (Hiltz, 1994). First, the role the instructor needed to take was different from the traditional classroom role. The instructor acted more as an active and dedicated facilitator rather than traditional teacher, as well as a consulting expert on the content of the course. Second, collaborative learning and student team-work were the educational methodology (Hiltz, 1994) which was shown in later studies to be a key factor in making distance courses as good or better than face-to-face courses (Hiltz and Wellman, 1997).

These results indicate that distance courses can be as effective as face-to-face courses when using any of the traditional measures, such as exams and grades. However, these traditional measures may be inadequate to measure of many of the benefits observed in classes that utilize computer-mediated communications technology for a number of reasons:

- Due to social pressures, students tend to be more concerned with how other students view their work quality than how the professor views it. They are significantly more motivated to participate in a meaningful way when their fellow students can view their contributions.
- When equality of communications is encouraged, students cannot get away with being passive or lazy. The transcript or electronic recording of the discussions shows who is and is not participating. It is visible to both the instructor and other students that someone is being lazy. (In fact, students seem to be more concerned with what the other students will think of their performance than what the professor will think.)

- The scope of what the outstanding students learn becomes even more noticeable.
- The performance of students at the lower end of the distribution is improved. The communications systems permit them to catch up, because they are able to obtain a better understanding of the material with which they are most uncomfortable or have the least background knowledge.
- The instructor can become more aware of his/her successes or failures with individual students because of the reflective nature of the student contributions to the discussion.

While these conclusions need confirmation through long-term longitudinal studies of student performances, the marketplace is also providing confirmation of the beliefs held by many experienced in teaching these classes. We are seeing that collaboratively oriented programs offer a solution to the problems, which are inherent in traditional correspondence courses.

Students benefit from the ability to electronically store lectures alone or in chunks integrated into other material on the Web. Electronic storage of lectures gives all students the power to choose freely whether they want to attend a face-to-face class or take the same course remotely. Traditional face-to-face students can later hear a lecture missed due to illness or travel. Students with English as a second language can listen to a lecture multiple times. Face-to-face students who have to travel or fall sick can use the same tapes to catch up and/or review material prior to exams.

In our view a student in a face-to-face class that is not augmented by a collaborative learning approach and by asynchronous group communications technology is not getting as good of an education as the distance student who has those benefits. It is the face-to-face student who may be suffering from the segregation of the college system into separate face-to-face and distance courses. These observations about the past and the present lead to some speculations about the future.

THE CHANGING NATURE OF ACADEMIC COURSES

A vital role of college education is to convey the mental model and pragmatics of a subject matter. To ensure learning of complex subject matter, an instructor needs to communicate his/her mental model and accompanying problem-solving approaches. The model, or structure and dimensions to understand and organize the material, along with the pragmatics of the subject matter become the student's starting point. These allow him/her to build the details, acquire new knowledge, and apply information in new situations. The more pragmatic the course content is, the more important it becomes for the instructor to convey his or her mental models and assess student assimilation.

Communication among students and with the instructor is particularly valuable in courses with high pragmatic content. By discussing and comparing their interpretations, student can reinforce their understanding and reduce their conceptual errors. Mixing distance and face-to-face students in the same discussion space results in having students with a great deal of work experience virtually mingle with those who have had none. Often undergraduate students who take distance courses are working, and most of those in the face-to-face courses are not working. Mixing both a distance and a face-to-face class leads to a better balance of backgrounds. Students, who have work experience, may have had an example from their life that illustrates a theoretical concept presented in the class. When a student shares this experience, it reinforces the importance of the concept and encourages other students to pay attention to the presentation.

Creative, interactive software programs accompanied with background tutoring can effectively teach students to master the skills currently taught in many undergraduate courses. When these courses are automated, the costs incurred are far below typical college tuition. In the future, colleges and universities will not be able to continue to charge current tuition costs for introductory courses that are largely skill oriented. For example, there are many stand-alone and Web-based software programs that offer introductory programming courses, as well as skills in many other areas. These courses are comparable to college courses and some are even based upon a textbook used on some college campuses. They are available for a few hundred dollars. The major difference is that they do not carry college credits.

At the other extreme we have the usual "skimming of the cream" that occurs with every change in technology. Today, private firms are willing to invest a millions dollars in single multimedia, largely automated courses to sell to industry that can afford to pay thousands of dollars of tuition for each student. When all the downsizing of outdated professionals was occurring, one of the first things to go was the elaborate investment these companies had made in internal course offerings and video classrooms and networks. Some companies had created their own internal college and claimed their employees did not need outside education. This type of thinking, like the concept of "just-in-time learning," which may be a euphemism for "teach them only what they need to know in order to do their current job," does not prioritize student growth and learning. Unfortunately, some institutions of higher education are no longer certain about whether their client is the student or industry. Until an enlightened consumer marketplace emerges, many transient but inferior offerings will be available on the market.

With the recent recession we find some institutions, like Duke, have cancelled their online executive MBA, for which they were charging twice the tuition of the on-campus offering. The recession has had the obvious results of discouraging and terminating a number of private sector efforts to enter the distance education market. The cream has largely turned sour.

Now that we have a national and international competitive market in courses, those colleges that accept skill knowledge from unaccredited sources, such as training courses and work experience, will obtain a market edge over those that do not. In fact, the student population will begin to expect institutions of higher learning to accept courses from any accredited institution.

In addition, institutions with clearly stated credit transfer policies will also obtain a marketplace edge. Individual courses, as well as total programs, will be the basic units in a national and international marketplace for higher education. There are no longer geographical monopolies on higher education. Only consortiums based upon real cooperation among the participating institutions will succeed. Many current attempts to market only the separate offerings of the participating institutions, or to impose added layers of administration between the courses and the students, are doomed to a marketplace failure.

The single noncommercial web site that focuses on distance education utilizing group communications is that of the Society for Asynchronous Learning Networks (http://www.aln.org). There are numerous group conferences for educators and administrators, as well as a newsletter and journal.

ROLE OF THE FACULTY

In any meaningful educational program, the major responsibilities for a given course should be the responsibility of a faculty member, including:

- Course design

- Choice and creation of course materials and assignments
- Approval and mentoring of instructors, adjuncts, and Ph.D. students who will also teach the course
- Performance monitoring of other instructors, adjuncts, and Ph.D. students
- Course and material updates

The technology allows senior professors or department chairs to effectively evaluate and mentor all instructors of particular courses, whether they are teaching traditional classroom courses or distance courses. The ability to review whole class discussions after the class is over gives senior faculty the ability to evaluate distance instructors hired to teach previously developed courses, as well as to review on-site instructors and junior faculty. Thus, they can improve and extend their mentorship and apprenticing relationships.

While educational institutions are rapidly developing programs for large populations of distance students, it does not appear that universities are creating tenured faculty lines that can be occupied by remote faculty. When additional faculty are needed to teach distance courses, instructors, rather than tenure track faculty, are often sought. Since the success of distance courses is largely dependent upon the capabilities of the instructor responsible for the particular course, the value of instructors able to master teaching at a distance will rise within institutions of higher education.

The technology we are using for distance education can allow faculty members to live anywhere they want to. Unique benefits will be available to outstanding teaching faculty. For example, one of the best full-time instructors for NJIT, which is located in beautiful downtown Newark, is a mother with two small children who never has to be on campus. She is teaching other instructors how to teach remotely. Similarly, a University of Colorado accounting professor, on sabbatical in

Thailand, is able to teach a course in the Distance MBA program.

There have been a few master programs where some or all of the instructors are located anywhere in the world. It is technically feasible for those wanting to escape winter cold to teach in places such as Hawaii that we could only dream about. The technology makes it feasible, but various administrative policies, unions, insurance companies, benefit programs, etc., have not yet caught up to the technology. There is increasing emphasis by accrediting agencies on treating remote instructors the same as faculty are treated. This is likely to bringing about a greater degree of equality between instructors and tenured track faculty. The outcome is uncertain, but it may mean that the costs for remote and traditional classes will equalize so that the profit margin in online classes will not be quite so high.

Role of the Technology

Some functions of technology that can facilitate this function are:

Asynchronous discussions: In the online environment, students can take as much time as they need to reflect on a discussion and polish their comments. This improves the quality of the discussion and changes the psychology and the sociology of communications. Students can address topics in the sequence they chose rather than in a predefined order. This leads to the development of different problem-solving strategies among the individual members of the class.

Instructor control of online conference and roles: With online course conferences (many per course), instructors control the membership of each, assign roles and enable other instructors to monitor conferences for joint teaching exercises involving more than one course. Groups within courses are able to set up private online conferences for team and collaborative work group assignments. Joint editing of items facilitates team work.

Question and answer communication protocol: Instructors are able to ask questions during discussions. They can control who views the answer and prevent other students from seeing the answer of the others or engaging in the resulting discussion until they have entered their answer. In studies of Group Decision Support Systems, it has been shown that asynchronous groups in an online Delphi mode generate many more ideas than unstructured discussions or face-to-face groups of similar size (Hee-Kyung et al., 2003). This area has proven to be a valuable tool in forcing equal participation. Use of question-and-answer communication protocol can be used to force each student to independently think through their answer without being influenced by the other students.

Anonymity and pen name signatures: When students with work experience are part of a discussion, they can use their real life experiences to illustrate the concepts the professor is presenting. Such comments from fellow students, rather than the professor, often make the instructor's message more meaningful to the students. A student confirming the theory presented by a faculty member through real life examples is more effective in making a point than "dry" data from an instructional article. Furthermore, students can talk about disasters in their companies with respect to decisions in any area and provide detail, including costs, when they are not identified and the anonymity of the company they work for is preserved. Also, the use of pen names allow individuals to develop alternative personas without divulging their real identity and is extremely useful in courses that wish to employ role playing as a collaborative learning method.

Membership status lists: The monitoring of activities, such as students' reading and responding to communications, allows the professor to know what each individual has read and how up-to-date each student is in the discussion. This allows the instructor to detect when a student is falling behind. Student collaborative teams can make sure that everyone in the team is up to date. Furthermore, students can easily compare their frequency of contributions relative to other students in the course.

Voting: Instant access to group and individual opinions on resolutions and issues are enabled by voting capabilities. This is useful for promoting discussion and the voting process is continuous so that changes of views can be tracked by everyone. Voting is not used to make decisions. Rather, its function is to explore and discover what are the current agreements and disagreements or uncertainties (polarized vs. flat voting distributions) so that the class can focus the continuing discussion on the latter. Students may change their votes at anytime during the discussion.

Special purpose scaling methods: These useful methods show true group agreements and minimize ambiguity. Currently we have a system which allows each student at the end of the course to contribute a statement of what they think is the most important thing they have learned in the course and then to have everyone vote by rank ordering all the items on the list. The results are reported using Thurstone's scaling, which translates the rank order by all the individuals to a single group interval scale. In this interval scale if 50% prefer A to B and 50% prefer B to A, the two items will be at the same point on the scale. It has been surprising what some of the results have been in some courses. For example, in a management of Information Systems course the concept of "runaway" software projects was felt to be twice as important as any other topic. The professor was quite surprised by this result until he began to realize that the students were using this concept of a mental model in which to integrate many of the other things they had learned.

Information overload: This occurs when enthusiastic discussions by students that are meant to augment the quality of the learning process aug-

ment only the quantity of the number of comments, instead, leading to the problem of "information overload." Currently this phenomenon limits the size of the group that can be in a single CMC class. Online discussions allow individuals to enter comments whenever it is convenient for them, without waiting for someone else to finish the point they were trying to make. This makes it physically possible and also very likely that a great deal more discussion will take place and much more information will be exchanged among the group than if only one person can speak at a time, as in the face-to-face classroom environment. Anything that reduces the temptation of some students to "contribute" comments or messages that have nothing to do with the meaningful discussions underway will increase the productivity of the discussion without information overload setting in. Among such functional tools the computer can provide are:

- **Class gradebooks:** This eliminates a tremendous amount of electronic mail traffic that would become very difficult for an individual instructor to manage with a large class.
- **Selection lists:** The instructor can set up lists of unique choices so that each student may choose only one item and others can see who has chosen what. This is very efficient for conveying individualized assignments and reduces a large portion of communications.
- **Factor lists:** Members of a class or group can add ideas, dimensions, goals, tasks, factors, criteria, and other items to a single, shared list which may then be discussed and modified based upon that discussion and later voted upon.
- **Notifications:** Short alerts notify individuals when things occur that they need to know about. For instance, students can be notified that a new set of grades or vote distribution has been posted, eliminating

the need for individuals to check for these postings. People can attach notifications to conference comments from a select list that provides alternatives like: *I agree, I disagree, I applaud, Boo!* Such appendages reduce significantly the need to provide paralinguistic cues of reinforcement as additional separate comments.

- **Calendars, agendas or schedules:** Students have access to a space to track the individual and collaborative assignments and their due dates. These are listed in an organized manner that links detailed explanations for each assignment, as well as questions and answers related to the assignments.

THE STATE OF THE TECHNOLOGY

The technology available today includes at least 250 versions of group communication software. However, some of them may not survive the recession. There are a growing number of software packages for course management. The online learning product landscape is changing at a rapid pace as companies are acquiring their competitors to expand functionality. A recent article gives an excellent summary of the popular platforms and the evolving nature of eLearning (Gray, 2002).

There are only a few of these that have wide usage and they are beginning to raise their prices to capitalize on their popularity. Most of these packages charge a fee per user, which is not the desirable fee structure for the customer. Many of the older conferences systems charge on a per server basis and it does not matter how many students one has. It is far cheaper to spend more on the hardware and a get a more powerful server. Also, the course management systems do not provide many of the useful software features one would like to have for group communications. Given the way prices are going, it might be better to pay some of the undergraduate students to

educate some of the faculty on how to create their own web sites and have their own pages for their courses that they update and maintain directly. This also has desirable long-term consequences in raising the ability of the faculty in this area. Once you have committed all your content to one vendor's system, you are a captured customer and will have to pay whatever they want to charge. Right now, software development is undergoing rapid evolution and no customer should put themselves in the box of only being able to use one vendor. If it is clear you are using a number of vendors, you may even be able to get some breaks on pricing and will certainly get the top level of service when each of them knows there is an alternative service readily available to the customer. In the coming decade, one can expect major upgrades of these software systems every few years and the best one today may not be the best one tomorrow.

COURSE DEVELOPMENT AND DELIVERY TECHNOLOGY

Unfortunately, many faculty do not know how to use the technology to design a successful course. As the historical record shows, it is a mistake when transferring an application to computers to just copy the way it used to be done onto the computer. Utilizing the methodology of collaborative learning is the key to designing courses using group communications technology. Simple systems, which attempt to impose a discussion thread on top of what is electronic mail technology, allow the student or the teacher only to view one comment at a time. This approach does not allow an individual to grasp the totality of any complex discussion. Only by placing the complete discussion thread in a single scrolling page can a person review and understand a long discussion. They can browse the discussion and cognitively comprehend it without having to perform extra operations and loose their cognitive focus. Users of such simple systems cannot generate a large complex discussion and have no way of realizing that complex discussion is even possible.

When online discussions are successful, they can easily go from enthusiastic wonderful discussions to information overload. To maximize the power of the technology to facilitate collaborative learning, critical development directions for the future should include:

- Tailorability of communication structures by instructor
- Tailorability of communication protocols by instructor
- Anonymity and pen name provisions
- Delphi method tools and the availability of scaling and social judgment (voting methods)
- Tools for collaborative model building
- Powerful information retrieval capabilities
- Tailorability by instructor of application-oriented icons and graphical components
- Tools for the analysis of alternative diagrams

Instructors also need to allow students to extend the discourse structure and to vote on the significance of incidents of relationships among factors in the problem domain by using Group Decision Support processes. The system should allow students to not only develop their own conceptual maps for understanding a problem, but also to detect disagreements about elements of the conceptual map and the meanings of terms. This is valuable preparation for problem solving in their professional life, a process that requires removing inherent ambiguities and individual meanings in the language used to communicate about a problem with others from diverse backgrounds.

Routines should be included that are based upon both scaling and social judgment theories which improve the ability of larger groups to quickly reach mutual understanding. Currently, few tools exist in current systems that support the use of collaborative model building, gaming, and Delphi exercises. The current generation of software does not often include the functions of anonymity and pen names.

Course instructors need to have complete control over course communication structures and processes and should be able to use their recently acquired knowledge for future offerings of the course. Currently, systems lack the needed integration of functions to easily evolve the changes in both the relationships and the content in a given field. A long-term advantage of teaching in the collaborative electronic environment is that the students create useful material for future offerings and can aid the instructor in monitoring the new professional literature.

Future technology will allow faculty to organize their material across a whole set of courses into a collaborative knowledge base available to the faculty teaching those courses. This would allow students and faculty would be able to create trails for different objectives and weave the material in that knowledge base to suit a group of students or a set of learning objectives. Individual learning teams would be able to progress through a degree program's knowledge base at the rate best for them, rather than setting the same timeframe for all learning teams or faculty teams. Faculty, individuals or teams would take responsibility for a specific domain with in the web of knowledge representing a degree program.

Collaborative technologies are changing the concept of what constitutes a course. Program material could be an integrated knowledge web based largely on semantic hypertext structures. Over time, the domain experts, the faculty, would continue to develop and evolve their parts of the web and wait for learning groups, composed of any mix of distance and regular students sharing the same learning objectives and needs.

Current vendor systems focus on the mass market and concentrate on tools to standardize and present course content. Group communication tools are usually just disguised message servers that offer only a discussion thread capability and little more, certainly not the complex capabilities discussed above. Vendors have not yet recognized the primary importance of group communications and how faculty members can guide and facilitate the process and be available for consultation as needed. Based upon the conceptual knowledge maps they design, faculty should be encouraged to develop content structures that are characteristic of their subject matter. In the end, faculty should have the ability to insert group communication activities anywhere in their professional knowledge base (e.g., question/answers, discussion threads, lists, voting, etc.).

EDUCATIONAL CONSUMERISM

Most of today's distance education course offerings give the illusion of difference by placing materials on the Web, instead of providing them through the mail. These online course offerings still use the correspondence course model. These offerings typically include e-mail systems for one-on-one communication between an instructor and an individual student, but do not include effective group communications featuring course content and delivery methodologies reworked for a distance-learning environment. E-mail is better than nothing, but no one would claim that e-mail is preferable to a face-to-face college course. The typical consumer of distance education does not understand the difference between courses with only simple e-mail systems and courses that have introduced sophisticated group communication processes.

Students in the United States pay the equivalent of the price of a used or new car every year to attend college. A student and his family make a major financial investment for a college educa-tion. Reports evaluating new and used automobile models and car-buying comparison web sites enable people to obtain detailed information on any model of car for little or no money. We predict the emergence of a successful "consumer report" organization for distance learning, similar to the college guides appearing each year, that will provide details about individual courses and instructors in different programs, unfiltered and direct from other students. (Some guides have already appeared, but some of them make money by charging the schools to list their programs.) There is, as of yet, nothing comparable to *Consumer Reports* or even the yearly *U.S. News & World Reports* independent rating of universities. Examples of sources now available are *Bears' Guide to the Best Education Degrees by Distance Learning* by John Bear, Mariah Bear, Tom Head, Thomas Nixon Ten Speed Press (ISBN: 1580083331; September 2001) or *Guide to Distance Learning Programs 2003* (*Peterson's Guide to Distance Learning Programs*, 2003) Petersons Guides (ISBN: 0768908191Bk&Cd-Rom edition; October 2002).

Ultimately, these sources will support more intelligent consumerism about college education. Since distance education eliminates a college's geographical monopoly, colleges will be forced to be much more sensitive to consumer pressures than they have been in the past. Since no educational institution or organization has had the foresight, so far, to do this, there is now a commercial Web firm that sells books, student services and other products and has committed to putting up a recommendation system to evaluate any distance course anywhere and have the results made Web accessible.

Today, university faculty are not rewarded as much on the quality of instruction as on their research and external funding. To obtain tenure and promotions, their instruction quality merely has to be acceptable and new faculty cannot afford to prioritize exceptional, innovative teaching. The problem is often with university administrations that focus their attention more on efficiency rather than effectiveness when it comes to teaching.

INSTRUCTORS AND REWARDS

Administrations place most of their priority on competitive research and sponsored funds. Many face a rude awakening, as they are less in touch with the fact that the educational process is undergoing an unanticipated, unexamined, fundamental change. Some of these institutions will not realize until it is too late to change entrenched attitudes and bureaucratic processes fast enough to compete in the new competitive environment. During the next decade, many institutions may fail as a result of intense international and national competition.

In the future, the underlying factor of success will be the faculty's commitment to excellence in teaching and the quality and talent of the instructors. Many institutions will be forced to reevaluate their faculty incentives and the relative importance of teaching and research. Marketplace mechanisms will also force this reevaluation, as the relative quality of teaching for each institution becomes more visible to the public and prospective students. These pressures will force the threshold for acceptable teaching quality definitely to rise.

As the quality of distance learning increases, the view of distance education courses as inferior to traditional classes will disappear. It will become the talent of the instructor and his or her facility that will determine success. Additional organizational layers of intermediaries will doom a program to failure. For example, the majority of students taking distance courses in the future will be regular students who schedule a mix of distance and face-to-face courses to accommodate their schedule, family commitments, work commitments and their desire to complete their education in a timely manner. Their course ratings do not distinguish distance courses and face-to-

face courses, but distance courses are even rated higher in many ways.

ALTERNATIVE VERSIONS OF THE FUTURE

Students of the future will have many choices, the spectrum of which can be illustrated by examples of two students at extremes:

The Positive Future for the Student

After careful consideration of my options, I decided to turn down a scholarship from an Ivy League university to go to eU (Electronic University). I wanted to continue learning my family's business and did not want to move away from my fiancée. I was able to easily make up for the cost of the eU tuition, which was about 25% of the traditional school's, even with the recent 15% tuition reduction which that school just announced. I was convinced when I discovered that eU was rated as highly as the Ivy League school for the quality of its courses. The response of more than three million students in the "Learning Consumer Database" made the results for most of my courses statistically significant to the .05 significance level. Also, I found the comments of the professors about their courses much more extensive in the eU course ratings. Hardly any of the other school's professors responded.

All of that encouraged me to check the resumes of most of the eU professors. The results were quite surprising. Most of them are retired from other universities where they had tenure before they came to eU. They are all paid the same salary of $150,000 and work out of their homes all over the world. One of them wrote that 95% of his time and effort is devoted to instructional activities because eU has eliminated all committees except one to determine the departmental curriculum.

The eU classes usually range from 20 to 50 students with a great deal of emphasis on class discussion and collaborative work. The profile of students shows that over 70% are working in professional areas related to their degree programs. As a result, class discussions are high caliber. I got to actually eavesdrop on some ongoing courses once I had submitted an application.

I can take my exams at the local community college, which has a franchise from eU, and I can also use their sophisticated multimedia computers. eU accepts courses from any college and university that has accreditation for the same degree, so I can use another distance program when the course I need is closed at eU or not offered that semester, without pre-approval. With the three-semesters-per-year program, I can move a lot faster than in most two-semester programs.

I am concerned, however, with getting in, as I am fresh out of high school and they take very few students like me. Their rejection rate is much higher than most Ivy League schools'. I have tried to convince them that my four years of part time work in the family business should be counted. I hope that helps.

The Negative Future for the Student

I have decided to apply to eU.com rather than Harvard. I really must spend the time learning the family business and my fiancée has told me in no uncertain terms that long separations are not in the cards. Oh well, it is a lot cheaper than Harvard and a lot of those video lectures were prepared by top notch professors at places like Harvard and the University of Chicago. They claim having a professor from Harvard on video is far better than just any old professor in a classroom. Most of the instructors for technology courses are from industry and I am told that if you get one from the company you are interested in working for and do a good job, you are more likely to get a job offer in the future. Courses in other areas seem to be mostly those tapes and automation. They require a computer joystick for the educational

software packages, so the school cannot be too bad, plus major Hollywood studios produce their multimedia software.

It does worry me that their tuition jumped by 20% in our area as soon as our local community college went out of business. I did not realize their tuition was geographically dependent. Their software costs are quite high since each course uses unique packages, including the ebooks generated by the professors. These materials seem be undergoing constant revision, but I suspect that is so the prior year's material cannot be sold in a secondary market among the students. Even though the average course size is one thousand students, eU does have these small discussion sections of 50 to 100 students run by the course graders. So, at least you can get help when you need it. Still, some courses use automated graders and I am not clear how that works, yet.

I was told the compositions in the first writing course and the programs in the first computer course are completely graded by the computer without the need for any human to look at them. An intelligent system not only designs the exam so that every exam is unique to every student in the course, but also uses your past performance profile to tailor the exam to your performance level. This allows even C students to get high point scores so they can feel good about themselves and show good results to their parents, who are probably financing their studies. Students are classified as Outstanding, Above Average, or Average, and then receive grades within those categories. Everyone has a chance to get a lot of A grades.

They sent me this funny form with their acceptance letter, where I must promise to not divulge any of my experiences in courses to any data collection process not approved by eU.com or they can deny me any future access to my records and rescind my degree. I don't understand the reason for that one at all. Oh well, I have no real choice, given my situation.

SUMMARY

In the first scenario above, the student will receive the same quality of education whether he studies on campus or at home. He will participate with a group of his peers and will establish a network of relationships to utilize throughout his career. He can also get to know his instructors and his fellow students well.

In the second scenario, the student will participate in a distance-learning program set up like a mass production process. This is a clear second choice, apparently forced on the student by circumstances and costs. This option sacrifices the quality of education for the ultimate efficiencies and mass delivery of courses.

The real variable, which will decide between these future alternatives, is whether higher education institutions integrate all their face-to-face students in the same communication environment, prioritizing collaboration for all students and rewarding faculty who introduce new technology in this way. Regardless of what is written down, in most universities the rewards for faculty are inextricably linked with research and external funding and instruction needs only to be acceptable to obtain promotion and tenure. Innovation in education and exceptional teaching are not prioritized for young faculty at many institutions.

This is clearly a problem with administrations. While administrations focus their attention on competitive research and sponsored funds, the educational process is undergoing an unanticipated, unexamined, fundamental change. The next decade will bring some rude awakenings. Because of the time needed to change attitudes and bureaucratic processes, some of these awakenings may occur too late. Competition in instruction on an international and national basis will become the principle determinant of institutional success or failure in the next decade. We are entering a free marketplace era for the enterprise of education at the university level. The Web is the first communication system where consumer reaction to experiences with alternatives is cheap and easy to collect, organize and provide. One of the key premises underlying the concept of a free market is the free flow of relevant information and that is going to happen for individual courses, as well as degree programs.

The most important factors for future success will be the quality and talent of the instructors and their commitment to excellence in learning. Many institutions may well have to reassess the relative imbalance in faculty rewards between teaching and research. In addition, marketplace mechanisms will make the quality of teaching more visible to the public and prospective students. We can expect the threshold for acceptable teaching quality to rise. Regular students will opt for distance participation in some of their courses, not only because it is convenient, but also because they perceive no loss of quality. As long as both versions of the courses utilize the same technology and learning methodology this is going to be true.

Ultimately, the fundamental changes that could ensure the future success of university and college level institutions may well have to come from the accreditation agencies in realizing that we are evolving in a competitive marketplace and that it is their role to ensure that the consumer has access to the information needed to make fair market decisions. Ensuring that courses in accredited degree programs can always be transferred among accredited institutions and that accreditation might have to be assigned to individual faculty as well as individual degree programs will most likely be a part of that evolutionary process.

NOTE

A great deal of recent evaluation studies are beginning to confirm our earlier findings based upon extensive and large scale studies at such places as SUNY, Drexel, Penn State and others. Some of these may be found in the Journal of ALN (http://www.aln.org) and on the ALN Evaluation Community web site (http://www.alnresearch.org).

REFERENCES

Cho, H. K., Turoff, M. and Hiltz, S. R. (2003, January). The impacts of Delphi communication structure on small and medium sized asynchronous groups. In *HICSS Proceedings*, IEEE Press.

Discenza, R., Howard, C. , & Schenk, K. (eds.). (2002). *The Design and Management of Effective Distance Learning Programs.* Hershey, PA: Idea Group Publishing.

Gray, S. (2003, August). Moving — elearning vendors take aim in the changing environment. *Syllabus,* 16(1), 28-31.

Harasim, L., Hiltz, R., Teles, L. and Turoff, M. (1995). *Learning Networks: A Field Guide to Teaching and Learning Online.* MIT Press.

Hiltz, S. R. (1993). Correlates of learning in a virtual classroom. *International Journal Of Man-Machine Studies,* 39, 71-98.

Hiltz, S. R. (1994). *The Virtual Classroom: Learning Without Limits via Computer Networks, Human Computer Interaction Series.* Intellect Press.

Hiltz, S. R., & Turoff, M. (1993). *The Network Nation: Human Communication via Computer.* MIT Press (original edition 1978).

Hiltz, S. R., & Turoff, M. (2002, April). What makes learning networks effective. *Communications of the ACM,* 56-59.

Hiltz, S. R., & Wellman, B. (1997, September). Asynchronous learning networks as a virtual classroom. *Communications of the ACM,* 40(9), 44-49.

Howard, C., & Discenza, R. (1996). *A Typology for Distance Learning: Moving from a Batch to an On-line Educational Delivery System.* Paper to be presented at the Information Systems Educational Conference (ISECON) in St. Louis, Missouri, (October 1996).

Howard, C., & Discenza, R. (2001). The emergence of distance learning in higher education: A revised group decision support system typology with empirical results. In L. Lau (Ed.), *Distance Education: Emerging Trends and Issues.* Hershey, PA: Idea Group Publishing.

McIntyre, S. and Howard, C. (1994, November). Beyond lecture-test: Expanding focus of control in the classroom. *Journal of Education for Management Information Systems.*

Nelson, T. H. (1965). A file structure for the complex, the changing and the indeterminate. In *ACM 20th National Conference Proceedings* (pp. 84-99).

Turoff, M. (1995, April). A marketplace approach to the information highway. *Boardwatch Magazine.*

Turoff, M. (1996). Costs for the development of a virtual university. *Journal of Asynchronous Learning Networks,* 1(1).

Turoff, M. (1997, September). Virtuality. *Communications of ACM,* 40(9), 38-43.

Turoff, M. (1998, Spring). Alternative futures for distance learning: The force and the darkside. *Online Journal of Distance Learning Administration,* 1(1).

Turoff, M. (1999, January/March). Education, commerce, & communications: The era of Competition. *WebNet Journal: Internet Technologies, Applications & Issues,* 1(1), 22-31.

Turoff, M., & Hiltz, R. S. (1986). Remote learning: Technologies and opportunities. *Proceedings,*

World Conference on Continuing Engineering Education.

Turoff, M., & Hiltz, R. S. (1995). Software design and the future of the virtual classroom. *Journal of Information Technology for Teacher Education,* 4(2), 197-215.

Turoff, M., Hiltz, R., Bieber, M., Rana, A., & Fjermestad, J. (1999). Collaborative Discourse Structures in Computer Mediated Group Communications. Reprinted in *Journal of Computer Mediated Communications on Persistent Conversation,* 4(4).

Chapter 8.6
Awareness Design in Online Collaborative Learning:
A Pedagogical Perspective

Curtis J. Bonk
Indiana University, USA

Seung-hee Lee
Indiana University, USA

Xiaojing Liu
Indiana University, USA

Bude Su
Indiana University, USA

ABSTRACT

Collaboration in online learning environments is intended to foster harmonious interactions and mutual engagement among group members. To make group performance effective, it is essential to understand the dynamic mechanisms of online groupwork and the role of awareness supporting dynamic online collaboration. This chapter reviews the nature of online collaboration from the standpoint of task, social, and technological dimensions and reconceptualizes the importance of awareness support into these three dimensions of online collaboration. Further, this chapter suggests key knowledge elements in each type of awareness. Detailed pedagogical examples and technological features for awareness support for online collaboration are proposed.

INTRODUCTION

In the midst of the emergence of advanced Web technologies, groupwork has arisen as one of the most promising and innovative practices in online as well as face-to-face teaching and learning. For decades, educators have been arguing that a technologically sophisticated learning environment can provide support for online inquiry and knowledge-building in a learning community. However, decisions about which technologies to use and the ways they that are used for collaboration greatly impact the quality and depth of computer-supported learning. Effective communication and productive groupwork across time zones and geographic distances are highly dependent on whether the technological tools are used in conjunction with appropriate pedagogical guidelines and technological support.

Groupwork in real world situations often relates to working together for a key part of a project. In educational settings, groupwork often entails learners building collective knowledge via dialogues while working together. The process of groupwork produces unexpected synergistic ideas between group members, the intense discussion or debate of ideas, and creative final products that extend far beyond the talents of any one individual.

Many groupware tools and course management systems offer a variety of advanced features, but most of these serve the function of delivering communication rather than supporting group activities (Kirschner & Van Bruggen, 2004). Given that deep understanding of peer interactions within a shared workplace impacts the success of group performance (Gutwin & Greenburg, 1999), problems within current groupwork environments often lie in their lack of support of intricate group tasks and dynamic group processes.

In order to make group performance successful, different types of awareness support as well as instructional design focused on groupwork are critical for online interactions (Dias & Borges,

1999; Kirsch-Pinheiro, Lima, & Borgers, 2003). Given that awareness is defined as "an understanding of the activities of others, which provide a context for your own activity" (Dourish & Bellotti, 1992, p. 1), any contextual information provided by instructional design and systems supports helps coordinate the collaborative work process. In other words, awareness can assist in building harmonious interactions by allowing learners to be aware of basic information such as what is going on, what the assigned task or learning goal is, and what and how they can work together in online environments. Importantly, a sense of awareness of others, of the assigned tasks or activities, and of the multiple communication modes available plays a crucial role in ordinary work processes. Given that the success of online collaboration is contingent on group member interactions in the learning process with support of awareness information, awareness should be a major concern when designing groupware tools and systems.

Up to now, system development approaches for supporting awareness in online learning environments have not addressed all of the dimensions of online collaboration. A key reason for these limitations is that most studies in this area have been approached from system development or technological perspectives, rather than from pedagogical ones. As a result, such studies have failed to explore the dynamic mechanisms involved in the online collaborative process.

In response, in this chapter, we discuss the importance of awareness as a way of supporting the process of online collaborative learning, and introduce some related previous studies. Second, we present a framework of awareness support for online collaboration. In detail, the framework describes how we can design awareness, both pedagogically and technologically, for online collaboration. At the same time, the conceptual framework illustrates how different technological applications can support pedagogical strategies that facilitate collaborative teaching and learning. From the standpoint of human factors research, this chapter suggests a set of focal points wherein pedagogical and technological supports may foster awareness within online collaboration.

LITERATURE REVIEW

Dimensions of Online Collaboration

Learning is truly a social endeavor. Learning typically takes place when students work in a group, though the size and duration of such a group drastically varies. Given that meaningful group learning experiences typically do not occur naturally, many researchers and scholars have suggested more intense investigations of issues related to creating meaningful group learning environments.

Broadly defined, group learning is an activity in which two or more people learn or attempt to learn something together (Dillenbourg, 1999). In a work setting, groupwork can be defined as an activity wherein learners pursue possible solutions related to a mutual problem (Roschelle & Teasley, 1995). Meanwhile, in an online environment, a "group" refers to "people with complementary competencies executing simultaneous, collaborative work processes through electronic media without regard to geographic location" (Chinowsky & Rojas, 2003, p. 98). In effect, the online group can be dispersed anywhere and can contact each other regardless of time and place.

In particular, in online learning environments, the value of groupwork lies as much in the process of learning as it does in the ultimate product(s). Accordingly, groupwork is a dynamic and complex force that generates outputs and products that are difficult for learners and instructors to predict. A variety of input variables in groupwork (e.g., expectations, experiences, cultural backgrounds, assigned tasks, etc.) influence one another and produce different learning results or outcomes. This dynamic learning process within a group becomes a part of the learning experience. Thus,

it is important that instructional designers and practitioners consider how to design learning environments in ways that support groupwork both pedagogically and technologically.

On the other hand, the features of online collaboration can be well described using the framework of online collaborative learning, which is comprised of three dimensions: (1) task, (2) social, and (3) technological dimensions (Carabajal, LaPointe, & Gunawardena, 2003). In face-to-face collaboration, much attention is made within the task dimension regarding integrating good strategies for effective task performance. Instructors and practitioners generally do not need to consider social and technological dimensions because simultaneous contact in the same space creates social climates naturally. Moreover, technological tools are minimally used, or at most, are supplemental devices in the support of collaborative work. However, due to geographic and time-related distances, online collaboration brings new situations and contexts to both instructors and learners. Thus, instructors need to recognize the multiple dimensions of online collaboration as well as the various roles needed to create successful online collaboration. Key aspects of the task, social, and technological dimensions in online collaboration (Lee, Bonk, Magjuka, Su, & Liu, in press) are summarized next.

• **Task dimension:** Group members accomplish group tasks through mutual engagement, idea negotiation, and the clarification of the key group goals or tasks. In the collaborative learning process, group members benefit from the group process through giving and receiving feedback and other assistance, exchanging information and resources, and challenging each other's thinking and reasoning patterns (Johnson & Johnson, 1996). This collaboration process fosters cognitive dissonance and conflict among group members, and, thus, encourages new knowledge construction. Due to the loss of explicit contextual cues in online environments, online collaborative groups may have more difficulty in reaching consensus than traditional residential groups. In this case, developing a shared mental model is important for motivating students to pursue high task performance. Setting mutual goals, achieving consensus, and providing groupwork guidelines are critical steps to assist students in effective group processes of completing a common task (Carabajal et al., 2003).

- **Social dimension:** The social dimension of online collaboration concerns the socio-emotional needs of group members and interpersonal relationships between group members. The richness of in-person communication comes from cues such as facial expression, eye contact, body gestures, and feelings of closeness. These types of cues often get lost or substantially reduced in online environments. Given the limited social cues, it is vital to explore various techniques to enhance the social dimension of groupwork in online environments. In addition, a rich social atmosphere helps minimize feelings of isolation in an online learning context, and cultivates a sense of community. Wegerif (1998), for instance, points out that group members are prone to be defensive, nervous, and reluctant to collaborate unless a sense of community is present. Social communications also promote mutual understandings and help build group norms. Such group norms often lead to group discipline which is especially important in online collaboration for enhancing group effectiveness and efficiency. More importantly, social engagement helps build trust among group members, which is essential for providing a social context that is conducive to virtual group collaboration.
- **Technological dimension:** The technological dimension is another distinct feature of

online collaborative groups. Collaborative technologies often require the modification and transformation of traditional pedagogical approaches by both online instructors and students to new patterns of communication and learning. The appropriate choices of technologies as well as effective uses of them have important consequences for the quality and depth of online collaborative group processing and task performances (Carabajal et al., 2003; Duarte & Snyder, 1999). With a plethora of technologies currently available for online collaborative group members, the importance of identifying technologies that optimize the performance of online groups cannot be overstated (Durate & Snyder, 1999). One critical factor to consider in terms of the effective use of technologies is its potential to facilitate a shared dialogue and information exchange among group members (Carabajal et al., 2003). Technologies, when used appropriately, can also facilitate socio-emotional well-being of the group entity as a whole as well as increase the social awareness among group members or participants that enhances the effectiveness of collaboration and coordination among instructors and students (Durate & Snyder, 1999).

Awareness in Online Collaboration

While awareness is a widely-known concept in the fields of computer science and human computer interactions, many researchers have pointed out that there is no overall comprehensive depiction of awareness in the prevailing research literature (Gutwin, 1997). Schmidt (2002) claims that awareness is perhaps not really a distinct concept due to its diverse and sometimes even contradictory definitions and usage.

Awareness can be defined in several ways, but, as noted earlier, awareness is generally considered as an understanding of the activities of others and their learning context (Dourish & Bellotti, 1992). This definition underscores that individual contributions are important to the group activity as a whole, and to assessing individual actions with regard to group performance.

The concept of awareness has recently gained increasing attention among researchers in the study of computer-supported environments, a field that considers awareness crucial for creating effective online collaboration. In a shared face-to-face workspace, effective groups naturally develop a shared knowledge and practice through group members continuously checking on the actions of other people as well as the evolution of group products. Such a shared mental model of a group requires minimal conscious effort (Gutwin & Greenberg, 1999). However, in online environments, knowledge of others is not available unless the system makes it explicit. Knowledge sharing is the key to group interaction. Lack of this awareness when collaborating online (e.g., awareness of group members and their associated skills, collaboration tasks and goals, role within the group, group progress, resources available, and knowledge gaps) creates continuous breakdowns in the flow of knowledge, and it has a negative impact on learning (Prasolova-Førland & Divitini, 2002).

Several researchers, in fact, have identified awareness support as a factor in effective group performance (Kirsch-Pinheiro et al., 2003). They argue that awareness can be a key mediator for effective group communication and coordination. Awareness can assist in simplifying the need for verbal communication, in assisting with coordination, in providing a context for assistance and anticipation, and generally in helping people recognize opportunities for getting closer as a team or workgroup (Gutwin, 1997).

Although the concept of awareness has been widely used in designing groupware and information systems, the use of such design concepts is rare in existing educational collaborative tools and systems. It is not an exaggeration to say that currently popular course management systems do not adequately maintain awareness to support group process and performance. In response, in this chapter, we extend the concept of awareness in the context of online collaborative learning environments, and, thus, propose a framework of awareness design to support online collaboration.

HOW TO ENHANCE AWARENESS IN ONLINE COLLABORATION

In the above sections, we have pointed out that in online collaborative environments, learners should recognize what is happening and know how to act and respond appropriately. In this section, we propose a framework of awareness design in online collaborative learning environments from the perspective of task, social, and technological dimensions of collaboration. First, we present how previous studies address awareness design from these three dimensions of online collaboration. Second, we address key elements of knowledge that constitute awareness, which support three different dimensions of online collaboration. The key elements of knowledge will be addressed with four "w" words and one "h" word: who, what, where, when, and how. Finally, based on the implications of previous studies and key knowledge elements of awareness design, we propose a substantial number of pedagogical examples and technological features for enriched awareness support.

Awareness Design for Task Dimension of Online Collaboration

Implications from Literature Review

A learning task is designed to achieve specific learning goals. A successful collaborative learning process involves setting task goals, coordinating efforts, and evaluating the progress of collaboration (Carroll, Neale, Isenhour, Rosson, & Mc-Crickard, 2003). The task dimension of online collaboration requires learners to be aware of task goals and structure as well as other people's activities in relation to progress toward task goals. Such understanding is essential for group members to maintain the awareness of overall situations to accomplish common collaboration goals.

The literature has identified several characteristics of awareness including four categories of task awareness. First, task-oriented collaborations operate on shared goals and plans. Well conceptualized goals stimulate effective strategies, timelines, and collective activities for effective group performance (Marks, Mathieu, & Zaccaro, 2001). In online environments, it is critical that all members of a group have a sense of goal awareness, which, according to Endsley (1995), is a shared understanding of the goal and plans.

Second, a complex project involves a variety of activities taking place in different time periods. According to research on activity awareness, knowledge about the task related activities of other group members (Jang, Steinfield, & Pfaff, 2002) falls under the rubric of task awareness. To coordinate with other members of the group in a learning task, it is important to be aware of other people's situations. Different types of collaboration tasks require task-specific awareness. In real time collaboration situations, when several learners operate simultaneously on a learning task, they need detailed information about their group members' activities. In a long-term task, when group members' actions are interdependent, they usually need to be aware of the progress, completions, and actions of others (Gutwin, 1997).

Third, to simply learn the status of others' activities may not be enough for effective coordination. In addition to knowing about group member activities, one must be aware of how others in the group perceive the general situation (Endsley, 1995) and interpret particular situations and activities. To enhance such awareness, a group needs regular communicative actions to support perspective awareness.

Fourth, another aspect of task awareness is to understand the resources that are available or potentially available for online collaboration (Espinosa, Cadiz, Rico-Gutierrez, Kraut, Scherlis, & Lautenbacher, 2000). Whether in its initial planning stage or group progress stage, the process of monitoring available resources critical to the tasks, along with effective communication of this information, leads to situational awareness (Marks et al., 2001).

Key Knowledge Elements for Task Dimension

Based on the awareness literature, we propose that task awareness for online collaboration should reflect the nature of the group task, awareness of group progress, awareness of group perceptions, and awareness of resources. Awareness within the task dimension can revolve around the four "w" and one "h" words: who, what, where, when, and how.

Awareness of the nature of group tasks. Group members need to build a shared mental model of the task goals and structure:

- What is the goal of the task?
- What is the structure of the task?
- What steps must we take to complete the task?

Awareness of group progress. Group members need to know the activity status of others and the overall situations of tasks in relation to the common goal:

- When do we have to finish the sub-goals and final goals of the task?
- How much have we done as a group?
- What more do we have to do?
- Who is doing what?

Awareness of group perceptions. Group members need to be aware of others' perceptions of group activities and group process:

- What are they thinking?
- Why do they think this way?

Awareness of resources. Group members need to have knowledge of available resources to complete their own tasks and assist in other members' tasks:

- What resources (tools/materials/skills/expertise) are needed to complete the task?
- What resources are available for the task?

Pedagogical Examples for Task Dimension

Listed next are a few pedagogical examples and possible groupware system features that support task awareness in online collaboration.

Setting mutual goals and expectations. Besides being encouraged to do initial goal-setting, online groups also need to have a codified document related to their overall goal statement and task structure. For example, in a groupware space, a graphical view of a project timelines that includes group deadlines for goal or sub-goal accomplishment can serve as a situational status reminder (Carroll et al., 2003).

- **Encouraging role assignment and task responsibilities.** In a collaborative learning task, the assigning of explicit roles to learners is an effective way of making learners aware of their task roles in the group. Groupware can use icons or avatars to indicate explicit roles that learners have been assigned (Gutwin, 1997).
- **Encouraging group discourse.** A group needs regular communicative actions to support perspective awareness. Regularly

scheduled online conversations or explicit comments regarding tasks or artifacts provide awareness of other group members' understandings and will contribute to a shared understanding of the group's purpose.

- **Providing work flow guidelines where a procedural map of task and sub-tasks is presented.** Monitoring groupwork flow is important to make effective collaboration. A workflow system that can present the tasks as a sequence of actions following a defined order is one example of a practice that keeps the group focused on task goal.
- **Reflecting group progress periodically.** Besides setting up regular meetings to update group progress and coordinate actions, many system features in groupware assist in awareness of group progress. To improve activity awareness, a system log of all tasks or activities is helpful for group members to understand what has happened (Jang et al., 2002). In addition, an activity notification system can send e-mail notices to the group to keep group activities up-to-date (Jang et al., 2002). A chronological visualization of project artifact histories and evolution is helpful to catch group members up with group progress on different fronts (Carroll et al., 2003). A simple calendar in a groupware system that displays the critical dates and provides detailed descriptions of the events might also serve the purpose of process awareness and assist in task progress (Jang et al., 2002).
- **Getting and sharing available resources.** A resource analysis should be conducted at the initial stage of groupwork. Group members are encouraged to share their expertise and personal resources that contribute to the group goals. Groups need to keep up-to-date on the availability of new information resources as well as emerging themes or patterns within those resources. A desig-

Table 1. Awareness design for task dimension of online collaboration

Key Elements	Pedagogical Examples	Technological Features
Awareness of the nature of group tasks • What is the goal of the task? • What is the structure of the task? • What steps must we take to complete the task? *Awareness of group progress* • When do we have to finish the sub-goals and final goals of the task? • How much have we done as a group? • What more do we have to do? • Who is doing what? *Awareness of group perceptions* • What are they thinking? • Why do they think this way? *Awareness of resources* • What resources (tools/materials/skills/ expertise) are needed to complete the task? • What resources are available for the task?	• Setting mutual goals and expectations • Encouraging role assignment and division of responsibilities • Encouraging cognitive discourse • Providing work flow guidelines where a procedural map of task and sub-tasks is presented • Reflecting on group progress periodically • Getting and sharing available resources	• Calendar • Graphical view of project timelines • Avatars or icons for role structure • Reflection note for group progress • Activity notification • Activity log • Electronic workflow guidelines • Group note pad • Private groupwork space or folders

nated group folder or private groupwork space in groupware facilitates the sharing of resources within the group.

A summary of awareness design for the task dimension in online collaboration is shown in Table 1.

Awareness Design for Social Dimension of Online Collaboration

Implications from Literature Review

In the context of collaborative learning, awareness support for the social dimension is defined as "the awareness that students have about the social connections within the group" (Gutwin, Stark, & Greenberg, 1995, p. 2). Several types of awareness that have been identified in the literature can be categorized under the larger umbrella of social awareness support. For example, informal awareness, which is concerned with who is around and whether they are available, is a part of social awareness (Gutwin, 1997). On the other hand, some scholars (Caselles, François, Metcalf, Ossimitz, & Stallinger, 2000) point out that social awareness should not merely be concerned with superficial awareness such as the existence of another entity. Non-superficial awareness that involves understanding and trust should also be a concern. Non-superficial social awareness is crucial to nurturing a context that is conducive to collaboration for group members where differences and conflicts can be resolved through mutual understanding and trust. This point can be further supported by social-cultural theory that highlights the importance of member relationship mutuality and the significance of interaction in collaborative learning (Sfard, 1998). To provide a comprehensive picture of social awareness, we can discuss it in relation to the emergence of online communities, group norms, and member trust.

Key Knowledge Elements for Social Dimension

Awareness support for the social dimension reflects the knowledge about a sense of learning community, group norms, and member trust. Similar to other kinds of awareness, social awareness also can revolve around the four "w" and one "h" word(s) mentioned earlier.

Awareness of a sense of community. To minimize feelings of isolation, members need to sense the existence of other members and need a sense of belonging to their group.

- Who is around?
- Who are they?
- How much do I know them?
- How much do they know me?
- How can I get in touch with them?
- When are they available?

Awareness about group norms. To be efficient, group members need to follow the explicit and implicit rules that have evolved over the period of collaboration.

- What are the appropriate ways to approach the group?
- What role should I take in this group?
- What roles will other members of the group assume?
- When/where is the appropriate time/location to work with them?
- How should I interact with other group members?

Awareness about member trust. To be effective, groups should demonstrate high levels of trust among members to decrease possible misunderstandings and unnecessary conflicts.

- What do I expect from my group members?

- What do they expect from me?
- How comfortable do I feel when I communicate with others?
- How comfortable do they feel when they communicate with me?
- How much do my group members trust me?

Pedagogical Examples for Social Dimension

Enhancing social awareness in the context of online collaborative learning can be approached from both technological and pedagogical aspects. Some pedagogical examples to support social awareness in online collaborations are listed below.

- **Encouraging discourse among group members:** A number of tactics can encourage discourse in virtual collaboration. For instance, instructors can read online discussions once in a while to provide feedback or even grade the quality and quantity of the discussions. In-depth and frequent communications among members definitely contribute to their mutual understanding.
- **Setting up regular communication patterns:** One approach to online collaborative learning is that group members attempt to minimize the time that they spend on communication since most of them have full-time jobs or other responsibilities. Asking group members to establish a regular communication pattern from the beginning can help establish a routine that will facilitate effective and perhaps seamless group communication.
- **Establishing and clarifying group norms:** Making the group rules clear—especially clarifying the implicit ones—can help members predict others' actions, and, thus, help support awareness.
- **Encouraging role negotiations:** Asking group members to assign roles during col-

Table 2. Awareness design for the social dimension of online collaboration

Key Elements	Pedagogical Examples	Technological Features
Awareness of a sense of community • Who is around? • Who are they? • How much do we know each other • When and whether are they available? *Awareness about group norms* • What are the appropriate ways to approach the group? • What roles should I and others play? • When/where is the appropriate time/location to work with other group members? • How should I interact with them? *Awareness about member trust* • What do group members expect from each other? • How comfortable are we to communicate with each other? • How much do my group members trust me?	• Encouraging social discourse among group members • Setting up regular communication patterns • Establishing and clarifying group norms • Encouraging role negotiations • Arranging peer teaching and peer evaluations • Arranging group reflection opportunities	• Threaded discussion spaces • Weekly logbook • Displaying multiple screens during synchronous collaboration • Virtual meeting place such as a "coffee house" where learners can bump into each other like in real life situations • Build-in artifacts that reflect member identity and activities that help trace collaboration patterns • Annotation and brainstorming tools, desktop videoconferencing, electronic whiteboards, Knowledge management portals and tools, mentoring exchange systems, translation tools, virtual classrooms or online presentation tools (Bonk, Wisher, & Nigrelli, 2004)

laboration can help clarify who is supposed to do what.

- **Arranging peer teaching and peer evaluations:** Asking group members to provide feedback and generally assist each other will create communication opportunities through which members can deepen their mutual understanding.
- **Arranging group reflection opportunities:** To enhance social cohesion and member trust, instructors may provide group reflection activities so that misunderstandings and conflicts can be discussed and solved among group members.

A summary of awareness design for the social dimension in online collaboration is shown in Table 2.

Awareness Design for Technological Dimension

Implications from Literature Review

Awareness support for the technological dimension of online collaboration links the task and social dimensions of online collaboration. Technological awareness allows effective collaboration. Minimal research has considered awareness support from the technological dimension, when compared to the attention paid to supporting awareness from both the task and social dimension standpoints.

In the context of collaborative learning, we define technological awareness through two aspects. Not surprisingly, the first part is the general knowledge of technology tools used in groupwork. The other key part is workspace awareness, defined as up-to-date knowledge about other learners' interactions with the shared technological workspace, including past, current, and future traces of activity in the groupware (Gutwin, 1997). While knowledge of online pedagogy is vital, the effectiveness of online collaborative environments greatly depends on how technological tools are applied for learning and how learners use them to make their groupwork meaningful. The general knowledge of technological tools is a critical aspect for group effectiveness but is often neglected in practice. For example, group members may have failed to take advantage of the potential interactive tools in a groupware system due to insufficient orientation sessions related to those tools. Technological tools should function as cognitive tools (Gao, Baylor, & Shen, 2005; Jonassen & Carr, 2000; Teasley & Roschelle, 1993), rather than as just communication or message delivery tools. This cognitive tool focus implies that the technological dimension needs to help expand learners' capacity to create, filter, share, and represent collective knowledge.

In addition, in online learning environments, a group workspace is sometimes the only central space that links different task activities of group members. The artifacts in a group space represent the shared repertoire of the community. Group members leave traces of their actions on the artifacts, reflecting various activities and resource ownerships (Prasolva-Førland & Divitini, 2002). The awareness of others' interactions within a group space not only facilitates effective coordination in the space but also facilitates task and social awareness for collective goals.

Key Knowledge Elements for Technological Dimension

Key knowledge of technological awareness covers awareness of general technology tool use and awareness of the groupwork space. Similar to other types of awareness, technological awareness cues can revolve around the key "w" and "h" words noted earlier.

Awareness of general technological tool uses. Group members need to have a good understanding of different tools for group communications.

- What technologies are available for the group?
- How do they function?
- How much time and energy do they require to learn and to use?
- How compatible are the available technologies that I have access to with the technologies of others in my group?

Awareness of groupwork space. To foster effective use of tools for collaboration, group members need to be aware of the status of others' interactions with the technological space.

- Where have they been in the workspace?
- What have they already done in the workspace?
- Where are they now?
- What are they doing in the workspace?
- Are they online or off-line at this moment?

Pedagogical Examples for Technological Dimension

Some pedagogical examples to support technological awareness in online collaborations are listed next.

- **Understanding what tools are available:** Learners need to be aware of the facility or system that they have for online groupwork. In effect, they should be aware of what technological tools and features are available for groupwork, including hardware and software. For instance, is there a

Table 3. Awareness design for the technological dimension of online collaboration

Key Elements	Pedagogical Examples	Technological Features
Awareness of general technological tool uses • What technologies are available for the group? • How do they function? • How much time and energy do they require to learn and to use? • How much are my technologies compatible with others? *Awareness of group workspace* • Where have they been? • What have they already done in the workspace? • Where are they? • What are they doing in the workspace? • Are they online or offline at this moment?	• Understanding what tools are available • Orienting to features and functions of tools and how to get support to use them • Understanding the usefulness of tools for certain groupwork • Matching appropriate tools with groupwork • Making a joint decision on types of tools used for group communication • Keeping group members aware of actions in groupwork space	• Tutorials • Tool use guidelines • Activity notification • Activity log • Annotations • Private groupwork space/folders

group workspace or a place to post shared documents? Are there threaded discussions or synchronous chat tools?

- **Orientating features/functions of tools and how to get support for their use:** One concern raised by online educators is that many learners do not know about the current technological tools that are embedded in course management systems or groupware systems, their pedagogical purpose or potential, how much time and energy they require to learn and to use, and whether different technologies are compatible with others. This lack of technology knowledge blocks learners' effective use of online collaborative environments. As a result, many learners need to acquire this knowledge while collaborating online. Instructor orienting tasks and activities as well as available system supports are highly valuable in fostering this knowledge and awareness as well as general system use.
- **Understanding the usefulness of tools for certain groupwork:** Setting expecta-

tions for technology requirements up front can help learners to have an idea of how to use given technological tools for their groupwork. One suggestion is to list relevant technologies with their pros and cons to help learners understand the usefulness of tools.

- **Matching appropriate tools with groupwork:** Knowledge of how to use technological tools, when to use them, and for what purposes they should be used, will help determine the effectiveness of online collaboration.
- **Making a joint decision on types of tools used for group communication:** Group members need to discuss and reach a consensus on the types of tools used in the group and be consistent in their use.
- **Keeping group members aware of actions in groupwork space:** Group members should make it explicit when their actions in the groupwork space could impact others' activities. Some groupware features help support this awareness. Activity notifica-

tions and logs in a groupware system or courseware usually track the changes within the artifacts and provide a history of actions in the groupware space. Data mining tools can help visualize such activity patterns and usage statistics, thereby providing more insights on group member participation and functioning.

A summary of awareness design for the technological dimension in online collaboration is shown in Table 3.

SUMMARY

Collaboration is not simply a collection of individual work. Optimal collaboration pursues mutual engagement for problem solving, rather than just communication transmissions or simple message delivery. Collaboration in online learning environments requires that learners have an appropriate sense of understanding of themselves, their peers, the required activities, and other important situational context information.

This chapter revisits the concepts of "awareness," which has been studied among computer science and is now being introduced into vari-

ous educational fields. In this chapter, we have examined the importance of awareness in online collaboration and presented an awareness framework from a pedagogical perspective, rather than a groupware development perspective. In particular, we addressed how to support awareness from the standpoint of the task, social, and technological dimensions of collaboration.

Yet, there are several concerns to be considered. First, though task awareness has been proposed as one important principle for designing effective groupware systems to support collaboration and coordination, several issues also have been raised regarding embedding features to support task awareness. For example, information overload is a well-known phenomenon resulting from over-designed task awareness tools and features (Kirsch-Pinheiro et al., 2003). When task awareness is not designed to accommodate individual needs, awareness information can also be extremely distracting and harmful to individual activities. On the other hand, the design of awareness tools should be complementary to other communication and coordination tools preferred by collaborators. Without such considerations, awareness tools may be redundant and need extra effort from collaborators in using awareness supports (Jang et al., 2002).

Second, individuals working in a group inevitably use other members' visible or predictable actions to pace and plan their own actions (Ackerman & Starr, 1995). Therefore, supporting social awareness can help group members to understand their overall situations better and make more suitable decisions about their own actions. However, the support of social awareness itself does not guarantee positive effects in group collaboration. Building social awareness definitely is a precondition for creating cohesive relationships among students. At the same time, providing social awareness activities or tasks with learners during online groupwork can be another burden for learners to address or take care of in addition to conducting task assignments.

Figure 1. Relationship of three dimensions for awareness support in online collaboration

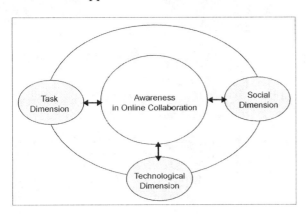

Careful determination of what social awareness information to provide and appropriately timing its delivery during virtual teaming or different stages of group collaboration would significantly impact task performance.

Third, technological awareness is definitely the most salient and probably the most significant portion of awareness in online collaboration. Still, support for technological awareness has received scant attention compared with other aspects of awareness.

As shown in Figure 1, awareness support from task, social, and technological dimensions positively connects to the effectiveness of online collaboration. Considering three dimensions are closely interrelated with one another, online educational practitioners, instructional designers, and instructors should consider balancing support and embedded cues when designing awareness in online learning environments.

In conclusion, the current groupware tools and course management systems provide some pedagogical examples and technological features for awareness support. The pedagogical examples and technological features discussed in this chapter contribute to online learning environments functioning not just as virtual spaces but also as cognitive tools. As online groupware and Web collaboration continue to emerge and educational theory and use related to such tools simultaneously evolves, there will be a plethora of interesting research and development avenues to enhance online collaborative group functioning and success. In addition, there will be interesting windows on previously hidden or unknown aspects to human learning and social interactions as well as countless opportunities for research that explores these innovative forms of group functioning and human learning and performance. These are exciting times for becoming aware of how online collaborative groups function, negotiate roles, foster new insights, and ultimately succeed or fail. It is definitely time for those in education to become more aware of online collaboration awareness.

ACKNOWLEDGMENT

This work was supported by Korea Research Foundation Grant (KRF-2003-037-B00071).

REFERENCES

Ackerman, M., & Starr, B. (1995). Social activity indicators: Interface components for CSCW systems. In *Proceedings of the 8th ACM Symposium on User Interface Software and Technology (UIST'95)* (pp. 159-168). Retrieved April 16, 2006, from http://portal.acm.org/citation.cfm?id=215969&col=GUIDE&dl=ACM&CFID=69623 356&CFTOKEN=139753 &ret=1#Fulltext

Bonk, C. J., Wisher, R. A., & Nigrelli, M. L. (2004). Learning communities, communities of practice: Principles, technologies, and examples. In K. Littleton, D. Miell, & D. Faulkner (Eds.), *Learning to collaborate, collaborating to learn* (pp. 199-219). Hauppauge, NY: Nova Science Publishers.

Carabajal, K., LaPointe, D., & Gunawardena, C. N. (2003). Group development in online learning communities. In M. G. Moore, & W. G. Anderson (Eds.), *Handbook of distance education* (pp. 217-234). Mahwah, NJ: Lawrence Erlbaum Associates.

Carroll, J. M., Neale, D. C., Isenhour, P. L., Rosson, M. B., & McCrickard, D. S. (2003). Notification and awareness: Synchronizing task-oriented collaborative activity. *International Journal of Human-Computer Studies, 58*(5), 605-632.

Caselles, A., François, C., Metcalf, G., Ossimitz, G., & Stallinger, F. (2000). Awareness and social systems. In G. Chroust & C. Hofer (Eds.), *Social systems and the future* (pp. 31-42). Österreichische Studiengesellschaft für Kybernetik, Reports of the Austrian Society for Cybernetic Studies, Vienna, Australia.

Chinowsky, P., & Rojas, E. (2003). Virtual teams: Guide to successful implementation. *Journal of Management in Engineering, 19*(3), 98-106.

Dias, M. S., & Borges, M. R. S. (1999). Development of groupware systems with the COPSE infrastructure. In *Proceedings of String Processing and Information Retrieval Symposium & International Workshop on Groupware* (pp. 278-285). Retrieved April 6, 2006, from http://csdl.computer.org/dl/proceedings/spire/1999/0268/00/02680278.pdf

Dillenbourg, P. (1999). What do you mean by collaborative learning? In P. Dillenbourg (Ed.), *Collaborative-learning: Cognitive and computational approaches* (pp. 1-19). Oxford: Elsevier.

Dourish, P., & Bellotti, V. (1992). Awareness and coordination in shared workspaces. In *Proceedings of the ACM Conference on Computer-Supported Cooperative Work (CSCW'92),* Toronto, Ontario (pp. 107-114). New York: ACM Press. Retrieved April 16, 2006, from http://portal.acm.org/citation.cfm?id=143468&coll=GUIDE&dl=GUIDE&CFID=73906796&CFTOKEN=24042431&ret=1#Fulltext

Duarte, D., & Snyder, N. (1999). *Mastering online collaborative groups: Strategies, tools and techniques that succeed.* San Francisco: Jossey-Bass.

Endsley, M. R. (1995). Toward a theory of situation awareness in dynamic systems. *Human Factors, 37*(1), 32-64.

Espinosa, A., Cadiz, J., Rico-Gutierrez, L., Kraut, R., Scherlis W., & Lautenbacher, G. (2000). Coming to the wrong decision quickly: Why awareness tools must be matched with appropriate tasks. In *Proceedings of the SIGCHI Conference on Human Factors in Computing Systems CHI 2000, 2(1)* (pp. 392-399). New York: ACM Press.

Gao, H., Baylor, A. L., & Shen, E. (2005). Designer support for online collaboration and knowledge construction. *Educational Technology & Society, 8*(1), 69-79.

Gutwin, C. (1997). *Workspace awareness in real-time distributed groupware.* Unpublished dissertation transcripts. University of Calgary, Alberta, Canada.

Gutwin, C., & Greenberg, S. (1999). *A framework of awareness for small group in shared-workspace groupware* (Tech. Rep. No. 99-1). Saskatoon, Canada: University of Saskatchewan, Department of computer science.

Gutwin, C., Stark, G., & Greenberg, S. (1995). Support for workspace awareness in educational groupware. In J. L. Schnase & E. L. Cunnius (Eds.), *Proceedings of Computer Supported Collaborative Learning (CSCL'95)* (pp. 147-156). Mahwah, NJ: Lawrence Erlbaum Associates.

Jang, C., Steinfield, C., & Pfaff, B. (2002). Virtual group awareness and groupware support: An evaluation of the GroupSCOPE system. *International Journal of Human-Computer Studies, 56*(1), 109-126.

Johnson, D. W., & Johnson, R. T. (1996). Cooperation and the use of technology. In D. H. Jonassen (Ed.), *Handbook of research for education communications and technology* (pp. 1017-1044). New York: Simon and Schuster Macmillan.

Jonassen, D. H., & Carr, C. S. (2000). Mindtools: Affording multiple knowledge representations for learning. In S. P. Lajoie (Ed.), *Computers as cognitive tools, volume II: No more walls: Theory change, paradigm shifts, and their influence on the use of computers for instructional purposes* (pp. 165-196). Mahwah, NJ: Lawrence Erlbaum Associates.

Kirsch-Pinheiro, M., Lima, J. V., & Borges, M. R. S. (2003). A framework for awareness support in groupware systems. *Computer in Industry, 52*(1), 47-57.

Kirschner, P., & Van Bruggen, J. (2004). Learning and understanding in virtual teams. *Cyberpsychology & Behavior, 7*(2), 135-139.

Lee, S., Bonk, C. J., Magjuka, R. J., Su, B., & Liu, X. (in press). Understanding the dimensions of virtual teams. *International Journal on E-learning.*

Marks, M. A., Mathieu, J. E., & Zaccaro, S. J. (2001). A temporally based framework and taxonomy of group processes. *Academy of Management Review, 26*(3), 356-376.

Prasolova-Førland, E. (2002). Supporting awareness in education: Overview and mechanisms. In *Proceedings of the International Conference on Engineering Education (ICEE 2002)* (Manchester, UK). MUIST. Retrieved April 16, 2006, from http://www.idi.ntnu.no/grupper/su/publ/ekaterina/ICEE2002.pdf

Prasolova-Førland, E., & Divitini, M. (2002). Supporting learning communities with collaborative virtual environments: Different spatial metaphors. In *Proceedings of the IEEE International Conference on Advanced Learning Technologies (ICALT 2002)* (Kazan, Russia). IEEE Press. Retrieved April 16, 2006, from http://lttf.ieee.org/icalt2002/proceedings/t605_Icalt114_End.pdf

Roschelle, J., & Teasley, S. (1995). The construction of shared knowledge in collaborative problem solving. In C. E. O'Malley (Ed.), *Computer supported collaborative learning* (pp. 69-97). Berlin: Springer-Verlag.

Schmidt, K. (2002). The problem with 'awareness': Introductory remarks on 'awareness in CSCW'. *Computer Supported Cooperative Work, 11*(3-4), 285-298.

Sfard, A. (1998). On two metaphors for learning and the dangers of choosing just one. *Educational Researcher, 27*(2), 4-13.

Teasley, S. D., & Roschelle, J. (1993). Construction a joint problem space: The computer as a tool for sharing knowledge. In S. P. Lajoie, & S. J. Derry (Eds.), *Computers as cognitive tools* (pp. 229-258). Hillsdale, NJ: Lawrence Erlbaum Associates.

Wegerif, R. (1998). The social dimension of asynchronous learning networks. *Journal of Asynchronous Learning Networks, 2*(1), 34-49. Retrieved September 4, 2005, from http://www.sloan-c.org/publications/jaln/v2n1/pdf/v2n1_wegerif.pdf

This work was previously published in Advances in Computer-Supported Learning, edited by F. Neto, pp. 251-273, copyright 2007 by Information Science Publishing (an imprint of IGI Global).

Chapter 8.7
On the Convergence of Formal Ontologies and Standardized E–Learning

Miguel-Ángel Sicilia
University of Alcalá, Spain

Elena García Barriocanal
University of Alcalá, Spain

ABSTRACT

Current efforts to standardize e-learning resources are centered on the notion of a learning object as a piece of content that can be reused in diverse educational contexts. Several specifications for the description of learning objects—converging in the LOM standard—have appeared in recent years, providing a common foundation for interoperability and shared semantics. At the same time, the Semantic Web vision has resulted in a number of technologies grounded in the availability of shared, consensual knowledge representations called ontologies. As proposed by several authors, ontologies can be used to provide a richer, logics-based framework for the expression of learning object metadata, resulting in the convergence of both streams of research towards a common objective. In this article, we address the practicalities of the representation of LOM metadata instances into formal ontologies, discussing the main technical and organizational issues that must be addressed for an effective integration of both technologies, and sketching some illustrative examples using modern ontology languages and a large knowledge base.

INTRODUCTION

The increasing interest in Web-enabled education and training (often referred to as e-learning) has fostered a growing interest in the definition of specifications and reference models for digital educational contents, in an effort to standardize them (Anido et al., 2002). The objectives of such standardization efforts include facilitating their interchange, their composition, and, ultimately, their mass customization (Martinez, 2001). At the same time, the vision of a Semantic Web (Berners-Lee, Hendler, & Lassila, 2001) has resulted in a renewed interest in the provision of shared, consensual knowledge representations (Davis, Shrobe, & Szolovits, 1993) for the annotation of documents, or, in a more general sense, of knowledge assets. The fact that e-learning and Semantic Web technologies somewhat intersect has been raised in several recent research papers—Hoermann et al., 2003; Sicilia and García, 2003a; and Stojanovic, Staab, and Studer, 2001—that focus on the convergence of two key concepts: learning objects and ontologies. On the one hand, learning objects are an approach to instructional design centered on the notion of reusability (Wiley, 2001) and understood as the capability for a digital content element (a learning object) to be used in several different learning situations, possibly in combination with other contents that were not originally designed for the same context (Sicilia & García, 2003b). Learning object metadata records are used to describe technical requirements, educational properties, and other kinds of information about learning objects, all of them in a standardized format, thanks to the availability of the IEEE LOM standard (IEEE, 2002) and several other conforming specifications. On the other hand, ontologies are logics-based consensual knowledge representations that are advocated as a means to annotate Web resources (or electronic resources in general) to provide them with semantic, machine understandable meaning, thus becoming enablers for knowledge management tools and processes (Fensel, 2002a). Consequently, it also provided that learning objects are a specific kind of digital resource (with an explicit instructional intention), ontology-based annotations are a candidate for expressing learning object metadata records as summarized in Sicilia and García, 2003a.

Since ontology formalisms are based on specialized logics (Baader et al., 2003) carefully designed for expressiveness and computational efficiency, they could be used to provide a richer (Kabel, 2001), semantic enabled computation framework for metadata-based e-learning. In fact, the use of formal ontologies to express metadata not only preserves current principles and practicalities that are applied to standardized metadata (Duval, Hodgins, Sutton, & Weibel, 2002), but also provides a richer framework for its realization and its subsequent use by automated tools. Principles like modularization and extensibility are addressed by the use of open XML-based formats prepared for the Web (Fensel, 2002a), while the principle of refinement is formally defined by the logical interpretation of subsumption. The practicalities of building application profiles for particular usages can be realized by means of constructing specialized ontologies from more general ones, adding specialized terms that restrict cardinalities, value spaces, or relationships among metadata entities. In addition, the satisfiability requirements of a given ontology can be used as a means to assess the completeness of metadata records.

But the integration of formal ontologies with the paradigm of learning objects poses several problems that have not been addressed yet. These problems can be categorized roughly in technical and organizational issues. Technical issues include the practicalities of expressing the structure, properties, and prospective contexts of the use of learning objects as description-logics expressions, and thus entail a notion of what a complete and consistent metadata record should be. Organizational issues are those that would eventually be caused by the adoption of ontology-based learning objects in organizations, as part of an integrated value process (Lytras, Pouloudi, & Poulymenakou, 2002b). These issues include the need for specific user interfaces, the provision of intraorganization conceptualizations coherent with shared ones, and new approaches for the assessment of the quality of metadata records.

This article is intended as an attempt to provide an initial framework for further research regarding the integration of ontology-based annotation in the practice of designing and describing learning objects, adhering to the LOM standard as the high-level reference model of common metadata elements. To do so, LOM metadata elements are examined from the viewpoint of expressing them in description logics, and surrounding practical issues that may interfere with the adoption of such an approach. The issues described here can be considered a specific aspect of the more general issue of the convergence of e-learning and knowledge management that has been analyzed elsewhere (Lytras, Pouloudi, & Poulymenakou, 2002; Maurer & Sapper, 2001).

The rest of this article is structured as follows. The second section describes related work and the motivation for the integration of formal ontologies with learning objects. The third section examines some of the principal technical and organizational issues that arise when trying to use formal ontologies to express LOM metadata elements with an educational intention. Specific metadata examples using the OpenCyc knowledge base and the W3C-OWL[1] language are provided as illustrations in the fourth section. Finally, conclusions and future research directions are sketched.

FORMAL ONTOLOGIES AS A LANGUAGE FOR LEARNING OBJECT METADATA

The use of logics-based languages for metadata description lies at the heart of the vision of a Semantic Web. But using such languages requires the understanding of specific annotation semantics or, alternatively, the provision of tools that hide these complexities to the average user by means of a carefully devised user interface. In addition, the semantics of current learning object metadata specifications are provided mainly through natural language explanation, so that the use of logic-based approaches requires some kind of support to provide a shared machine-understandable form to annotations. In this section, we review previous works that explicitly address the use of formal ontologies for the purpose of describing learning objects, and then, the main benefits of such approach are summarized.

Related Work

A recent literature review concerning the topic of this article can be found in Lytras, Tsilira, and Themistocleous (2003). Here we will comment only on some of the approaches that are relevant for our subsequent discussion. Stojanovic et al. (2001) provide a comprehensive list of high-level benefits of using Semantic Web technology for e-learning, and they also provide a structure made up of three ontologies that addresses the description of the topics of the learning material (*content*), the form of presentation (*context*), and the composition of learning materials (*structure*). In addition, they provide a number of concrete examples illustrating the use of ontologies to provide semantically richer metadata annotations. But the concepts in ontology are not mapped to the metadata structure of the existing specification, and the concept of a learning object is not made explicit, so that semantic convergence remains still open. The IMAT project (Kabel, 2001) uses an ontology, including syntactical, semantic, and educational elements to facilitate the edition of training materials. But despite the fact that many of the metadata items could be considered analogous to elements in current specifications, no attempt is made to map them. Project k-MED (Hoermann et al., 2003) provides a knowledge-based architecture that links LOM-described contents—in the so-called MediaBrickSpace—to concepts in an ontology called ConceptSpace. Consequently, it maintains two different description languages—the ontology for semantic description and plain LOM—to describe the rest of the aspects. Although such separation of concerns is perfectly compatible with current learning object specifications, there is no reason to restrict the use of the ontological language only to a part of metadata items. Since LOM clearly differentiates the physical representation (i.e., the syntax or representation language) from the semantic interpretation (i.e., the conventions about the meaning of each particular data item), ontology description languages can be used as a substitute for other representation formats like plain XML to take advantage of the richer logics-based structures described by Stojanovic et al. (2001).

An explicit representation of learning objects as instances inside an ontology is addressed by Sicilia and García (2003a) and Qin and Finneran (2002), but a comprehensive discussion of the LOM metadata is not provided. In addition, Nilsson, Palmér, and Naeve (2002) discuss several high-level issues for e-learning metadata, and Dhraief, Nejdl, and Wolf (2001) use the recently developed LOM-RDF binding[2] as the storage format for a repository of learning materials.

But despite the considerable number of projects using Semantic Web technologies to build e-learning related functionalities, the issue of integrating concrete learning content specifications with concrete ontology description languages still remains open. Such an undertaking will be required to come up with a *lingua franca* to semantic annotation of learning objects that represents a strong agreement between e-learning specifications and Semantic Web languages. More concretely, the LOM standard represents the less common denominator of agreement in learning contents description, so that its mapping to ontology languages can be considered the more urgent step in the convergence of these two initiatives.

Benefits of a Logic-Based Metadata Approach

The benefits of using ontology description languages to express learning object metadata are of two fundamental types. On the one hand, those languages provide richer knowledge representation formalisms (Davis, Shrobe & Szolovits, 1993) for metadata descriptions than using plain text, XML bindings, or even RDF. Here, the benefits are the result of using description logics instead of simply using structured data in XML format or using RDF, which is a less expressive language than DAML or OWL. On the other hand, the use of ontologies eventually may produce synergies with the technological advances that are taking place under the overall label of Semantic Web. The most prominent of such synergies may come from the availability of shared, consensual ontologies on many domains along with tools to develop systems that exploit them for diverse intelligent behaviors. But such infrastructure today are far from being a reality, since large ontology engineering efforts are only beginning to produce relevant results in terms of the open availability of large consensual conceptualizations, and little Semantic Web content is currently available (Benjamins et al., 2002). Consequently, nowadays a more realistic view (to which we adhere in this article) is that of focusing on the role that Web-enabled ontologies can play as a better representation language for the implementation of advanced learning object handling and delivery systems, both open or proprietary.

The use of ontologies for describing learning contents enables the extension of current specifications with additional relations and axioms among metadata items, without breaking their original semantics. For example, negative assertions about a learning object—like "this content is not appropriate for K-12"—could be represented through axioms, and the inverse of some relationships can be stated, as described in Stojanovic et al. (2001). In addition, the fact that annotations make reference to terms in large logical conceptualizations enables the construction of software agents that perform some kind of reasoning to initiate subsequent processes from fine-grained personalized information filtering to the overall scheduling of learning processes, given the knowledge gap of a business unit.

Ontologies can be used also for purposes that exceed the scope of the current LOM specification, although they are out of the scope of this article. For example, links between learning objects can be provided a specific and semantic-rich intention by the use of link-type ontologies (Sicilia et al., 2003a), and ontologies would eventually play a central role in the modeling of learner profiles (Dolog & Nejdl, 2003).

USING FORMAL ONTOLOGIES FOR THE ANNOTATION OF KNOWLEDGE ASSETS

Some of the practicalities of the integration of IEEE LOM metadata with formal ontologies have been addressed in some scattered research works described above, including Sicilia and García (2003a), and Hoermann et al. (2003). A draft RDF-LOM mapping is currently available, but a comprehensive account of the issues that arise when expressing LOM metadata elements through ontology description languages has not been provided. In addition, the organizational issues (derived from the technical requirements) that surround taking such approach to metadata as part of a knowledge management system have not been discussed yet. In this section, we address the most prominent of both kinds of issues. It should be noted that here we adopt an understanding of ontologies as a formal semantic account, represented through terminological logics, according to the clarifications provided by Lytras, Tsilira, and Themistocleous (2003).

Technical Issues

When looking for a way to integrate learning object descriptions into ontological support, the first issue to be solved is the logical form of learning object metadata. Assuming a conventional description logics database (Baader et al., 2003), there exists learning objects that should be treated as instances (also referred to as individuals or facts), so they are elements of the assertional A-box that are categorized by one or several terms in the terminological T-box. At a minimum, the term *learning object* should be used to denote the class of all resources, holding a mandatory, essential property storing its URI as a means to uniquely identify the resource, as described in Sicilia and García (2003a). This fact is apparently uncontroversial, but the rest of the technical apparatus is a matter of design and engineering. Here, we'll provide the results of our past experiences in engineering ontologies aimed at building learning systems, providing the rationale for each design decision.

From the just described basic structure, instances belonging to any existing ontology can be linked with assertions to any learning object, but in order to enable an uniform annotation, some form of standardized learning object predication ontology is required, defining a uniform semantic (or rhethoric) intention for each type of assertion involving an arbitrary ontology. Such approach is not new but has been used in diverse hypermedia research systems in the form of—taxonomies of link types—(Kopak, 1999), one of the most cited ones being the one crafted by Trigg (1986).

Figure 1 depicts the main components described so far, so that the learning object ontology would describe the structure (e.g., parts of a learning object) and main characteristics of the learning objects and their metadata, while the learning object predication ontology is used as a conventional means of linking learning objects with any arbitrary public or proprietary ontology. Some intentions are already present in some way in existing metadata standards (e.g., the classification-purpose LOM element enables the specification of prerequisites and learning objectives), but others can be borrowed from library science and hypermedia (e.g., pointers to refutations or analogies) to enhance the opportunities for subsequent intelligent behaviors based on explicit and well-defined semantic intentions, possibly bearing some degree of strength, as described by Sicilia, García, Aedo, and Díaz (2003a). Ontology language and interoperation protocol issues for establishing such links reasonably can be expected to be solved, thanks to the OWL effort (Fensel, 2002) and the emergence of Web services as an Internet-based communication means, including the ontological support for the description of the services themselves (Gaio, Lopes, & Botelho, 2003).

From a technical viewpoint, two additional ontologies are required to support learning process life cycles (Lytras, Pouloudi, & Poulymenakou, 2002) both for authors or knowledge providers and for learners. The first one should deal with learning processes (e.g., courses, occasional provision of support for work operations, or even other personal information like notes or commentaries). The second one should deal with the knowledge creation processes themselves, including, for example, descriptions of the stage in which each learning object is inside the cycle of an organization, like evaluation or initial annotation. The former ontology enables the tracking and exploitation of the actual learning experiences that have a place in the organization, while the latter provides support for the learning object lifecycle.

Figure 1 shows an example of a common subclass of learning objects labeled "exercises." Creating subclasses of the learning object class will eventually enhance the composition capabilities of software modules automating learning-object-based activities. Surprisingly, learning object typing has not been exploited so far in current metadata standards. Although specifications allows for declaring the kind of resource a given

learning object is, they neglect the interesting fact (possibly for simplicity's sake) that some metadata items can be exclusive of specific types of learning objects. For example, a type virtual reality interaction may be annotated with details of the required interaction appliances (e.g., helmet and gloves) and also with other characteristics like the requirement for conversation with avatars and other specific issues. Another more specific example may be a learning object acting as a surrogate for an individual (e.g., an instructor or expert), so that metadata descriptors may include information to physically contact him or her, or its timetable.

In addition, ontological support should consider the diverse characteristics of metadata elements, since they would call for different treatment in tools and processes of knowledge creation. Table 1 describes several possible and overlapping categories of metadata elements, according to their plausible interpretations, their degree of definiteness, their obligatory nature, and their complexity.

Many metadata elements, due to the nature of the information they are intended to capture, may allow for imperfect values. By imperfection, we mean any of the multiple facets of imprecision or uncertainty (Smets, 1997). Such imperfection may be unavoidable, or it may be convenient in certain situations. For example, the allowance for imprecise dates, durations, and/or judgments can be used to obtain more realistic estimations as described in Palomar, Sicilia, and García (2002). In addition, the meaning and usage of some metadata items may be subject to several distinct interpretations or variants of a single one. A simple example is that of *Description*, which may be provided in two machine-undistinguishable fashions, as stated in the standard: "This description need not be in language and terms appropriate for the users of the learning object being described." Finally, metadata items vary in their data complexity, so that some of them may allow for values of arbitrary complexity (i.e., formulas

Figure 1. Main ontological partitions and some examples

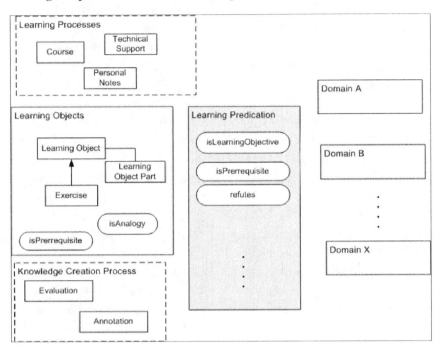

Table 1. Possible categories of existing metadata elements

Category	Description	LOM Examples
Precise	Element whose values are free of indefiniteness or uncertainty.	*Identifier* [1.1] that allows the specification of unique labels.
Imprecise	Elements that allow for some degree of vagueness, imprecision or any other kind of information imperfection.	Regions and historical periods [1.6] have not sharp boundaries, and linguistic values like "medium" and "low", as used in *Interactivity Level* [5.3] and others are inherently vague.
Unambiguous	Elements for which a single interpretation is normally used.	*Lyfecycle* [2] provides a description of the history of changes of the learning object.
Interpretable	Elements that admit more than one interpretation or significant variants of a single one.	*Description* [1.4] can be targeted to learners but also to learning experience designers.
Straightforward	Simple data types.	All of the items in the *General* category.
Complex	Structures of arbitrary complexity and level of composition.	*Structure* [1.7] denotes the "shape" of the learning object, thus determining a kind of information structure.

in description logics that require the appropriate parsing and handling software).

Some cases that arise in practice point out the necessity of crossing the A-box and T-box ontological levels by linking an instance to a given term, which cannot be accomplished with some of the current metadata editors. The most common example is that of providing a metadata item that links a learning object with its topic, so that the topic is a term and not a concrete instance of the term. A possible compromise solution for this situation—used in the application described in Sicilia, García, Aedo, and Díaz (2003b)—is that of reifying terms as special instances so that they can be linked at the assertion level, but it would be preferable to use an ontology language that allows for that kind of meta-level crossing situations.

As a consequence of the technical issues discussed, the description of ontologies for learning processes should encompass a wide range of modeling constructs that have not been put together to date and would require a laborious and effort-consuming standardization process, more ambitious in depth and coverage than current approaches to metadata (Duval, Hodgins, Sutton, & Weibel, 2002). Some examples of those constructs will be given in the examples provided in the next section of this chapter.

Organizational Issues

For this approach to become a reality in organizations, new tools and practices targeted to knowledge workers would be required. A recent study (Pagés et al., 2003) points out that current learning object annotation practices, as done in prominent repositories, mainly produces unstructured annotations (i.e., free text oriented towards human reading) so that structured metadata items are often short and incomplete, being most of the significant information in free textual form. As has been sketched previously, the annotation of learning objects through ontologies requires a more disciplined approach than that of typing some descriptive specification-conformant fields and providing textual descriptions intended for the human reader. In order to enhance metadata annotation practices, new editing tools are required that provide a convenient usability explicitly targeted to knowledge production tasks, automating as many of the aspects of such process as possible. The *Semantics for Learning* tool described by—Lytras, Pouloudi, and Poulymenakou (2002b) is an example of these kinds of novel tools. In addition, new ontology-based search processes would be required to enable efficient annotation practices, so that research relevant to ontology-based querying like Sicilia et al. (2003b) and García and Sicilia (2003)—is also relevant for this purpose. Without intending to be exhaustive, Table 2 provides a list of tool features that should be made available for an efficient and effective interaction.

But tools alone are not enough to guarantee that properly annotated learning becomes available. New practices and shared guidelines would be required for the sake of uniformity and reus-

Table 2. Features desirable for ontology-based learning object annotation tools

Feature	Description
Interoperability	Able of discovering and working with ontology repositories and also internal ontology databases, by means of standardized languages and protocols. The *Annotea* project (Kahan *et al.*, 2002) represents an important step in that direction.
Search	Ontology-based metadata creation processes are ontology-search intensive processes, so that this kind of functionality should be given a special research priority. Relevant work includes (Sicilia *et al.*, 2003b; García & Sicilia, 2003; Papazoglou, Proper, & Yang, 2001).
Personalization	Due to the expected large size of many ontologies, personalization would be needed to make more efficient the interaction with the tool, e.g. by providing direct access to the user's most frequent used terms or adapting to the style of each individual. Existing research on adaptive hypermedia (Brusilovsky, 2001) can be applied to that feature.
Collaboration	Knowledge creation processes can be made more effective in cooperative or collaborative settings. An example of such mechanisms can be found in (Dodero, Aedo & Díaz, 2002)

ability. This entails novel paradigms for learning object descriptions that are able to overcome the neutrality of current specifications with respect to how metadata must be created and what are the required, recommended, and optional fields. Some recent work in that direction has borrowed techniques from the software engineering community (Sánchez & Sicilia, 2003; Sicilia & Sánchez, 2003) in an attempt to provide stricter metadata languages oriented toward machine-understandability. From the viewpoint of organizations, this would entail managerial implications similar to those described by Anderson (2001), since a significant increment in knowledge-worker effort is to be expected due to the increasing demand for meta-information.

In addition, since the adoption of ontology-based learning objects eventually requires more disciplined annotation practices, more research would be needed regarding quality standards for learning objects. Quality notions should include reusability as a key aspect that is intimately connected with the quality and accuracy of metadata records, as described by Sicilia and García (2003b). In this way, the provision of metadata can be considered a measurable value-adding process that enhances the usability and reusability of learning objects by specifying a semantically rich context enabling any form of automated or manual knowledge management process.

Last, but not least, the vision of ontology-enabled learning objects depicted thus far requires a previous investment in engineering and the availability of shared, consensual ontologies that would eventually serve as a basis for interorganization integration. Current ontology repositories like DAML[3] are insufficient, both in diversity and coverage, and consortia and institutions are only beginning to standardize their terminologies. In addition, the consensus approach to create ontologies can be considered controversial due to the amount of coordinated effort required and the need to reconcile multiple views into a single one. In any case, while organizations wait for those

shared conceptualizations to be available—intra-organizational ontologies—based on current standards to the larger possible extent—appear as the solution for internal knowledge management processes. But again, this entails a large amount of resources that should be properly forecast and financed.

EXAMPLES OF INTEGRATING IEEE LOM WITH FORMAL ONTOLOGIES

The issues discussed must be translated to concrete implementations to prove the feasibility and usefulness of using ontologies to express learning object metadata. In what follows, we briefly summarize concrete case studies as representative examples of typical architectures of software systems supporting LOM inside an ontological structure.

Integrating LOM in OWL-RDF Applications

LOM categories (e.g., General, Lifecycle, Meta-Metadata, Technical, Educational, Rights, Relation, Annotation, and Classification) are simply groups of related metadata elements, so that they can be mapped to namespaces, just as it is done in the LOM-RDF mapping. The rest of the elements require an explicit design as classes and properties with their associated restrictions. In what follows, some examples of modeling LOM as an OWL-RDF ontology are provided. To do so, the terminology of the OILEd[4] ontology editor is used, since it was used to formalize the entire LOM specification.

Assuming that the learning object class described above is defined, a number of LOM items can be modeled as simple *datatype* properties, or as properties connected to structures that, in turn, are simply datatype property holders. For example, the title element (number 1.2 in LOM)

Equation 1.

```
<owl:Class rdf:about="http://www.uah.es/lo#Learning_Object">
  <rdfs:subClassOf>
    < owl:Restriction>
        <owl:onProperty rdf:resource=" http://www.uah.es/LOM_1#identifier"/>
        <owl:allValuesFrom>
           <owl:Class rdf:about=" http://www.uah.es/LOM_1#Identifier"/>
        </owl:allValuesFrom>
     </owl:Restriction>
  </r dfs:subClassOf>
  <rdfs:subClassOf>
      <owl:Restriction>
        <owl:onProperty rdf:resource=" http://www.uah.es/LOM_1#title"/>
        <owl:cardinality
rdf:datatype="http://www.w3.org/2001/XMLSchema#nonNegativeInteger">1
        </owl:cardinality>
        </owl:Restriction>
  </rdfs:subClassOf>
  ...
```

can be modeled as a cardinality restriction (only one title is allowed) of the corresponding string datatype property attached to the learning object class. The identifier element (1.1), which holds catalog/entry pairs, can be modeled by an additional identifier class with catalog and entry as mandatory properties. Description, keywords, and other simple elements can be modeled in a similar way. Equation 1, the OWL/RDF fragment, illustrates these simple mappings as subclass restrictions.

Elements like coverage and language can be modeled similarly as simple strings, but it is preferable to model languages, cultures, regions, and the like as instances of their respective classes, so they are connected to other terms, enabling reasoning on those contexts. For example, dialectal variants can be identified, and bordering regions can be retrieved to search for other related learning objects.

Other elements require more elaborate translations. This is the case of the structure element

(1.7) that describes the underlying organizational structure of the learning object. This value can be used to explicitly describe additional learning object structures. For example, hierarchical learning objects can be translated to the UML structure depicted in Figure 2, in which the property child in a learning object subclass can be used to build a tree of arbitrary depth and breadth. This formal structure, consisting of a property restriction, enables free composition of structures and also permits software to handle different structures differently. For example, the status of the learning object could be propagated to its constituents, and annotations of composites could be interpreted also for its parts.

The learning resource type also can be modeled by declaring subclasses of the learning object term. Once again, different types of objects can be provided with different properties and restrictions, informing potential software modules about specialized structures. The following is an example.

Figure 2. Terms for describing explicit learning object tree structures

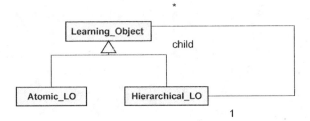

The classification category (9) provides in its purpose element a rich annotation means that somewhat represents a streamlined version of the learning object predication ontology showed in Figure 1. Here, the use of ontologies enhances the utility of classification, since complex logical formulas can be used for annotations—as described in Sicilia and García (2003a), to formulate composite prerequisites with or and and logical connectives—and taxonomies of competences and skills also can provide a context for the learning object. An example of such applications to human resource selection is provided in Sicilia, García, and Alcalde (2003). In addition, more detailed ontologies of link types can be used as a refined version of the list of purposes provided in LOM (Sicilia, García, Aedo, & Díaz, 2003a), also introducing the possibility of making links among learning objects as a generalization of the relation category (7).

Integrating LOM in OpenCyc

Cyc is a large knowledge base of common sense terms (Lenat, 1995), with over 100,000 atomic terms and an associated efficient inference engine using a variant of predicate calculus called CycL. OpenCyc[5] is the open source version of the Cyc knowledge base. Cyc is structured into locally consistent contexts called microtheories and provides tools to import an export to the DAML Semantic Web language.

When using a large knowledge base like Cyc, the first required task is that of locating relevant knowledge to locate our new definitions in connection with them. Consequently, we first should classify learning objects. They can be considered instances of ComputerFileCopy in the sense that at least when referenced inside a concrete system,

Equation 2.

```
(#$genls #$LearningObject #$ComputerFileCopy)
(#$isa #$ElQuijote #$LearningObject)
(#$argIsa #$lomLanguage 2 #$HumanLanguage)
(#$lomLanguage #$ElQuijote #$SpanishLanguage)
```

Equation 3.

```
(#$genls #$BiographicalDescription #$LearningObject)
(#$isa #$AznarBio #$BiographicalDescription)
(#$lomAbout #$AznarBio #$JoseMariaAznar)
(#$lomA   uthor #$MarianoRajoy #$AznarBio)
(#$holdsIn (#$ TimeIntervalInclusiveFn (#$YearFn 1996)(#$YearFn 2003))
(#$presidentOfCountry #$Spain #$JoseMariaAznar))
```

Equation 4.

```
(#$isa #$text1 #$LearningObject)
(#$isa #$text2 #$LearningObject)
(#$containsAnalogy #$text1 #$text2 #$high)
```

they are uniquely identifiable physical objects that contain information (hence an InformationBearingThing) readable by computers. Then, LOM metadata can be attached to LearningObjects as CycL sentences (see Equation 2).

Since constants like *HumanLanguage* and *SpanishLanguage* are provided by OpenCyc, assertions related to them are available to perform queries or any kind of knowledge base processing. Predications about learning objects can be done through CycL relations, only taking into account the pertinent arity and type restrictions, if any. For example, a type of learning object containing a biography can be represented as shown in Equation 3.

The above definitions allow a software module searching for an overview of contemporary Spanish history to locate Aznar's biography and put it in the context of his governing years. Other information may be used to decide to include it or not, or to combine it with other learning objects providing a contrasting perspective. In the example, the biography was written by one of Aznar's collaborators, and such fact could be encoded in OpenCyc also, thanks to its extensive provision for asserting facts about geopolitical organizations.

In addition, using sentences as shown in Equation 4 creates pointers to other learning objects containing analogies with a given degree of similarity.

For modularity and efficiency reasons, a domain-specific microtheory for learning objects could be defined. Learning processes in general naturally belong to the *HumanActivitiesMt* microtheory that encompasses statements both about individual and organizational activities. As illustrated in the examples, OpenCyc provides a large knowledge base that is ready to be used by connecting learning objects with existing assertions.

CONCLUSION AND FUTURE RESEARCH

The use of formal ontologies to describe reusable learning objects provides a better support for the development of intelligent tools, since the semantics of ontology definition languages are richer than those of RDF and also of simple information structuring XML schemas. But widespread adoption of ontology-based learning objects would not come without a cost, since a significant effort and novel tools and metrics would be required to properly annotate an organization's knowledge assets; to a level of detail that enables their automated handling. This calls for further research in standardized practices and evaluation criteria that are not currently addressed by specifications and standards.

Here, we have outlined the translation and use of LOM metadata into formal semantic descriptions. Concretely, the use of modern Web-enabled ontology languages has been summarized, and an illustration of the benefits of the integration of learning object descriptions has been provided through OpenCyc examples. More comprehensive learning object specifications, including the description of learning processes, should be addressed in the future as part of a far-reaching research agenda that has been outlined elsewhere (Lytras, Tsilira, & Themistocleous, 2003).

Future research should follow two main directions. On the one hand, technical ontology engineering research should aim to develop richer and more flexible learning object metadata schemas and supporting run-time services. On the other hand, the managerial and human factors issues regarding the adoption of such technology should be studied, and appropriate tools and practices should be devised to facilitate their use.

Important ontology engineering issues include the integration of e-learning specifications with Semantic Web languages and with existing large knowledge bases. The integration with OpenCyc has become an important issue now that it is being considered for IEEE standardization as part of a shared upper ontology. Such integration would eventually come up with richer semantic interpretations enabling the expression of run-time constraints in LMS, converging with stricter specification approaches that have begun to appear recently, like the design by contract approach (Sánchez & Sicilia, 2003).

Related managerial issues include the effective and consistent use of ontologies in any kind of knowledge management process and their integration in internal processes of knowledge validation and review. Only through this kind of managerial support will organizations be able to come up with a knowledge-based infrastructure of enough quality to enable a new generation of tools and systems exploiting the advantage of richer representation formalisms. In order to achieve this goal, new knowledge-edition tools are required that integrate seamlessly with existing business applications and that place minimal burden on everyday activities while maintaining consistency. Further research in human-computer interaction in that context would be required to attain this goal.

REFERENCES

Anderson, E.G. (2001). Managing the impact of high market growth and learning on knowledge worker productivity and service quality. *European Journal of Operational Research, 134*(3), 508-524.

Anido, L.E., Fernández, M.J., Caeiro, M., Santos, J.M., Rodríguez, J.S., & Llamas, M. (2002). Educational metadata and brokerage for learning resources. *Computers & Education, 38*(4), 351-374.

Baader, F., Calvanese, D., McGuinness, D., Nardi, D., & Patel-Schneider, P. (eds.) (2003). *The description logic handbook: Theory, implementation and applications.* Cambridge.

Benjamins, V.R., Contreras, J., Corcho, O., & Gómez-Perez, A. (2002). Six challenges for the Semantic Web. *Proceedings of the 2002 International Semantic Web Conference.*

Berners-Lee, T., Hendler, J., & Lassila, O. (2001). The Semantic Web. *Scientific American, 284*(5), 34-43.

Brusilovsky, P. (2001). Adaptive hypermedia. *User Modeling and User Adapted Interaction: Ten Year Anniversary issue, 11*(1/2), 87-110.

Davis, R., Shrobe, H., & Szolovits, P. (1993). What is a knowledge representation? *AI Magazine, 14*(1), 17-33.

Dhraief, H., Nejdl, W., & Wolf, B. (2001). Open learning repositories and metadata modeling. *Proceedings of International Semantic Web Working Symposium.*

Dodero, J.M., Aedo, I., & Díaz, P. (2002). Participative knowledge production of learning objects for e-books. *The Electronic Library, 20*(4), 296-305.

Dolog, P., & Nejdl, W. (2003). Challenges and benefits of the Semantic Web for user modelling. *Proceedings of the Workshop on Adaptive Hypermedia and Adaptive Web-Based Systems.*

Duval, E., Hodgins, W., Sutton, S.A., & Weibel, S. (2002). Metadata principles and practicalities. *D-Lib Magazine, 8*(4).

Fensel, D. (2002a). Ontology-based knowledge management. *IEEE Computer, 35*(11), 56-59.

Fensel, D. (2002b). Language standardization for the Semantic Web: The long way from oIL to OWL. *Proceedings of the 4th International Workshop on Distributed Communities on the Web* (pp. 215-227).

Gaio, S., Lopes, A., & Botelho, L. (2003). From DAML-S to executable code. In B. Burg et al. (Eds.), *Agentcities: Challenges in open agent environment* (pp. 25-31). Springer-Verlag.

García, E., & Sicilia, M.A. (2003). Designing ontology-based interactive information retrieval interfaces. *Proceedings of the Workshop on Human Computer Interface for Semantic Web and Web Applications* (pp. 152-165).

Hoermann, S., Seeberg, C., Divac-Krnic, L. Merkel, O., Faatz, A., & Steinmetz, R. (n.d). Building structures of reusable educational content based on LOM. *Proceedings of the 15th Conference on Advanced Information Systems Engineering — Workshop on Semantic Web for Web-based Learning.*

IEEE Learning Technology Standards Committee (2002). Learning object metadata. *IEEE 1484, 12*(1).

Kabel, S. (2001) The added value of ontology-based instructional markup. *Proceedings of AI-ED* (pp. 496-499).

Kahan, J., Koivunen, M.R., Prud'Hommeaux, E., & Swick, R.R. (2002). Annotea: An open RDF infrastructure for shared Web annotations. *Computer Networks, 39*(5), 589-608.

Kopak, R.W. (1999). A proposal for a taxonomy of functional link types. *ACM Computing Surveys, 31*(4).

Lenat, D.B. (1995). Cyc: A large-scale investment in knowledge infrastructure. *Communications of the ACM, 38*(11), 33-38.

Lytras, M., Pouloudi, A., & Poulymenakou, A. (2002a). Knowledge management convergence: Expanding learning frontiers. *Journal of Knowledge Management, 6*(1), 40-51.

Lytras, M., Pouloudi, A., & Poulymenakou, A. (2002b). Dynamic e-learning setting through advanced semantics: The value justification of a knowledge management oriented metadata schema. *International Journal of e-Learning, 1*(4), 49-61.

Lytras, M., Tsilira, A., & Themistocleous, M.G. (2003). Towards the semantic e-learning: An ontological oriented discussion of the new research agenda in e-learning. *Proceedings of the Ninth Americas Conference on Information Systems.*

Martinez, M. (2001). Successful learning—Using learning orientations to mass customize learning. *International Journal of Educational Technology, 2*(2). Retrieved June 17, 2003: http://www.outreach.uiuc.edu/ijet/v2n2/martinez⁻

Maurer, H., & Sapper, M. (2001). E-learning has to be seen as part of general knowledge management. *Proceedings of ED-MEDIA.*

Nilsson, M., Palmér, M., & Naeve, A. (2002). Semantic Web meta-data for e-learning: Some architectural guidelines. *Proceedings of the 11th World Wide Web Conference* (WWW2002).

Pagés, C., Sicilia, M.A., García, E., Martínez, J.J., & Gutiérrez, J.M. (2003). On the evaluation of completeness of learning object metadata in open repositories. *Proceedings of the Second International Conference on Multimedia and Information & Communication Technologies in Education* (mICTE03).

Palomar, D., Sicilia, M. A., & García, E. (2002). Modeling and interchange of enhanced life-long learning profiles. *Proceedings of the First International Conference on Information and Communication Technologies in Education* (pp. 1309-1314).

Papazoglou, M.P., Proper, H.A., & Yang, J. (2001). Landscaping the information space of large multi-database networks. *Data & Knowledge Engineering, 36*(3), 251-281.

Qin, J., & Finneran, C. (2002). Ontological representation of learning objects. *Proceedings of the Workshop on Document Search Interface Design and Intelligent Access in Large-Scale Collections.*

Sánchez, S., & Sicilia, M.A. (2003). Expressing preconditions in learning object contracts. *Proceedings of the Second International Conference on Multimedia and Information & Communication Technologies in Education* (m-ICTE03).

Sicilia, M.A., & García, E. (2003a). On the integration of IEEE-LOM metadata instances and ontologies. *IEEE LTTF Learning Technology Newsletter, 5*(1).

Sicilia, M.A., & García, E. (2003b). On the concepts of usability and reusability of learning objects. *International Review of Research in Open and Distance Learning, 4*(2).

Sicilia, M.A., García, E., Aedo, I., & Díaz, P. (2003a). Using links to describe imprecise relationships in educational contents. *International Journal for Continuing Engineering Education and Lifelong Learning (IJCEELL)* (in press).

Sicilia, M.A., García, E., Aedo, I., & Díaz, P. (2003b). A literature-based approach to annotation and browsing of Web resources. *Information Research Journal, 8*(2).

Sicilia, M.A., García, E., & Alcalde, R. (2003). Fuzzy specializations and aggregation operator design in competence-based human resource selection. *Proceedings of the WSC8 Conference (Springer series).*

Sicilia, M.A., & Sánchez, S. (2003). Learning object "design by contract." *Proceedings of the 5th WSEAS Telecommunications And Informatics Conference.*

Smets, P. (1997). Imperfect information: Imprecision-uncertainty. In A. Motro, & P. Smets (Eds.), *Uncertainty management in information systems: From needs to solutions* (pp. 225-254). Kluwer Academic Publishers.

Stojanovic, L., Staab, S., & Studer, R. (2001). E-learning based on the Semantic Web. *Proceedings of the World Conference on the WWW and Internet (WebNet 2001).*

Trigg, R.H., & Weiser, M. (1986). TEXTNET: A network-based approach to text handling. *ACM Transactions on Office Information Systems, 4*(1), 1-23.

Wiley, D.A. (ed.) (2001). *The instructional use of learning objects.* Bloomington, IL: Association for Educational Communications and Technology.

ENDNOTES

1. http://www.w3.org/TR/2002/WD-owl-ref-20020729/
2. http://kmr.nada.kth.se/el/ims/metadata.html
3. http://www.daml.org/ontologies/
4. http://oiled.man.ac.uk/
5. http://www.opencyc.org

This work was previously published in the International Journal of Distance Education Technologies, Vol. 3, No. 2, pp. 13-29, copyright 2005 by Idea Group Publishing (an imprint of IGI Global).

Chapter 8.8
Guerilla Evaluation:
Adapting to the Terrain and Situation

Tad Waddington
Accenture, USA

Bruce Aaron
Accenture, USA

Rachael Sheldrick
Accenture, USA

ABSTRACT

This chapter provides proven strategy and tactics for the corporate evaluator. Topics include: adopting a performance-based operating model (the V-model) to shift focus from training for activity to training for results; using the V-model to plan and structure communication; leveraging modern measurement and statistics to save time and money (e.g., item response theory, sampling procedures, regression); leveraging available data to calculate training ROI (return on investment); determining when to hire or contract skills and knowledge; using technology to save time and money; and making the most of your available applications.

INTRODUCTION

Most corporate evaluators confront an assortment of decisions and trade-offs between the prescribed models of their discipline and the demands of the current business and technological situation. These exigencies demand flexibility in measurement and evaluation approaches. Adapting successfully to the situation and the reality of the corporate terrain often requires creative or "guerilla" tactics. In this chapter we share some of the tactics that have served us well in our endeavors to conduct effective evaluation in the larger corporate system within which we operate. For us, guerilla evaluation means adapting to two primary domains or constraints:

1. **Terrain:** The unique demands of the modern corporate training environment. Successfully navigating the corporate terrain requires tactical planning and communication, and a good map. Our map is a systems development model adopted from our corporate methodology and repurposed for evaluation and performance improvement work. This model guides our work as well as our communication plans. It helps us identify the unique characteristics and requirements of each stakeholder group, and deliver the information that each group needs in a timely manner.

2. **Situation:** Constraints on resources of time, money, people, skills, and technology. We respond to these situational constraints with a second set of guerilla tactics based on lessons learned: leveraging data and statistical skills, using contractors and consultants, and maximizing our use of available technology.

It is important to note that what we describe in this chapter is based on our experiences as a small team with a large agenda. Our group of four is the evaluation team for the central learning organization of Accenture, a leading global management consulting and technology services organization, with more than 75,000 people in 47 countries. Our team's defined mission probably reflects that of other corporate evaluation teams:

To be an integral partner within our organization and provide valued information that enables us to improve the products and programs necessary to build capabilities within our company.

In essence, our mission is to use evaluation to create and drive value.

Our collective background includes graduate training in evaluation, statistics, and measurement, as well as applied measurement and evaluation in the public and private sectors. In recent years our business context has become more challenging (but more interesting) and our strategies have evolved in response. This chapter will describe and illustrate a few of the important challenges confronting today's corporate evaluator, and what can be done to meet these challenges with resourcefulness, flexibility, and creativity.

NAVIGATING THE TERRAIN

Evaluation within a business context and the discipline of measurement as practiced and taught within academic settings are different enterprises and require different approaches. In our experience, successful corporate evaluators place themselves "in the trenches" and frame their work in the business context, rather than attempt to organize their efforts around lofty theory. As we'll describe later, academic training and acumen in statistics and measurement are critical levers for adding value, but in the corporate arena, the alignment of evaluation strategies with the business needs of the organization is of primary importance. Two strategies we have employed for the purpose of adapting to the business terrain are:

1. Adopting a performance-based operating model
2. Planning and structuring communication effectively in a business context

As is usually the case in adapting to any terrain, it helps to have a map. Ours is a model that guides our evaluation work and relates our work to the goals that the organization values. It is also vital for organizing our communication with stakeholders about our processes and results.

Creating a Map of the Terrain

It is widely purported that the Human Resource Development (HRD) field is evolving of necessity from its traditional training paradigm to a performance-based approach. In this approach, the objective of the training function is no lon-

Table 1. Paradigm shift in training evaluation

Training for Activity	Training for Results
No business need for the program	Program linked to specific business needs
No assessment of performance issues	Assessment of performance effectiveness
No specific measurable objectives for behavior and business impact	Specific objectives for behavior and business impact
No effort to prepare the work environment to support transfer	Environment prepared to support transfer
No efforts to build partnerships with key managers	Partnerships established with key managers and clients
No measurement of results or cost benefit analysis	Measurement of results and cost benefit analysis
Planning and reporting on training is input focused	Planning and reporting on training is output focused

ger the activity of training but the production of performance and business results (Stolovich & Keeps, 1999; Fuller & Farrington, 1999). Naturally, a shift in the training enterprise requires a shift in the evaluation of that enterprise. Phillips (1997) describes the characteristics of this shift, as shown in Table 1.

If an organization has not emphasized this shift, the evaluator has a unique opportunity to help clients and colleagues make the transition to performance-based solutions by leveraging the evaluation function in pursuit of ROI (return on investment) data. Because the bottom-line results of ROI analyses are attractive to today's clients, requests for ROI results are common. In addressing these requests, an evaluator can demonstrate that ROI is integral to a larger performance-based framework that involves needs analysis and identification of the relevant business issues and results. In effect, the evaluator can use ROI to sell a performance-based approach.

Consider, for example, a typical request for the ROI of a course or program. Such requests are not usually based on deep experience in HRD ROI methodology, but rather on the client's desire to know what bottom-line return will be achieved on the investment made in a learning solution. An evaluator engaging this client can demonstrate the need and value of a performance-based approach by illustrating how determination of ROI for the learning solution must be based on the measurable business results that are at the core of the client's need and central to the client's request for a solution. By structuring this conversation clearly, the evaluator can test the links between the learning or performance solution that's on the table, and the ROI and business results that should be realized. Often, these links have not been clearly defined, which puts the client's return at risk. A clear performance model will help structure these conversations with clients. By guiding the client through this process of mutual discovery, the evaluator enters the realm of performance-based HRD through ROI. In fact, doing this extends the evaluator's role to that of performance consultant, and small efforts such as these can help an organization transition from a training mode to a performance mindset.

As HRD organizations journey from an activity-based training culture to a performance-based and results-based approach, the evaluation model should likewise shift to a performance-based framework. The evaluation model should address each of the following fundamental objectives:

1. Measure the extent to which training meets the business goals and needs of the organization.
2. Link HRD intervention efforts to measurable results and estimation of return on investment (ROI) of training dollars.
3. Provide valuable information about training to the key decision makers for the purpose of continuous improvement.

In our effort to meet these requirements, we adapted a model from Accenture's core business methodology to map out the process for evaluating learning assets. Referred to as the V-model, it bridges our local business context, best practices in evaluation, and the shift in emphasis from traditional training activity to performance improvement approaches. The original V-model is an industry best practice for quality control in the development of applications, providing a structured development framework comprising verification, validation, and testing processes. The HRD adaptation of the model is shown at a basic level of detail in Figure 1.

The model proceeds down the left side of the V, moving from broad analysis activities (starting with the business need and definition of business requirements) through finer detail specification and specific design decisions. This essentially is a process of analysis and decomposition of broader objectives into more granular requirements, with verification at each step to validate links to the preceding set of requirements. During this analysis phase, the business need is translated into human performance requirements and solution designs. Development of the intervention occurs in the middle of the process, at the bottom of the V. Then, proceeding up the right side, each step signifies a specific set of measurements or products for its left-side counterpart. The process unfolds upward, finally resulting in Level 4 evaluation metrics (Kirkpatrick, 1994) and their monetary transformation into ROI estimates. In this reflective fashion, business results and ROI are measured in a manner that matches the business needs that were the original drivers of the problem or opportunity.

The key to the V-model is its symmetry. During the analysis and design stages (left side of the V), the model requires simultaneous development of metrics that reflect and link plan with result, gap with measurement, and ultimately, investment with return on investment. This 'testing' process, as well as the validation and verification processes inherent in the V-model, ensure the integrity of the system and provide tight linkages between the analysis-design phase and the measurement phase. When implemented properly, this results in performance solutions that are tied to the business needs that defined the original problem or opportunity, and metrics that are tied to performance objectives. Because it is adopted from our internal business culture, the V-model has been effective both for organizing our work and communicating our processes and results to stakeholders. It orients our clients in evaluating learning and performance interventions, guiding us from point A to point B.

Figure 1. V-model

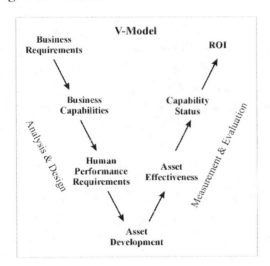

Using the Map as a Guide to Communication

The V-model not only maps the terrain and guides evaluation work, it also helps in planning communication. The importance of communication in corporate evaluation seems underrated. The corporate evaluator must realize that in a real-world business context, the value of the evaluation effort is transaction-based and depends on the communication and reception of information. The evaluator's work is complete only upon an effective information transaction, with results delivered in a language that the audience understands and at a time that is useful for decision making. Communication and dissemination of evaluation results therefore require as much attention as any other aspect of the evaluation effort and should be planned, designed, delivered, and managed.

In order to be deliberate and effective, communication planning should be mapped to the overall evaluation model. The V-model is useful for identifying the level and characteristics of target audiences for evaluation results. These audiences are identified for each set of evaluation results on the right side of the V-model (e.g., effectiveness of a course or program, its impact on business capabilities in the field, and return on investment). When mapped to the corresponding front-end processes of the performance model (the left side of the V), the identified target groups represent those whose needs, assumptions, and expectations have been clarified and incorporated into evaluation planning. These front-end requirements provide the basis for reporting results. Figure 2 displays the types of audiences within our organization that correspond to each of the levels of the V-model.

In addition to focusing results and reporting, identifying these groups allows us to consider their unique characteristics and needs. For example, at the bottom of the V, where courses are developed, instructional designers and project

teams need rich, detailed information at the level of individual learning objectives as they pilot and revise a course. It is important for evaluators working with these project teams to report and interpret patterns, and to integrate qualitative and quantitative data to provide solid formative information for continuous improvement. At the other extreme, near the top of the model, executives typically need a much bigger picture of how the course or program is delivering results. If the project has been selected for ROI evaluation, they will want monetary estimates, and evaluators should be prepared to provide details on the method used and the basis of the results. But, these stakeholders will typically need a more comprehensive and global summary of the course or program effectiveness across levels. Simple indices or scorecard presentations will summarize the big picture.

Similarly, mapping the communication plan onto the larger model allows evaluators to anticipate the flow of evaluation information between target audiences, and effectively plan the structure and timing of dissemination. This involves evaluating the consistency of messages released to each level of the target audience, and considering how each audience might use the information in reporting to others.

Piloting the Communication

An evaluator should feel confident that the information planned for delivery exactly answers the questions posed at each level of the model and for each stakeholder group. If necessary, the communication can be piloted before releasing results. Ideally, this should be done on the front end of the process while working with stakeholders to define needs and requirements. An easy way to do this is by producing mock reports with simulated data. This helps ensure that the right kind of information is being given in the right way. It can also save time on the back end of the project, because

Figure 2. V-model communication map

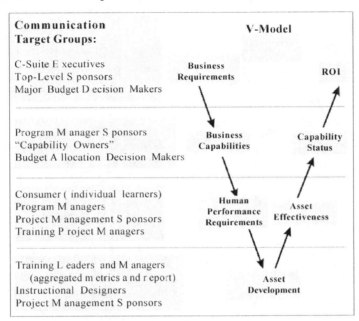

the simulated data can simply be replaced with actual data, and the tables and graphs refreshed. Because stakeholders use evaluation information to make decisions, piloting the communication will help clarify that the right questions are being addressed in order to make those decisions, that the right data will be collected, and that the best analysis method will be applied to the data.

Timing the Communication

When communicating evaluation results in today's business environment, timing is nearly everything. When developing the communication plan and working with stakeholders, it is important to assess their context to determine exactly when they will need the information. Of course, their answer is often "yesterday," but the reality is that various pressures drive stakeholder needs, and a pretty-good evaluation today may be worth much more than a perfect one tomorrow. The stakeholders' external pressures should be identified in order to determine the most appropriate time to

disseminate information. Well-crafted evaluation, delivered at the right time to the right decision makers, is the fundamental achievement of the corporate evaluator's work. Conversely, it is very frustrating to deliver results of a well-crafted and labor-intensive evaluation effort to stakeholders who have already made their decisions. They will respond to their pressures, make a decision for better or worse, and proceed to other pressing issues. If the decisions aren't optimal because they didn't have good data, everyone loses.

To Give is to Receive

Since knowledge is said to be power, one might be tempted to hoard it. We have found, however, that the opposite is true—the value of our work increases with the size of our audience. We choose to publish evaluation reports to an internal intranet site with open employee access. Organization productivity is dependent on collaborative efforts among strategic partners and allies throughout the organization, and these partners depend on information. Making information easily available serves to decrease transaction costs for program improvements within the organization and increase the entrepreneurial efforts of a wider audience to add value.

Users are not likely to misuse or inappropriately slant evaluation output if we succeed in delivering a cohesive and meaningful presentation of results and follow the communication plan. We have also found it helpful to identify early on those who used and found value in the reports in order to share their positive experiences with colleagues.

ADJUSTING TO SITUATIONAL CONSTRAINTS

Another major challenge for corporate evaluators is doing more with less and making the most of available resources. By effectively leveraging available resources, the evaluation effort can succeed in its goal to demonstrate the relationship between performance improvement interventions and bottom-line impact. This, in turn, will help improve the quality of solutions and ensure that training and performance improvement expenditures are seen as investments, not merely costs. Demonstrating the value of performance improvement interventions helps protect budgets for important training and evaluation projects during challenging periods in the economic cycle and ultimately ensures the health of the human resource system in the organization. We have developed four strategies for coping with the frequent lack of resources encountered during evaluation efforts in a business context:

1. Leveraging statistics to save time and money
2. Leveraging available data to deliver value
3. Hiring or contracting needed skills
4. Using technology to save time and money

Leveraging Statistics to Save Time and Money

With skills in statistics and measurement, an evaluator has a wider range of tools for doing good work. Following are four examples from our experience.

Case 1: The Fickle Flexibility of Raw Score Ratings

We were asked to evaluate an ongoing and extremely expensive annual event for our company's leadership. Three years of data showed a statistically significant *decline* in ratings of the event, but qualitative comments suggested that people thought the event was getting better each year. In resolving this conflicting information, we applied item response theory (IRT) analysis techniques, which allow raters and rated items to be separately and objectively measured on a common scale, independent of any particular sample of respondents. Because unique identifying information had been collected along with the ratings, we knew how each person had rated the event and could take into account individual differences in how the event was rated (e.g., some people are easier "graders" and tend to give higher ratings). With these biases factored out, we found that the event was actually receiving significantly higher ratings each year. Investigating further, we discovered that each year the event planners had been inviting fewer managers (who tend to be more positive in their ratings) relative to higher-level executives (who are more severe raters). By systematically eliminating the easy "graders," they had accidentally biased the data against themselves. Resolving this conflict of information (decreasing ratings but positive comments) gave the managing partner for the event better information with which to make decisions. The importance of having reliable and valid information was made clear when she told us, "You've just earned your pay for the year."

Case 2: Using Sampling to Save Thousands of Dollars

Another tactic we used in the analysis for this annual event was to sample wisely. The event was attended by several thousand people whose hourly bill rate was extraordinarily high. Previous evaluators had surveyed every participant with a lengthy survey. If each survey took 10 minutes to complete and 3,000 people completed them, the cost to the company was $250,000—a very expensive way to gather information. We needed information on the quality of the event, but we did not need to know what every participant thought of every aspect of the event. The large size of the event allowed us to gather information on different parts of the event from different participants and still have a reliable measure of the event's quality. In other words, we sampled both people and items. A few items remained common to the four different surveys sent to participants so that all of the items could be calibrated with IRT techniques, ensuring psychometrically valid measures, independent of the sample used to collect the data. By sampling both items and people, the length of each survey was cut to two minutes and the number of people completing the survey to 1,000. The total estimated opportunity cost thus dropped to $17,000, saving the company $233,000 while still delivering valid results.

Case 3: Providing Valid and Reliable Measures—Even When You Can't Choose the Items

We also used item response theory both to develop valid measures of our company's e-learning products and to avoid a series of political battles. Our e-learning group had developed millions of dollars worth of training products, but no evaluation data had been collected. When the organization's leadership required that comprehensive evaluation be put in place, the product managers took different approaches and selected different items to collect evaluation data. While all of these items consisted of general evaluative statements reflecting perceived value of instruction, the selection criteria in many cases appeared to be the manager's perception that the item would produce high ratings (since subtle manipulations of item wording and rating scales can influence the raw score ratings). When our team was asked to provide a comprehensive evaluation of these products, some resistance was evident within the development group (as is often the case when evaluation is perceived as externally mandated). In working through the political issues and compromises, each development manager was still allowed different selections of items for their products' evaluation surveys, rather than being constrained to a common set of identical items. We created an item bank that consisted of general evaluative statements reflecting perceived value of instruction. These items were calibrated using IRT techniques since a data matrix could be built in which some people rated multiple products and some products were rated by the same items. A common measure was derived across all products that was both valid and reliable, and allowed legitimate comparisons between courses. Because the items were psychometrically calibrated, it did not matter which specific items appeared on any product evaluation. For any combination of items, we could calculate an overall measure of value that allowed us to directly compare the e-learning products. Even though we lost the battle (managers controlled which items appeared on the surveys), we were able to win the evaluation war (regardless of the items chosen, we were able to provide valid and reliable measures).

Case 4: Using Data to Drive Course Design

One current project also leverages IRT. To date we have data from more than 75,000 evaluations of roughly 3,000 courses. Each course is coded on variables such as course length, format (e.g., classroom or e-learning), learning objective level (e.g., skill building or knowledge dissemination), and course-design variables (e.g., goal-based scenario). We will be able to develop an objective calibrated measure of instructional value as the dependent variable and use multiple regression to identify those aspects of course design (described above) that predict the perceived value of courses. This information should be useful to course developers and will further enable "data-driven design." This should help answer the question, "For this audience and with these constraints, how should we build this course?"

Leveraging Available Data

Depending on their skill sets and experience with data analysis, corporate evaluators might be asked to perform various analyses and answer questions beyond the scope of the planned evaluation of specific courses. Tackling these broader research questions usually involves working with extant data, and these data from various sources can often be used creatively to deliver exciting results. Three examples from our experiences demonstrate this theme.

Case 1: A Unique Way of Calculating Training ROI

Most people think about the value of training in terms of how a specific course gives a specific person a specific skill that allows that person to accomplish a specific task. While this is not inaccurate, our work on training ROI revealed larger effects of training—effects in recruiting, chargeability, bill rates, and retention. Ignoring these effects risks underestimating the broader value of the training program. The benefits gained from taking a single course may be fleeting, but the benefits gained from taking many courses can be pervasive. We chose, therefore, to look

at how training histories (rather than individual courses) affect broad business concerns—how to attract and keep good people and make them more productive during their tenure. Our work on training ROI has been recognized for its innovation; in June 2002 we won the Corporate University Xchange Excellence Award in the category of Measurement.

To calculate the ROI of training, we analyzed over a quarter-of-a-million human resource records on all employees who have been with the company. We began by calculating net per-person Contribution [(Chargeable hours * Bill rate) – (Cost rate * total hours)], the individual's margin over his or her career with the company. We used regression to determine how training histories (e.g., hours spent in training) affected Contribution, making sure to remove the effect of potential biases in the data such as career level, work experience, inflation, and business cycles. The results of the analysis showed that people who take the most training (those in the top 50th percentile) were both more billable and charged higher bill rates than those who had taken less training. These employees were also more likely to stay with the company longer. We based our analysis on actual data collected within the company and were able to document a 353% ROI for training expenditures.

Case 2: If You Scratch My Back, I'll Scratch Yours

A common constraint in corporate evaluation is the availability and quality of data. Evaluators are the quintessential knowledge workers, often required to generate meta-knowledge (i.e., measuring what the organization knows). Evaluation involves creating knowledge from information and intelligence from data. In addition to the data generated through course evaluations and knowledge/skill assessments, other existing sources of data generated or collected by internal organizations that might be of use to the training organization include employee or customer satisfaction data, and marketing or market research data. Often the data needed to answer the business question being asked are not accessible. Helping the owners of the data realize value by providing evaluation expertise and methodology is a valuable tactic in gaining access to these additional sources of data. For example, gaining access to human resources records was essential for our ROI analysis. By assisting HR in an analysis of personnel performance ratings and the promotion process, we not only added more value to our organization, but were able to gain access to the data we needed for our own analyses.

Case 3: Know the Nature of the Task

Ideally, evaluators are asked to evaluate a project by providing objective summative or formative feedback. It is not uncommon, however, to find that what is actually expected is a kind of positive, quantified advertisement. Stakeholders might be influenced by the "drive for 5s" approach to evaluation, where the primary object of conducting evaluation is to achieve a maximum "grade" on a five-point rating scale. Such motivations are understandable, and typically are not informed by a sense of the importance of constructing items that produce variance and balancing item difficulty (i.e., the ease with which an item can be rated highly). In general, data will be more accurate and informative when there is increased variance in responses, which cannot occur if everybody gives a rating of a 5. To balance the need for good measurement with good advertising, evaluators can include both types of items—tougher questions that provide good measurement information and easier items that provide good press. Both types of items can be used ethically, if you are mindful that consequential validity—the degree to which the intended actions and inferences based on the data are supported by the data—is maintained. A course that gets a lower percentage of 5s on a difficult item may be an outstanding course, but

a statistically naive stakeholder might not take into account the difficulty of the item and decide to cut funding for the course because of its "low" score. Paradoxically, the "advertising" items (a lot of 5s), though statistically spurious, might allow for the right decision to be made (e.g., this course had a lot of 5s so I'll continue to fund it). The paradox is that sometimes you can only get the right decision by providing bad data, a truly guerilla tactic.

We have been successful in persuading people to use items with greater item difficulty by convincing them that it is in their long-term interest to have these tougher items. Suppose 95% of participants rate a program as some form of "Excellent." The sponsors of the program may increase the funding for the program for the next year and expect higher results for the increased investment. Because the ratings are already near the upper limit (100%), they cannot get much higher, and future results may be misinterpreted as a lack of progress. While course sponsors might prefer "feel-good" items for advertising purposes, this must be balanced with the high-quality measurement provided by tougher, more valid questions.

Hiring or Contracting Skills and Knowledge

Even under the best of circumstances, an evaluator at times might not have the time or level of skill necessary to complete a project. Multiple projects may be imposing serious time demands, or there may be an opportunity to deliver value that requires familiarity with new techniques or processes. When such demands or requirements are too great, assistance can be sought in external sources, such as university students or consultants. Both sources can provide the skills needed (and provide the opportunity to develop those skills within the evaluation team) and free up time for the evaluators to complete more pressing or valuable projects.

It is useful to maintain contacts with local universities, as students can be an excellent resource; they are also generally more cost effective than external consultants. In addition, if you have positions opening, the experience can be a useful trial period for the organization and the student to assess the fit. We have had great success using graduate students from a nearby Ivy League school. These contractors are paid more than they could make on campus and less than the typical cost for employees (a tactic that keeps both our company and the students happy), and we strictly follow the law for employing temporary contract workers. These contractors can be invaluable in consolidating research relevant to current projects, providing statistical assistance, and handling time-consuming tasks.

The internal staff should define the target problems and break them down into manageable steps. The contractors then can work on the tasks and develop potential solutions. With this assistance, the evaluation staff can work on more projects and deliver valuable results more quickly than would be possible otherwise. The internal staff might also proactively identify areas that would be of interest to the organization and create reports based on the work of the contractors. For instance, one of our contractors read and performed content analysis on over two million words of qualitative comments from various surveys that we have in place and provided a well-synthesized qualitative summary that was quite useful and well received by many stakeholders.

Another type of external consultant that can be leveraged is a highly specialized expert with recognition in the field. These consultants are hired because of their specialized evaluation expertise and the unique value they can deliver to an evaluation project. For example, an external expert with deep expertise and experience in calculating the ROI of human performance interventions can be contracted for assistance with such analyses. Unlike lower cost contrac-

tors that require some level of supervision, these higher-level consultants work more like partners, and different strategies are needed to manage and leverage these relationships.

In any contracting relationship, make the scope of the project clear. Define the expectations for work, roles, and responsibilities. In setting the parameters of the relationship, a long-term view should be maintained in order to take full advantage of the partnership. To the extent possible, opportunities and expectations for knowledge transfer should be built in, so that at the end of the project, the evaluation team has gained new competencies in the expert's domain. The scope of this knowledge transfer will vary according to that domain and the project circumstances, but such experiences are among the best professional development opportunities around, and they should be capitalized on. The ultimate goal should be the ability to tackle the same type of project independently as a result of the experience.

A number of tactics can be used to reach this goal. An evaluation consultant's typical service might include tasks that an evaluator or evaluation team can assume (e.g., production of survey instruments, data collection, or internal communication with project stakeholders). Relieving the external consultant of these tasks can free up time and an expectation can be set for collaboration on evaluation design or for more frequent meetings to discuss problems and options.

It also helps to do some research on the area of expertise prior to engaging the consultant. Initially, an evaluator should have enough background to know that the consultant's particular expertise is a good match for the organization's performance issue, and that this expertise or resource is not available internally. Investing time up front, and demonstrating some knowledge and expertise while setting expectations with the consultant, will earn their confidence and assist in the development of a close working relationship that will result in deeper learning.

A final report or document should be developed that has sufficient depth and documentation to provide guidance for conducting similar projects in the future. This should include whatever elements are deemed helpful for reproducing the processes. Such elements might include sufficient background on the theoretical bases of the approach, description of the processes involved, detailed evaluation design or planning documents, and copies of instruments used. As an expert in the field, the consultant should have no trouble generating this sort of document. It can probably be synthesized easily from the body of her work, which could include books and professional papers. Also, given the consultant's direct engagement in the evaluation project and organization, she is ideally suited to tailor this body of work to an evaluator's particular context.

Using Technology to Save Time and Money

Evaluating courses can be a time-consuming task, particularly when the evaluations are developed and distributed manually. Because of the time involved in designing, developing, distributing, and analyzing the results of a course evaluation, it may not be possible to evaluate every course that a company offers its employees. By leveraging technology, however, the task of evaluation cannot only become more manageable, but time and resources may also be freed up to allow more intensive and valuable data analyses that further benefit the company. The following examples illustrate this use of technology.

The Leap from 200 Customized Surveys to 3,000 Standardized Surveys

Before June 2001, course evaluations at Accenture were done on a case-by-case basis, and evaluations were developed specifically for individual courses. The responsibility of deciding whether a course should be evaluated usually fell to the course sponsor. If the decision was not made early enough in the course development process, the course was not evaluated until well after it was released, or it was simply not evaluated at all. When these post-course surveys were created, the design process often included considerable iterative input from the course sponsor and others in an effort to provide quality client service. However, the expectations of the sponsors had to be balanced with survey design standards (i.e., questions and rating scales that provided reliable and valid metrics), and many hours were often spent in discussions over the wording of particular items or a number of questions. These individual course surveys were distributed on an ad hoc basis via e-mail, and the reports that presented the survey results were created and updated manually.

While customized surveys provided value, the process behind such evaluations (from survey design to distribution to reporting) was hardly efficient. The time needed to support course evaluations left little room for reflection on evaluation processes, and no time to work on more interesting and business-relevant projects, such as training ROI. By leveraging technology, however, the course evaluation process was greatly streamlined. A system was created that not only provided evaluation services for *all* of the learning assets provided by Accenture, but also allowed us the chance to provide more valuable evaluation services to the company at large, rather than limiting it only to individual course owners.

In June 2001, Accenture rolled out myLearning, a personalized, Web-based portal that gave its employees immediate access to all information related to company-sponsored learning assets. The myLearning portal includes:

- A Course Catalog and Learning Management System (LMS), which provide a list of required and recommended courses based on an employee's role within the company and gives employees the ability to register for courses online.

- A Decision Support Center, which provides a comprehensive and immediate snapshot of learning metrics, including expenditures per area and course evaluation ratings.
- A Course Evaluation System, which provides "5-star" ratings and detailed course reviews for every course to guide employees in their learning decisions.

The 5-star ratings and course reviews provided by the Evaluation System are based on participant responses to end-of-course surveys. These surveys are automatically distributed by the LMS upon course completion; data are stored in a SQL server, which warehouses data, feeds 5-star ratings and course reviews back into the LMS for publication to end users, and transfers learning metrics data to the Decision Support Center (DSC) reports.

Rather than creating individualized surveys for every course, which would be time consuming to create and of limited utility due to lack of standardization, we created four distinct surveys based on the following types of assets:

- Instructor-Led Training (e.g., classroom training)
- Virtual Training (e.g., Web-based seminars)
- Online Course (e.g., CBT)
- Books/Publications

We selected survey questions based on the results of an item response theory analysis of over 10 million data points culled from six years of in-depth classroom evaluations (our earlier custom evaluations), test scores, follow-up surveys, faculty ratings of student learning, and supervisor ratings. Though brief, the 5-star rating system is a psychometrically reliable and valid proxy for the previously completed in-depth course evaluations. As might well be expected, a few clients were not pleased with the systemic changes to evaluation processes. Sponsorship and support from key leaders was essential in articulating the new value that was to be realized from adopting the widespread changes.

Not only are surveys distributed automatically, but reporting is also provided automatically, via myLearning's Decision Support Center (DSC). The DSC integrates information from a variety of sources—including course evaluations, accounting, and global personnel tracking (all of these data have been stored on an SQL server to facilitate integration)—to make immediately available the information that learning stakeholders need to make better education investment decisions. The reports are pre-formatted as Excel pivot tables that can be customized by user groups so that they may examine their learning programs from a variety of perspectives and make appropriate modifications.

Because the reports are linked to the SQL server that houses evaluation data, the results of those evaluations become immediately available when surveys are returned. And, because the reports are also linked to usage and financial information, the reports become more valuable to multiple groups. Accenture executives can quickly create reports that provide an organizational summary of educational spending and effectiveness, learning sponsors can easily manage the day-to-day learning activities of their employees, and course owners can determine which courses provide the greatest benefit, based on course selections, evaluations, and expenditures.

This universal and automatic evaluation system has several advantages that drive continuous improvement:

- Surveys are distributed automatically, rather than manually.
- All learning assets are now evaluated, rather than just a handful of courses. In its first year, the 5-star rating system collected over 75,000 evaluations on more than 3,000

courses. This much data allows for the data-driven course design mentioned above.

- To help Accenture employees make better and more informed learning asset selections, the 5-star rating and participant comments associated with the course are posted in the course catalog.
- The 5-star rating system allows Accenture to use statistical process control methods on all of its courses in order to weed out those that aren't performing up to standard.
- With myLearning Decision Support, Accenture learning stakeholders have on-demand access to all measurement data, course usage information, and financials.

Online Surveys

The myLearning LMS delivers surveys through the Internet; participants are either sent a link to the survey in an e-mail message or click a button in their transcript to open the survey in a new browser window. There are many benefits to creating online surveys, most notably their ease of use. Because online surveys are accessible via the Internet, they can be completed at any time and from nearly any computer. Rather than distributing the surveys manually, participants can be given a link to the survey. A link to the survey can also be added to the website for a specific course or organization, making it easy to find and complete.

Although Accenture uses an LMS to distribute its online surveys, an LMS is not necessary to develop online surveys. All that is needed is web space that can be accessed by everyone who should complete the survey, a person who can develop an HTML form (i.e., the actual survey), a repository to collect the data (e.g., a database or XML file), and a spreadsheet or other analysis package to analyze the data. Even if these things are not available, it is still possible to create and distribute online surveys. Many companies provide services for the creation and storage of an online survey (search online for "online survey solutions" or "create online survey"), as well as basic analysis and reporting services as the data are collected.

"High-Tech" Paper Surveys

Developing a learning management system that incorporates evaluation and reporting is just one way to make the most out of technology. While an LMS can incorporate a highly efficient evaluation system, it requires a large investment of time and resources to develop. There are many other ways to make evaluations more efficient and valuable without requiring such an investment. Even paper surveys and processing can leverage technology well. At times, paper surveys are the best alternative for survey administration—for instance, when those taking the surveys do not have access to the Internet or a computer, or clients insist on immediate and tangible administration of the questionnaire. But hand-entering data from paper surveys can be an onerous and error-prone task, that ties up valuable resource hours.

Nonetheless, if the paper survey relies on a selected-response item format, it can use "bubble" responses (e.g., fill-in-the-bubble formats found on standardized tests), and can be scanned into a computer and processed using software that can read optical marks. Such software does not require "bubble sheets" (e.g., standardized testing forms) in order to process the surveys; even surveys created in Microsoft Word can be scanned and processed. This software automatically codes and enters the data into a spreadsheet, which can then be saved into many other common database formats, such as Microsoft Excel or Access. With a scanner and scanning software, even paper surveys can be a high-tech and efficient way to provide evaluation services.

Another alternative for paper survey data processing is professional data entry. Data entry companies are fairly common and provide a cheap and flexible (i.e., variable-cost as opposed to fixed-cost) way to handle data entry. Usually if you ship them the data overnight, they can e-mail you hand-entered results within a day or two for a reasonable price. Because they are experts, the error rates are much lower than having an in-house non-expert enter data.

Making the Most of Applications

One of the best ways to make the most of technology is simply to make the most out of your applications. One of the most widely used applications, Microsoft Excel, is also one of the most under-used, in terms of the flexibility and power it offers. Three of the most powerful features this program has to offer are formulas, pivot tables, and macros. (Visual Basic for Applications is also a powerful tool, and is available beginning with Excel 2000, but can be more difficult to learn. VBA gives users a way to add highly interactive and dynamic features to their reports. Many books, as well as the Excel help menu, are available to help users who are interested in VBA for Excel.) With these tools, you can create appealing and informative reports quickly and easily. You can also create surveys that look and act much like HTML surveys do, with features such as drop-down menus, checkboxes, and option buttons.

Excel's features make it a reasonable option for automating reports. If the data can be stored in an Excel worksheet (not to exceed 65,536 rows or 256 columns), a blank report template can be created with all of the formulas, charts, and tables already entered and formatted. A blank worksheet can be reserved to hold the data, and the formulas, table cells, and chart sources can be entered such that they reference the appropriate cells on the data sheet. Of course, this means that some thought will have to go into how the data sheet will look (both when the data are compiled and in the report template), and how the tables and/or charts should look. Once the template is set up, however, a few macros can be recorded to automatically update the report with the click of a

button once the data are copied into the reserved data worksheet.

If your evaluations produce too much data to be stored on an Excel worksheet, Excel can still be used to generate customizable reports to meet the needs of your clients. Pivot tables and charts can be created that draw upon data stored in an external database, such as Microsoft Access or SQL server, by using ODBC connections (they can also use data stored in an Excel worksheet). Pivot tables are an excellent way to create interactive, modifiable tables that can be easily refreshed and updated as new data come into your database. They offer a variety of summary functions to present your data—including means, standard deviations, counts, and percentages—and also give you the flexibility to create your own summary functions; for instance, a formula can be created that divides one variable in your dataset by another (e.g., course price by number of training hours), without having to add the formula to your original data. Filters allow you to display different summaries of the data by limiting the report to only those categories you are interested in. A report on course evaluations can include filters based on student demographics so that results from a particular geographic location or office can be quickly accessed. With a few clicks of the mouse, the report can drill down to increasing levels of detail (e.g., from broad industry category, to a specific industry within the category, to an office location). Pivot charts can similarly present an interactive and modifiable report in chart format.

Excel can be used not only for reporting, but also to create the actual surveys used in course evaluation. Forms can be created that have a look and functionality quite similar to HTML surveys, are easy to create, and can be distributed and collected via e-mail. The Control Toolbox and Forms toolbars provide multiple options to create dynamic survey questions. Drop-down menus, checkboxes, text boxes, and lists are just a few of the choices available. Multi-page surveys can easily be created with macros programmed to direct the user from worksheet to worksheet. A "thank-you" page (or other announcement) can appear at the end of the survey to thank users for their participation. Best of all, the data do not have to be hand-entered in a spreadsheet when the surveys are returned. The survey questions can be linked to a data sheet hidden behind the survey such that the data are automatically collected in a spreadsheet as a user completes the survey.

Excel is described here only to provide a few examples of how a commonly used application can make course evaluation easier. This is certainly not to suggest that Excel is the only application that can or should be used. Many applications have a great deal of flexibility and power that can be used for data collection and reporting, but often this power is unrecognized or underutilized. While it might take some time, it is important to become familiar with all of the features available in the applications you use. You may be surprised how much time-consuming work is minimized by simply making the most of your everyday applications.

CONCLUSION

Every evaluation team should exist to add value to their organization. But, as Peter Drucker (2001) says, quality isn't what you put into your work; quality is what others get out of your work. Therefore, in order to deliver quality, the corporate evaluator must creatively navigate the business terrain. Adopting an appropriate performance model is key to such navigation. We use the V-model to link training to business goals, link intervention efforts to measurable results and ROI, and support continuous performance improvement. The V-model also helps plan effective communication by identifying which decision-makers should receive what type of evaluation results and when. It makes our work more valuable, because our work is tied directly to business goals and it makes our communications more effective, because they are targeted to the appropriate audiences and timed to be valuable in aiding decisions. This is important, because evaluation work that is not understood by the organization or that is not delivered when it is needed for decision-making will not be valuable in business no matter how valid and reliable the results.

An evaluator not only needs to be aware of and adaptive to the terrain, or business context, but must also have the strategies and flexibility necessary to adapt to situational constraints. Evaluators are often faced with constraints on the money available for evaluation work, the data available for analysis, and the time or skills available to complete projects. Statistical and technological tools can give an evaluator the ability to provide quality work more efficiently (e.g., using sampling to send out fewer and shorter surveys, creating automated processes). If data exist that would improve the quality of evaluation work, but access to the data is limited or denied, using evaluation or statistical skills to help another group achieve quality results can add value to the organization and provide access to the needed data. When time or skills are at a premium, hiring contractors or external consultants can free up time to tackle larger projects and create an opportunity for skill development within the evaluation team.

In summary, guerilla evaluation tactics allow you to adapt to rugged terrain and to changing situations, thereby creating and driving value in your organization. As such, guerilla evaluation requires deliberate innovation, and deliberate innovation is the first step toward innovation delivered.

REFERENCES

Drucker, P.F. (2001). *The essential Drucker.* New York: HarperCollins Publishers.

Fuller, J. & Farrington, J. (1999). *From training to performance improvement: Navigating the transition*. San Francisco: Jossey-Bass Pfeiffer.

Kirkpatrick, D.L. (1994). *Evaluating Training Programs*. San Francisco: Berrett-Koehler Publishers.

Phillips, J.J. (1997). *Return on investment in training and performance improvement programs*. Houston, TX: Gulf Publishing.

Stolovich, H.D. & Keeps, E.J. (1999). *Handbook of human performance technology: Improving individual and organizational performance worldwide* (2nd ed.). San Francisco: Jossey-Bass Pfeiffer.

This work was previously published in Instructional Design in the Real World: A View from the Trenches, edited by A.-M. Armstrong, pp. 136-159, copyright 2004 by Information Science Publishing (an imprint of IGI Global).

Chapter 8.9
Standards?
What and Why?

Phil Long
Information Services and Technology, USA

Frank Tansey
Technology Consultant, USA

ABSTRACT

Specifications define the nature of the interconnections between the distinct parts of complex learning systems, but not their boundaries. Next generation CMS tools are emerging from standards discussions that challenge current e-learning systems design boundaries. They raise the prospect of a complex but smoothly functioning set of components and services that aggregate in ways that best serve individual communities of users. Users need to engage in the process to express their requirements for e-learning software. These building blocks, produced by a small number of organizations, are establishing the framework that will enable CMS environments to become vastly different than the CMS you might now be using.

INTRODUCTION

Our exploration of next-generation course management systems begins with the important and somewhat hidden efforts to develop e-learning specifications and standards. These building blocks, produced by a small number of organizations, are establishing the framework that will enable CMS environments to become vastly different than the CMS you might now be using. The environment that emerges from well-defined specifications is a landscape that makes the current boundaries set by course management systems both artificial and limiting. The logical outcome of this work is a complex but smoothly functioning set of components and services that aggregate in ways that best serve individual communities of users. Specifications define the nature of the interconnections between these distinct parts of a complex learning system but not their boundaries. The result is a future world where we'll look back on this discussion of CMS software as a quaint footnote in the development of more robust educational technologies for teaching.

COMMON NEEDS

Specifications and standards arise from the need to promote technical, syntactical, and semantic interoperability. This need is important in relation to metadata, content, databases, or repositories, designs for learning, vocabularies, learner profiles, assessment, expression of competencies, and networking protocols. Standards and specifications make the "abilities" (Nissi, 2003) of e-learning possible. These abilities include:

- **Interoperability:** Systems work with other systems, within and between institutions or organizations. Content developed in one system is not restricted to that system by proprietary encoding or protocols.
- **Reusability:** Learning objects or resources are easily used in different curricula, learning settings, and for different learner profiles.
- **Manageability:** The system tracks information about the learner and the content.
- **Accessibility:** A variety of learners, with different learner profiles such as educational and physical needs, easily access and assemble the content at the appropriate time.
- **Sustainability:** The technology evolves with the standards to avoid obsolescence.

Why are Specifications Important?

Specifications enable people to focus on a problem by providing a shared vocabulary of words and ideas. They represent a current "state of the art" consensus among developers and architects of educational software about a particular data structure, functional behavior, or service that is important for an online learning system. They are intended to capture agreement in the face of change. As such, they provide a hedge against the risks of this volatile environment. To achieve the best return on investment, these systems must be sustainable, flexible, scalable, and interoperable with new learning technologies.

Specifications and Standards Live in the Background

A key advantage of an effective standard or specification is that, with proper implementation, the standard becomes largely invisible. In this state, the standard is a building block for features that differentiate one product from another.

Take, for example, a typical electronic device you use every day. When you purchase a clock radio or a microwave oven, you focus on the features of the device. You want good sound from your radio or a small size for your microwave. You don't think about the plug that you will insert into the wall to power the device. Plugs and electrical sockets have been standardized, as have voltages and currents. You are not expected to think about these factors to use each device you have purchased. If you needed an adapter for each electrical item, you would think twice about every purchase.

The same advantage of key standards applies to CMS systems. If your content had a standard "plug" for all CMS systems, your world of content choice would be greatly expanded. Similarly, if all CMS systems could communicate with the system used by the registrar's office to exchange key student information, your class roll would always be up to date, and grade submission would be virtually finished when you posted your grades to the CMS. Such is the promise of specifications, yet unrealized.

A CLOSER LOOK AT SPECIFICATIONS AND STANDARDS

Laying the Foundation

Specifications lay a foundation on which learning technologies should be built. They represent common agreement among communities with expert knowledge in a particular domain. Specifications are the first step on the road to standards. Standards are created by accredited bodies that give their stamp of approval to specifications placed before them. Standards represent in some sense an end point in the process. The expectation is that by the time a specification is recognized by an accrediting body, such as the Institute of Electrical and Electronics Engineers (IEEE), it has achieved a degree of stability in which change is measured and modifications follow a strictly delineated process.

What Type of Specifications?

The world of specifications is complicated in educational software because it is often viewed from differing and contradictory perspectives. This complication results in breaking down the problems that course management systems are supposed to address in wildly different ways. It's no wonder that the outcome is expressed in incompatible, divergent, and generally impoverished models for software-enabled e-learning. The problem is exacerbated by the fact that course management systems don't exist in a vacuum. They must integrate with other systems that provide necessary services for students, including library systems, student information systems, and authentication systems. Without considering the ecology of the whole e-learning landscape, there is little likelihood that these related services can be efficiently leveraged or well-integrated into course management system environments.

Software architects call this process of breaking a complex system down into its logical parts "factoring" the problem space. If the functional domains that need to be represented in the course management system are clearly identified, then it becomes easier to describe the components of the system. Describing the components carefully results in a model, and only then can important specifications be mapped to the e-learning system.

Smyth, Evdemon, Sim, and Thorne (2004) have provided a set of principles on which course management systems, or any technology-mediated learning system, should be built. They identify three patterns that appear commonly in the design of e-learning systems. These represent a consistent factoring of the educational domain for online learning. These three elements are: (1) data representation, (2) communications, and (3) interfaces.

Data Representation

The most well-defined area of specifications in the course management world applies to the description of data used by it. This makes sense, as describing the data in a learning system is essential to being able to move data from place to place. This need drove early specifications activity under the rubric of data exchange. Having agreed-upon data representations permits information to move between course management systems in such a way as to preserve meaning and structure.

Without defined and agreed-upon specifications for data, interoperability is impossible. Creating these specifications means achieving agreement in the structure, meaning, and language used to describe the data in e-learning systems. The specifications that have been developed to date for defining data include, for example, the Dublin Core metadata suite, IMS Content Packaging, and the ADL Shareable Courseware Object Reference Model (SCORM).

Communications

Course management systems are frequently self-contained, stand-alone, or at best, hand-tailored to fit into an existing enterprise environment. Defining how two or more systems communicate to achieve generic agreement on the framework to be used is a critical step to facilitate systems integration.

Examples of the problem and benefits of an agreed-upon solution are on your desktop and used every day. You select an e-mail client that has features you find useful. Regardless of your e-mail client, you can send and receive mail successfully because the programs follow common e-mail protocol specifications, such as POP or IMAP. This process works because both the mail client and the mail server have implemented a common protocol. Additional functionality, such as specifying a method of authenticating mail, might be added, but this is only an enhancement to the specification that is the basis of e-mail exchange.

Interfaces

The most striking contribution of Smyth, Evdemon, Sim and Thorne's (2004) work is the clarity they provide in defining the roles of interfaces. Interfaces can be thought of as a contract between system components that describes what a given component does and what it expects of the other components. The authors describe two fundamental approaches to establishing the responsibilities for either providing a service to an e-learning application, or using a service provided from somewhere else, either within an e-learning system, or from another system entirely (e.g., a campus authentication system).

Applications that are in the category of service consumers tend to be dependent on programming language. An interface between service consumers and providers insulates the consuming application from details that are specific to the service provider. This allows the service provider to choose technologies that are optimal for it, leaving the service consumer unaffected by such changes. This approach also permits the consuming application to select among different implementations of a service.

Let's look at an example. Most e-learning systems don't permit the option to store files in multiple locations, for example, locally, centrally, or remotely. If an e-learning application implemented multiple file interfaces, it could store files locally if not attached to a network, or centrally and mirrored to a remote server if it were. The user then might be able to select where the data should be stored, if preferred. These multiple interfaces coexist and provide functionality as well as abstraction between the consuming application and the service provider. The Open Knowledge Initiative (OKI) has developed a suite of such consumer-oriented, open-source interface definitions (OSIDs) as candidate specifications for e-learning developers.[1]

Provider-oriented interfaces are independent of language choices, often implemented over networks where the interface is not on a local user's machine. Web services are built using these interfaces to maximize the availability of the service with minimal dependence on the local consuming application. Web Service Definition Language (WSDL) is an example of a provider-oriented interface to an application that might be incorporated in an e-learning system.

Smyth, Evdamon, Sim and Thorne (2004) make the crucial observation that these interfaces are not mutually exclusive—they offer specific capabilities that can be leveraged by the builders of e-learning systems. Interfaces of either type may be implemented within an application on a single machine or across a complex network topology. What's important to the observer is this framework of data structures, communications protocols, and interfaces. e-learning system developers should be able to describe their work and their systems from this perspective, and the corresponding specifications or specification candidates that they are implementing. It should be clear from the above that proprietary implementation of service interfaces (Web services or any other) should be avoided.

Specifications Are Not Standards

More often than not, you hear about standards (rules or models) rather than specifications (detailed descriptions of work to be done). We hear about Web "standards" when referencing the World Wide Web Consortium, or W3C. There are W3C standards (for example, the standards for using URL and HTTP), but many of W3C's contributions are closer to specifications and guidelines. In the CMS world, there are few standards but many specifications, some of which are evolving to standards. Because the CMS world is evolving so rapidly, there has yet to be sufficient time to mature many specifications into standards.

The function of a specification is to provide a sufficiently detailed initial description to implement a defined scope of work. It is based upon and promotes early implementations but is a bit of a moving target. As implementation and technologies progress, specifications evolve to respond to detailed requirements.

Standards, on the other hand, are more static and become the rule. That is not to say that the standards never change. They do, by becoming a new standard, not just a new version of the same standard. The granular review involved in the standard process theoretically eliminates the need for multiple versions of standards. To call a specification a standard is a premature freezing of work. It is important not to propose a standard until the work is fully matured and accepted.

A CLOSER LOOK AT SPECIFICATIONS

Specifications Address Real Issues

Successful specifications or standards address real issues. It is unlikely that any specification group would devote limited resources to trivial or nonconsequential work. There must be substantial consensus that an issue is worthy of a specification for the process to begin. Specifications may address a seemingly small part of a larger issue; if one considers a single specification within a larger context, then each specification is a building block.

Specification development is hard work. It is rarely done in isolation; in fact, competitors frequently sit at the same table. It requires strong technical skills, a deep understanding of topic area, the capability to reduce extensive requirements to essential features, and willingness to compromise. On top of this, it is frequently necessary to produce results in a short time frame. This is the backdrop for most specification efforts.

Let's examine a single specification as an example. The IMS Content Packaging Specification details a set of rules for packaging course material and identifying the material in the package. It is not concerned with how the content will run on another system, what the content is, or how it is sequenced in a course. The scope of the specification is limited, but packaging is a real problem that virtually all course management systems must address.

A course consists of content that is packaged for a particular need. For example, content packaging takes content material, exercises, and exams and bundles them into a course. All items cited are examples; there is no rule on what should be included in a content package. An implementation of this specification operating within a single CMS would be an improvement over a proprietary implementation since it provides the opportunity for future interoperability. However, it may be of limited value if the CMS remains isolated.

However, the impact of the content packaging specification is greatly enhanced when a content package can be exchanged from one CMS to another. Packaging and exchange is a real issue, especially on campuses that support more than one CMS and should be of crucial importance to any faculty member thinking of changing institutions. Any potential vendors or developers should be called upon to explain how their implementation addresses this need. While some vendors may view this as creating an undesirable exit strategy for a potential customer, the real point is that it can hinder access to one's own intellectual property and should be a prerequisite for considering the system's acquisition in the first place.

Specifications Help Avoid Reinventing the Wheel

The development of course management systems has followed the pattern common to early-stage technology innovations. There is a period of wide and rapid development, when diverse strategies are pursued, emphasizing differentiating feature sets and unique underlying designs. Commonly agreed specifications, if acknowledged at all, are implemented with unique "extensions" that further distinguish a given product. Course management systems developers, faculty and student users, instructional technologists, and information systems professionals have lived through this period of exhilarating change and frustration. It's important to recognize that implementing new functionality as proprietary interpretations of specifications impedes interoperability and slows progress in e-learning development.

Real value will be derived from paradigm-shifting processes that, like systems theory, look at a new level of integration to provide benefits otherwise not possible but fundamentally resting on work that has already been done at lower levels. Specifications need refinement, which can

be achieved only through use and feedback. It's important, as new ideas and people engage in this work of building better and more robust course management technologies, to focus these efforts not only on the right level of abstraction, but also in the right direction.

Specifications Promote Opportunities for Adoption

The consensus process involved in specification development and standard setting promotes the opportunity for adoption in multiple implementations. Once vendors or in-house developers become aware of, or involved in, an open specification or standard effort, the advantages of implementing such an approach quickly overwhelm proprietary solutions. While there may initially be resistance to dropping a solution developed in-house, specifications and standards promote adoption of key technologies. In the CMS world, if key underlying common needs are addressed through specifications and standards, developers are free to focus on key features and services that address the significant needs of users. Standards and specifications enable wider adoption of e-learning.

Specifications Create Flexibility

If we started our exploration of specifications with common needs, we end it with flexibility. In the beginning, we had a series of common problems that needed to be addressed. Take just two of the common issues encountered—course portability and the two-way exchange of enterprise information. Initial solutions to these challenges are often unique, requiring comparable effort to repeatedly solve the same problems across the universe of CMS implementations.

As specifications are developed to address the problem of moving course content from one system to another, or exchanging selected data between different CMS packages, limited resources can be liberated. These limited resources can then be better used, for example, to develop new content or to expand the range of services supported on the CMS. Greater flexibility is created, and, in turn, the CMS effectiveness is increased.

Similarly, if it becomes necessary to move to a new CMS environment, carefully implemented content packaging specifications already in place offer the flexibility to pack instructional materials and transfer them to a newly procured CMS. The key to this flexibility is a set of underlying specifications that provide solutions to common needs. If, however, the CMS used had yet to implement broad specification support, flexibility is limited, and once again it's necessary to create or procure tools for a specialized migration. Addressing underlying common needs with specification-based approaches is a critical component for next-generation course management systems.

Specifications Need Not Be Perfect

In the specification development world, technical perfection is low on the list of priorities. That is not to say specification working groups don't strive for this goal, but perfection in the midst of rapid technological change is unrealistic and fleeting, at best, and unnecessary. What is more important is a specification that can be implemented by the largest group of developers, a specification that provides the basic functionality required in a particular area. A necessary goal of a successful specification is the ability to develop future capabilities without losing prior capabilities. It must be backwards-compatible to be forward-looking.

Specification development frequently requires compromise from firmly held positions. In some cases, a vendor may believe that the right solution has already been reached and that anything less than its solution is inappropriate. Unless the vast majority of participants are willing and able to accept a single solution, then compromises must be made to achieve broad-based acceptability of the specification.

Specifications Can Be Good Enough

If a specification does not need to be perfect, what is "good enough"? Following our consensus model, most specifications focus on workable, acceptable solutions. Let's examine the concept of "good enough" in the context of a theoretical specification.

Involved in the process are a number of developers, each with his or her own particular set of development tools. As the specification is scoped, it is agreed that the issue is a core function for most e-learning environments. Several developers advance potential solutions to this issue. Developer A proposes an extremely elegant solution that takes full advantage of a proprietary technology. Developer B advances its own proprietary approach that, while not as elegant as Developer A's, is more inclusive of other developers. Developer C proposes a new approach that is more technology agnostic.

In this specification process, the key factor is finding a solution that all developers can implement that will permit development of a usable and scalable implementation on all platforms. The specification process must reach consensus on an acceptable solution to all—a solution that is good enough.

Resulting compromises may initially seem to decrease functionality. However, with this specification in place, developers can turn their attention to building on top of the specification to differentiate their CMS, knowing that the core need has been addressed.

SPECIFICATION DEVELOPMENT

The Specification Cycle

The specifications development process is essentially one of iterative work. The model in Figure 1 is followed by one of the central specifications

Figure 1. Generic specifications development process

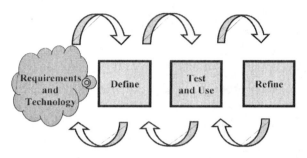

development bodies, IMS Global Learning Consortium, and is illustrative of this work.

Creating specifications is a community process. Once a need has been identified, there are at least three strategies by which a new specification can be derived: adopt the work of others as is, modify an existing specification to better meet current needs, or start from scratch and create something new. Rarely does the first strategy, adopting others' work without modification by the community, make sense, as specifications are a response to a problem of one sort or another identified by a community.

Modify Existing Specifications

Groups within a larger community often work on a common local approach to a problem with the expectation that other groups will see the value of their perspective and adopt it as their own. This approach to adoption and its challenges is often a driving force behind larger efforts to push for a common specification. A good example is Education Modeling Language, or EML.

EML was developed at the Open University of the Netherlands and released in December of 2000. The goal of EML is to address differences in the ways that learning processes are described by creating a common approach to the documen-

tation of teaching strategies and materials. This process then facilitates consistency, comparability, and reuse of learning materials distributed and "played" in electronic environments.

Like many similar efforts within a large and diverse community, the creators of EML found that adoption of their approach to describing learning activities failed to reach a critical mass, and overall progress toward a common method to describe learning designs was stymied.

To push for wider adoption for a learning design specification, the EML team engaged in an international specifications development forum, the IMS Global Learning Consortium, through the work group formed therein called the IMS learning design work group. They contributed their 1.0 version of EML as a candidate for consideration toward a new specification. Use case scenarios were written and compared. Through this community process, the original EML candidate specification was modified, in some ways substantially, and was ultimately released as the IMS Learning Design (see *www.learningnetworks.org*, for example), and the improvement of the specification itself (see *www.imsglobal.org/learningdesign/*) in February of 2003.

Standing on the Shoulders of Specifications

Arriving at an agreed specification for a part of the e-learning puzzle sets the stage for important subsequent work. The outcome of achieving the learning design specification has spawned a wide range of projects extending the design toward tools that use it. This is a benefit from achieving consensus in a specification; it generates practical tools that advance the discipline through broader engagement of the community of developers, instructors, faculty, and students. The experiences gained from these projects will generate valuable information to guide the next iteration of the specifications development process.

There are a number of "players" in the specification community that deserve mention here. The IMS Global Learning Consortium is the successor to the IMS Project, formed by the National Learning Infrastructure Initiative of Educause in 1997. After its initial three-year charge was fulfilled, the project spun off into an independent organization. IMS produces a range of specifications to support critical e-learning infrastructure needs. Key IMS specifications include content packaging, question and test interoperability, learner information package, enterprise specification, and simple sequencing. IMS has an active development process, and new specifications, as well as enhancement to released specifications, are produced on a regular basis. The organization includes vendors and consumers of e-learning products with a significant representation from academia. IMS specifications are developed within the organization and are then freely distributed to the public. Several IMS specifications have been adopted as key components of other initiatives such as ADL/SCORM and the Open Knowledge Initiative (OKI). In addition to specification development, IMS offers workshops and developer support. They also conduct briefing on their efforts and the general state of e-learning specifications.

OKI defines an open and extensible architecture for general-purpose infrastructure with a specific focus on learning technology targeted to the higher education community. OKI provides detailed specifications for interfaces among components of a learning management or other environment and open-source examples of how these interfaces work. The OKI architecture is intended for both commercial product vendors and higher education product developers. It provides a stable, scalable base that supports the flexibility needed by higher education, as learning technology is increasingly integrated into the education process. OKI defines an architecture that precisely specifies how the components communicate with each other and with other systems.

The architecture offers a standardized basis for development with proven, scalable technologies encouraging the development of specialized components that integrate into larger systems. OKI service interface definitions (OSIDs) make up the basic elements of the architecture. OKI is building a base XML generation, or "gen," code that can emit various object-oriented language bindings, as well as providing Java versions of these application programming interfaces (APIs) for use in Java-based systems. OKI's partners and developer community are providing open-source examples and reference implementations that use the APIs. OKI began as an independent project funded by the Andrew W. Mellon Foundation and continues through engagements with IMS, Centre For Educational Technology Interoperability Standards (CETIS), and other organizations worldwide.

The Aviation Industry CBT Committee (AICC) is an industry consortium that focuses on the development of computer-based training specification. With its roots in the aviation industry, the organization has narrower focus than other specification initiatives. However, the organization is mindful that it is important to create specifications that have broader application across the e-learning environment. Thus, AICC has produced a number of important specifications, referred to as "AICC Guidelines and Recommendations" (AGR), that have been adopted by other e-learning segments. The AICC offers both a self-regulated and independently certified compliance program for products against one or more of the nine AGRs.

The Advanced Distributed Learning (ADL) Initiative is an initiative lead by the U.S. Department of Defense in collaboration with other government agencies, industry, and academia. The primary goal of the project is to develop the framework for a learning environment that supports the interoperability of learning tools and course content meeting the needs of all initiative participants. One such reference implementation is the Sharable Content Object Reference Model (SCORM). To create the SCORM, the ADL documents, validates, promotes, and sometimes funds the creation of specifications and standards from other sources. Through a series of "co-labs," a wide range of SCORM implementations are tested. A principle of the initiative is to promote collaboration in the development and adoption of tools, specifications, guidelines, policies, and prototypes. The SCORM is a reference model that defines the interrelationship of course components, data models, and protocols so that learning content objects are sharable across systems that conform with the same model. The SCORM contains a collection of specifications adapted from global specification bodies and consortia to provide a comprehensive suite of e-learning capabilities enabling interoperability, accessibility, and reusability of Web-based learning content.

The Dublin Core Metadata Initiative focuses on a single goal, making it easier to find information. Through a series of working groups, Dublin Core develops interoperable online metadata standards that support a broad range of purposes and business models including specialized vocabularies that are defined to meet the needs of specific populations. Through a metadata registry and educational outreach, Dublin Core metadata is promulgated to and between the various metadata communities.

A FURTHER EXPLORATION OF STANDARDS

If standards are different from specifications, where are the differences? A key difference is the maturity of a specification. Specifications reflect an initial early consensus on an approach to addressing a technical issue. It is possible that multiple specifications might be developed to address a similar issue. That was the case with metadata, as several groups developed varying specifications to address the discovery of learning objects. The feedback loop is a key component to moving a specification to a standard. As specifications are implemented, issues emerge that require refinement of the specification. Sometimes these issues require changes in the actual specification; other times, issues force changes to the implementation. In either case, the specification is progressing through a maturation process, the end point of which is a declared standard.

When a specification is proposed as a standard, the submission may be from a specification body, or it may be from one of the developers who have implemented the specification. In either case, standards review is the final polishing of a single specification or the melding of multiple specifications into an agreed-upon whole.

Standards Trail the Marketplace

By the time a specification reaches a standards body, it may significantly trail implementations and acceptance in the marketplace as the standards process is a more deliberative process than specification development. Specifications are frequently developed in six months or less from the initial proposal to the adoption of version one of the specification. As specification implementations occur, the developer community is validating the specification through implementations. Over time, successful early adoption spurs more extensive adoption of specifications. The key for any specification is to promote early adoption and validation of the approach to a perceived problem the specification addresses.

Customers and developers often feed off each other in the early phases. As we pointed out earlier, specifications are developed in response to customer and developer needs. If customers are aware of a specification that addresses their particular needs, they frequently press their developer to implement a solution that has a specification-backed approach. Similarly, developers frequently find commonly adopted specifications a market advantage.

Thus, in fairly short order, the marketplace may coalesce around an implemented specification long before any standards body is ready to review the proposed solution. Going back to our electrical plug analogy, the plug has been specified, developers are including it in their products, and customers have grown to expect a common plug, but there is not yet the universal acknowledgment of our plug as the solution. At this point, the specification is running ahead of a standard in the marketplace.

Standards Reflect Further Consensus

Standards organizations are well aware of the specification development process. They understand that specifications are rooted in early agreement on approaches to key issues. For a standards organization, there is a responsibility to make sure that an adopted standard is more than a consensus — it must be a comprehensive solution. During the standards review process, the boundary of a submitted specification may expand to encompass related specifications, or the scope of the standard may be pared down.

As the standards body moves through its process, a primary outcome is the promotion of final consensus. Participants who have yet to implement a proposed standard are now taking a closer look at the details of the proposal. Issues are being clarified, and in some cases, further implementations are being validated. No proposal moves through a standards body quickly, without protracted inspection and consensus building.

The Role of Standards

Standards play an important role when multiple specifications address essentially the same issue. The International Organization for Standards (ISO) and IEEE are representative standards bodies at the end of our continuum. Frequently, specification organizations submit proposals to organizations like IEEE or ISO after the marketplace has had some time to implement and fully exercise a specification. The typical standards process will involve the review of any existing specifications and the formulation of a draft standard from the range of available options. Sometimes this draft standard will refine one selected specification; other times, the draft may include pieces of multiple specifications. Only after extensive deliberation will a standard be approved.

Melding Specification and Standards Development

Trying to accelerate the standards process, IEEE's Learning Technology Standards Committee (LTSC) began working on a metadata standard while several specification groups were still developing competing metadata specifications. In this role, the LTSC became another specification effort rather than waiting for the marketplace to settle out. The LTSC began to develop its own standard after efforts to combine the work of Dublin Core, IMS, and ARIADNE failed. Eventually, LTSC, IMS, and ARIADNE contributed their work to the LTSC. From these contributions and the standards process, the Learning Object Metadata standard was forged. Sometimes it is better to let the specification community do its work; sometimes the standards process needs a jump-start.

IMPLICATIONS FOR USERS

It is easy to think specifications are for technical staff members, who focus on arcane details deep within a CMS. In most cases, however, the requirements of users drive the specification process. This can be seen by exploring a number of user issues and their implications for specification- or standards-based solutions in next-generation course management systems.

To some, it might seem that the CMS market-place is coalescing into just a few vendors. The reality is that there are still a significant number of solutions available. One should anticipate next-generation course management systems will be more plentiful. Whether one believes that the market is contracting or expanding, portability is a key issue for most users. Unless one wants to rely on a single vendor to continue to meet e-learning requirements, the ability to move content easily from one solution to another is a high priority. Understandably, at procurement time, looking to your next CMS is not a high priority, but it should be a consideration in any procurement.

Maintaining Multiple Course Management Systems

In the early days of course management systems, it was not uncommon for a CMS to simply appear running on computers in faculty offices or on departmental servers. In this period, the prolif-eration of e-learning systems on campuses was significant. Low initial costs, high enthusiasm among new online learning advocates, and rapid development were major drivers of this prolifera-tion. Where there were commercial offerings, they were characterized by proprietary solutions, with each CMS implementing key features in its own specialized fashion.

As course management systems moved from the early adoption stage to broader campus usage, the need to maintain multiple solutions became a serious problem. Instructional technology services were frequently flooded with requests to imple-ment similar content and resources on multiple CMS platforms. Soon the bloom of excitement from new and evolving technology faded. Those actively utilizing a CMS in this period will re-member the issues it presented.

There were decidedly different responses to this situation. On some campuses, the selection and support of the CMS became an enterprise-wide issue. Many campuses limited the choices

and support to one or perhaps two solutions. Departments that were satisfied with their prior choice were sometimes forced to abandon their option.

EDUCOM's National Learning Infrastruc-ture Initiative (NLII) observed this problem and concluded that there was another way to address the issue through the development of e-learning specifications. NLII created the IMS Project, which evolved into the independent IMS Global Learning Consortium. One of its early goals was to create specifications that facilitate the interoperability of key components within course management systems.

Specialized Needs

To some, implementing specifications and standards is a limitation on choice, forcing all to implement the same capabilities. However, specifications organizations recognize that there are specialized needs, and further recognize that there may be more than one specification in the same domain that must be supported. Thus, specifications and standards need not only restrict but can also support specialized needs.

Sometimes the specialized needs of a constitu-ency or constituencies result in more than one specification addressing a particular need. One example of this is metadata, or information to de-scribe the development and uses of particular data. There are several widely recognized metadata specifications and standards. For example, Dub-lin Core Metadata, the LTSC's Learning Object Metadata (LOM), and the Metadata Encoding and Transmission Standard (METS) all offer different approaches to metadata specification.

To an individual not interested in metadata, the functional abilities of these approaches look similar. When procuring a CMS, multiple imple-mentations for metadata specifications should be supported. This will provide the greatest likeli-hood that future systems will support the needs of your community.

Specification Groups Want to Respond to Real Needs

Specification development efforts are dependent on good user input. Widely implemented specifications begin with user requirements and well-documented use cases. These two inputs become the benchmark for the specification development process. Think carefully about this. The technical staff developing a specification is critically dependent on users to describe their needs. If users skip this important opportunity for input, they may doom themselves to receiving technically sound specifications that misunderstand their needs. Similarly, the limited resources available for specification development may be misdirected to issues that do not address user needs.

Consequently, most specification development efforts are constantly seeking user input as they cannot effectively perform their task without user participation. Some of the most valuable specification development efforts extend user participation throughout the specification creation process. That way users can validate approaches as they surface. As a specification is released in draft form, both technology developers and users are critical to the validation of the proposed specification.

Users Have a Forum for Input

Developing a specification involves a variety of roles. These include system architects and developers as well as users. A common comment from user constituencies, especially academia, is that specifications are not fully addressing the needs of the constituency. One of the most frequently heard pleas from specification groups is for requirement documents and use cases. This circular reference problem is broken when users provide their active input into the process. It is important to emphasize that users focus their contributions on the requirements and use cases rather than on architectural or technical solutions.

Compliance

Before we launch into some of the key players in specification development, it is important to add a word about specification compliance. As specifications are developed and then adopted or supported by a vendor, there is an expectation that someone has certified these specifications. This concept, known as compliance testing, is a costly process. Frequently, specifications have no formal compliance process, so care must be taken to clearly understand claims of compliance. While it is possible for a vendor to fully conform to a specification, there may not be full compliance certification because of a lack of testing.

This lack can lead to significant problems during implementation when the scope of the compliance is tested against real-world needs. Imagine, for example, that a company has implemented a specification that is fully operational and functions as expected within the test system. However, during implementation, it is discovered that the specification, as implemented, fails to scale to the volume of the transactions the system generates. A similar problem can develop in a multivendor environment when all vendors indicate compliance with a specification but the various implementations do not interoperate. System one cannot exchange information with system two even though the specification indicates that information and data exchange is required. These problems can be addressed in formal compliance testing. However, if there is no formal compliance testing, it is essential to define the expectations and to understand a vendor's implementation of the specification. Ways to resolve specification compliance issues should be clearly spelled out in any procurement.

Specification development groups are mindful of these issues and do work to promote solutions to these problems. However, consumers of course management systems that rely on these specifications should be prepared to pressure vendors and specification development efforts to clearly define reasonable expectations for self-claims of compliance. In the case of the specifications groups, active participation in the process will be key to pressuring for accurate claims of compliance.

THE ROLE OF SPECIFICATIONS AND STANDARDS IN THE RFP PROCESS

The importance of a well-thought-out request for proposal (RFP) cannot be emphasized enough. When it comes to specifications and standards, it is not sufficient to include only simple statements such as "must support the XYZ specification." This vague requirement leads to an evaluation process that is little more than checking boxes for proclaimed specification compliance. In this scenario, vendors can respond to such a requirement in the affirmative with only minimal implementation.

This is not a criticism of the vendor community. A user may ask for support of the XYZ specification, the vendor may have implemented it, but the implementation may not meet the user's needs. Or, in some cases, specification support is available, but the core product may largely ignore the specification, and there may be significant expenses to having a meaningful implementation of the specification.

While the vendor may be technically meeting the bid requirements, it is the campus or user that will suffer the consequences. The fault in a vague RFP is a lack of understanding on the part of the campus as to the full implications of required specifications. Failing to fully understand the implication of the specifications leads to a lack of specificity in the RFP. The fiscal, time, and functional penalties for this approach are significant. Properly articulating both the scope of the requirement and the importance of the specifications and standards compliance is critical to a successful RFP.

To overcome this potential problem there are some important details to include in an RFP. In a general sense, and particularly in the case of specifications and standards, it is helpful to build a use case for the key aspects of an RFP. The use case should include functionality and scaling metrics to illustrate use patterns. For example, a campus might need to use the same content on more than one CMS, or there might be a need to assure an easy transition of content to a next-generation CMS. A thorough use case would help the campus and the vendor understand the scope of the users' expectations and needs.

Especially in the latter case, moving to a next-generation CMS and adopting a specification-based approach would provide the greatest probability of packaging and moving content. Remember, while the new vendor is likely to support the import of content from a competing CMS, they are less likely to be interested in helping move it to the next system.

A second improvement to any RFP would be the careful selection of key specifications and standards. In some cases, a particular organization is addressing a range of e-learning issues via specifications, while other organizations are specializing in a single specification. It is important to do your homework to understand the various specifications available to you and the vendor community. Most specification organizations will point to vendors that have implemented a particular specification. A sample implementation may also be available to help you evaluate a specification for inclusion in your RFP.

Later we will be suggesting some specifications that should be high on a user's list of potential specifications, but even that list should only be a starting point for consideration. For example, there are multiple metadata specifications with slightly different approaches. Users should take the time to understand the implications of choosing a particular specification. This includes functional and technical implications, as well as the integration between specifications and the new CMS.

Table 1. Specifications and standards

Specification	URL
Dublin Core Metadata	http://dublincore.org/documents/
EduPerson	http://www.educause.edu/eduperson/
IEEE Learning Object Metadata	http://ltsc.ieee.org/wg12/index.html
IMS Specifications	http://www.imsglobal.org/
Metadata Object Description Language	http://www.loc.gov/standards/mods
MPEG	http://www.chiariglione.org/mpeg/standards.htm
OKI Authentication, Authorizations, Filing, and Digital Repository	http://sourceforge.net/project/showfiles.php?group_id=69345&package_id=68278
MPEG7	http://www.chiariglione.org/mpeg/standards/mpeg-7/mpeg-7.htm
SOAP	http://www.w3.org/TR/soap/
WSDL	http://www.w3.org/TR/wsdl

Consider an RFP to be an opportunity to communicate to the vendor community about which specifications are important. This is important with existing specifications, but is also critical to the development of emerging specifications.

Early on, vendor support of specification efforts was influenced by the simple inclusion of the "must support XYZ specification" statements. Both the vendor and user community should be well beyond these types of simple statements. Inclusion of detailed use statements and awareness of current and emerging specifications demonstrates to the vendor community the importance of supporting bedrock specification efforts.

KEY SPECIFICATIONS AND STANDARDS

The specifications and standards shown in Table 1 should be considered during the procurement of a next-generation course management system. Specifications and standards are under continual development. It is important that users track the specifications and standards under active development.

CONCLUSION

E-learning specification and standards organizations are setting the stage for the next generation of course managements systems. Users need to engage in the process to express their requirements for e-learning software. Users who rely solely on the vendor community to integrate specifications and standards into proprietary course management systems are likely to remain in the current realm of course management systems.

Institutions and users interested in next-generation systems have a number of opportunities to both advance their progress and assure themselves and their institutions that they are investing in the future rather than the past. Tracking key specification efforts as an observer is a first step. Understanding the issues being addressed and matching them to user needs provides an important benchmark. During the procurement of the next CMS at an institution, users should take extra care to have potential vendors fully articulate their positions on adopting and complying with key e-learning specifications and standards. Users should develop their own use cases and compare them against the vendor's implementation. Users must fully understand how the vendor will address these issues.

Institutions capable of more fully engaging in the specification effort are encouraged to join the key organizations described in this chapter, providing requirements and use cases as the basis for development. These institutions should also participate in the work groups that create the specifications and standards that will enable next-generation course management systems.

REFERENCES

DIN (German Institute for Standardization). (2004). *Learning sequence.* Available at http://eduplone.net/products/learningsequence/

DIN. (2000). *Economic benefits of standardization: Summary of results final report and practical examples.* Available at http://www.din.de/set/aktuelles/benefit.html

Duval, E. (2004). Learning technology standardization: Making sense of it. *Computer Science and Information Systems,* 133-143. Available at http://www.comsis.fon.bg.ac.yu/ComSISpdf/Volume01/InvitedPapers/ErikDuval.pdf

elive Learning Design. LD-Suite. Available at http://learningnetworks.org/forums/showthread.php?s=&threadid=202

European Committee for Standardization. http://www.cenorm.be/cenorm/index.htm

Nissi, M. (Ed). (2003) *Making sense of learning specifications and standards: A decision maker's guide to their adoption* (2nd ed.). The Masie Center, Learning Technology & e-Lab Thinktank. Available at http://www.masie.com/standards/s3_2nd_edition.pdf

Reload: Reusable eLearning Object Authoring and Delivery. Available at http://www.reload.ac.uk/

Smythe, C., Evdemon, J., Sim, S., & Thorne, S. (2004). *Basic architectural principles for learning technology systems* (DRAFT). Available from Alt-i-Lab Topical Working Sessions, http://www.imsglobal.org/architecture.pdf

Why e-Learning Standards? Available at http://careo.prn.bc.ca/losc/mod3t1.html

ENDNOTE

[1]   http://sourceforge.net/okiproject

This work was previously published in Course Management Systems for Learning: Beyond Accidental Pedagogy, edited by P. McGee, C. Carmean, and A. Jafari, pp. 14-38, copyright 2005 by Information Science Publishing (an imprint of IGI Global).

Chapter 8.10
A New Taxonomy for Evaluation Studies of Online Collaborative Learning

Lesley Treleavan
University of Western Sydney, Australia

ABSTRACT

In this chapter, the literature of online collaborative learning (OCL) is extensively reviewed for contributions to evaluation. This review presents a new taxonomy for evaluation studies of OCL, identifying studies of students' experiences, studies of instructional methods and sociocultural studies. Studies that focus on evaluating students' experiences engage approaches from phenomenology and ethnography to explore students' perceptions of collaborative learning. Instructional method studies attend to evaluation of the tools, techniques and outcomes. Sociocultural studies emphasize the socially constructed nature of the teaching and learning processes and are concerned, therefore, with evaluation in its social context. The sociocultural studies fall broadly into three clusters: pedagogical studies, linguistic studies and cross-cultural studies. The analysis highlights the need for theory-driven empirical evaluation of OCL. Accordingly, three theoretical frameworks for OCL evaluation are discussed. Emphasis is placed on a Communicative Model of Collaborative Learning, developed from Habermas' Theory of Communicative Action, for its contribution to evaluating what takes place within the social context of students' communicative practices that is productive of collaborative learning in an online environment.

INTRODUCTION

Collaborative learning, especially as it can now be supported by computer-mediated communication, is receiving significant attention by those concerned with developing higher education students' capabilities to meet the increasingly complex challenges of working in a postmodern world. Students indicate that they often enjoy and benefit from collaborative learning with and from their peers, when equity and group dynamics are appropriately addressed. Teachers are engaging new online communication tools and assessing different instructional design techniques to enable better collaborative learning and to improve student learning outcomes (Bonk & Dennen, 1999; Freeman, 1997). Higher education institutions are investing in Web-based learning systems and e-learning support units (Sheely, Veness & Rankine, 2001) often in the hope of saving costs associated with face-to-face delivery of instruction and in attempts to capitalize on globalized higher education markets.

But is collaborative learning actually happening in these new electronically-mediated spaces? What is the nature of that collaborative learning and, most importantly, how can we evaluate the quality of this collaborative learning? How do we know it is worth doing now, and continuing to improve? For whom is evaluation undertaken: students, teachers, educational developers, designers, e-learning specialists or those who stand beyond participation, yet hold the purse strings? What would we evaluate for such different audiences? And would such differently targeted evaluation be undertaken in the same way? Does a cautionary call need to be sounded against those in higher education whose agendas are shaped by assumptions that technologically advanced, flexible delivery, necessarily equates with learning, collaborative or otherwise? This chapter sets out to explore some of these questions, while raising others and, hopefully, provoking still more.

Whereas earlier chapters in this volume have defined and explored the nature of online collaborative learning (OCL), this chapter examines the evaluation of OCL. Evaluation has been defined by Gunawardena, Carabajal and Lowe (2001) as:

A systematic and purposive inquiry that includes the collection, analysis and reporting of data relating to the efficiency, appropriateness, effectiveness, and value of operational characteristics and outcomes of a procedure, program, process or product. (p. 3)

Such a definition needs to be applied to perhaps four questions: Why are we evaluating? For whom are we evaluating? What are we evaluating? How will we evaluate these aspects?

Evaluation of OCL requires attention not only to processes of collaborative learning, but also to the means through which they are achieved—computer-mediated communication. Furthermore, the goal of collaborative learning is understood here, as Ronteltap and Eurelings (2002) state, to "create a situation in which productive interactions between learners can be generated" (p. 14). This notion of productivity, what the learners produce together, necessarily involves evaluating the different contexts, processes and outcomes that facilitate and support such productivity. Evaluation in these new contexts challenge traditional approaches to evaluation and require new theoretical frameworks to guide analysis and interpretation.

Three components of collaborative learning have been identified by Brandon and Hollingshead (1999) as collaboration, communication and social context. While communicative processes have been examined in linguistic studies and the social contexts of these new online learning communities have been the subject of many sociological studies, the majority of the literature reports case studies that only evaluate collaborative learning implicitly, as they focus principally on its perceived benefits in achieving learning outcomes.

Much of our understanding as practitioners, who are deeply committed to the value of collaborative learning as we observe our students engage more fully when we incorporate it into our programs, and as researchers, who are attempting to study those transformative shifts facilitated by collaborative learning processes, rests powerfully on our tacit knowledge (Tsoukas, 2003) of what actually happens within these shared, interactive online spaces. What is therefore required is a theoretical framework that enables explicit and systematic investigation of what takes place within the social context of students' communicative practices that is productive of collaborative learning. In this way, the integration between the three components—collaboration, communication and social context—are investigated holistically, rather than reduced to their three constituent components.

Accordingly, this chapter argues that OCL may be effectively and usefully evaluated from within the students' online communicative practices. Three theoretical models that facilitate evaluation from such contextualized locations are presented: the collaborative learning model (Ronteltap & Eurelings, 2002), the interaction analysis model (Gunawardena, Lowe, & Anderson, 1997) and the communicative model of collaborative learning (Cecez-Kecmanovic & Webb, 2000a). The latter is advanced here as an appropriate theoretical framework for theory-driven empirical evaluation of OCL, based on its application in a number of empirical studies elsewhere (Treleaven & Cecez-Kecmanovic, 2001; Treleaven, 2003a; Treleaven, 2003b). This chapter is, therefore, structured into three parts. First, in the major part, the literature of OCL is extensively reviewed for its contributions to evaluation. This wide-ranging review is presented within a new taxonomy for evaluation studies of OCL. Second, from this review, the relative absence of rigorous evaluation and appropriate theoretical models informing evaluation in the expanding body of OCL research is highlighted. Three theoretical frameworks, notably those of Ronteltap and Eurelings (2002), Gunawardena, Lowe and Anderson (1997) and Cecez-Kecmanovic and Webb (2000a) are elaborated. Particular attention is paid to the communicative model of collaborative learning (Cecez-Kecmanovic & Webb, 2000a) for its evaluation of OCL by examining students' linguistic acts as they are produced on electronic bulletin boards. Third, the conclusion points to the value of a theoretical model, such as the communicative model of collaborative learning, as a pedagogical tool for effectively evaluating not only students' collaborative learning but also for further improving and testing the design of OCL.

LITERATURE REVIEW OF OCL EVALUATION

This section reviews the nature and extent of the literature in the area of evaluating OCL (Mason, 1992; Alavi, 1994; Gunawardena, Carabajal, & Lowe, 2001). Clustering distinct approaches within the literature demonstrated clearly that no single study fits neatly into one mutually exclusive category. Nevertheless, it is useful to identify three trends in the literature concerned with evaluating OCL: studies of students' experiences, instructional method studies and sociocultural studies. In turn, the focus of each domain is slightly different, respectively: students' perceptions, tools and techniques, and teaching and learning processes in a social context. This range of evaluative literature has extended significantly in the last decade, beyond the surveys (online and hard copy), case studies, empirical experiments and quantitative analyses (messages sent, logons, replies, threads), identified in Mason's 1992 review of methodologies for evaluating computer conferencing.

Studies of Students' Perceptions and Experiences

It is notable that in much of the literature on OCL, there is an implicit focus on evaluation. Often such implicit evaluations are descriptively orientated toward the students' experience of learning based on phenomenographic theories (Gosling, 2000). For these studies borrow implicitly from ethnographic methods as they focus on the general value of online learning in a wide range of settings, while they also borrow from phenomenology as they investigate the students' experiences and perceptions of their online learning (Sanders & Morrison-Shetlar, 2001; Sullivan, 2001; Gallini & Barron, 2002; Askov & Simpson, 2001; Weiner, 2002). Fewer, however, are focused on the value of their collaborative learning.

Illustrative of phenomenological studies are Kitchen and McDougall's (1999) examination of graduate students' perceptions of the educational value of their collaborative learning in a course delivered via computer-mediated communication. Their research is based on interviews and analysis of students' communication practices, in order to determine the most important aspects associated with their collaborative experiences. Although students report enjoying the convenience and opportunity for collaboration, some also report dissatisfaction with the instructional strategy and the delivery medium. McAlpine (2000) examines the stimulus that students receive from the online learning approach in a Master's course and the value they perceive in learning from colleagues and lecturers in a collaborative way. McIsaac, Blocher and Mahes (1999) investigate student and teacher perceptions of interaction in online computer-mediated communication and suggest principles for distance educators to incorporate into online classes, such as providing immediate feedback, participating in the discussions, promoting interaction and social presence, and employing collaborative learning strategies. Salmon (2000), in her practical guide to e-moderating students' learning, draws on such evaluative approaches.

Instructional Method Studies: Tools, Techniques and Outcomes

Second, other more targeted investigations evaluate the effectiveness of different online learning tools and techniques for collaborative learning, and thereby student outcomes, in a wide range of settings (for example, Collings & Pearce 2002; Chen, Liu, Ou, & Lin, 2001; Chang, 2001; Nakhleh, Donovan, & Parrill, 2000; Chalk, 2000). Typifying this approach is Ravenscroft and Matheson's (2002) evaluation of two collaborative dialogue games, finding that they produce significant improvements in students' conceptual understanding. However, since these tools are differentially successful depending on the nature of the conceptual difficulties experienced by the learners, they conclude that developments in collaborative e-learning dialogue should be based on pedagogically sound principles of discourse. Bonamy, Charlier and Saunders (2001) focus on evaluation of products and how these can be useful "bridging tools." Nevertheless, they also see a role for evaluation beyond this specific focus on tools. In their project with a network of teachers, researchers and learners from different European institutions, Bonamy et al. (2001) adopt an approach to evaluation that is characterized by an evaluation distinguishing between individual experience and institutional contexts and an evaluation using monitoring data and qualitative validation. In the search for new knowledge in the area of changing practices in learning, their useful study points to the importance of attending to evaluation explicitly.

Relatedly, another cluster of studies analyze the link between online instructional techniques and different student outcomes (such as Grossman, 1999; Redding & Rotzien, 2001; Ellis & Cohen, 2001). Much of this work is evaluative and of interest for teachers wanting to improve course designs. For example, Alon and Cannon (2000) analyze an Internet-based learning forum that aims to link student teams in international collaborative learning projects. The purpose of the forum is to empower students to participate in setting learning goals and learning processes, and enabling instructors to be closer to the student in the learning process. The article reports on the experience of one college that used the forum and discusses likely outcomes that may emerge from using Internet-based experiential projects in the classroom. Nevertheless, it is not the concern of such research to focus on whether it is OCL, in particular, rather than some other features of the course design or implementation, that is contributing to student outcomes.

Sociocultural Studies: Socially Constructed Pedagogy Mediated by Language

The sociocultural studies fall broadly into three clusters: pedagogical studies, linguistic studies and cross-cultural studies. What they share is an understanding that learning occurs in a social context. As such, evaluation of learning needs to take into account the nature of the learning environment and social interaction processes, as well as the tools and techniques employed. What distinguishes the sociocultural studies from studies that give more emphasis to "instruction" via tools and techniques of "delivery" is recognition that meaning, and thus learning, is socially constructed. So the social context, especially, but not only, its cultural and cross-cultural demographics within which meaning is produced, becomes an integral part of the research.

Pedagogical Studies

Many pedagogical studies examine online learning communities and collaborative learning from a sociological orientation. They adopt the perspective that learning is a social process and cannot be seen independently of the social context (Tu & Corry, 2001; Kumpulainen, Salovaara, & Mutanen, 2001; Ronteltap & Eurelings, 2002). Tu and Corry (2001) emphasize the importance of the construction of online learning communities. Their study examines an online learning community drawing on Goffman's (1974) self-presentation and Short, William and Christie's (1976) social presence theory. They critique the limitations of research into online communities for the short-term period of study and analytic focus on products at the expense of the individual engaged. A further study by Tu (2002) examines the learner's perception of social presence in three computer-mediated communication systems. Using a social presence and privacy questionnaire, Tu identifies three dimensions of social presence

as social context, online communication and interactivity.

Some very useful, systematic work has been undertaken by Ronteltap and Eurelings (2002), who analyze collaborative interactions for quantity (what types of learning issues generate most interactions?) and quality (what types of learning issues generate the highest level of information processing?), using a model that is discussed later in the chapter. Their highly structured experimental research examines the productivity of small group learning and the level of cognitive activity. Thus they classify documents according to whether they are low order cut and paste, summarizing or higher level original postings that require more processing of information with synthesis, interpretation, reflection and referencing to other contributions. They note that although tools and functionality are important, these do not necessarily produce interactions of a quality that lead to learning.

Other studies, such as Kumpulainen et al.'s (2001) examination of sociocognitive processes, evaluate multimedia learning while paying attention to the social interaction that takes place. Their investigation is facilitated in this regard by the use of a wide range of interactive media for their research (video and audio recordings, online observations, interviews, questionnaires and assessments of students' poster displays). Thus, they are able to summarize, in case-based analytic descriptions, the nature of students' navigation processes, social interactions and cognitive activities in the multimedia context. The authors conclude that more attention has to be paid to the design of instructional situations and pedagogical supports for multimedia-based learning.

Linguistic Studies

Research into dialogue and online discussions (Weasonforth, Biesenbach-Lucas, & Meloni 2002; Collot & Belmore, 1996; Yates, 1996), like much of the literature on OCL, has not given much attention to developing a theoretical model for the analysis of communicative practices in collaborative online learning (Gunawardena et al., 1997). Again, they focus on identifying features of the dialogue and discussions, examining the processes that take place online and sometimes comparing them with offline processes (Johnson & Johnson, 1996; Curtis & Lawson, 1999). Furthermore, as Jones (1998) argues, evaluation techniques, such as content analysis and single features of computer-mediated communication (Mason, 1992; Henri, 1992), do not include relevant data like casual chatting or ad hoc activity in student texts.

Nevertheless, many linguistic studies are located broadly within a theoretical framework that supports attention to the socially constructed and socially mediated processes of learning and meaning production. These studies draw on an ethnographic approach. For example, Jones (1998) explores online group work using ethnographic techniques, and argues that this approach stresses the social context, refusing "a priori categories" of analysis, in order to concentrate specifically on the learning process.

This focus on the social processes mediating online communication has been researched by Tidwell and Walther (2002). They explore the exchange of personal communication and effects of communication channels on self-disclosure, question asking and uncertainty reduction between online partners.

A study by Hron, Hesse, Cress and Giovis (2000) raises significant questions for OCL and for evaluation, in particular. Their experiment tests the use of two different methods of dialogue structuring to keep the conversation coherent within virtual learning groups. Implicit structuring of collaborative exchanges elicited discussion on the subject matter and on key questions, arising from learning already undertaken together. In contrast, explicit structuring of dialogue by the designer provided additional rules for discussion,

encouraging students to argue and participate equally. Both implicit and explicit structuring of discussion facilitate stronger orientation to subject matter and less "off-task talk," than in unstructured groups. Nevertheless, a post-test was not able to distinguish any differences in knowledge between the groups using the different dialogue approaches. Their finding points to the difficulty of such comparative evaluation, for it relies heavily on the effectiveness of the assessment of student learning.

The implication of such a statement is to question what we seek to evaluate and how we assess the kinds of knowledge that pedagogically oriented approaches may produce. In the context of knowledge (Tsoukas, 2003), can we indeed assess students' knowledge adequately and thus usefully evaluate the learning environments, behaviors and mechanisms that are employed in their online courses? It is for this reason that evaluating from within, where collaborative learning is enacted and productivity displayed, that the CMCL discussed below is so appealing.

Cross-Cultural Studies

As Web-based delivery of learning is employed for more diverse student populations, studies that evaluate OCL by taking into account demographic differences and cross-cultural factors are of increasing importance. With the range of flexible delivery of asynchronous learning programs set in global markets for higher education, the demographics of students are increasingly more diverse in age, culture, ethnicity, language, work experience and familiarity with technology (for example, Bates, 2001; Lauzon, 2000; Warschauer; 1998, Herring, 1996). Gunawardena, Nolla, Wilson, Lopez-Islas, Ramirez-Angel and Megchun-Alpizar (2001) conducted a cross-cultural study of group process and development in online conferences between participants in Mexico and the United States. In their survey, they identify significant differences in student perception of the norming and performing stages of group development. The groups also differed in their perception of collectivism, low power distance, femininity and communication. Some of the challenges in cross-cultural studies, noted by Gunawardena et al. (2001) are finding equivalent samples for quantitative research and developing qualitative research that conceptualizes identity issues, so as to move past simplistic stereotyping and better understand how people define themselves.

Evaluating the impact of these cultural differences within and between groups becomes important in the context of OCL. For the dynamics of social interaction, and hence collaborative learning, are demonstrably affected by such differences. Evaluation studies in this specific context challenge pedagogical assumptions and highlight issues of generalizability across different cohorts and individual differences.

THEORETICAL FRAMEWORKS FOR EVALUATING OCL

Given the untheorized nature of much of the evaluation that is undertaken both implicitly and explicitly (Arbaugh, 2000), the findings of Alexander and McKenzie's (1998) investigation into innovation in information and communication technology (ICT) education in Australia are not surprising. They found that commonly used evaluation tools/forms at universities are inadequate for the purpose of improvement, and that ICT educators usually do not have the skills to implement educational designs that could improve their teaching. Following their recommendation to develop better ways of evaluating whether educational programs meet their intended aims, the Australian Universities Teaching Committee (AUTC) funded an action learning evaluation project conducted by the Computing Education Research Group at Monash University (Phillips,

2002). The learning-centered framework (for evaluating computer facilitated learning) adopted in this staff development intervention is based on work by Alexander and Hedberg (1994) and Bain (1999) and encompasses four stages — analysis and design, development, implementation and institutionalization. Arguably, the outcomes of such projects directed toward developing evaluation skills and wide-reaching change in practices need to be informed by, and may be enhanced by, empirical research into OCL evaluation, which is soundly conceptualized and theory-driven.

Toward addressing this gap in the empirical OCL research to date, three theoretical models from the literature of OCL are now discussed. Each of these models take into account the social interaction processes involved in OCL and have been developed by Ronteltap and Eurelings (2002), Gunawardena et al. (1997), and Cecez-Kecmanovic and Webb (2000a).

The first two frameworks are briefly presented before the latter is discussed as one approach that is proving useful in terms of theorizing, and then evaluating, students' communicative practices and knowledge creation from within the collaborative learning spaces.

Collaborative Learning Model

A major contribution to identifying a framework for what requires evaluation in OCL has been developed by Ronteltap and Eurelings (2002). In their model, they distinguish the learning environment that in turn mediates the learning behavior engaging with a learning mechanism. These three dynamics and their inter-relationships are represented below in the schema (Figure 1) developed to illustrate their model. These elements, which can be employed generally for both design and analysis, are utilized to plan and evaluate their own study. The learning environment includes the curriculum and its learning materials, assessment processes and learning

Figure 1. Collaborative learning model (Schema based on Ronteltap & Eurelings, 2002)

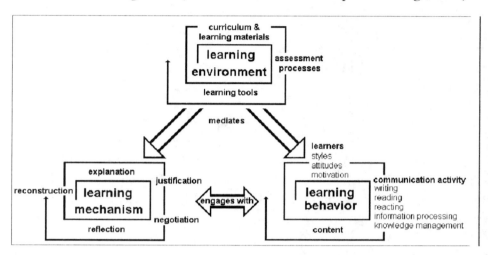

tools. The learning behavior refers to all aspects relevant to studying the interactions between learners, and thus includes learners (their learning styles, attitudes, motivation), activity prior to and during communication (writing, reading, reacting, information processing and knowledge management) and content. The learning mechanisms are generated by communicative activity and include explanation, justification, negotiation, reflection and reconstruction.

In Ronteltap and Eurelings' (2002) first application of the model to an experiment, their work points to how higher levels of information processing are achieved through issues related to practical learning. Their model is also helpful in articulating the transformative effects of asynchronous online collaboration as the opportunity for collaborative application of knowledge, reflection and restructuring of that knowledge by participants. The future directions of their work toward a theory of mediated action as a framework for analyzing the inter-related processes within the model holds potential for further theoretical developments applied to evaluating OCL.

Similar work, based on systems theory and earlier grounded theory, by Gunawardena and her colleagues (Gunawardena et al., 2001; Runawardena et al., 1997) parallels the systems, interactivity and outcomes of Ronteltap and Eurelings' (2002) framework. However, Gunawardena et al. (1997) applied their model for content analysis that dealt with only one part of this social interaction. In this later work they emphasize content, context, collaboration and control, and the interrelations between them. Like Alexander and Hedberg (1994), Housego and Freeman's (2000) conceptualization of Web-based learning operating at different levels adds the very important institutional dimension to a general evaluative framework.

Interaction Analysis Model

Recognizing that the development of theory on which to base collaborative learning in higher education is much needed, Gunawardena et al. (1997) developed an interaction analysis model for examining the social construction of knowledge in computer conferencing.

This model was designed to help answer two evaluation questions: namely, was knowledge constructed within the group by means of the exchanges among participants? and, second, did individual participants change their understanding or create new personal constructions of knowledge as a result of interaction within the group? (Gunwardena et al., 1997). Given that the professional development participants engaged online for only one week, the model and methodology itself is of more interest here than the substantive findings from its application.

Their model outlines five phases in the co-construction of new knowledge within which different types of cognitive activity (such as, questioning, clarifying, negotiating and synthesizing) takes place. The five phases are identified as sharing/comparing, dissonance, negotiation/co-construction, testing tentative constructions and statement/application of newly constructed knowledge.

Communicative analysis of the phases of knowledge co-creation pays attention not to the threads and the specificity of messages but to the broader evaluation of whether engagement in a collaborative forum successfully produces new knowledge. The researchers themselves have subsequently claimed that this model, focusing as content analysis tends to do on one aspect of the learning process itself, has not adequately engaged the issues of evaluation (Gunawardena et al., 2001). Nevertheless, their approach to investigating knowledge creation and sharing online may well be worth building on.

The Communicative Model of Collaborative Learning

The theoretical framework that shapes Cecez-Kecmanovic and Webb's (2000a) communicative model of collaborative learning is developed from Habermas' (1984) theory of communicative action. Despite critiques of Habermas' value for

communication in teaching (see Heslep, 2001), other studies have productively drawn on his theory. For example, Gosling (2000) evaluates participants' perceptions of negotiated learning against the tutor's declared philosophy of "the ideal speech conditions." Ekstrom and Sigurdsson (2002) investigate international collaboration in nursing education through the lens of Habermas' theory, analyzing the influence that politics and economics have on active communication and the potential benefits of shared meaning and understanding achieved by interaction and discourse.

The communicative model of collaborative learning (CMCL) is based on three assumptions, according to Cecez-Kecmanovic and Webb (2000a): first, that collaborative learning is enacted and mediated by language; second, that collaborative learning involves processes of social interaction and third, that acts of communication or language acts function as social interaction mechanisms through which collaborative learning and knowledge co-creation processes may be produced.

The CMCL (Table 1) identifies and classifies language acts as constituents of collaborative learning along two dimensions: the dominant orientation of learners and the domain of knowledge. First, the model identifies an orientation to learning (shown as a desire to know and to interact with others to increase mutual understanding and construct knowledge cooperatively), then an orientation to achieving ends (shown in students' primary motivation to achieve a goal, such as passing the subject, getting a good mark or getting the best mark in the class) and an orientation to self-representation and promotion (shown in students' attempts to impress others by portraying a particular self-image).

Second, the model differentiates between language acts that refer to different domains of knowledge, such as those related to subject matter and any substantive issues (theory, application, problem solving, etc.); linguistic acts addressing norms and rules that regulate the conduct of interactions and interpersonal relations in the collaborative learning process; and linguistic acts addressing personal experiences, desires and feelings by which students express themselves and shape both their individual and collective sense of self and of their learning processes.

Using this 3x3 matrix, the CMCL enables classification of linguistic acts produced in specific learning situations. Communicative analysis based on this model takes into account both the knowledge domain (subject content, norms and rules, and personal experience) that a specific linguistic act refers to and, at the same time, its orientation (learning, achieving ends and self-representation) that shapes the productivity within the learning situation. Accordingly, the nature of the linguistic act and its productivity, and how it contributes and what it enables (in the flow of linguistic acts with and between postings) in the construction and maintenance of collaborative learning processes can be analyzed and differentiated. From such analysis, the type and extent of collaborative learning taking place can be identified.

To illustrate, the specific linguistic act may be of the same type, e.g., subject matter, however, what it actually produces depends on the student's orientation. A student oriented to learning may seek mutual understanding with other students on a topic, collaboratively creating new knowledge. Whereas, a student oriented to achieving a passing grade may attempt to fulfill course requirements instrumentally, without necessarily engaging in collaborative learning. A student oriented to self-representation may present their posting on a subject in such way as to try and make an impression on some members of their cohort, neglecting the ongoing processes of developing understanding that lead to knowledge sharing and creation. It is important to note, therefore, that interpretation of

a linguistic act is always within the context of the learning situation itself and the flow of linguistic acts constituting that learning process.

While there is insufficient space in this chapter for more specific illustration, the CMCL has been applied, tested and developed as a theoretical framework in a collaborative action research study (Reason & Bradbury, 2001) of OCL in a Web-enhanced learning environment. The empirical investigation of students' communicative practices in an undergraduate management subject has been documented in a number of research papers, focusing on different aspects of the study (Treleaven & Cecez-Kecmanovic, 2001; Treleaven 2003). Central to this empirical investigation has been the deployment of the CMCL to analyze sets of student postings on an electronic bulletin board for the dominant modes of linguistic acts, and the flow between these postings as knowledge is created collaboratively. The first phase of the study traced the development and productivity of a collaborative learning space on an electronic bulletin board, by examining the numbers and types of postings made throughout the semester. Student demographics (gender and English as a second language) related to participation were also investigated. A number of methodological and pedagogical implications raised in this first action research cycle informed improvements in the Web-mediated learning design of the subject. The second phase examined the revised design, evaluating the extent and nature of the collaborative learning facilitated by more structured online discussion. Accordingly, the pedagogical implications for learners and for designers were the major outcomes of the second action research cycle. These two cycles are brought together elsewhere (Treleaven, in preparation) within the newly developed taxonomy here for examining the evaluation of OCL.

Table 1. CMCL (Source: Cecez-Kemanovic & Webb, 2000a, b)

Dominant orientation to \ Knowledge domains	SUBJECT MATTER (1)	NORMS AND RULES (2)	PERSONAL EXPERIENCES, DESIRES AND FEELINGS (3)
LEARNING (A)	**A1** - Linguistic acts about subject matter raised in order to share views and beliefs, to provide arguments and counter-arguments leading to mutual understanding and knowledge creation	**A2** - Linguistic acts that establish norms and rules regarding interaction and collaboration; cooperative assessment of legitimacy, social acceptability and rightness of individual behavior	**A3** - Linguistic acts expressing personal views and feelings about learning process and other learners, aimed at sharing experiences and increasing mutual understanding
ACHIEVING ENDS (B)	**B1** - Linguistic acts that raise or dispute claims and provide arguments about subject matter, with an intent to frame attention, influence others and achieve goals	**B2** - Acts of changing or interpreting norms and rules about the interaction process, so as to suit a particular student interests and goals (may be at the expense of others)	**B3** - Acts expressing personal experiences in a way that influences other learners and instructors, so as to help achieve goals (e.g., emphasizing personal success)
SELF-REPRESENTATION AND PROMOTION (C)	**C1** - Raising or disputing claims and arguments as a performance on a stage that serves personal promotion (often neglecting an ongoing argumentation process)	**C2** - Raising or disputing claims about norms/rules or their violation in order to attract attention and establish oneself as a distinguished student (e.g., a leader, an authority, etc.)	**C3** - Linguistic acts expressing personal experiences and feelings that project an impression of importance in a group or of a key role in a situation (e.g., domination)

TOWARD MORE RIGOROUS EVALUATION

This chapter asked what we need to evaluate and how this has been done in the literature of OCL. The chapter detailed a new taxonomy of evaluation studies, identifying those which focus on student experiences; those concerned with tools, techniques and outcomes; and those that emphasize the socially constructed nature of language. In particular, the chapter showed how we are in a better position now, than 10 years ago, to evaluate whether collaborative learning is taking place in these online learning environments.

Most importantly, the chapter addressed the issue of rigorous theoretical conceptualization that good teaching and learning practice deserves. It demonstrated how evaluation has been approached implicitly, without theory-driven empirical investigation, even in pedagogical studies. When the technological and pedagogical developments of the last decade are considered, there have been huge advances in opportunities for designing and delivering Web-mediated learning. It is arguably time for evaluation to make concomitant advances. Consideration was therefore given to three models with a sociocultural perspective that pays attention to the importance of the context of social interaction processes in which OCL takes place.

Application of one of these models, the CMCL developed by Cecez-Kecmanovic and Webb, has demonstrated in empirical studies that it is possible to evaluate OCL from within the social interaction space. The CMCL is a valuable pedagogical framework that enables useful and fine analytical distinctions to be made regarding students' orientation to collaborative learning online. Furthermore, the model enables the productivity of the OCL space to be evaluated by tracing how collaboration and knowledge co-creation is generated (and may be inhibited) within Web-mediated environments.

The CMCL model enables evaluation that is both descriptive as well as functional, and, significantly, can be applied indicatively to improve the design of Web-mediated collaborative learning environments. Thus, designers do not need to wait until student perceptions, tools and outcomes have been evaluated post-hoc. Each part of the design phase can be tested before its implementation.

Nevertheless, let us not assume that because evaluation can point to valued student experiences, effective instructional methods or even successful pedagogical outcomes, that all will be well for students and those who design, facilitate, monitor and evaluate for this innovative mode in higher education. Online learning and flexible delivery is becoming increasingly politicized. With the pressures currently on higher education, we are yet to see the emergence of evaluation research that demonstrates how courses can be delivered more efficiently via flexible delivery. Nonetheless, the interpolation of managerialism into our pedagogical research is not far off. Such assumptions concerning the instrumental efficiency of Web-mediated learning and its entrepreneurial potential are being widely voiced in the expansion of flexible delivery and online learning. It cannot be too long before we are required, or funding is made available, to research the economics of online delivery modes. Whether such research will be able to retain an emphasis on pedagogy remains to be seen. In looking to other areas of higher education from which to make inferences, there is little reassurance available. It may well be those institutions that make, or can afford to make, substantial investment in and commitment to the provision of quality education will lead and inspire those that are less well endowed and less courageous in embracing and supporting change.

ACKNOWLEDGMENT

The assistance of a University of Western Sydney Seed Grant is gratefully acknowledged.

REFERENCES

Alavi, M. (1994). Computer-mediated collaborative learning: an empirical evaluation. *MIS Quarterly, 18,* 159-174.

Alexander, S., & Hedberg, J. G. (1994). Evaluating technology-based learning: Which model? In K. Beattie, C. McNaught, & S. Wills (Eds.), *Interactive Multimedia in University Education: Designing for Change in Teaching and Learning, Vol. A59*, pp.233-244. Amsterdam: Elsevier B.V. (North Holland).

Alexander, S., & McKenzie, J. (1998). *An Evaluation of Information Technology Projects for University Learning.* Canberra, Australia: Committee for Committee for University Teaching and Staff Development and the Department of Education, Employment, Training and Youth Affairs.

Alon, I., & Cannon, N. (2000). Internet-based experiential learning in international marketing: The case Globalview.org. *Information Review, 24*(5), 349-356.

Arbaugh, J. B. (2000). How classroom environment and student engagement affect learning in Internet-based MBA courses. *Business Communication Quarterly, 63*(4), 9-26.

Askov, I., & Simpson, M. (2001). Researching distance education: Penn State's online adult education med degree on the World Campus [Electronic version]. Retrieved November 1, 2002 from http://www.avetra.org.au/PAPERS%202001/askov.pdf.

Bain, J. D. (1999). Introduction. *Higher Education Research and Development, 18*(2), 165-172.

Bates, T. (2001). International distance education: Cultural and ethical issues. *Distance Education, 22*(1), 122-136.

Bloom, B. (1956). *Taxonomy of Educational Objectives: The Classification of Educational Goals, by a Committee of College and University Examiners.* New York: Longmans, Green.

Bonamy, J., Charlier, B., & Saunders, M. (2001). "Bridging tools" for change: Evaluating a collaborative learning network. *Journal of Computer Assisted Learning, 17*(3), 295-305.

Bonk, C. J., & Dennen, V. (1999). Teaching on the Web: With a little help from my pedagogical friends. *Journal of Computing in Higher Education, 11*(1), 3-28.

Brandon, D. P., & Hollingshead, A. B. (1999). Collaborative learning and computer-supported groups. *Communication Education, 48*(2), 109-126.

Cecez-Kecmanovic, D., & Webb, C. (2000a). A critical inquiry into Web-mediated collaborative learning. In A. K. Aggarwal (Ed.), *Web-Based Learning: Opportunities and Challenges* (pp. 307-326). Hershey, PA: Idea Group Publishing.

Cecez-Kecmanovic, D., & Webb, C. (2000b). Towards a communicative model of collaborative Web-mediated learning. *Australian Journal of Educational Technology, 16*(1), 73-85.

Chalk, P. (2000). Webworlds-Web-based modelling environments for learning software engineering. *Computer Science Education, 10*(1), 39-56.

Chang, C. (2001). A study on the evaluation and effectiveness analysis of Web-based learning portfolio (WBLP). *British Journal of Educational Technology, 32*(4), 435-458.

Chen, D.-G., Liu, C.-C., Ou, K.-L., & Lin, M.-S. (2001). Web-learning portfolios: A tool for supporting performance awareness. *Innovations in Education and Training International, 38*(1), 19-32.

Collings, P., & Pearce, J. (2002). Sharing designer and user perspectives of Web site evaluation: A cross-campus collaborative learning experience. *British Journal of Educational Technology, 33*(3), 267-278.

Collot, M., & Belmore, N. (1996). Electronic language: A new variety. In S. C. Herring (Ed.), *Computer-Mediated Communication – Linguistic, Social and Cross-Cultural Perspectives* (pp. 13-28). Amsterdam: John Benjamins.

Curtis, D., & Lawson, M. (1999, July). *Collaborative online learning: An explanatory case study.* Paper presented at the HERDSA Annual International Conference, Melbourne, Australia.

Ekstrom, D. N., & Sigurdsson, H. O. (2002). An international collaboration in nursing education viewed through the lens of critical social theory. *Journal of Nursing Education, 41*(7), 289-295.

Ellis, T., & Cohen, M. (2001). Integrating multimedia into a distance learning environment: Is the game worth the candle? *British Journal of Educational Technology, 32*(4), 495-497.

Freeman, M. (1997). Flexibility in access, interaction and assessment: The case for Web-based teaching programs. *Australian Journal of Educational Technology, 13*(1), 23-39.

Gallini, J. K., & Barron, D. (2002). Participants' perceptions of Web-infused environments: A survey of teaching beliefs, learning approaches, and communication. *Journal of Research on Technology in Education, 34*(2), 139-156.

Goffmann, E. (1974). *Frame Analysis: An Essay on the Organization of Experience.* Cambridge, MA: Harvard University Press.

Gosling, D. (2000). Using Habermas to evaluate two approaches to negotiated assessment.

Assessment & Evaluation in Higher Education, 25(3), 293-304.

Grossman, W. M. (1999). On-line U. *Scientific American, 28*(1), 41.

Gunawardena, C., Carabajal, K., & Lowe, C. A. (2001, April). *Critical analysis of models and methods used to evaluate online learning networks*. Paper presented at the Annual Meeting of the American Educational Research Association, Seattle, Washington, USA.

Gunawardena, C. N., Lowe, C. A., & Anderson, T. D. (1997). Analysis of a global online debate and the development of an interaction analysis model for examining social construction of knowledge in computer conferencing. *Journal of Educational Computing Research, 16*(4), 397-431.

Gunawardena, C. N., Nolla, A. C., Wilson, P. L., Lopez-Islas, J. R., Ramirez-Angel, N., & Megchun-Alpizar, R. M. (2001). A cross-cultural study of group process and development in online conferences. *Distance Education, 22*(1), 85-121.

Habermas, J. (1984). *The theory of communicative action – Reason and the rationalisation of society* (Vol. I). Boston, MA: Beacon Press.

Henri, F. (1992). Computer conferencing and content analysis. In A. R. Kaye (Ed.), *Collaborative Learning Through Computer Conferencing: The Najdeen Papers*, pp.117-136. Berlin: Springer-Verlag.

Herring, S. C. (ed.). (1996). *Computermediated Communication – Linguistic, Social and Cross-Cultural Perspectives*. Amsterdam: John Benjamins.

Heslep, R. D. (2001). Habermas on communication in teaching. *Educational Theory, 51*(2), 191ff.

Housego, S., & Freeman, M. (2000). Case studies: Integrating the use of Web-based learning systems into student learning. *Australian Journal of Educational Technology, 13*(3), 258-282.

Hron, A., Hesse, F. W., Cress, U., & Giovis, C. (2000). Implicit and explicit dialogue structuring in virtual learning groups. *British Journal of Educational Psychology, 70,* 53-64.

Johnson, D. W., & Johnson, R. T. (1996). Cooperation and the use of technology. In D. H. Jonassen (Ed.), *Handbook of Research for Educational Communications and Technology* (pp. 1017-1044). New York: Simmon and Schuster Macmillan.

Jones, C. (1998). Evaluating a collaborative online learning environment. *Active Learning, 9,* 31-35.

Kitchen, D., & McDougall, D. (1999). Collaborative learning on the Internet. *Journal of Educational Technology Systems, 27*(3), 245-258.

Kumpulainen, K., Salovaara, H., & Mutanen, M. (2001). The nature of students' sociocognitive activity in handing and processing multimedia-based science material in a small group learning task. *Instructional Science, 29*(6), 481-515.

Lauzon, A. C. (2000). Distance education and diversity are they compatible? *The American Journal of Distance Education, 14*(2), 61-70.

Mason, R. (1992). *Computer Conferencing: The Last Word.* Victoria, British Columbia, Canada: Beach Holme.

McAlpine, I. (2000). Collaborative learning online. *Distance Education, 21*(1), 66-80.

McIssac, M., Blocher, J. M., & Mahes, V. (1999). Student and teacher perceptions of interaction in online computer mediated communication. *Educational Media International, 36*(2), 121-131.

Nakleh, M. B., Donovan, W. J., & Parrill, A. L. (2000). Evaluation of interactive technologies for chemistry Websites: Educational Materials for Organic Chemistry Web site (EMOC). *The Journal of Computers in Mathematics and Science Teaching, 19*(4), 355-378.

Phillips, R. A. (2002). *Learning-centred evaluation of computer-facilitated learning projects in higher education: outcomes of a CUTSD Staff Development Grant, Staff Development in Evaluation of Technology-based Teaching Development Projects: An action inquiry approach.* Retrieved on April 20, 2003 from the web site: http://www.tlc.murdoch. edu.au/project/cutsd01.html.

Ravenscroft, A., & Matheson, M. P. (2002). Developing and evaluating dialogue games for collaborative e-learning. *Journal of Computer Assisted Learning, 18*(1), 93-102.

Reason, P., & Bradbury, H. (eds.). (2001). *Handbook of Action Research: Participative Inquiry and Practice.* London: Sage Publications.

Redding, T. R., & Rotzien, J. (2001). Comparative analysis of online learning versus classroom learning. *Journal of Interactive Instruction Development, 13*(4), 3-12.

Ronteltap, F., & Eurelings, A. (2002). Activity and interaction of students in an electronic learning environment for problem-based learning. *Distance Education, 23*(1), 11-22.

Salmon, G. (2000). *E-Moderating: The Key to Teaching and Learning Online.* London: Kogan Page.

Sanders, D. W., & Morrison-Shetlar, A. (2001). Student attitudes toward Web-enhanced instruction in an introductory biology course. *Journal of Research on Computing in Education, 33*(3), 251-262.

Sheely, S., Veness, D., & Rankine, L. (2001). Building the Web interactive study environment: mainstreaming online teaching and learning at the University of Western Sydney. *Australian Journal of Educational Technology, 17*(1), 80-95.

Short, J. A., Williams, E., & Christie, B. (1976). *The Social Psychology of Telecommunications.* London: John Wiley & Sons.

Sullivan, P. (2001). Gender differences and the online classroom: Male and female college students evaluate their experience. *Community College Journal and Practice, 25*(10), 805-818.

Tidwell, L. C., & Walther, J. B. (2002). Computer-mediated communication effects on disclosure, impressions, and interpersonal evaluations: Getting to know one another a bit at a time. *Human Communication Research, 28*(3), 317-348.

Treleaven, L. (2003). A tale of two evaluations: Better practice for learning collaboratively online. In C. Bond & P. Bright (Eds.), Learning for an Unknown Future, Research and Development in Higher Education, Vol. 26, pp. 547-556. Sydney: Higher Education Research and Development Society of Australasia.

Treleaven, L. (in preparation). Reframing the class as a "virtual organisation": A study of pedagogy, technology and innovation in management education.

Treleaven, L. (in press). Three approaches to evaluating online collaborative learning: A collaborative action research study.

Treleaven, L., & Cecez-Kecmanovic, D. (2001). Collaborative learning in a Web-mediated environment: A study of communicative practices. *Studies in Continuing Education, 23*(2), 169-183.

Tsoukas, H. (2003). Do we really understand tacit knowledge? In M. Easterby-Smith & M. A. Lyles (Eds.), *The Blackwell Handbook of Organizational Learning and Knowledge Management.* Blackwell, Oxford, pp. 410-427.

Tu, C. (2002). The measurement of social presence in an online learning environment. *International Journal on Elearning, 1*(2), 34-45.

Tu, C., & Corry, M. (2001). A paradigm shift for online community research. *Distance Education, 22*(2), 245-263.

Warschauer, M. (1998). Technology and indigenous language revitalization: Analyzing the experience of Hawaii. *The Canadian Modern Language Review, 55*(1), 139-159.

Weasonforth, D., Biesenbach-Lucas, S., & Meloni, C. (2002). Realizing constructivist objectives through collaborative technologies: Threaded discussions. *Language, Learning and Technology, 6*(3), 58ff.

Weiner, C. (2002). A new Alternative: adolescent students study in cyberspace. *Dissertation Abstracts International, 63*(1-A), 155.

Yates, S. C. (1996). Oral and written linguistic aspects of computer conferencing: A corpus based study. In S. C. Herring (Ed.), *Computer-Mediated Communication—Linguistic, Social and Cross-Cultural Perspectives* (pp. 29-46). Amsterdam: John Benjamins.

Chapter 8.11
Staffing the Transition to the Virtual Academic Library:
Competencies, Characteristics, and Change

Todd Chavez
Tampa Library at the University of South Florida-Tampa, USA

Change brought about by innovations in computing technologies has fundamentally altered the nature of work in academic libraries. In his description of the term informatica electronica, Gilbert (1998) suggests that despite the way technology is changing how library staff do their work, it should not change the emphases on traditional services to patrons, such as accessing and retrieving information. This chapter also focuses on human changes that accompany the migration from print to electronic collections, from traditional to online services, and from the academic research library of a decade ago to the virtual library of today and tomorrow.

INITIAL CONSIDERATIONS

The most important management decision to be made remains staffing the academic research library (Tennant, 1998). Historically, this has been a rather straightforward process, including the selection of a pool of candidates, each possessing similar experiences, skills, and competencies. A senior librarian would chair the search committee, with a selection of existing staff. Following one or more interviews, and perhaps a presentation, the library would solicit employment references, make the decision, tender the offer of employment, and the new employee would begin work.

In a nationwide survey, over 4,000 human resources professionals identified the two most significant issues facing their organizations (KnowledgePoint, 2001). Seventy-nine percent of the respondents stated that recruitment of qualified employees was their greatest challenge into the near future while 51% identified retention. Further elements contributing to the challenges of recruitment and retention included compensation, the need to demonstrate value for the employee, and poor management. Seventy-one percent of the human resources professionals stated that their employees cited improved communication as the most important factor contributing to retention rates. They also identified poor selection skills and practices as contributing to difficulties (KnowledgePoint, 2001).

Clearly, academic libraries are not exempt from many of the same pressures facing the respondents to the survey. In the past, it was possible to identify the specific skills and experiences that were desirable in an employee and either hire an individual with those skill sets or train an existing employee. Given the pace of change in today's academic library, this requires that library administration possess a crystal ball to predict which knowledge base and skills will remain important in the future (Tennant, 1998).

TECHNO-CHANGE AND THE CHANGING NATURE OF ACADEMIC LIBRARIES

Lynch and Smith (2001) reported on the results of a content analysis of 220 job announcements over a 25-year period (1973-1998) in College and Research Libraries News. Their research focused on the specific job characteristics listed in the position advertisements. They posited that position announcements in the News were probably representative of current trends and job requirements of the profession as a whole. Several significant trends were reported in this study.

The authors found that few traditional job elements persisted throughout the job announcements. First, although the requirement for a Master's degree in Library Science (MLS) from an American Library Association (ALA) accredited program in Library and Information Science was the most persistent (present in 80% of the advertisements), there has been a decline in M.L.S. requirements, particularly among the largest academic research libraries where special-

ized degrees are often required (Lynch & Smith, 2001). Association of Research Libraries (ARL) salary surveys for the period 1985 to 1998 reveal that a growing percentage of the professionals in these libraries were without the MLS (Lynch & Smith, 2001). Although the authors state that the knowledge, skills, and abilities formed from a library and information science (LIS) education continue to dominate the academic library workforce, an equally valid interpretation is that the ARL institutions are functioning as harbingers of future trends.

Lynch and Smith (2001) also found that computing technologies as they relate to library and information science were incorporated into all jobs and thus were present in all position announcements (emphasis added). The authors conclude that new hires alone cannot meet the academic library's increasing need for technological proficiency; rather, that the institutions must invest in a systematic program of continuing education and training.

In addition, Lynch and Smith discuss the increasing incidence of requirements for instructional experience, emphasizing a desire for teaching skills and knowledge of learning theories and methodologies and a growing and recent emphasis on departmental and unit team environments. Coupled with a concurrent emphasis on behavioral skills, such as effective oral and written communication, flexibility, and creativity, Lynch and Smith conclude that organizational cultures are changing. However, the changing emphasis on teams and increasing solicitation for behavioral skills supporting team organization and interaction is challenged by an apparent contradiction: position announcements for administrative jobs do not reflect the changes in organizational structure implied by the non-administrative position advertisements. What this apparent "disconnect" means for future organizations is not explored, but one may assume that some future crisis will emerge to challenge the existence of two divergent sets of expectations.

There are specific examples of the changing nature of work in the academic library. Nofsinger (1999) suggests that changes and innovations in computing technologies compel a systematic requirement for training and retraining for 21st century reference librarians in the following core competencies: reference skills, subject knowledge, communication skills, interpersonal abilities, knowledge and skills in technology, critical thinking skills, supervisory and management skills, and commitment to user services. For the cataloging side of the profession, Wendler (1999) cites the explosion in electronic publishing and the concomitant requirement for metadata as the impetus underlying the challenges to the cataloger's ability to order the chaos. It is clear that developments in computing technologies are changing the very nature of the academic library's mission and thus the staff's work.

Support staff is not immune to the effects of rapid technological change. Librarians tend to share many common competencies gained through the experience of graduate education in the discipline. This is not the case with paraprofessionals, who come to the academic library with a plethora of skills and experiences, diverse both in content and in level of accomplishment. Sheffold (2000) suggests that paraprofessional training and continuing education are quite often the first areas impacted by budget reductions. Organizationally, support staff are often left to operate the desks during important meetings and training opportunities for professional staff. Thus, the impact of change upon support staff is particularly serious.

Reporting on a case study of change within an academic library, Farley, Broady, Preston and Hayward (1998) characterize change as occurring on three levels: organizational, technological, and human. They caution the administrator to ensure that the concerns of all staff are examined and addressed prior to implementing change because "the negative impact of change on staff, even if successfully managed, must not be underestimated"

(p. 151). Positing that academic librarianship has changed more over the last few decades than in its entire history, the authors cite four areas in which the change has been dramatic: economics, technology, higher education, and organization (p. 153).

HUMAN CHANGES

Technological change is the one constant for the academic library engaged in the transition from traditional format resources and services to future electronic collections and services. Nevertheless, the human dimension may well dwarf the technologically derived sources of change in terms of long-term impact upon the academic library. The most significant of these human changes include considerations of the changing demographics of the work force and management's response to these fundamental factors.

The demographic profile of a "typical" academic librarian (Bell, 1999; Cooper & Cooper, 1998) is white, female, and 45 years of age. Her undergraduate training is likely to be either in the arts and humanities or in the social sciences, with some graduate-level coursework in these disciplines. Regardless of where this typical librarian works, she is probably from the "reference side" of the profession. She possesses approximately 13 years of professional experience and earns $43,000 per year. This librarian is a member of a group who typically retires by age 63 (Matarazzo, 2000).

Consider the incoming library school graduate. At an average age of 36 years, this librarian is solidly "Generation X" (i.e., an individual born between 1961 and 1981). Contrasting significantly with the earlier generations, 'Xer's' are skill-focused, survivalists in orientation, used to rapid and unending change, and technologically competent (Cooper & Cooper, 1998, p. 20). Administrators and managers who are unaware of or unwilling to embrace these generational differences are positioning themselves for future difficulties. The generational changes between librarians have import in such areas as organizational culture, reward systems, training requirements and methods, and budget.

Morgan (2001, p. 58) highlights the importance of incorporating the human factor in any attempt to manage or adapt to change. He states, "What sometimes gets forgotten in all this [concern for change] is the human element involved in what is likely to be a heavily technology-driven future … Research suggests that 90% of change initiatives that fail do so because human factors are not taken into account." Morgan's human factors include communication, staff involvement, and generational dynamics.

With human communication a recurring theme in much of the change literature, consider the effects of email in today's libraries. Hierarchical communication is dead. It is no longer necessary to make an appointment with the Dean of Libraries to place an idea or complaint directly on his or her desk. Lubens (2000) concludes that not only has email had a positive effect on staff productivity, it increases the staff's understanding of the organization. More importantly, it promotes good communications practices allowing staff to have immediate access to people and vital information to deal with change.

RESPONSE TO CHANGE

In an indictment of academic librarians' recognition of the fundamental results of the technological change experienced over the past decade, Herring (2001) accuses librarians of reaching "stasis," of creating or contributing to their own unemployment by becoming comfortable and complacent with the minimal adaptations made to date. He describes several external trends that threaten to make academic libraries irrelevant: 1) the "everything's-on- the-Internet" challenge; 2) competition with commercial information pro-

viders; 3) failure to be proactive in technological developments and innovations (i.e., allowing technology to drive library services and collections); and 4) a fundamental "disconnect" with the new generation of information consumer. Although Herring identified technological change as the catalyst underlying libraries' own undoing, he simultaneously makes it clear that obsolescence is not guaranteed – academic library professionals can make changes to the seemingly inevitable death of the library.

In an examination of the effects of technological change on academic library staff, Poole and Denny (2001) surveyed professional and paraprofessional personnel in 28 Florida community college libraries. They found that respondents were overwhelmingly positive about the changes that accompanied technological innovations, e.g., approximately 69% of the staff enjoyed the changes as contrasted to less than five% reporting that they disliked computers. From questions designed to assess the ability of training efforts to keep pace with the rate of technological change, Poole and Denny concluded that training needs were sufficient in Florida's two-year colleges. However, they identified the lack of management's commitment to involve staff in planning and decision making as a significant area of contention.

Herring (2001) states it is highly unlikely that anyone in the academic library community needs to be convinced of the challenge of change. Hudson (1999) suggests that managers first become clear as to the appropriate concepts to employ in this situation. She distinguishes between change that she defines as relating to a specific situation, and transition which is a psychological construct: change is "…a gradual process, internal to the individuals who are going through it …Transition is the process people go through to internalize the change" (p. 36).

Farley et al. (1998) identify four areas of human resource management that would minimize the negative effects of change: communication and information sharing, staff involvement and participation, training and development, and job design. Management should recognize that: 1) the traditional organizational structure in academic libraries is the opposite of what is needed to manage change and facilitate transition, and 2) "people are an organization's greatest asset but it would seem that few organizations truly believe this or act as if they do" (p. 162).

In addition to staff participation, Morgan (2001) champions adoption of a managerial style characterized by: 1) flat organizational models; 2) teamwork and project management; 3) strong links between library and institutional parent; and 4) a spread of accountability. He emphasizes developing and fostering a strategic awareness in all staff, ensuring that everyone understands the strategic goals they serve in the course of their daily work and getting a handle on the tendency for technology to drive people as opposed to the reverse (p. 60). Green, Chivers and Mynott (2000) similarly emphasize communication, developing peer relationships, staff involvement in decision making, appropriate recognition and reward, training, and staff development as being critical to effective, positive management of change and as motivational tools.

JOB SKILLS AND COMPETENCIES: ARE THEY PRACTICAL?

Should managers and human resources professionals serving the staffing needs of academic libraries rely upon job competencies – either formal (published) or informal (anecdotal) – to make selection and hiring decisions? Once an employment decision is made, can managers productively use these same competencies to evaluate and promote staff? These questions are deceptively simple. The answer to either query depends upon whether the competencies employed are statements of job skills, including lists of specific technological proficiencies, or by

contrast, are stated in terms of desirable personal characteristics.

In an article describing desirable skills and competencies for librarians in the new millennium, Tennant (1999) lists knowledge of imaging technologies, optical character recognition, markup languages, cataloging and metadata, indexing and databases, user interface design, programming, Web technology, and project management. Although it is certain that different readers will argue for the continued validity of one or more of the skills listed, how relevant are all the skills that are listed in 2002? Arguably, skills with markup languages may no longer be essential given the quality of editing applications that accomplish the markup function for users who have word processing skills. At the USF Tampa Library, the experience is that such technologies as interface design and programming are best outsourced to individuals or organizations whose skills are at the "bleeding edge" of currency.

To illustrate the difficulty in employing skill lists for selection purposes, consider this recent example. After an extensive planning period, library management decided to establish a Geographic Information Systems (GIS) Research and Data Center. In January 2001, library management initiated a nationwide search for a qualified GIS Librarian to manage the center. The position announcement was carefully crafted to reflect the needs of the center with due consideration of minimal competencies. The library selected a candidate that matched the knowledge base and skills and assigned a start date of May 1, 2001. However, the following month, ESRI, the premier designer of GIS software applications used by all academic units at the USF Tampa campus, announced a major development in version eight of their GIS software application. In July 2001, the new GIS Librarian went to ESRI training to learn the new format. The lesson here? Skills in a particular application are good today and obsolete tomorrow. Flexibility and willingness to accept change are critical for success.

Now consider an extreme example of futuristic predictions regarding technological change in academic libraries and the competencies that would be required to bring this change to fruition. Gillett (1998) suggests that libraries use nanotechnology to produce information on demand from templates on a molecular level. In essence, he envisions "the library as factory"–a future as repositories of information templates in infinite variety. Based upon this view of the future, what are the job skills and competencies that academic librarians must possess to succeed in the world of "molecular information?" A shift from lists of job skills to competencies appears imperative.

The Association of South Eastern Research Libraries' Competencies for Research Librarians (Perez et al., 2000) outlines five competencies that define what is best in a research librarian. Perez et al. state that the successful research librarian possesses such attributes as "intellectual curiosity, flexibility, adaptability, persistence, and the ability to be enterprising" (p. 3). Woodsworth (1997) focuses on both particular technologies and personal characteristics when reviewing competencies for librarians who are best viewed as elements of a "global digital information infrastructure." These parallel Nofsinger's (1999) competencies for the reference librarian of the 21st century and Tennant's (1998) admonition regarding the importance of hiring and selection in academic libraries. Tennant also distinguishes between skills and traits and goes on to list such personal characteristics as capacity to learn quickly and constantly, flexibility, and innate skepticism as critical to the librarian of the future.

PERSONAL CHARACTERISTICS

Lynch and Smith (2001) noted the increasing incidence of behavioral characteristics in the position announcements they analyzed. Tennant (1998), Sheffold (2000) and Morgan (2001) have alluded to the importance of such personal characteristics as flexibility, enabling skills, and risk taking. Perhaps of more immediate importance, does the profession have a firm grasp on those personal characteristics and behaviors that are counterproductive to the transition from the traditional to the online environment?

Hudson (1999) suggests that conflict and stress are inevitable but need not be disastrous. It is important to recognize that failure to adapt to change need not be solely limited to unhappiness, lost workdays, or retention problems. The conflict and stress associated with change adaptation difficulties can cause violence or other unacceptable behaviors. Staff struggling to adapt to change and/or protect their own self-interests become stressed. Stress breeds conflict: stress, and conflict can result in abusive behaviors. This is the sobering side of the nature of change; it is a side of change that academic library managers must consider or be remiss in their responsibilities.

Bullying

Hannabus (1998) suggests that bullying is widespread in the workplace. Bullying takes a variety of forms: physical assault; gossip and rumor-mongering; ridiculing arguments in meetings; public criticism; overloading individual workers with assignments; denying annual or sick leave; abusing internal processes designed to alleviate management-worker tensions (e.g., grievances). Hannabus (1998) characterizes the classic bully as an individual with low self-confidence and low self-esteem, i.e., someone who is fearful that his or her inadequacies (perceived or real) will become evident. Bully-victim behaviors are symptoms of a more significant and pervasive problem (Hannabus, 1998).

Dealing with bullying requires a concerted effort by many in the organization. First, the victim must acknowledge that he or she is being bullied. Once a victim believes that he or she fully understands the dynamics of the situation, a meeting with management is in order to determine if the problem is one of bullying. Once that determination is made, the library should take assertive action. Bullies need to be confronted about their behavior since they need to understand the effect their actions have upon both the victim and the organization. Violent bullies should be removed from the workplace. Counseling sessions for all concerned may also be productive. To illustrate the importance of eliminating bullying, consider the following example. When a project team of volunteers was to be disbanded and integrated into the larger library system, most members of the team understood the need to make this change and actively facilitated the transition. However, one individual (opposed to change) publicly criticized colleagues, disrupted meetings, and threatened to file grievances, thus making the group's efforts to transform as difficult as possible.

Passive-Aggressive Behavior

Another common workplace phenomenon is the growing incidence of passive-aggressive behavior in the face of transition. McIlduff and Coghlan (2000) make it clear that passive-aggressive behavior is much more than a mere strategy adopted by individuals faced with the uncertainties that accompany change on this level. Described as "a pervasive pattern of passive resistance to demands for adequate social and occupational performances, beginning by early childhood and present in the functioning of the person in a variety of contexts" (p. 717), the term "passive" is the key to understanding the disorder. Passive-aggressive behaviors are exhibited in ways that do not directly offend other parties involved but do accomplish the intended goal of "getting back at authority figures for perceived ill treatment or injustice" (p. 718). These behaviors typically surface as a resistance to demands for performance.

Individuals immersed in passive-aggressive behaviors typically perceive change as threatening or unnecessary; assess the impact of the proposed change as a threat to themselves, the

organization, and/or the clientele served; and respond to the change by dodging, opposing, or resisting the thrust of the initiative. Interventions, done in either a one-to-one setting or within a team environment, include: 1) calm but assertive communications describing the reason for intervention; 2) genuine efforts to understand the passive-aggressive individual's context; and 3) resolution to work through the problem, however long it may require (McIlduff & Coghlan, 2000). Toleration of passive-aggressive behavior will doom an organization to fail in its efforts to move forward.

Organizational Fit

First and foremost, academic libraries must attract, retain, and train staff to understand organizational culture. Staff must be capable of understanding the organizational culture, both as productive members and as prospective members seeking to join "the team." Accurate assessment of an organization's culture is critical to an employee's potential for success and a source of added stability to the organizational unit (Sannwald, 2000).

Ethics of Workplace Behavior

A second trait essential to successful adaptation to this changing environment is best described as a fully internalized and "automatic" sense of the ethics of workplace behavior, which is not the same as the principles of conduct that govern librarianship. The ethics of workplace behavior are personal rules of engagement that are designed to ensure integrity in all actions (Caville & Hoskins, 2001). Two examples of the ethics of workplace behavior are: "It is ethical to positively change the organization; it is unethical to damage it. It is ethical to go above and beyond expectations; it is unethical to do anything less" (pp. 11-13). Clearly, desirable staff are those who possess similar internal ethical standards.

Leadership

Metz (2001) argues that academic libraries are suffering from a general lack of leadership capable of leading in a discontinuous future. Defining a "discontinuous future" as one lacking sequence and cohesion, Metz challenges academic library leaders to recognize the significance of the transformation from print repositories to portals to electronic collections and thus develop new mind sets that value differences, redefine and eliminate historical limitations, manage expectations, and think discontinuously. He also stresses the importance of being a generalist possessed by many of the personal characteristics described in this chapter and simultaneously cautions against security in specialized skills.

Creativity

In a side-by-side comparison of inventories of desirable personal characteristics (Tennant, 1999, 1998; Oberg, 2000; Wilson, 1999), many similarities are immediately apparent: capacity to learn quickly (and constantly), skepticism, public service orientation, enabling skills, appreciation for colleagues, risk-taking philosophy, and so forth. Without exception in either form or meaning, one of the most desirable traits is creativity.

Creativity may well be the most essential of the personal characteristics discussed in this chapter. Defined as the "ability of providing an original or inventive response to a problem," (Yong, 1994), creativity is possible only when such traits as flexibility, risk taking, enabling, and comfort with change are present. More importantly, creativity can be assessed both during the selection process and after employment (Williams, 2001). Creative people possess four characteristics: 1) problem sensitivity ("the ability to identify the "real" problem"); 2) idea fluency ("the ability to generate a large number of ideas from which to choose"); 3) originality ("new ways to adapt existing ideas to new conditions"); and 4) flexibility ("ability to consider a wide variety of dissimilar approaches to a solution") (Yong, 1994, pp. 17-18).

Apart from attracting and selecting creative personnel is the matter of how to address existing employees, individuals who possess a rich and irreplaceable knowledge of the organization and are thus important to the successful continuance of the academic library's mission. Certainly, library management cannot simply abandon these individuals in an unswerving search for creativity, but we can train them. Williams (2001) endorses a program of creativity training including creative problem solving, creative self-statement (enhances creative performance), and "synectics," a brainstorming technique in which the user seeks to make the strange familiar and the familiar strange. It is clear that academic libraries can incorporate creativity training into their organizational repertoire, but the challenge is to ensure that negative behaviors such as bullying and passive-aggressive behavior do not combine to make the effort irrelevant. This is a particular challenge in an organization that is unsuccessfully dealing with the generational dynamics described by Cooper and Cooper (1998).

Yong (1994) and Williams (2001) also emphasize the need to ensure that managers receive training in appropriate methods for managing creative people, including such tools as role-playing and behavioral modeling. Among the areas wherein management may effectively and productively promote creativity are organizational culture and structure, work group design and use, job design, social support for the creative process, and recognition and evaluation.

CONCLUSION

It is clear that staff remains a library's most important asset in successfully transforming the traditional academic library into a 21st century organization. The need to devise selection strategies to attract the best personnel, to implement management practices and organizational structures conducive to retaining productive and creative staff, and to initiate training for valued existing personnel cannot be emphasized enough. While libraries cannot – and should not – abandon current hiring and selection practices in a wholesale manner, continued reliance on traditional lists of job skills separated into minimum and preferred qualifications will not facilitate employing the most desirable personnel available for a given function. The "life-time" employment system currently utilized in the majority of academic institutions may well work in direct opposition to elimination of the undesirable traits (bullying and passive-aggressive behavior) in favor of a creative workforce.

FUTURE ISSUES

To accomplish the goal of attracting and retaining creative personnel to academic libraries, it is essential that libraries broaden their perspectives on selection practices to incorporate measures of creativity heretofore unknown in the traditional academic search process. While it is unethical to wantonly disregard the rules of the organization, it is equally important to recognize that many institutional selection processes are antiquated leftovers from an age wherein diversity was the primary goal in selection. In this milieu, such considerations as creativity and flexibility take a backseat to representative candidate pools, and diverse search committees appear to be more essential than the ability of the membership to contribute to the selection process. It is important to recognize the wider implications of a desire for diversity: diversity of education, diversity in thought, and diversity in approach to change.

REFERENCES

Bell, C. (1999). Y210, succession planning in libraries: Finding the common ground. PNLA Quarterly, 64(1):20-21.

Caville, P., & Hoskins, A. (2001). Raising the bar on workplace behavior. PNLA Quarterly, 65(3):10-14.

Cooper, J. R., & Cooper, E. A. (1998). Generational dynamics and librarianship: Managing Generation X. Illinois Libraries, 80(1):18-21.

Farley, T., Broady-Preston, J., & Hayward, T. (1998). Academic libraries, people and change: A case study of the 1990's. OCLC Systems and Services, 14(4):151-164.

Gilbert, B. (1998). The more we change, the more we should stay the same: Some common errors concerning libraries, computers, and the Information Age. In M. T. Wolf, P. Ensor, & M. A. Thomas (Eds.), Information imagineering: Meeting at the interface (pp. 3-11). Chicago: ALA.

Gillett, Stephen L. (1998). Nanotechnology: The library as factory. In M. T. Wolf, P. Ensor, & M. A. Thomas (Eds.), Information imagineering: Meeting at the interface (pp. 219-227). Chicago: ALA.

Green, J., Chivers, B., & Mynott, G. (2000). In the librarian's chair: An analysis of factors which influence the motivation of library staff and contribute to the effective delivery of services. Library Review, 49(8):380-386.

Hannabus, S. (1998). Bullying at work. Library Management, 19(5):304-310.

Herring, M. Y. (2001). Our times they are a-changin', but are we? Library Journal, 126(17):42-44.

Hudson, M. P. (1999). Conflict and stress in times of change. Library Management, 20(1):35-38.

KnowledgePoint, Inc. (2001). Survey finds recruitment and retention the top issues facing employers. Library Personnel News, 14(2):8.

Lubens, J. (2000). "While I was busy holding on, you were busy letting go": Reflections on e- mail networks and the demise of hierarchical communication. Library Administration & Management, 14(1):18-21.

Lynch, B. P., & Smith, K. R. (2001). The changing nature of work in academic libraries. College & Research Libraries 62(5):407-420.

McIlduff, E., & Coghlan, D. (2000). Reflections: Understanding and contending with passive aggressive behaviour in teams and organizations. Journal of Managerial Psychology, 15(7):716-736.

Matarazzo, J. M. (2000). Library human resources: The Y2K plus 10 challenge. Journal of Academic Librarianship, 26(4):223-224.

Metz, T. (2001). Wanted: Library leaders for a discontinuous future. Library Issues, 21(3):1-6.

Morgan, S. (2001). Change in university libraries: Don't forget the people. Library Management, 22(1/2):58-60.

Nofsinger, M. N. (1999). Training and retraining reference professionals: Core competencies for the 21st century. Reference Librarian, 64:9-19.

Oberg, L. R. (2000). How to make yourself indispensable: A survivor's guide for the 21st century. Library Mosaics, 11(1):14-15.

Perez, D., Drake, M., Ferriero, D., & Hurt, C. (2001). Competencies for research librarians.

ASERL Web. [Electronic Resource]. Retrieved 11/14/2001 from: http://www.aserl.org/statements/competencies/competencies.htm

Poole, C. E., & Denny, E. (2001). Technological change in the workplace: A statewide survey of community college library and learning resources personnel. College & Research Libraries, 62(6): 503-515.

Sannwald, W. W. (2000). Understanding organizational culture. Library Administration & Management, 14(1):8-14.

Sheffold, D. (2000). Support staff professional development: Issues for the coming millennium. OLA Quarterly, 5(4):7.

Tennant, R. (1999). Skills for the new millennium: Personal characteristics essential to digital librarians. Library Journal, 124(1):39.

Tennant, R. (1998). The most important management decision: hiring staff for the new millennium. Library Journal, 123 (3):102.

Wendler, R. (1999). Branching out: Cataloging skills and functions in the digital age. Journal of Internet Cataloging, 2 (1):43-54.

Williams, S. (2001). Increasing employees' creativity by training their managers. Industrial and Commercial Training, 33 (2):63-68.

Wilson, A. (1999). Ride the wave of the future: Be flexible in your job. Library Mosaics, 10 (6):7.

Woodsworth, A. (1997). New library competencies: Roles must be defined within the context of a global digital information infrastructure. Library Journal 122(7), 46.

Yong, L. W. (1994). Managing creative people. Journal of Creative Behavior 28(1),16-20.

This work was previously published in Building a Virtual Library, edited by A. Hanson & B. Levin, pp. 212-222, copyright 2003 by Information Science Publishing (an imprint of IGI Global).

Chapter 8.12
Knowledge Management Trends:
Challenges and Opportunities for Educational Institutions

Lisa A. Petrides
Institute for the Study of Knowledge Management in Education, USA

Lilly Nguyen
Institute for the Study of Knowledge Management in Education, USA

ABSTRACT

While the pressure of public accountability has placed increasing pressure on higher education institutions to provide information regarding critical outcomes, this chapter describes how knowledge management (KM) can be used by educational institutions to gain a more comprehensive, integrative, and reflexive understanding of the impact of information on their organizations. The practice of KM, initially derived from theory and practice in the business sector, has typically been used to address isolated data and information transfer, rather than actual systemwide change. However, higher education institutions should not simply appropriate KM strategies and practices as they have appeared in the business sector. Instead, higher education institutions should use KM to focus on long-term, organization-wide strategies.

INTRODUCTION

Knowledge management (KM) can be used by educational institutions to gain a more comprehensive, integrative, and reflexive understanding of the impact of information on their organizations. Specifically, the practice of KM, initially derived from theory and practice in the business sector as described in the previous chapter, provides a framework to illuminate and address organizational obstacles around issues of information use and access (Davenport, 1997; Friedman &

Hoffman, 2001). Yet introducing the concept of KM into the educational arena from the business sector has been a slow and often underutilized process. This is partially due to the fact that KM is a multi-layered and systems-oriented process that requires organizations to rethink what they do and how they do it (Brown & Duguid, 2000; Senge, 1990). Additionally, educational institutions are traditionally hierarchical with silo-like functions, making cross-functional initiatives difficult to implement (Friedman & Hoffman, 2001; Petrides, McClelland, & Nodine, 2004).

However, educational institutions can perhaps learn from KM efforts in the business sector, in terms of the limitations and drawbacks associated with KM. In fact, there are several compelling reasons why educational institutions have not, and perhaps should not, simply re-appropriate KM, as popularized by the business sector, into their own organizations. For example, in the business sector, there has been an appeal to focus on information technology and systems as solutions to problems of knowledge transfer and knowledge sharing (Hovland, 2003; Huysman & de Wit, 2004). Coupled with a profit motive, KM as it exists in the business sector is often limited in its ability to create far-reaching organizational change (Hammer, Leonard, & Davenport 2004). Furthermore, recent trends in the field also fail to fully distinguish between data, information, and knowledge (Huysman & de Wit, 2002). Consequently, organizations merely address singular and isolated data and information transfer, rather than actual systemwide and organization-wide change.

These particular limitations are especially salient now as higher-education institutions face an increasing number of challenges that have forced them to rethink how they are accountable to external demands, as well as how to improve internal accountability. Rather than focus on micro-level information-sharing activities, implementing KM strategies and practices requires these educational institutions to examine the larger context of information sharing within the organization, specifically how their people, processes, and technology function within it. As such, neither data-sharing activities nor technological implementation should be viewed as the ultimate objective and final stage of a KM strategy. Instead, KM practices necessitate strategies that build upon current practice, leading to more comprehensive and organization-wide changes in knowledge practices and actions.

How then can educational institutions translate isolated sharing activities into long-term learning? This chapter illustrates how KM strategies and practices enable higher-education institutions to distinguish between data, information, knowledge, and action and how this iterative cycle can help organizations assess their available resources—that is, their people and processes along with their technology. In turn, this chapter demonstrates how KM can help educational institutions place themselves on the path toward continuous learning and organizational reflexivity.

CONCEPTS AND THEORIES

An overview of KM practices in the business sector demonstrates an overwhelming focus on simplified solutions, specific applications, and singular information-transfer activities. Recent accounts suggest that KM has seen limited impacts in the private sector due to overemphasis on technological hardware and software (Hammer et al., 2004; Hovland, 2003; Huysman & de Wit, 2004). This may be due in part to the fact that it is often easier to persuade organizations to acquire new technology tools than to modify or redesign existing organizational processes (Coate, 1996).

However, these particular approaches to KM are less likely to embrace a systematic approach to how organizations function. By focusing too narrowly on isolated information-sharing activities, organizations are prematurely confined and prevented from engaging in a more integrative approach to KM. These information-sharing activities, which some might argue are wrongly classified as KM, may include electronic search and retrieval, document management, and data warehousing systems. These examples demonstrate important yet isolated occurrences of information activities and practices. However, these practices are often implemented disassociated from a larger organization-wide strategy. Secondly, and perhaps more importantly, the interpretation of these as KM does not acknowledge a vital distinction between information and knowledge. It is this delineation that pinpoints the incremental process behind the implementation of KM strategies and practices: Information is data with contextual meaning, data that has been categorized, or subjected to a process of sense-making and interpretation. Knowledge is information that is put into action through the process of problem-solving, decision-making, feedback processes, and so on (Davenport, 1997).

Therefore, developing policies and processes that fundamentally support and organizationally align information-sharing activities to each other is one of the first steps an organization must take to embrace and develop successful KM strategies. Often, an organization will try, yet fail, to implement an entire host of activities related to data collection and information access, only to find that the necessary organizational conduits for information sharing and new knowledge creation are not in place. How an organization shares information, along with the incentives and rewards to do so, and a culture that supports information-based decision-making are all key components that need to be in place before KM can be successfully implemented.

People, Processes, and Technology

KM strategies and practices come to embody the interactions between people, processes, and technology. These three—people, processes, and technology—all function as an integral part of the ongoing dynamics as organizations struggle to meet their information needs. First, it is people, not systems or technology, who "know." Thus, it is people who manage the policies, priorities, and processes that support the use of data, information, and knowledge. KM strategies and practices seek to engage different groups of people across various levels of an organization in the process of collective sense-making and decision-making. Whether these groups are formal or informal, a KM strategy includes supporting individuals in coming together to share information to address their collective needs.

Likewise, self-evident processes or embedded, day-to-day work practices can greatly affect the exchange and sharing of information within any organization. For example, it may be common practice within an organization for decision-making authority to be exercised only at the most senior level. These kinds of decision-making processes can create barriers to ownership, in which individuals are not provided with the appropriate incentives to make their own decisions and changes, let alone use data and share information. By uncovering these processes, KM strategies and practices can help identify knowledge gaps, and thus enable people to obtain the information they need and encourage them to share it with others, sometimes creating new knowledge and improved decisions. In highlighting patterns of information use that might not be evident otherwise, KM practices encourage a certain level of organizational reflexivity, which allows organizations to better understand themselves, in turn leading to more informed decision-making.

Rather than situating technology as the focal point, KM practices approach technology as an essential resource that is necessary for changes in organizational process to occur, but not sufficient. Recent trends in KM may grant technology disproportionate authority in how organizations share information. However, technology and information systems are neither the driver of information sharing, nor are they tangential to the process. Instead, technology is of equal importance in its ability to impact how information flows throughout an organization. Therefore, KM is the combination of people, processes, and technology that come together to promote a robust system of information sharing, while guiding organizations toward ongoing reflexivity and learning.

In summary, recent KM trends in the business sector often do not explicitly address all of the organizational resources necessary to implement KM, namely, the people and processes as well as the technology. To some, KM is used as a phrase to describe the technology that is used to manage an organization's data, such as data on monthly sales figures or a database of successful sales strategies. However, the way that these information systems are used is fully contingent on the strategies and policies employed by the organization, and does not constitute KM on its own. It is not uncommon to hear a claim that a vendor has developed "knowledge management software," rather than "developing software that could be used to help an organization implement KM strategies and practices." Although this distinction may appear to split semantical hairs, we argue that these types of technology present only one part of a larger whole within organizations, but they often do not address the necessary steps to become an organization that uses information and knowledge to develop continuous learning throughout.

Data–Information–Knowledge–Action

KM strategies and practices are predicated on the distinction between information and knowledge. Other research in KM makes this distinction to highlight that information undergoes a series of processes that transform it into knowledge as it flows and is exchanged among individuals within an organization (Davenport & Prusack, 1997; Drucker, 1998; Wilson, 2002). To further refine this notion, we assert that information and knowledge need to be further delineated. As such, we propose four stages that comprise the KM cycle: data, information, knowledge, and action. Data are the facts and quantitative measures that are available within any organization. When groups or individuals take data and contribute their own interpretation and categorization, data can be transformed into information. In turn, knowledge is the resulting understanding that allows people to share and use this information that is now available to them. Once this knowledge is applied to make specific decisions or address problems, it is transformed into an action. Each component of the cycle builds upon the preceding element, feeding back and connecting actions and decisions and new learning, which eventually translates back to new questions that are informed by data once again.

There is a certain set of activities and practices that typically takes place in each part of the cycle, where each component builds upon the one before it, making it an iterative process of change or improvement. Data activities in the KM cycle can include accessing data by departmental request, or retrieving data directly from information systems and placing them within personalized spreadsheets. Information activities may include analyzing data to find patterns, problems, and discrepancies, or aggregating and disaggregating data, writing reports, or discussing findings from the data with colleagues. Knowledge activities entail formal and informal discussion and collaboration to address issues and problems in the context of the data and information. It is important to note that the knowledge stage of the KM cycle encompasses a process of collective sense-making, which includes ongoing discussion, collaboration, and feedback, thus shifting individual data and information practices into the organizational environment. The last stage of the cycle is then implementation of changes and action that result from the iterative process.

Therefore, organizations that simply engage in the collection and distribution of data are

Figure 1. The data-information-knowledge-action cycle

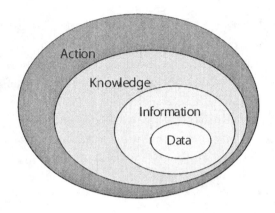

engaged in data management activities only. However, knowledge management is more than the mere aggregation of data management practices. KM practices include the management of the infrastructure that supports the data–information–knowledge–action cycle, as well as the implementation of the process. In these examples, we see then that KM activities and practices bring together all four components of the cycle: data, as well as information, knowledge, and action. In turn, KM strategies embrace practices at every stage of this cycle, and integrate the people, processes, and technology within the organization. It is important to note that each stage of the KM cycle is not mutually exclusive. An organization that fully adopts KM strategies and practices also demonstrates activities within each component of the KM cycle simultaneously. Engaging in the knowledge stage of the KM cycle also includes individuals engaging in data and information activities. In fact, KM practices necessitate that individuals simultaneously engage all three stages of practice, data, information, and knowledge as they implement changes and action (see Figure 1).

Thus, the KM cycle demonstrates the dynamic qualities of KM strategies and practices. Their simultaneous, ongoing, and cyclical na-

ture further highlights the necessary feedback and iterations that serve as the foundations for ongoing reflexivity and learning. As such, KM practices demonstrate how knowledge is most valuable not when stored in static repositories, but when exchanged across groups of people, used and applied to inform actions and change. KM strategies and practices can help organizations better identify their information-sharing and knowledge-generating activities, which, in turn, can help organizations capitalize on the iterative nature of knowledge-sharing activities.

CURRENT CHALLENGES FOR KM IN HIGHER EDUCATION

Increasing pressures and demands for data on student success have translated into an increased call for reliable information regarding critical outcomes in higher education. Due to rising public accountability pressures and strains on fiscal resources, many legislators have begun to demand information that can be directly linked to academic outcomes. As a result, these institutions are faced with requirements to provide accurate data and information around a growing number of issues and outcomes. In order to do so, the institutions are now re-evaluating their own knowledge strategies and practices.

However, these processes of re-evaluation have proven to be challenging. To begin with, the information technology infrastructure at many higher-education institutions is problematic. Rather than having one robust and integrated system, educational institutions more often maintain several information systems that support various functions throughout the organization, some of which are antiquated legacy systems. In addition to this fragmented information technology infrastructure, there are often inconsistent priorities around data collection, which can result in inaccessible or unreliable data. These characteristics translate into disparate data silos throughout the organization, redundant data gathering, and information hoarding, the cost of which is an impaired ability to sustain knowledge development, growth, and effective decision-making (Petrides et al., 2004). In an increasingly performance-driven climate, this only exacerbates these already problematic and costly practices.

Furthermore, cultural issues associated with information hoarding and overall disincentives for sharing and cross-functional cooperation can undermine KM implementation strategies in educational institutions. In a climate of accountability, data and information can appear threatening as well as politically charged, particularly when programs or other initiatives are under fiscal strain. Nevertheless, educational institutions can minimize these potentially negative consequences by developing KM strategies under a set of policies that explicitly encourage change and progress rather than penalize mistakes. A culture that is intolerant of mistakes can severely impede KM initiatives (Davenport & Prusack, 1997). The psychological instability that can arise is a very real challenge that can curtail any change initiative. As such, when implementing a KM strategy, educational institutions are better served by fostering an environment that reduces the sense of fear and retribution that individuals within the organization may face, for example, as they uncover data and information that may support unpopular opinions.

KM practices also require long-term strategies and commitments in order to fully realize their benefits. While educational institutions have tentatively begun to incorporate KM strategies, they will benefit from gaining a better understanding of the current limitations of these recent approaches to KM in the business sector, such as the narrow focus on seemingly easier-to-address solutions—for example, creating a data warehouse from which to extract student data. In microscopically fixating on specific information solutions, many current trends in KM do not help these institutions build the capacity to

sustain long-term organization-wide change, but instead limit the potential that information and knowledge sharing can have.

While KM researchers may recognize the importance of distinguishing between data, information, and knowledge, KM practitioners in the private sector have not necessarily taken into account these distinctions. In this particular conception of KM, knowledge is then simply used as an overarching term for all three—data, information, and knowledge. Subsequently, many of the products, repositories, and exchange activities that are currently termed KM prove to merely support data and information, rather than actual knowledge. Doing so runs the risk of prematurely curtailing the necessary feedback mechanisms for continuous organizational learning.

However, it becomes much more difficult to address systemic barriers to knowledge sharing. The desire to find narrow and short-reaching solutions is often rooted in a compartmentalized understanding of the nature of organizational barriers to information sharing, even though these problems are more than technological. These problems include people's prevailing attitudes, beliefs around knowledge sharing, and systematic and structural disincentives to share and exchange. For example, the politics of information are often heavily embedded in organizational culture and structure, which complicates efforts to change processes that could be used to potentially support and drive knowledge sharing and creation. Recent evolutions of KM do not necessarily take into account the organizational cultures and structures that serve as barriers to data sharing, information sharing, and eventually knowledge sharing. Furthermore, these recent developments in KM fail to acknowledge the evolving and iterative qualities of knowledge. Knowledge is only useful when it is shared, transmitted, or acted on in some capacity. During these exchanges, knowledge undergoes an ongoing and continual cycle of change from data, information, knowledge, and action. However, these distinctions are lost as KM practitioners attempt to find solitary solutions to problems of data and information.

If these attempts at KM remain truncated and narrowly focused on simplified solutions, specific applications, and singular knowledge-transfer activities, these tools can only marginally improve an organization's use of information and knowledge and do not address the deep-rooted processes and strategies necessary to overcome these barriers. Information technologies and applications only incrementally improve an organization's ability to facilitate data sharing and information exchange. As such, these approaches demonstrate a bounded set of limitations that ultimately prevent organizations from overcoming their current obstacles and diminish their ability to build a self-sustaining and long-term organization-wide system, thus undermining the very benefits KM practices have to offer.

Therefore, we suggest that educational institutions should not simply appropriate KM strategies from the business sector and apply them to their organizations. If KM is being implemented poorly, does that mean it should be done away with completely? Or does it hold its own as a concept worth striving for? The current limitations and drawbacks of KM in the private sector should serve as a warning for educational institutions. These organizations should be careful not to prematurely fragment their KM practices and focus on narrow applications and solutions. Instead, higher-education institutions stand to benefit from an approach that incorporates a more long-term and inclusive strategy to their knowledge activities. As such, improved methods of data and information sharing need to be coupled with embedded and long-term KM strategies in order to address the organization-wide factors that can either impede or promote an ongoing culture of research, reflexivity, and long-term organizational learning. If the evolutionary qualities of knowledge management—as it evolves from data, information, and knowledge—cross through multiple groups of people within an organization,

as well as traverse the three key organizational resources available—that is, people, processes, and technology—then the dynamic process that guides successful KM strategies and practices is more readily supported and maintained.

OPPORTUNITIES FOR KM IN EDUCATION

Educational institutions demonstrate a great need for improved knowledge-based systems. We already find that there are many formal and informal administrative processes, information-sharing patterns, work incentives, information silos, and other work practices that have flourished over time, yet these can also critically impede organizational and systematic information flow and knowledge exchange. KM strategies and practices can begin to integrate these disjointed systems. For example, the use of information maps and audits can initially be used to obtain a bird's-eye view of the current processes and practices, and their corresponding strengths and weaknesses. This type of initial diagnosis proves to be important for implementing KM in order to identify the most appropriate entry point for change. The cyclical quality of KM encourages organizations to take an honest and reflexive stance on what is already going on in their organization. KM requires that educational institutions candidly address their current patterns and processes, and only from this position begin to capitalize on the opportunities that KM strategies and practices can offer. This process of organizational re-evaluation and reflexivity proves to be the most difficult challenge for educational institutions. At the same time, the process offers the ideal opportunity for these institutions to integrate KM to promote sustainable learning within their organizations in order to meet these external demands as well as improve organization-wide effectiveness.

Higher-education institutions can begin to translate these strategies into action by identifying their information shortages and needs, including finding out where people are already requesting more data and information. These institutions can also start by identifying groups of people who already maintain synergistic relationships of collaboration and sharing within the institution. In fact, educational settings already demonstrate many information-sharing activities in effect, such as existing formal or informal communities of practice. However, to sustain ongoing inquiry and continuous learning, educational institutions need to strategize as to how they will systemically embed these activities and practices within the very fabric of the organization. Taken individually, information-sharing activities can be used toward incremental improvement; however, when KM is adopted and executed as an organization-wide strategy, improved methods of data and information sharing can be used to continually promote the development of KM-based practices. This can help educational institutions become more informed in their decision-making as a whole. All of this helps to lay the foundation for a robust culture of inquiry and reflexivity, thus establishing the mechanisms for sustainable, long-term organizational learning.

Perhaps more importantly, student access and success are the likely benefactors of these KM practices. KM practices can promote organizational reflexivity in such a way that educational institutions better understand their own weaknesses and strengths, and can then allocate their resources to where they are most needed. As demands for accountability rise, educational institutions need to become much more adept at assessing students' needs along with their own institutional capabilities. KM practices can help bring these two together, that is, aligning institutional capabilities and resources to better address students' needs and thus student success. Subsequently, educational institutions that engage in KM practices for continuous learning at the organizational level also engage in promoting continuous learning for their students.

OPPORTUNITIES FOR CONTINUOUS LEARNING

In conclusion, to fully realize the potential of KM, educational institutions will need to change the focus of KM from isolated knowledge-sharing activities to long-term, organization-wide strategies. Thus, KM practices can help educational institutions meet their goal of improved decision-making to advance student learning, allowing these institutions to begin to identify the value of programs and services that contribute to student access and success. This requires not only addressing information policies, but also taking a closer look at the institution's own processes and current practices to stimulate ongoing and constructive data use. Therefore, KM practices can be used to help educational institutions develop a sense of reflexivity across all levels of the organization, thereby providing these institutions with the means for a sustainable culture of inquiry and continuous learning.

REFERENCES

Brown, J.S. & Duguid, P. (2000). Balancing act: How to capture knowledge without killing it. *Harvard Business Review, 78*(5), 3-7.

Coate, L.E. (1996). Beyond re-engineering: Changing the organizational paradigm. In A. Kendrick (Ed.), *Organizational paradigm shifts* (pp. 1-18). Washington, DC: National Association of College and University Business Officers (NACUBO).

Davenport, T.H. (1997). *Information ecology: Mastering the information and knowledge environment.* New York: Oxford University Press.

Davenport, T.H. & Prusack, L. (1997). *Working knowledge: How organizations manage what they know.* Cambridge, MA: Harvard Business School Press.

Drucker, P. (1998). The knowledge-creating company. In Drucker, P. et al. (Eds.), *Harvard Business Review on knowledge management* (pp. 1-19). Cambridge, MA: Harvard Business School Press.

Friedman, D. & Hoffman, P. (2001). The politics of information. *Change, 33*(2), 50-57.

Hammer, M., Leonard, D., & Davenport, T. (2004). Why don't we know more about knowledge? *MIT Sloan Management Review, 45*(4), 14-18.

Hovland, I. (2003). Knowledge management and organisational learning, an international development perspective: An annotated bibliography, Working Paper 224. Retrieved October 7, 2004, from Overseas Development Institute (ODI) Web site, http://www.odi.org.uk/rapid/Publications/Documents/WP224.pdf

Huysman, M. & de Wit, D. (2004). Practices of managing knowledge sharing: Towards a second wave of knowledge management. *Knowledge Process Management, 11*(2), 81-92.

Petrides, L.A., McClelland, S.I., & Nodine, T.R. (2004). Costs and benefits of the workaround: Inventive solution of costly alternative. *The International Journal of Educational Management, 18*(2), 100-108.

Senge, P.M. (1990). *The fifth discipline: The art and practice of the learning organization*. New York: Currency.

Wilson, T.D. (2002). The nonsense of knowledge management. *Information Research, 18*(1), Paper 144. Retrieved October 7, 2004, from http://InformationR.net/ir/8-1/paper144.html

This work was previously published in Knowledge Management and Higher Education: A Critical Analysis, edited by A. Metcalfe, pp. 21-33, copyright 2006 by Information Science Publishing (an imprint of IGI Global).

Chapter 8.13
Online Academic Libraries and Distance Learning

Merilyn Burke
University of South Florida, USA

Bruce Lubotsky Levin
University of South Florida, USA

Ardis Hanson
University of South Florida, USA

BRIEF HISTORY OF DISTANCE LEARNING

Historically, distance learning or distance education began as little more than "correspondence courses," which promised an education in one's own home as early as 1728 (Distance Learning, 2002). By the 1800s the concept of distance education could be found in England, Germany and Japan (ASHE Reader on Distance Education, 2002).

In 1933, the world's first educational television programs were broadcast from the University of Iowa and in 1982, teleconferencing began (Oregon Community Colleges for Distance Learning, 1997), often using videotaped lectures, taped-for-television programs and live programming, adding a human dimension. Students and faculty were now able to interact with each other in real time, enhancing the learning process by allowing student access to teachers across distances.

ACADEMIC DISTANCE LEARNING & THE VIRTUAL LIBRARY

Distance learning can be defined by the fact that the student and the instructor are separated by space. The issue of time is moot considering the technologies that have evolved allowing real-time access. Today, universities around the world use various methods of reaching their remote students. With the use of technology, access becomes possible, whether it is from campuses to remote sites, or to individuals located in their own homes.

The development of course instruction, delivered through a variety of distance learning methods (e.g., including Web-based synchronous and asynchronous communication, e-mail, and audio/video technology), has attracted major university participation (Burke, Levin & Hanson, 2003). These electronic learning environment initiatives increase the number of courses and undergraduate/graduate degree programs being offered without increasing the need for additional facilities.

During the 2000-2001 academic year, the NCES (National Center for Education Statistics) estimated in the United States alone there were 3,077,000 enrollments in all distance education courses offered by 2-year and 4-year institutions, with an estimated 2,876,000 enrollments in college-level, credit-granting distance education courses, with 82% of these at the undergraduate level (Watts, Lewis & Greene, 2003, p. iv). Further, the NCES reported that 55% of all 2-year and 4-year U.S. institutions offered college-level, credit-granting distance education courses, with 48% of all institutions offering undergraduate courses, and 22% of all institutions at the graduate level (ibid, p. 4). It is clear that distance education has become an increasingly important component in many colleges and universities, not only in the United States, but also worldwide.

Although educational institutions create courses and programs for distance learners, they often omit the support component that librarians and accrediting organizations consider critical. It is recommended that courses be designed to ensure that students have "reasonable and adequate access to the range of student services appropriate to support their learning" (WICHE, Western Interstate Commission for Higher Education). Further, courses should incorporate information literacy skills within the course or in class assignments to ensure skills for lifelong learning (American Library Association, 1989; Bruce, 1997).

Distance learning (DL) students are unlikely to walk into the university's library for instruction on how to use the resources, from print to electronic journals, as well as services such as electronic reserves and interlibrary loan. The elements of any successful distance-learning program must include consideration of the instructors and the students, both of whom have needs that must be examined and served.

With imaginative use of technology, libraries have created "chat" sessions, which allow 24/7 access to librarians who direct students to the resources that are available online or through interlibrary loan. In addition, librarians assist faculty in placing materials on electronic reserve so that their students can access the materials as needed. Libraries have become more willing to provide mail services and desk top delivery of electronic articles to their distance learning students and, when that is not possible, refer their students to local libraries to take advantage of the interlibrary loan system. Online tutorials have been created to help students learn how to access these resources, while other libraries have specific departments that assist their distance education students and faculty. The role of the library in this process is one of support, both for the students and the faculty.

CHANGES IN DISTANCE LIBRARIANSHIP

Of all of the "traditional" library functions, such as materials provision, electronic resources, and reciprocal borrowing available to the distance learner, there remains a significant gap in service, that of reference. Although chat lines and other 24/7 services are available, these services simply do not provide the DL student the same quality of service that the on-campus student gets when he or she consults with a librarian in person. Newer versions of distance learning course software provide external links to resources, but do not yet include reference service by e-mail and live chat sessions in their basic packages. It will be the responsibility of the library to make these services easily available and known to the distant learner, whose contact to the institution may not include information about the library and its resources. Proactive planning by the library with those who are responsible for distance education can ensure that the students are made aware of what is available for them in the library.

Recently, libraries have been looking at e-commerce business models as a functional way to serve their clientele in reference services, as today's "customers" are savvier, and businesses have become more sophisticated in responding to customers' needs. Libraries can use these models to provide the services for DLs whose level of skills has risen with the increased use of the Internet. Coffman (2001) discusses the adaptation of such business tools as customer relations management (CRM) software, such as the Virtual Reference Desk, Webline, NetAgent, and LivePerson. These programs are based upon the "call center model," which can queue and route Web queries to the next available librarian. A quick visit to the LSSI Web site (Library Systems and Services, L.L.C, *http://www.lssi.com*) allows a look into the philosophy of offering "live, real-time reference services". LSSI's "Virtual Reference Desk" allows librarians to "push" Web pages to their patrons' browser, escort patrons around the Web and search databases together, all while communicating with them by chat or phone (*www.lssi.com*). Many of these systems provide the capability to build a "knowledge base" that can track and handle a diverse range and volume of questions. These collaborative efforts, with a multitude of libraries inputting the questions asked of them and creating FAQs (frequently asked questions lists), provide another level of service for the distance learner (Wells & Hanson, 2003).

These systems have great potential, and while they show tremendous possibilities, they need more work to make them more functional for library use. Chat sessions are problematic when the patron is using his or her phone line to connect to the computer, and libraries must look to the emerging technology to find solutions to such issues to prevent becoming obsolete.

Another direction is the development of "virtual reference centers," which would not necessarily have to be located in any particular physical library. Current collaboratives among universities have created consortial reference centers accessible anywhere and anytime. The reference center librarian could direct the student to the nearest physical resource or to an online full-text database based upon the student's educational profile (e.g., university, student status, and geographic location). Although the physical library may indeed become a repository for books and physical items, the reference component may no longer be housed within that particular building.

An example of support is Toronto's Ryerson Polytechnic University (Lowe & Malinski, 2000) infrastructure, which is based upon the concept that, in order to provide effective distance education programs and resources, there must be a high level of cooperation between the university, the departments involved, and the library. At Ryerson, the Continuing Education Department studied what types of support the students needed and identified technical, administrative, and academic help as three major areas of concern. Technical help was assigned to the university's computing services; administrative help was available on the Web and through telephone access, and academic help included writing centers, study skill programs, and library services. Ryerson's philosophy encompassed the concept that synchronization of all these components would assist in making the student's experience richer and give the student a higher degree of success

The library and the distance education unit worked to provide connectivity to resources that were important to the classes being taught online or at-a-distance. It is these types of library involvement that can make distance learning an even more successful and enriching experience. When a university system, as a whole, embraces a collaboration of all its components, both the students and the university reap the rewards.

CONCLUSION

Distance education will only continue to grow. In order to support this educational initiative, academic libraries must establish a supporting framework and commitment to those services traditionally provided by libraries such as lending books and answering reference questions in person or by telephone, plus new services such as "live chat" and desk top delivery of articles that are unique to the virtual environment. Faculty and students in distance learning courses depend on the academic library for their resources and services, and the library must be able to deliver materials to students or assist them in finding alternate sources in a timely manner. Libraries need to be able to identify their DL students using the necessary resources to verify information. Help desks, chat rooms, e-mail programs, and live reference all contribute to the support of the distance learning programs. Since DL students may never visit a library's physical facility, it is important to provide information on how best to access the library virtually.

Faculty members also require library support for their courses. For example, materials may be scanned and placed on the Web or videos may be "streamed" for online access. In order to digitize and make these items accessible, faculty need information on the correct use of copyrighted materials. It is also advisable to put into place an action plan to implement a program for distance learning and a method for assessing that program once it is in place.

FUTURE TRENDS

As distance learning continues to flourish, research will be needed to examine the effective implementation and ongoing management of distance education. While several issues emerge as salient, such as the social aspects of communication in the networked environment, and the integrity of Web-based course resources, it is the role of libraries in support of distance education that must be considered. Although much has been written about the social isolation of distance work, recent advances in groupware technologies have enhanced an individual's ability to stay connected for both work and social exchange through the use of synchronous and asynchronous remote communication (Li, 1998; Watson, Fritz et al., 1998). However, the increased use of technology suggests that formal and extensive training on both distance technology and team communications are necessary (Venkatesh & Speier, 2000).

Libraries, often overlooked in this process, have to be far more assertive in the distance learning process. Libraries can be a center of technical and administrative help along with the traditional academic role that they have normally held. The growing DL field allows librarians to re-define their roles, and request monies for advanced technological necessary to become as "virtual" as the classes being taught. In addition, to serve the ever-increasing DL population, library education must now include the course work that will provide future librarians the training necessary to serve this ever-expanding population.

REFERENCES

American Library Association. (1989). Presidential Committee on Information Literacy. *Final Report.* Chicago: American Library Association.

Bruce, C. (1997). *Seven faces of information literacy.* Adelaide, South Australia: AUSLIB Press.

Burke, M., Levin, B.L., & Hanson, A. (2003). Distance learning. In A. Hanson & B.L. Levin (Eds.), *The building of a virtual library* (pp.148-163). Hershey, PA: Idea Group Publishing.

Coffman, S. (2001). Distance education and virtual reference: Where are we headed? *Computers in Libraries, 21*(4), 20.

Distance Learning. (2002). 1728 advertisement for correspondence course. Retrieved March 8, 2002, from http://distancelearn.about.com/library/timeline/bl1728.htm

Li, F. (1998). Team-telework and the new geographical flexibility for information workers. In M. Igbaria & M. Tan (Eds), *The virtual workplace* (pp. 301-318). Hershey, PA: Idea Group Publishing.

Lowe, W., & Malinksi, R. (2000). Distance learning: Success requires support. *Education Libraries, 24*(2/3), 15-17.

Oregon Community Colleges for Distance Learning. (1997). *The strategic plan of the Oregon Community Colleges for Distance Learning, distance learning history, current status, and trends.* Retrieved March 8, 2003, from http://www.lbcc.cc.or.us/spoccde/dehist.html

Venkatesh, V., & Speier, C. (2000). Creating an effective training environment for enhancing telework. *International Journal of Human Computer Studies, 52*(6), 991-1005.

Watts, T., Lewis, L., & Greene, B. (2003). *Distance education at degree-granting postsecondary institutions: 2000–2001.* Washington, D.C.: National Center for Education Statistics. Retrieved from http://nces.ed.gov/pubs2003/2003017.pdf

Wells, A.T., & Hanson, A. (2003). E-reference. In A. Hanson & B.L. Levin (Eds.), *The building of a virtual library* (pp.95-120). Hershey, PA: Idea Group Publishing.

WICHE (Western Cooperative for Educational Telecommunications). (n.d.). *Balancing quality and access: Reducing state policy barriers to electronically delivered higher education programs.* Retrieved September 2, 2003, from http://www.wcet.info/projects/balancing/principles.asp

KEY TERMS

Chat: A real-time conferencing capability, which uses text by typing on the keyboard, not speaking. Generally between two or more users on a local area network (LAN), on the Internet, or via a bulletin board service (BBS).

CRM (Customer Relationship Management): This term refers to how a company interacts with its customers, gathers information about them (needs, preferences, past transactions), and shares these data within marketing, sales, and service functions.

Desk Top Delivery: Using electronic formats to send articles to users.

Distance Learning/Distance Education: Taking courses by teleconferencing or using the Internet (together with e-mail) as the primary method of communication.

Electronic Reserves: The electronic storage and transmission of course-related information distributed by local area networks (LANs) or the Internet. Also known as e-reserves; in addition to displaying items on a screen, printing to paper and saving to disk are often allowed.

Internet: A worldwide information network connecting millions of computers. Also called the Net.

Link-rot: The name given to a link that leads to a Web page or site that has either moved or no longer exists.

Next Generation Internet (NGI): Currently known as Abilene, the next generation Internet refers to the next level of protocols developed for bandwidth capacity, quality of service (QOS), and resource utilization.

Real-Time: Communication that is simultaneous; see **synchronous.**

Social Aspects of Communication: A social process using language as a means of transferring information from one person to another, the generation of knowledge among individuals or groups, and creating relationships among persons.

Streaming Video: A technique for transferring data as a steady and continuous stream. A browser or plug-in can start displaying the data before the entire file has been transmitted.

Synchronous and Asynchronous Communication: Synchronous communication is when messages are exchanged during the same time interval (e.g., Instant Messenger™). Asynchronous communication is when messages are exchanged during different time intervals (e.g., e-mail).

Virtual Library: More than just a means of collocating electronic resources (full-text materials, databases, media, and catalogues), a virtual library also provides user assistance services, such as reference, interlibrary loan, technical assistance, and so forth.

Voice over Internet Protocol (VoIP): A protocol that enables people to use the Internet as the transmission medium for telephone calls.

Web (World Wide Web): A global system of networks that allows transmission of images, documents, and multimedia using the Internet.

This work was previously published in the Encyclopedia of Information Science and Technology, Vol. 4, edited by M. Khosrow-Pour, pp. 2199-2202, copyright 2005 by Idea Group Reference (an imprint of IGI Global).

Chapter 8.14
Online Assessment and Instruction Using Learning Maps:
A Glimpse into the Future

Jim Lee
CTB/McGraw-Hill, USA

Sylvia Tidwell-Scheuring
CTB/McGraw-Hill, USA

Karen Barton
CTB/McGraw-Hill, USA

ABSTRACT

Online assessment, in its infancy, is likely to facilitate a variety of innovations in both formative and summative assessment. This chapter focuses on the potential of online assessment to accelerate learning via effective links to instruction. A case is made that detailed learning maps of academic progress are especially conducive to effective skill and concept diagnosis and prescriptive learning, contributing construct validity and precision to assessment results and coherence to instructional interventions. Item adaptive testing using learning maps and the paradigm of intelligent agents is discussed in the context of a vision of a seamless integration of assessment and instruction. The chapter is primarily speculative rather than technical.

A GLIMPSE INTO THE FUTURE

In this chapter the authors invite the reader to take a step back from the pressures of educational policy and politically driven educational reform movements to consider one possible direction of development of online educational assessment and instruction in the age of the Internet and advancing technology.

Instruction and assessment are closely related, integral aspects of the learning process. Instruction is the process by which learning is facilitated and guided; assessments are opportunities for learning, as well as feedback mechanisms that inform and have the potential to positively and dynamically affect instruction. It is only when this feedback provides relevant, timely information for enhancing instruction that its full potential can be realized. Online approaches to assessment are ideal for this purpose.

Imagine a classroom where students interact with an online assessment system on a weekly or monthly basis. The assessments are formative in nature (designed to inform the instructional process), as well as diagnostic, describing the specific skills a student has mastered and has yet to master in order to meet a prescribed educational standard. Linked to these diagnostic test results are instructional references and other supports to help the teacher remediate, sustain, or advance the student. With this support, the teacher is better able to meet the specific learning needs of each student and track all students' cumulative progress toward achievement of prescribed educational standards. Now imagine the same teacher delegating the teaching of some skills and concepts in each content area to learning software. As each student interacts with the software program, his or her knowledge state is continually assessed in order to customize the instructional inputs. Imagine further (if you do not mind a wildly speculative leap) an online learning environment, available in this same classroom via subscription, in which the construction of instructional inputs suggested by the ongoing assessment of the student's knowledge state is achieved by automated searches of public domain Web sites.

The argument of this chapter will be that detailed learning maps of academic progress are likely in this and coming decades to play a role in progress toward a vision of closely integrating assessment and instruction—either in the classroom, or in software, or in cyberspace.

It should be stated from the outset that there is no necessary distinction between the data required for sound formative assessment and sound summative assessment for accountability requirements. The cumulative records of student achievement, based on ongoing, detailed formative assessment, can potentially be aggregated and expressed in periodic summative reports. This chapter focuses, however, on the use of online formative test data as immediate feedback for learning and the design of instruction.

LEARNING MAPS

One of the foundations for online assessment advances may be the representation of academic learning sequences in detailed learning maps of a particular kind, defined here as networks of sequenced learning targets. Compared to other more general methods of ordering content (for example, by grade-level collections of knowledge indicators, by statistical results, or by the definition of emerging general stages of skill and knowledge), the detailed ordering of learning targets in learning maps is especially conducive to effective concept and skill diagnosis and prescriptive learning. Learning targets may be skills, concepts, or any discrete focus for a lesson. In addition to intentional learning targets, learning maps can include transitional knowledge states (errors or partially correct knowledge) that typically occur on the learning path.

Each learning target in a learning map may be represented graphically as a node in a network of nodes. These nodes are ordered in precursor/postcursor (learning order) relationships, as shown in Figure 1, which represents the learning path to an understanding of the causes of seasons on Earth. Figure 1 is a section of a larger learning map with links to other science strands. This section of the map focuses on the learning targets that need to be mastered in order to understand the causes of seasons on Earth. The shaded nodes are those that have been mastered by the student. This particular student appears to be making the mistake, common among students and the general population, of attributing seasons on Earth to the changing distance between Earth and the sun as Earth travels around the sun in an elliptical orbit. Before the student will have a fully correct understanding of the causes of seasons on Earth, he or she will need to master the learning targets in the unshaded nodes.

Although learning sequences traditionally have been mapped in a very general way in the scope-and-sequence charts of instructional materials and in state and district curriculum standards for kindergarten through grade 12, the detailed ordering of specific skill and concept development in learning maps is at a level of granularity that supports precise instructional intervention. It is likely—or so we argue in this chapter—that online assessments based on learning maps, currently in the experimental stage of development, will be tested more widely in classrooms throughout the United States (and probably elsewhere) in this and in coming decades.

In order to understand the power of learning maps for online assessment and for the attainment of significant academic improvement effects, the difference between a general ordering of learning objectives and the more precise ordering in a learning map should be discussed further. The ordering of learning targets in a learning map (with arrows pointing in the direction that learning proceeds) is based on the fact that skills and concepts are best taught and learned in a carefully constructed sequence. For example, a student cannot grasp the concept of seasons without a solid grounding in precursor concepts (axis of rotation, Earth's revolution, the different heating effect of direct versus indirect sunlight, etc.). The construction of this understanding must follow a path that takes into account the necessity of foundational knowledge before proceeding to the next step.

This does not mean, however, that all students follow the same path to proficiency. While some precursors of a concept or skill may be essential, others are merely facilitative and not necessary for all students. (For example, students can learn to carry in addition mechanically without understanding the concept of place value; however, understanding the concept of place value may facilitate learning how to carry in addition for some students.) For this reason, students may vary from each other in the way in which they progress through learning maps to a specified outcome; and the arrows that define the learning-order relationships in learning maps must have probabilities associated with them, based on data from thousands of students.

Precursor/postcursor relationships hypothesized by content experts and/or cognitive scientists to exist between nodes on a learning map must be continuously validated through ongoing test administrations on a variety of student populations. In this context, a node is validated as a postcursor of another node in a domain if, according to test data from an adequate student sample, there is a high probability that success on that node will not occur without success on the precursor node. Significant differences in the probability data from one subpopulation to the next may necessitate the use of different node orderings for distinct subpopulations.

LEARNING MAPS AND THE VALIDITY OF ASSESSMENT

Before we outline how learning maps potentially can support integrated online assessment and instruction solutions (the future vision), it is important to stress how learning maps can enhance the construct validity of the assessment dimension of these solutions. Traditional achievement tests try to estimate a student's level of possession of a construct such as "mathematics achievement." *Construct* is the psychometric term used to refer to something to be measured (such as mathematics achievement) that can only be inferred from actual performances (such as performance on a set of mathematics test items). Developing test items for a one-time or summative test that covers all the skills that define the broad and unobservable construct of mathematics achievement is, practically speaking, impossible. Therefore, tests only sample a subset of skills from the universe of possible skills that represent "mathematics achievement" (Crocker & Algina, 1986).

In an online diagnostic environment, where many different subsets of items can be administered over time, the sample of skills covered and the number of items given is collectively much larger[1] than would be possible on a one-time, summative test. This improved sampling, in itself, increases the likelihood that a test is valid, that it accurately measures what it claims to measure.

Whatever construct each test sets out to measure must be clearly defined and accurately measured. With clear definitions and accurate measurement, the assessment is considered valid in regard to the construct, or has some level of construct validity (Messick, 1989). With the construct operationally defined (via a learning map, for example), tests and test items can be conceived and developed as evidence of the construct via observable skills.

The construct validity of the assessment can be enhanced by carefully and systematically defining many of the skills and concepts that are components of the construct being assessed. Cognitive task analyses, an inherent component of instructional design (e.g., Anderson, 1995), can be used to define constructs as collections of learning targets. For example, Deno (1997) describes the task analytic approach as a way to take a task and break it into subtasks, illustrating in linear form the requisite skills for completing the overall task successfully. Such an approach has been used in many studies, including the development of computer-based performance assessments (Mislevy, Steinberg, Breyer, Almond, & Johnson, 1999) in which items are broken down into each skill and subskill necessary to access the item, respond to the item, and provide the correct answer to the item.

In a similar way, the construction of learning order sequences in learning maps entails a precise, highly granular decomposition and definition of larger units of knowledge into specific, small steps in skills and concept development. Here the applied question is: "What skills and concepts are necessary or helpful for learning X?" This question in turn is applied to the precursors of X, proceeding backward in this way until the network of learning targets that support understanding of X has been defined. In principle, this analysis is straightforward (or straightbackward); in practice, it is messy and difficult for a number of reasons. A key point, though, is that a learning map does not have to be perfect in order to be useful. The data from test administrations over time can contribute to the improvement of the map, including alterations in the ordering and granularity of learning targets. Equally important to keep in mind is the improvement in clarity and specificity possible with learning maps (like the map in Figure 1), as contrasted with the ambiguity, incompleteness, lack of grade-to-grade articulation, and wildly fluctuating granularity that are common in the statements describing what students are supposed to know and be able to do in many state and district curriculum standards documents (the content frameworks for most current assessments).

The careful definition and precise targeting of specific skills and concepts in learning maps help to ensure that irrelevant, non-targeted skills are not necessary for success on any test item. In other words, the concept or skill intended to be assessed by each item is all that is being measured. This more detailed and precise definition of each component of the construct to be measured might be called "construct amplification." Construct amplification supports the valid assessment of the achievement of *all* students.

For example, some students demonstrate their knowledge more effectively using graphics rather than words. If visual models and modes of understanding are neglected in the definition of the construct to be assessed, the resulting tests will not accurately assess the achievement of those students. For this reason, careful attention to the definition of constructs and their components is one of the key principles of universally designed assessments, tests designed according to guidelines that will enhance their accessibility and validity for all students, including special subpopulations (Thompson, Johnstone, & Thurlow, 2002).

Learning maps represent a construct such as mathematics achievement as a collection of hundreds, even thousands, of precisely defined learning targets ordered in a learning sequence. These learning-order hypotheses have to be validated theoretically by cognitive and curriculum experts, and empirically with many populations and subpopulations of students. The validated order of the learning targets adds to the evidence of construct validity. Once the learning order relations are established, mastery of some learning targets that were not directly assessed may be inferred from the mastery of other targets. For example, mastery of addition without regrouping may be inferred from mastery of addition with regrouping. In this way the assessment of components of mathematics achievement extends via inference to include learning targets not directly assessed by items in the test.

Figure 1. Example of a learning map: Understanding the causes of seasons

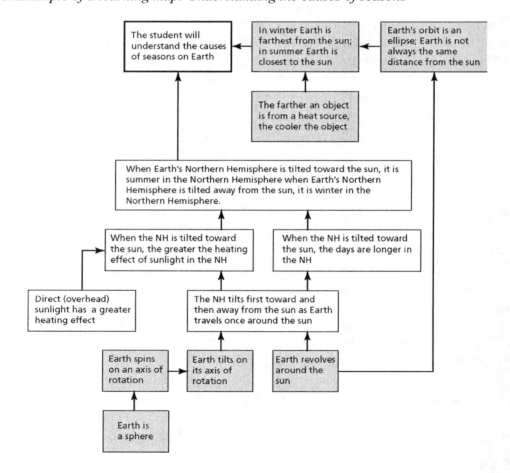

LEARNING MAPS AND THE COHERENCE OF INSTRUCTION

The characteristics of learning maps that enhance the validity of assessments can also influence the coherence of the instruction that follows from the assessment (the instruction dimension of the integrated assessment/instruction solution). As already indicated, the statements describing what students are expected to know and be able to do in each content area at each grade level often lack precision or completeness and therefore require interpretation and elaboration before they can become the foundation for the specification of items to be used in achievement testing. Learning-order considerations are observed in the most general way in these documents, but not at the level of acquisition of specific skills and concepts. Viewed longitudinally from grades K-12, the sequencing of learning targets in each content area may have critical gaps or violate learning order considerations. Learning materials and large-scale assessments, developed by textbook and testing companies and customized for each state based on the state curriculum standards document, tend to inherit deficiencies of detail and learning-order coherence in the standards document. In this way, the decomposition and ordering of knowledge in state standards documents paradoxically tend to maintain incoherence in education while overtly attempting to reduce it.

Referring again to Figure 1, one state's curriculum standards document, for example, requires that by grade 7, all students must understand the causes of seasons. However, in the same state there is no explicit reference in the standards document to several of the precursor concepts, either in the grade 7 standards or in any of the earlier grades. The learning map fills in the sequence of detailed learning targets involved in acquisition of this concept. Assessing the student on these learning targets is necessary for precise and accurate diagnosis of the student's current knowledge state, as well as prescription of a well-ordered instructional sequence to help the student understand the causes of seasons. The results of online diagnostic assessment of a student's knowledge of the learning targets in this map provide the teacher a useful blueprint for helping the student to achieve success on this grade 7 standard. In this way, learning maps can bring increased coherence and efficiency not only to the process of assessing students, but also to the instructional intervention that follows from the use of the assessment results.

LEARNING MAPS, ITEM-ADAPTIVE TESTS, AND INTELLIGENT AGENTS

The ability of learning maps to increase the efficiency of assessment can be exploited to its fullest potential when used to support item adaptive testing (a mode of testing in which the selection of the next test item is contingent on the student's responses to the previous test items) that utilizes the paradigm of intelligent agents.

The intelligent agent paradigm defines an agent as a system capable of undertaking *actions* in response to a series of *perceptions* (Russell & Norvig, 2002). In educational assessment, the presentation of an item to a student is one kind of action. A student's response to the presented item is one kind of perception. An intelligent assessment agent can be an online system whose actions are items and whose perceptions are student responses.

Interestingly, the perceptions of the agent, or student responses, need not be bound to a single assessment. The agent can remember the student's performance on past assessments, and begin subsequent tests or presentation of items in accordance with the agent's understanding of the student. This model can be extended even further, taking into account other perceptions (such as the student's history of interactions with instructional materials, assuming you have this data in a compatible format), producing an assessment even more precisely adapted to that particular student.

The agent's representation of the student's knowledge "state" can be based on the kind of learning maps already discussed. The knowledge state is perceived by the agent as a learning "boundary" between the mastered and unmastered targets in the learning map—the leading edge, so to speak, of the student's progressive mastery. Adaptive assessments as described herein would focus primarily on the learning targets just ahead of this boundary, where it is most likely that a student is currently progressing in understanding. The agent would tend to avoid presenting items well within the area of mastered concepts, since it already has evidence that success on these items is highly probable. The agent would also tend to avoid presenting items that assess learning targets well beyond the boundary, since it already has evidence that failure on these items is highly probable. The student's learning boundary (pattern of mastered nodes on the learning map) is in fact a collection of precise evidence for selecting each successive item to present to the student.

An intelligent item-adaptive assessment always uses the learning-order relationships in the learning map to select items with high "inferential value." (In the game of 20 questions, for example, asking whether the unknown object is "bigger than a breadbox" has high inferential value, because either a yes or a no response eliminates about half the objects in the room from consideration.) The inferential value of an item assessing a learning target in a learning map is related to the number of direct precursor and postcursor links it has to other learning targets in the map.

Whether the student responds correctly or incorrectly to the first item, the response (new evidence) results in a revised estimate of the student's knowledge state, which leads to intelligent selection of the optimal learning target to be assessed by the second item. This process continues until the student's current boundary in the map has been modified to include all the new learning targets that have been mastered.

Learning maps and the tests based on them can be designed so that wrong answers to test items provide evidence of transitional knowledge states (errors or partially correct knowledge) on the path to mastery of a learning target. For example, the map of development of understanding the causes of seasons (Figure 1) includes the misconception path taken by students who construct a naïve theory (personal misconception) regarding the cause of seasons based on the general idea that things are hotter the closer they are to a heat source. Consider the following multiple choice test item:

The temperatures in the Northern Hemisphere in winter are generally colder than in summer due to the fact that when it is winter in the Northern Hemisphere

A. it is more likely to snow
B. Earth is farther from the sun
C. there is more cloud cover blocking the sun's rays
D. the Northern Hemisphere is tilted away from the sun

Selection of answer choice B maps a student onto the misconception path in the learning map for understanding the causes of seasons.

Student misconceptions and errors are useful data in and of themselves, contributing to the accurate representation of the student's learning boundary and supporting the intelligent selection of subsequent items. The key point is that no piece of information—no perception, in agent terms—is lost to the system.

One of the interesting points about computer adaptive assessment is that by its very nature, it must continually make judgments about a student's mastery of a subject and also provide a continual update of assessment results, all while the assessment is in progress. Immediate feedback and storage of the pattern of student responses to test questions can become the basis for an ongoing calculation and prescription of the shortest instructional route to any prescribed educational goal or standard. Herein lies the potential of learning-map-based "intelligent" assessment (still perhaps decades away from realization) to support an extremely close integration of assessment and instruction in a responsive learning environment. Look again at Figure 1 describing the path to understanding the causes of seasons. The shaded learning targets on the map are nodes currently mastered by the student; the unshaded learning targets remain to be mastered before the student will have a scientific understanding of the causes of seasons.

Imagine that each of the unmastered targets in the learning map is linked electronically to a variety of appropriate online instructional activities, each teaching precisely the concept to be mastered. Once node mastery is demonstrated, a learning activity associated with the optimal next node can be tackled. The process can therefore proceed back and forth between learning activities and learning checks (assessment) until the understanding of the causes of seasons has the required depth, based on the sequence of concept development described in the learning map.

IN THE MEANTIME

Currently, in response to increased demand from state departments of education for online classroom formative assessment solutions, several assessment companies are offering online formative assessment programs in some strands of some content areas. What is motivating this investment? The increased availability of computers and online access in school systems through the United States are certainly enabling factors, but the primary motivational factor has been the apparent frustration of many educators with the traditional annual, on-demand summative assessment paradigm. Summative assessment results are intended ultimately to support gains in student learning *indirectly* via school improvement programs, identification of at-risk students, and so forth. However, a major frustration of many educators with this paradigm seems to be *the time lag between test administration and reports delivery.*

A second major frustration seems to be the *lack of diagnostic specificity in summative test reports.* Standardized norm-referenced tests report general information about student achievement, such as a percentile rating of how a student is functioning relative to peers in a given content area (e.g., mathematics), or in some general categories of instruction within the content area (e.g., measurement in mathematics). The specifics of what the student knows and does not know, at a level of detail that will support instructional intervention, are lacking in these reports. Hence, the call for online classroom formative tests that can provide a teacher with the right level of instructionally relevant information and instant turnaround time for receiving that information. Supporting the wisdom of the trend toward implementing online formative assessment solutions is a body of research suggesting that effective classroom formative assessment practices of all kinds can yield substantial academic improvement effects (Black, Harrison, Lee, Marshall, & William, 2003).

If significant achievement gains are perceived to accrue from this investment in formative assessment, testing companies will increasingly invest in the further development and elaboration of online formative assessment models and in ways to use this data for accountability purposes. The development of valid and reliable methods for aggregating classroom online formative assessment data for accountability purposes could have a watershed effect on the future of assessment. The substantial public commitment to and investment in accountability testing would then be available as a powerful support for the development of new generations of online assessment/instruction solutions.

CONCLUSION

Online assessment solutions with instructional links are likely to increase in popularity in coming years for the reasons cited above. Although detailed learning maps of the kind described in this chapter are difficult to construct and validate, we believe they can provide the most powerful content infrastructure for integrating assessment and instruction in the classroom, in software learning programs, and in the online environment. Compared to other more general methods of ordering content, they represent learning in a detailed way that enhances the efficiency, construct validity, and precision of the assessment as well as the coherence of the instructional intervention.

REFERENCES

Anderson, J.R. (1995). *Learning and memory: An integrated approach.* New York: John Wiley & Sons.

Black, P., Harrison, C., Lee, C., Marshall, B., & William, D. (2003). *Assessment for learning.* New York: Open University Press.

Commission on Excellence in Education. (1983). *A nation at risk: The imperative for educational reform.* Washington, DC: Commission on Excellence in Education.

Crocker, L., & Algina, J. (1986). *Introduction to classical and modern test theory.* New York: Holt, Rinehart & Winston.

Deno, S.L. (1997). "'Whether' thou goest: Perspectives on progress monitoring." In E. Kameenuii,

J. Lloyd, & D. Chard (Eds.), *Issues in educating students with disabilities.* Hillsdale, NJ: Lawrence Erlbaum.

Fuchs, L. (1998). Computer applications to address implementation difficulties associated with Curriculum-Based Measurement. In M. Shinn (Ed.), *Advanced applications of Curriculum-Based Measurement* (pp. 89-112). New York: Guilford Press.

Messick, S. (1989). Validity. In R.L. Linn (Ed.), *Educational measurement* (pp. 13-103). New York: American Council on Education/Macmillan.

Mislevy, R.J., Steinberg, L.S., Breyer, F.J., Almond, R.G., & Johnson, L. (1999). A cognitive task analysis with implications for designing simulation-based performance assessment. *Computers in Human Behavior, 15,* 335-374.

Phillips, S.E. (1994). High-stakes testing accommodations: Validity versus disabled rights. *Applied Measurement in Education, 7*(2), 93-120.

Russell, S.J., & Norvig, P. (2002). *Artificial intelligence: A modern approach* (2nd ed.). New York: Prentice-Hall.

Thompson, S.J., Johnstone, C.J., & Thurlow, M.L. (2002). *Universal design applied to large-scale assessments.* (Synthesis Report 44). Minneapolis, MN: University of Minnesota, National Center on Educational Outcomes. Retrieved from *http://education.umn.edu/NCEO/OnlinePubs/Synthesis44.html*

ENDNOTE

[1] The use of learning maps as the underlying content framework for any assessment can make it possible to generate and maintain the large item pools associated with online

diagnostic testing, even when the testing is customized to assess student progress toward proficiency on a given state's standards. By mapping standards and indicators in state and district curriculum frameworks to the learning targets in learning maps, the items used to assess standards in state A and the items used to assess standards in state B can be drawn from a common item repository based on the learning maps.

This work was previously published in Online Assessment, Measurement, and Evaluation: Emerging Practices, edited by D. Williams, S. Howell, and M. Hricko, pp. 227-237, copyright 2006 by Information Science Publishing (an imprint of IGI Global).

Chapter 8.15
Future Directions of Multimedia in E–Learning

Timothy K. Shih
Tamkang University, Taiwan

Qing Li
City University of Hong Kong, Hong Kong

Jason C. Hung
Northern Taiwan Institute of Science and Technology, Taiwan

ABSTRACT

In the last chapter, we discuss how advanced multimedia technologies are used in distance learning systems, including multimedia authoring and presentation, Web-based learning, virtual environments, interactive video, and systems on mobile devices. On the other hand, we believe pedagogic theory should be incorporated into the design of distance learning systems to add learning efficiency. Thus, we point out some suggestions to the designers of future distance learning systems.

INTRODUCTION

Distance learning, based on styles of communication, can be categorized into synchronized and asynchronized modes. The advantages of distance learning include flexibility of time and space, timely delivery of precisely presented materials, large amount of participants and business opportunity, and automatic/individualized lecturing to some degrees. Both synchronized and asynchronized distance learning systems rely on multimedia and communication technologies. Due to its commercial value, distance learning is becoming a killer application of multimedia and communication research. We discuss current distance learning systems based on the types of multimedia technologies used and point out a few new research directions in the last section.

MULTIMEDIA PRESENTATIONS AND INTERACTIONS

Authoring and playback of multimedia presentations are among the earliest applications of multimedia technologies. Before real-time communication and video-on-demand technologies, multimedia presentations were delivered to kids and distance learning students on CD ROMs. The advantage of multimedia presentation over traditional video tapes is due to interactivity. Multimedia presentations allow one to select "hot spots" in individualized topology. Techniques to realize this type of CD ROM presentations allow a rich set of media coding and playback mechanisms, such as images, sounds, and animations (including video and motion graphics). Successful examples include MS PowerPoint, Authorware Professional, Flash, and others.

With the development of communication technologies, multimedia computing focuses on efficient coding mechanism to reduce the amount of bits in transmission. Synchronization among media became important. Inner stream synchronization is implemented in a single multimedia record, such as the interleaving coding mechanism used in a video file, which includes sound track and motion picture track. Another example of inner stream synchronization and coding is to merge graphics animation with video stream (Hsu, Liao, Liu, & Shih, 2004). On the other hand, inter stream synchronization is more complicated since both the client (i.e., user) side and the server (i.e., management system) side need to work together. Inter stream synchronization allows packages (e.g., sound and image) to be delivered on different paths on a network topology. On the client side, packages are re-assembled and ensured to be synchronized. Another example of a recent practical usage of inter stream synchronization is in several commercial systems allowing video recording to be synchronized with MS PowerPoint presentation or Flash. Some systems (Shih, Wang, Liao, & Chuang, 2004) use an underlying technology

known as the advanced streaming format (ASF) of Microsoft. ASF allows users or programs to embed event markers. In a playback system on the client side, users can interrupt a vide presentation, or jump to another presentation section. The video presentation can also use markers to trigger another presentation object such as to bring up a PowerPoint slide (converted to an image) or another multimedia reference. In order to deliver a synchronized presentation, an ASF server needs to be installed on the server machine.

ASF provides a preliminary technology for video-on-demand (or lecture-on-demand). In order to support multiple clients, it is necessary to consider bandwidth allocation and storage placement of video records. Video-on-demand systems (Hua, Tantaoui, & Tavanapong, 2004; Mundur, Simon, & Sood, 2004 allow a video stream to be duplicated and broadcast in different topology on multiple channels, to support multiple real-time requests in different time slots. In addition, adaptive coding and transmission mechanism can be applied to video-on-demand systems to enhance overall system performance.

Video-on-demand allows user interactions to select video programs, perform VCR-like functions, and choose language options. Interactive TV (Liao, Chang, Hsu, & Shih, 2005) further extends interactivities to another dimension. For instance, the users can select the outcome of a drama, refer to specification of a commercial product, or answer questions pre-defined by an instructor. The authoring and playback system developed in Liao, Chang, Hsu, and Shih (2005) takes a further step to integrate video browser (for interactive TV) and Web browser. Thus, distance learning can be implemented on set-top box.

WEB-BASED DISTANCE LEARNING AND SCORM

Most multimedia presentations can be delivered online over Internet. And, Web browser is a common interface. HTML, XML, and SMIL are the representation languages of learning materials. Typically, HTML is used in the layout while other programming languages (such as ASP) can be used with HTML to retrieve dynamic objects. As an extension to HTML, XML allows user defined tags. The advantage of XML allows customized presentations for different Web applications, such as music and chemistry, which requires different presentation vocabularies. In addition, SMIL incorporates controls for media synchronization in a relatively high level, as compared to inner stream coding technologies. HTML-like presentations can be delivered by Web servers, such as Apache and MS IIS.

Although Web browsers are available on different operating systems and HTML-like presentations can be reused, search and reuse of course materials, as well as their efficient delivery, are key issues to the success of distance learning. In order to achieve reusability and interoperability, a standard is needed. The advanced distributed learning (ADL) initiative proposed the sharable content object reference model (SCORM) (The Sharable Content Object Reference Model, 2004) standard since 2000. Main contributors to SCORM include the IMS Global Learning Consortium, Inc., the Aviation Industry CBT (computer-based training) Committee (AICC), the Alliance of Remote Instructional Authoring & Distribution Networks for Europe (ARIADNE), and the Institute of Electrical and Electronics Engineers (IEEE) Learning Technology Standards Committee (LTSC). The SCORM 2004 (also known as SCORM 1.3) specification consists of three major parts

- **The content aggregation model (CAM):** Learning objects are divided into three categories (i.e., assets, sharable content objects (SCOs) and content organizations). The contents of the learning objects are described by metadata. In addition, CAM includes a

definition of how reusable learning objects are packed and delivered.

- **The run-time environment:** In order to deliver learning objects to different platforms, a standard method of communication between the learning management system (LMS) and the learning objects is defined.
- **The sequencing and navigation:** Interactions between users (i.e., students) and the LMS are controlled and tracked by the sequencing and navigation definitions. This also serves as a standard for defining learner profiles, as well as a possible definition for intelligent tutoring.

The SCORM specification clearly defines representation and communication needs of distance learning. To realize and promote the standard, a few SCORM-compliant systems were implemented (Chang, Chang, Keh, Shih, & Hung, 2005; Chang, Hsu, Smith, & Wang, 2005; Shih, Lin, Chang, & Huang; Shih, Liu, & Hsieh, 2003). However, common repository for SCORM learning objects, representation of learner records, and intelligent tutorial mechanisms to facilitate sequencing and navigation are yet to be identified. On the other hand, most existing SCORM-compliant LMSs fail to support the newest specification, except the prototype provided by ADL.

VIRTUAL CLASSROOM AND VIRTUAL LAB

Web-based distance learning supports asynchronized distance learning in general. Usually, distance learning programs rely on Web browsers to deliver contents, collect assignments from students, and allow discussion using chat room or e-mails. These functions can be integrated in a distance learning software platform such as Blackboard (http://www.blackboard.com/) and WebCT (http://www.Webct.com/). On the other hand, real-time instruction delivery can be broadcast using video channels, or through bi-directional video conferencing tools (Deshpande, & Hwang, 2001; Gemmell, Zitnick, Kang, Toyama, & Seitz, 2000). Real-time video communication requires sophisticated network facilities and protocols to guarantee bandwidth for smooth transmission.

In addition to online delivery of instruction, lab experiments can be realized using remote labs or virtual labs (Auer, Pester, Ursutiu, & Samoila, 2003). Remote lab uses camera and advanced control technologies to allow physical lab instruments to be accessed by students using Internet. Virtual lab may or may not include physical experimental instruments. Emulation models are usually used. In most cases, assessment of experiment outcomes from software emulation is compared with those from physical devices.

Virtual reality (VR) and augmented reality techniques can also be used in distance learning (McBride & McMullen, 1996; Shih, Chang, Hsu, Wang, & Chen, 2004). Most VR systems use VRML, which is an extension of XML for 3-D object representations. The shared-Web VR system (Shih, Chang, Hsu, Wang, & Chen, 2004) implements a virtual campus, which allows students to navigate in a 3-D campus, with different learning scenarios. Behaviors of students can be tracked and analyzed. The incorporation of game technologies points out a new direction of distance learning, especially for the design of courseware for kids. With wireless communication devices, ubiquitous game technologies can be used for mobile learning in the near future.

MOBILE LEARNING

Wireless communication enables mobile learning. With the capability of multimedia technologies on wireless connected notebook computers, PDAs, and even cellular phones, system developers are possible to implement distance learning systems on mobile devices (Meng, Chu, & Zhang, 2004; Shih, Lin, Chang, & Huang, 2004). The challenges of deploying course materials on small devices, such as cellular phones, include the limited display space, slow computation, and limited memory capacity. On a small display device, reflow mechanism can be implemented (Shih, Lin, Chang, & Huang, 2004). The mechanism resizes contents into a single column layout, which can be controlled using a single scroll bar on PDAs or cellular phones. To cope with small storage, prefetching technique on subdivided course contents can be used. Thus, the readers can download only the portion of contents of interesting.

To realize learning management systems on wireless network connected devices, a distributed architecture needs to be designed between the server and the client (e.g., PDA). SOAP is a communication protocol very suitable for the architecture. SOAP packages are messages that can be sent between a client and server, with a standard representation envelope recommended by the W3C (http://www.w3.org/). The advantages of the protocol include platform independency, accessibility, and implementation efficiency. In addition, in order to maintain the status of each individual learner, learner profiles needs to needs to be defined. Yet, SCORM contains only a preliminary description of learner profile definition. The representation of course contents should also consider how to enable small packages to be delivered on a remote request. Cashing mechanism and hand shaking protocol are important issues yet to be developed. In addition, in some occasion for situated learning, location awareness is necessary for situated collaborative learning.

On the other hand, synchronized distance learning on mobile devices requires efficient real-time streaming due to the limited bandwidth of current wireless communication systems (Liu, Chekuri, & Choudary, 2004. Even as 3G mobile communication technologies are available, smooth video streaming requires a broader channel and a robust error resilience transmission mechanism.

HYBRID INTERACTIVE SYSTEMS AND PEDAGOGICAL ISSUES

Whether learning activities are implemented on mobile devices or PC clients, efficient collaboration is the key issue toward the success of learning. A SCORM-based collaborative learning LMS is developed in Chang, Lin, Shih, and Wang (2005). The system allows learning activities among students to be synchronized based on the Petri net model. The instructor is able to supervise the collaboration behavior among a group of students. Whether or not it is SCORM compliant, a distance learning platform should support collaboration in either synchronized or asynchronized manner. At least, a CSCW-like system should be implemented to support the need of collaboration.

Recently, personalized Web information delivery has become an interesting issue in data mining research. A distance learning server is able to analyze student profiles, depending on individual behaviors. Learner profiles can be stored and analyzed according to traversal sequences and results from tests. This type of distance learning system is based on the self-regulation principles of the social cognitive theory (Bandura, 1986). A system using this approach should allow students to plan on their study schedule based on individual performance (Leung, & Li, 2003), while the underlying intelligent mechanism can guide students to a suitable study schedule, which can be reviewed by an instructor. To facilitate user friendliness, self-regulation can be incorporated with Web-based interfaces and mobile devices. To some degree of the usage of artificial intelligence (Shih & Davis, 1997), an intelligent tutorial system is able to generate individualized lectures (Leung & Li, 2003).

We realize that, it is possible to design an integrated learning environment to support the application of the scaffolding theory (Zimmerman & Schuck, 1989). Scaffolding, proposed by L. S. Vygotsky, was viewed as social constructivism. The theory suggests that students take the leading role in the learning process. Instructors provide necessary materials and support. And, students construct their own understanding and take the major responsibility. Between the real level of development and the potential level of development, there exists a zone of proximal development. This zone can be regarded as an area where scaffolds are needed to promote learning. Scaffolds to be provided include vertical and horizontal levels as a temporary support in the zone of proximal development. The scaffolding theory is essential for cognitive development. It also supports the process of social negotiation to self-regulation. There are three properties of the scaffold

- The scaffold is a temporary support to ensure the success of a learning activity.
- The scaffold is extensible (i.e., can be applied to other knowledge domains) and can be used through interactions between the learner and the learning environment.
- The scaffold should be removed in time after the learner is able to carry out the learning activities independently.

The scaffolding theory indicates three key concepts. Firstly, in the zone of proximal development, the relationship between the scaffolds providers and the receivers are reciprocal. That means that the instructor and students negotiate a mutual beneficial interactive process. Secondly, the responsibility is transferred from the instructor to the student during the learning process. Depending on the learning performance, the instructor gradually gives more control of the learning activities to the student for the ultimate goal of self-regulation. Finally, the interaction facilitates the learners to organize their own knowledge. Scaffolding also encourages the use of language or discourse to promote reflection and higher-order thinking.

Pedagogical principles are not multimedia technology. However, the developers of distance learning system should be aware of the concept.

SUMMARY

This chapter summarizes multimedia technologies for distance learning systems. While we were looking for the essential needs of professional educators and students, in terms of "the useful multimedia distance learning tools," we have found that lots of tools were developed by computer scientists. Most of these tools lack of underlying educational theory to show their usability. However, software is built for people to use. In spite of its advanced functionality and outstanding performance, any system will be useless if no one uses it. Thus, we believe the specification of a distance learning system should be written by educational professionals, with the help of computer scientists.

From the perspective of multimedia and Internet computing, there are a few challenging research issues to make distance-learning systems more colorful and useful. We highlight a few here

- **Interactive TV:** Video-on-demand technologies should be highly integrated with interactive TV and set-top box devices, which should be extended to incorporate different modals of interaction. A sophisticated bi-directional inter stream synchronization mechanism needs to be developed.
- **Standards:** The most popular standard is SCORM. However, the definition of user profile, federal repository, and adaptive techniques for mobile devices are yet to be investigated.
- **High communication awareness:** Video conferencing tools should be integrated with awareness sensors, to bring the attentions on interested video area to users.
- **Virtual and remote lab:** A standard development specification for creating virtual or remote labs is not yet developed. The standard should allow reusable lab components which can be assembled to facilitate different varieties of lab designs.
- **Adaptive contents for mobile learning:** Different mobile devices should have different functional specifications to guide a central server to transmit device and user dependent media for efficient learning.
- **Intelligent tutoring:** User profile dependent tutoring based on intelligent technology applied on Web technology should be used. Pedagogical considerations can be applied on intelligent tutoring.

Among the developed platforms for distance learning, an assessment mechanism, especially the one based on educational perspective, should also be proposed. It is the hope that the multimedia research community can work with educational professionals and the distance learning industry together, to develop a standard distance-learning framework for the success of our future education.

REFERENCES

Auer, M., Pester, A., Ursutiu, D., & Samoila, C. (2003, December). Distributed virtual and remote labs in engineering. *International Conference on Industrial Technology ICIT 2003,* Slovenia (pp. 1208-1213).

Bandura, A. (1986). *Social foundations of thought and action: A social cognitive theory.* Englewood Cliffs, NJ: Prentice-Hall.

Chang, F. C., Chang, W., Keh, H., Shih, T. K., & Hung, L. (2005). Design and implementation of a SCORM-based courseware system using influence diagram. *International Journal of Distance Education Technologies, 3*(3), 82-96.

Chang, W., Hsu, H., Smith, T. K., & Wang, C. (2005). Enhancing SCORM metadata for assessment authoring in e-learning. *Journal of Computer Assisted Learning, 20*(4), 305-316.

Chang, W., Lin, H. W., Shih, T. K., & Wang, C. (2005, March 28-30). Applying Petri nets to model SCORM learning sequence specification in collaborative learning. *Proceedings of the 19th International Conference on Advanced Information Networking and Applications,* Taiwan.

Deshpande, S. G., & Hwang, J. (2001, December). A real-time interactive virtual classroom multimedia distance learning system. *IEEE Transactions on Multimedia, 3*(4), 432-444.

Gemmell, J., Zitnick, L., Kang, T., Toyama, K., & Seitz, S. (2000, October-December). Gaze-awareness for video conferencing: A software approach. *IEEE MultiMedia, 7*(4), 26-35.

Hsu, H. H., Liao, Y. C., Liu, Y.-J., & Shih, T. K. (2004). Video presentation model. In S. Deb (Ed.), *Video data management and information retrieval* (pp. 177-192). Hershey, PA: Idea Group Publishing.

Kien, A., Hua, M. A., Tantaoui, & Tavanapong, W. (2004, September). Video delivery technologies for large-scale deployment of multimedia applications. *Proceedings of the IEEE* (Special issue on Evolution of Internet Technologies towards the Business Environment), *92*(9), 1439-1451.

Leung, E. W. C., & Li, Q. (2002). Media-on-demand for agent-based collaborative tutoring systems on the Web. *IEEE Pacific Rim Conference on Multimedia 2002* (pp. 976-984).

Leung, E. W. C., & Li, Q. (2003). A dynamic conceptual network mechanism for personalized study plan generation. *ICWL 2003* (pp. 69-80).

Liao, Y., Chang, H., Hsu, H., & Shih, T. K. (2005). Merging web browser and interactive video: A hypervideo system for e-learning and e-entertainment. *Journal of Internet Technology, 6*(1), 121-131.

Liu, T., & Choudary, C. (2004, October). Realtime content analysis and adaptive transmission of lecture videos for mobile applications. *Proceedings of the 12th ACM International Conference on Multimedia,* New York.

McBride, J. A., & McMullen, J. F. (1996, January). Using virtual reality for distance teaching a graduate information systems course. *Proceedings of the 29th Hawaii International Conference on System Sciences 1996* (Vol. 3, pp. 263-272).

Meng, Z., Chu, J., & Zhang, L. (2004, May). Collaborative learning system based on wireless mobile equipments. *IEEE Canadian Conference on Electrical and Computer Engineering CCECE 2004* (Vol. 1, pp. 481-484).

Mundur, P., Simon, R., & Sood, A. (2004, February). End-to-end analysis of distributed video-on-demand systems. *IEEE Transactions on Multimedia, 6*(1), 129-141.

The Sharable Content Object Reference Model. (2004). *ADL Co-Laboratory.* Retrieved from http://www.adlnet.org/

Shih, T. K., Chang, Y., Hsu, H., Wang, Y., & Chen, Y. (2004). A VR-based shared Web system for distance education. *International Journal of Interactive Technology and Smart Education (ITSE), 1*(4), 4.

Shih, T. K., & Davis, R. E. (1997, April-June). IMMPS: A multimedia presentation design system. *IEEE Multimedia* (pp. 67-78).

Shih, T. K., Lin, N. H., Chang, H., & Huang, K. (2004, June 27-30). Adaptive pocket SCORM reader. *Proceedings of the 2004 IEEE International Conference on Multimedia and Expo (ICME2004)*, Taipei, Taiwan.

Shih, T. K., Liu, Y., & Hsieh, K. (2003, July 6-9). A SCORM-based multimedia presentation and editing system. *Proceedings of the 2003 IEEE International Conference on Multimedia & Expo (ICME2003)*, Baltimore.

Shih, T. K., Wang, T., Liao, I., & Chuang, J. (2003). Video presentation recording and online broadcasting. *Journal of Interconnection Networks* (Special issue on Advanced Information Networking: Architectures and Algorithms), *4*(2), 199-209.

Zimmerman, B. J., & Schuck, D. H. (1989). *Self-regulated learning and academic achievement.* New York: Springer-Verlag.

This work was previously published in Future Directions in Distance Learning and Communication Technologies, edited by T. K. Shih, pp. 273-283, copyright 2007 by Information Science Publishing (an imprint of IGI Global).

Chapter 8.16
The Changing Library Education Curriculum

Vicki L. Gregory
University of South Florida, USA

INTRODUCTION

Libraries of the 21st century are very different places from those that existed at the beginning of the 20th century, and very different as well from the libraries of only 25 years ago. Library education has striven to keep pace with all the myriads of changes. Within the last 100 years, fortunately and necessarily in order to retain its relevance, professional library education and practice has evolved from the centrality of teaching and writing the "library hand" to providing modern curricula such as services for distance learners and Web-based instruction using course management systems such as Blackboard, WebCT, and so forth. Along the way, the library profession has often been first not only to accept but also to adopt and apply the technological innovations now common to modern civilization. Throughout, library educators have paved the way to the acceptance of innovation in libraries by instructing students to use and apply new technologies.

BACKGROUND

The revolutionary changes over the past 25 years in the educational curriculum for schools of library and information science, which are necessitated by the exponential expansion of computer-based technologies, requires an almost constant and continuous reexamination of the skills and expertise that need to be acquired by the next generation of librarians. Although much has changed in libraries, the core of who we are and what we are truly remains the same. Librarianship is and will continue to be a profession devoted to bringing users and information together as effectively and efficiently as possible. To meet that ideal, librarians have used technology to enhance and create services. In addition, it is important to meet emerging educational needs of our increasingly multicultural and diverse society. Librarians have recognized that changing expectations and lean budgets require organizations to call upon the talents of everyone (Butcher, 1999). And librarians have become more engaged in teaching and research in order to serve the needs of users better (Bahr & Zemon, 2000).

Importance of People and People Skills

Computer technologies and communication systems have had an undeniable impact on society as a whole and on our profession, but it is also critical to remember the importance of the individual and of the need for interpersonal skills in our profession, which at its heart remains basically a "people profession." We harness technology for a reason—to promote learning and the dissemination of information—and we do not simply revere technology for its own sake. With the aid of computer specialists, we could design the best information system imaginable, but unless it operates in a manner that is accessible to people, nobody will use it. The ability of librarians (whether through collecting, organizing, or retrieving information) to act as intermediaries between users and the world's information resources will, in my opinion, never become outdated (Gilbert, 1998). In sum, the rapid changes in all types of libraries and the burgeoning of new technologies for librarians to learn, while increasing the amount of information that students need to have under their "academic belts" if they are to enter successfully into a library career, nevertheless remain rooted in the need to carry out the traditional librarian roles—though hopefully faster, cheaper, smarter, and more effectively.

Preparing Students in Traditional Areas of Library Responsibilities

The traditional heart and sole of a library is and remains, of course, its collections—from the time of the great Alexandrine library of the Classical era, libraries have been, in essence, civilization's repositories of learning, and hence the materials through which learning is transmitted down the

generations. Current students preparing for the future (and indeed the present) electronic library cannot be permitted to overlook the continued, lasting importance of print publications in the library's carrying out of its role, but by necessity, they must be equipped to deal with the rapidly expanding world of digital medium. Thus, collection development courses must reflect an appropriately balanced approach, emphasizing the latest technology not as an end in itself, but rather as simply another tool to use in addressing the problems arising in acquiring adequate resources for a library collection in whatever format is most appropriate for the particular library and the "task at hand" (Thornton, 2000).

As librarians and information professionals go about the process of acquiring electronic information resources in carrying out their collection development role, they must also continue to recognize and care about the important questions that have always concerned libraries, in respect to questions of future accessibility and preservation of library resources. Electronic materials—with their typical provision to libraries only through a licensing regime rather than through outright purchase—present altogether different problems for the library than do print materials. Collection development and preservation must remain an important part of the library school curriculum, no matter how dominated the library may become with electronic materials (Kenney, 2002).

In most conceptions of the libraries of the future, reference librarians may expect to continue to play many of the same reference roles that they have traditionally performed in interacting with their libraries' users. Reference librarians will continue to serve in an intermediary role to assist users in finding needed information and providing important "value-added" services through the production of instructional materials and guides to information resources. However, many of these functions, out of necessity, will be performed in media other than those that have been traditionally utilized. Collaboration and instruction may be expected to take place in a Web-based "chat" environment or by e-mail, rather than through a face-to-face meeting over the reference desk (Abels, 1996; Domas, 2001).

Reference librarians of the future must therefore acquire teaching skills as well as informational skills. They will need to be able to teach information literacy skills as students discover that just finding some online information on a topic and pushing the "print" or "download" button is not enough. In the electronic information world, librarians must be prepared to evaluate resources in a somewhat more in-depth way than was necessary when they could often depend upon refereed print journals for the majority of their information (Grassian & Kaplowitz, 2001).

In addition to all the vagaries involved with the classification and cataloging of traditional print materials, technical services librarians today, and doubtless more so in the future, will have to be prepared to cope with all the exponential varieties and forms that electronic resources may take. Technical services professionals are increasingly dealing with so many different formats and kinds of materials that may defy classification and are often not traditionally cataloged; other approaches, such as indexing and abstracting techniques and the development of in-house library-constructed databases, as well as Webliographies, may be undertaken as methods of organizing the access and retrieval process. Future graduates planning a career in the technical services areas should place a much greater focus than is presently typically allowed for in most library school curriculums on the technological aspects of information provision. Concurrently, library and information science schools need to take steps to provide for the programs and/or the courses that will include building student skills in document creation for the digital library environment. Unfortunately, all this cannot be allowed to serve as a replacement for the traditional knowledge and skills involved in cataloging and classification. As a minimum, students will need to gain a hands-on

knowledge of the architecture of the infrastructure and databases behind a digital library. This means that LIS schools must develop additional specific courses, rather than trying to make room in the already overstuffed basic "organization of knowledge" classes that most schools currently offer (Vellucci, 1997).

FUTURE TRENDS

In the foreseeable future, it is probable that more and more instruction will be provided in a distance mode utilizing Web delivery, videoconferencing, and other technological means of providing instruction. A burden on many LIS faculty members at present is how to adapt a course, originally designed for a face-to-face classroom encounter to a Web-based encounter. Although the goals, objectives, and major assignments for a class might remain the same, the overall means of delivery puts more pressure on faculty members to devise new ways of delivering material (Gregory, 2003). Both virtual and print reserve materials may become problematic as distance from the home site increases. Compounding the traditional instructional component, there is the additional element of computer support on a 24-hour, 7-days-a-week basis (Young, 2002). Increasingly, when something goes wrong with the computer on a student's end, the faculty member is expected to be able to do computer troubleshooting over the telephone or by e-mail. It is common for programs and universities to provide technical support, but even so the faculty member usually gets caught up in the technical support problems, obviously much more so than when the class is taught in the traditional manner (Carey & Gregory, 2002; Newton, 2003, p. 418). Of course, when the academic computing staff person or the faculty member is unavailable, the next major organization on the campus that fields these questions is—you guessed it—the library. Librarians must be able to deal with technical, computing, or network issues and attempt to aid the beleaguered student (or faculty member). So although these issues primarily affect the teaching of library and information studies classes, they also have a major impact on the services demanded of the library (Barron, 2003).

CONCLUSION

The rapidly changing requirements in the educational curriculum of schools of library and information science resulting from the exponential expansion of computer-based technologies naturally result in a reexamination of the knowledge and skills that need to be acquired by the next wave of library and information professionals. Skills in the use of new technologies are not only important in professional work, but in the education process itself, as more and more LIS courses are being offered via the Web, with faculty and students utilizing course management software.

REFERENCES

Abels, E.G. (1996). The e-mail reference interview. *RQ, 35*(Spring), 345-358.

Bahr, A.H. & Zemon, M. (2000). Collaborative authorship in the journal literature: Perspectives for academic librarians who wish to publish. *College and Research Libraries, 61*(5), 410-19.

Barron, D.D. (Ed.). (2003). *Benchmarks in distance education: The LIS experience*. Westport, CT: Libraries Unlimited.

Butcher, K. (1999). Reflections on academic librarianship. *Journal of Academic Librarianship, 25*(5), 350-353.

Carey, J.O. & Gregory, V.L. (2002). Students' perceptions of academic motivation, interactive participation, and selected pedagogical and structural factors in Web-based distance education. *Journal of Education for Library and Information Science, 43*(1), 6-15.

Domas White, M. (2001). Digital reference services: Framework for analysis and evaluation. *Library & Information Science Research, 23*(3), 211-231.

Gilbert, B. (1998). The more we change, the more we stay the same: Some common errors concerning libraries, computers, and the Information Age. In M.T. Wolf, P. Ensor & M.A. Thomas (Eds.), *Information imagineering: Meeting at the interface* (pp. 219-227). Chicago, ALA.

Grassian, E.S. & Kaplowitz, J.R. (2001). *Information literacy instruction: Theory and practice*. New York: Neal-Schuman.

Gregory, V.L. (2003). Student perceptions of the effectiveness of Web-based distance education. *New Library World, 104*(10), 426-433.

Kenney, A.R. et al. (2002, January). Preservation risk management for Web resources. *D-Lib Magazine, 8*(1). Retrieved December 12, 2003, from www.dlib.org/dlib/january02/kenney/01kenney.html

Newton, R. (2003). Staff attitudes to the development and delivery of e-learning. *New Library World, 104*(1193), 312-425.

Thornton, G.A. (2000). Impact of electronic resources on collection development, the roles of librarians, and library consortia. *Library Trends, 48*(4), 842-856.

Vellucci, S.L. (1997). Cataloging across the curriculum: A syndetic structure for teaching cataloguing. *Cataloging and Classification Quarterly, 24*(1/2), 35-39.

Young, J.R. (2002). The 24-hour professor: Online teaching redefines faculty members' schedules, duties, and relationships with students. *The Chronicle of Higher Education, 38*(May 31). Retrieved from chronicle.com/weekly/v48/i38/38a03101.htm

KEY TERMS

Collection Development: The portion of collection management activities that has primarily to do with selection decisions.

Collection Management: All the activities involved in information gathering, communication, coordination, policy formulation, evaluation, and planning that result in decisions about the acquisition, retention, and provision of access to information sources in support of the needs of a specific library community.

Course Management System (CMS): Computer software system that provides a course shell with a number of integrated tools, which may include chat software, a threaded discussion board, online grade books, online testing, and other classroom functions.

Digital Libraries: Organized collections of digital information.

Distance Education: A planned teaching and learning experience that may use a wide spectrum of technologies to reach learners at a site other than that of the campus or institution delivering the course.

Information Literacy: An integrated set of skills and the knowledge of information tools and resources that allow a person to recognize an information need and locate, evaluate, and use information effectively.

Videoconferencing: Conducting a conference between two or more participants at different geographical sites by using computer networks to transmit audio and video data.

This work was previously published in the Encyclopedia of Information Science and Technology, Vol. 5, edited by M. Khosrow-Pour, pp. 2799-2802, copyright 2005 by Idea Group Reference (an imprint of IGI Global).

Chapter 8.17
E-Learning and New Teaching Scenarios:
The Mediation of Technology Between Methodologies and Teaching Objectives

Cecilia Mari
Università Cattaneo - LIUC, Italy

Sara Genone
Università Cattaneo - LIUC, Italy

Luca Mari
Università Cattaneo - LIUC, Italy

ABSTRACT

This chapter analyzes the reciprocal influences between various teaching methodologies supported by information and communication technology (ICT) and the teaching objectives that are pursued by means of these methodologies. The authors present the main characteristics of the conceptual model which has led to the definition of the teaching objectives and the results of the experience of the "eLearning@LIUC" project, where the validity of the hypotheses underlying the model has been tested through their application within concrete contexts. They believe that the presented model, with its analysis of the possible correlations between teaching objectives, teaching methodologies, and technological tools, can provide a new awareness of the opportunities offered by the adoption of ICT in teaching.

THE POINT OF VIEW

The use of tools based on information and communication technology (ICT), and on the Internet in particular, usually aims at pursuing economies of scale by reducing distribution costs and/or increasing the number of users. The focus is therefore more on reach—or quantity—than richness—or quality (Weigel, 2000). E-learning projects do not escape this tendency. Although direct interpersonal relationships are generally recognized more effective that those mediated by ICT, their lack of reproducibility makes them expensive, from the point of view of both the teacher, since each new edition requires the replication of many of her/his costs, and the learner, due to space and time bonds which demand her/his here-and-now presence.

This leads to a prevalence of strategies which tend to interpret e-learning as a tool for reducing the organizational costs of education rather than as a method for improving the quality of education. The common emphasis is on efficiency in the management of educational processes (D'Angelo, 2003), particularly with respect to the distribution of and access to teaching material and the remote interaction among the subjects involved in the process.

The Objectives of E-Learning

In the design of an e-learning project, two general objectives can be sought (Keeton, Sheckley, & Krejci-Griggs, 2002):

- **Efficiency:** in the attempt to reduce the space and time bonds of teaching processes, e-learning operates as a substitute for traditional education, thus increasing some of its quantitative features;
- **Effectiveness:** in the attempt to improve teaching processes, e-learning operates as a complement to traditional education, thus increasing some of its qualitative features.

For organizations whose business is education, as in the case of our university, and for whom e-learning can prove a good opportunity to pursue their own mission (Moore, 1993; Trentin, 2000; Piccoli, Ahmad, & Ives, 2001; Smith, Ferguson, & Caris, 2001; Syed, 2001), these poles represent the extremes of a continuum of options. It cannot

be assumed in fact that an organization aims at reducing costs without considering quality, nor that an improvement of quality is pursued without considering costs.

Our Objective

The main objective of the eLearning@LIUC experimental project, where the model here described takes its origins and has been extensively tested, is of exploiting e-learning as a means to offer students occasions for a more effective learning experience. The emphasis is thus on the dimension of learning more than on the dimension of teaching and, as a consequence, on the point of view of the learner (Huba & Freed, 1999; Weimer, 2002). At the basis of this choice there was the need to find integrative, and not substitutive, solutions to the existing offer: activities which could be carried out "together with", and not "instead of", the traditional ones, as literature has been suggesting already for some time now (Tsichristzis, 1999; Marold, Larsen, & Moreno, 2000).

The Cattaneo – LIUC University has several years experience in the use of ICT for didactics. Each course makes use of a dedicated site—with updated information about the course syllabus, the lesson plan, communications as well as downloadable teaching material—integrated in the management system of the University itself. These tools have proven very useful as a support to the management of didactics, but it is clear that they do not have a specific role in enhancing the learning, as for example the extension of the office hours of a university library can facilitate learning conditions for a student but does not obviously influence her/his learning style.

The Individual Roles in a Blended Solution

It is worth noting that a project which aims at improving the quality of teaching, and not the quality of the management process supporting teaching, has the same objectives as a traditional teaching situation. In particular, it specifically involves the work and competence of teachers as the subjects who activate and shape the educational experience. Several degrees of freedom are usually left to a teacher in the planning of a teaching situation: they are related for example to the relationships between teacher and students, the time dedicated to in-class lessons and to individual work, the adoption of a deductive or inductive approach, the degree of interactivity, the time dedicated to exercise and practice, … e-learning—in particular if, as in our case, the attempt is to adopt it in an integrative, and thus *blended*, way—allows to further increase the dimensions of this option space: besides the traditional scenarios, in fact, new ones are introduced because of the use of technological tools.

Due to all these factors, the competence of the teacher is crucial for the success of the project. While her/his proficiency in planning a traditional lesson is taken for granted, when it comes to the use of ICT the lack of established teaching models and the limited experience often make it difficult to really enhance the quality of teaching by properly exploiting the available ICT tools. This is due to a lack of awareness, and perhaps even of information, on which tool(s) could / should be used in which situations and for which objectives: there exists, on one hand, the prospect of great potential and, on the other, considerable complexity and little experience.

The descriptive framework we present here is aimed at bridging these dimensions (objectives, situations, tools) in the option space of the e-learning activities. As the inductive result of the several experiences of blended e-learning we have co-designed (together with the disciplinary teachers) and coordinated in our University in the last three years, such a framework has been already partly validated also as a tool for supporting the design of new activities.

Consistently with its premises and context, this work is therefore addressed to those who make

use of teaching methodologies, to help them plan teaching situations appropriately and consciously, taking the most advantage of the opportunities currently offered by ICT.

THE eLEARNING@LIUC PROJECT

In the context and with the objectives described, and on the general basis of the discussions which took place at the European Council of Lisbon in March 2000 (see http://europa.eu.int/comm/education/policies/2010/doc/info2004.pdf), at the beginning of 2001 LIUC University started a project of integrated experimentation in e-learning. It began with an internal call for proposals addressed to the teachers for the design and implementation of *e-learning seminars*. The condition was that they should offer innovative teaching and learning experience from the point of view of both the methodology and the tools adopted.

The Guidelines

Considering the lack of a firm theoretical framework on the pedagogy of e-learning, no strict rules were imposed on the teachers regarding either teaching objectives and methodologies, or technological choices, so that information about these aspects could be acquired in the course of the experimentation. Each teacher was however asked to create materials characterized by "specifically e-learning" features, according to some general guidelines, inspired by the idea of distinguishing "specifically e-learning activities" from both:

- what we have called "e-distribution", currently well exemplified by the usual opportunity of making traditional materials available in some electronic format downloadable from a Web site;
- distance learning.

The guidelines, as presented to the teachers, were (Mari et al., 2002):

- if the innovation of the project is found in the way the material is accessed, and not in the way it is used, then the project is about e-distribution, not e-learning (and therefore it is not suitable for the call for proposal);
- if the material can be transferred onto a paper support without a loss in its effectiveness, then the project is about e-distribution, not e-learning (and therefore it is not suitable for the call for proposal);
- if the teaching experience is characterized only by the presence of a forum and/or on-line tutors, then the project is about distance learning, not e-learning (and therefore it is not suitable for the call for proposal.

And positively, a project is about e-learning when tools are used that change the way people learn.

New Dimensions in Learning

The challenge was then to create teaching materials and use them in a way which could bring about this change and positively affect the students' learning process. In this sense, the use of technologies makes it possible to emphasize some aspects which can enrich the experience of learning with new dimensions:

- **Multimediality:** the use of multimedia elements makes learning easier because it helps students to focus and keep their attention on complex contents, thanks to the activation of different senses (Jacobson, 1994; Laeng, 1996; Maragliano, 1998; Guttormsen & Krueger, 2000);
- **Hypertextuality:** the hypertext, structured as a manifold system of non-linear relationships among texts, allows students to follow their own personal paths and to create new

ones each time (Calvani & Varisco, 1995; Colazzo & Molinari, 1996);

- **Interactivity:** interactive components make it possible to work with the material in a *learning-by-doing* approach, which brings about a higher involvement, a deeper understanding and a better retention of the subjects (Dede, 1990; Johnson, Johnson, & Smith, 1998).

Blended Learning

We also tried to emphasize the characteristic of *blending*, that is, "mixed" didactics, e-learning and traditional, using ICT tools which would contribute to make teaching processes more flexible (Saunders et al, 2000). The basic hypothesis is that the way teaching processes are traditionally managed is unnaturally rigid in its structure, because it is carried out in two distinct phases: first the teacher transfers information to the class and then, but only afterwards, a control on the quality of learning follows. Moreover, the phase which is really crucial for learning tends to go "one way", from the teacher (who speaks, or whose texts are read) to the students. Anybody with some experience in teaching, on the contrary, knows perfectly well how important the direct contact with the students is for the success of the teaching itself, just as it is known that a good way to learn something is to teach it to somebody else. A good teacher looks for a feedback from the students, and a good student tries to learn actively.

Active Learning

With the aim of exploring the possibility of activating interactive teaching processes where evaluation is carried out during the learning, that is, in a two-way approach, we have structured a use of e-learning that softens some distinctions which are traditionally considered as unmodifiable, in particular:

- between class activities and individual activities;
- between learning of theoretical knowledge and practice with this knowledge;
- between moments when teaching material is used and moments of interaction with teachers and/or other students;
- between the production of teaching material and its utilization.

The Projects

Operatively, we have worked within the frame of *learning-by-doing* supported by ICT tools with the creation of "e-learning projects". Each of them consisted of a 12/15 hour seminar, whose contents were generally, but not necessarily, an integration of the program of an institutional course; seminars were open to a variable number of students (20 to 40) and held by one or more teachers in classrooms equipped with a network of connected computers.

In order to enable ourselves and the students to experiment with the possibilities and limits of e-learning in the best way, we have chosen to work with contents relating, rather than to the typical disciplines to which e-learning is applied (ECDL and foreign languages), to specialized disciplines which are qualifying in the students' university curriculum, for the faculties of Economics, Law and Engineering at LIUC.

Today, after more than three years of experimentation, about one hundred seminars have been carried out, and they have involved a total of more than 2000 students.

THE FIRST RESULTS OF EXPERIMENTATION: A PROPOSAL OF E-LEARNING SCENARIOS

We aim at providing teachers with ideas about possible uses of the tools and the various types of teaching materials, to be used according to the pursued objectives. Our starting point was, consequently, our concrete experience in the eLearning@LIUC project.

From the list of the seminars and the teaching tools we have used, we have identified a sublist of the possible ways of utilization in different contexts. For each element of this sublist, we have identified the corresponding teaching objectives.

Our work, for example, may start from questions like: what is the teacher's objective when s/he chooses to use a forum to open a discussion about a case study s/he intends to analyze? What is the interest of introducing an exercise on a subject, through a simulator or a business game, before explaining the related theory?

The Framework

Starting from these considerations, we have derived a framework which shows, for each experience in our analysis, the relationship between the e-learning tools, their different uses, the type of interaction, and the teaching objectives that were pursued.

We have noticed that, within this classification, the different *types of interaction*, made possible by the technological tools, among the roles involved (teacher, students, computers) are the following:

teacher—student/s;
student/s—student/s;
student—computer;
both face-to-face and distance.

The *teaching objectives* (Bloom, 1956; Calvani & Rotta, 1999; Badii & Truman, 2001) we have identified as possibly common to the different contexts, which contribute to the major meta-objective of *helping students to learn better*, appear to be:

- teaching to gather, organize, and analyze information (Ausubel, 1998; Trentin, 1998);
- awakening a critical mind (Wilson, 1996; Ausubel, 1968);
- stimulating active participation and collaboration among students (Tinzmann et al., 1990; Gokhale, 1995; Palloff & Pratt, 1999; Cenarle & Biolghini, 2000);
- encouraging the practice of what has been learned in theory (deductive approach) (Aster, 2001; Johnson et al., 1998);

- fostering a *learning-by-doing* approach (inductive approach) (Gross, 1993; Schank, 2000).

Table 1 reports in detail what has been said up to now, in particular:

- the columns *OBJECTIVES* and *TYPE of INTERACTION* contain the items included in the previous lists;
- the column *TOOL* reports the software tools developed and used in the seminars (please

Table 1. The framework

OBJECTIVES	TYPE of INTERACTION	TOOL	USE
Teaching to gather, organize and analyze information	Teacher — Student/s	Forum	Distance – to communicate and exchange material between a lesson and the following
			Distance – to manage FAQ
	Student — Computer	Animations	Face-to-face or distance – to support students in the application of procedures
		Film	Face-to-face – to support theory through the viewing of a real case
		Glossary	Face-to-face or distance – used by students to get deeper and clarify key concepts
		Multimedia Presentation	Distance – as self learning, to revise and get deeper into the contents presented during the lessons
		Hypertextual Structure	Face-to-face – to organize themes and related contents, at students' disposal for navigation
			Face-to-face – to show links and connections among teaching resources and comment a possible navigation path together
			Face-to-face – to give an overview of the subject, presenting teaching resources through conceptual maps
		Test	Distance – as self evaluation
			Face-to-face – to discuss results together
Awakening a critical mind	Teacher — Student/s	Forum	Face-to-face – to open the discussion on specific theoretical themes
	Student — Computer	Film	Face-to-face – to support theory through the presentation of a real case
			Face-to-face – to discuss starting from a real situation and drawing general conclusions
		Multimedia Presentation	Distance – as self learning, to revise and get deeper into the contents presented during the lessons
		Hypertextual Structure	Face-to-face – to show links and connections among teaching resources and comment a possible navigation path together
			Face-to-face – to give an overview of the subject, presenting teaching resources through conceptual maps

continued on following page

Table 1. continued

OBJECTIVES	TYPE of INTERACTION	TOOL	USE
Stimulating active participation and collaboration among students	Teacher — Student/s	Virtual Community	Face-to-face – to manage the class real time during exercises and practice
			Face-to-face – to allow exchanges of documents and files real time among teachers, students, and groups of students during guided practice
		Forum	Face-to-face – to open the discussion on specific theoretical themes
			Face-to-face – to give assignments and collect students' feedback and works
	Student/s — Student/s	Virtual Community	Face-to-face – to allow exchanges of documents and files real time among teachers, students, and groups of students during guided practice
		Forum	Distance – to communicate and exchange material between a lesson and the following
	Student — Computer	Film	Face-to-face – to discuss starting from a real situation and drawing general conclusions
		Simulator/ Business Game	Face-to-face – to practically apply what has been presented from a theoretical point of view
			Face-to-face – to start from a concrete problem and then go back to the related theory
		Test	Distance – as self evaluation
			Face-to-face – to discuss results together
OBJECTIVES	**TYPE of INTERACTION**	**TOOL**	**USE**
Encouraging the practice of what has been learnt in theory (deductive approach)	Student - Computer	Animations	Face-to-face or distance – to support students in the application of procedures
		Exercise	Face-to-face or distance – to practically apply what has been presented from a theoretical point of view
		Film	Face-to-face – to support theory through the presentation of a real case
		Simulator/ Business Game	Face-to-face - to practically apply what has been presented from a theoretical point of view
Fostering a learning-by-doing approach (inductive approach)	Student - Computer	Exercise	Face-to-face or distance – to practically apply what has been presented from a theoretical point of view
			Face-to-face – to start from a concrete problem and then go back to the related theory
		Film	Face-to-face – to discuss starting from a real situation and drawing general conclusions
		Simulator / Business Game	Face-to-face – to start from a concrete problem and then go back to the related theory

notice that this list does not aspire to completeness and does not include all software tools which could be used for e-learning, but only those which were actually used in the course of our project);

- in the column *USES* we have described "how" the tools were used, also specifying the details about time and space (face-to-face/distance, synchronous/asynchronous).

SOME CONSIDERATIONS, FOR A SYNTHESIS

The first element which emerges from the analysis of the table is that different uses for the same software tool correspond to different teaching objectives. To properly structure a teaching path it is thus not enough to specify, as it is usually done, that a certain tool will be adopted; on the contrary, the choice of the specific way this tool will be used is crucial. When planning a teaching situation with particular objectives in mind, these can be connected to the objectives in the list above; the table is to be read from the first column, and the educational context can be structured each time according to the different needs, in terms of interaction and uses of the tools.

The guidelines of this table will lead the teacher to the creation of courses and lessons which, taking advantage of the possibilities offered by technology, will prove potentially more effective from the point of view of the learner. Moreover, such experience will increase awareness among teachers about the opportunities at their disposal thanks to the use of ICT.

CASE STUDIES

Some significant examples follow, which show the application of this approach in the phase of planning together with the teachers.

Case: planning and execution of the seminar *Business strategy and policy* (12 hours, 25 students).

Teaching objectives: stimulating active participation in the lesson and awakening a critical mind.

When planning the seminar, one of the objectives suggested by the teacher was that of fostering the students' skills for analysis and critical thinking, starting from a considerable amount of connected documents.

In this case we decided to develop a hypertextual structure which would allow the reading not only of the documents, but also of the relations existing among them, according to a conceptual map which made them explicit.

Some themes for discussion were connected to the map nodes, and they were activated by the teacher in the classroom through the use of a forum. The students contributed with their answers and comments starting from the teacher's cue. In this way, the material provided became the starting point for a new elaboration by the students, based on their personal thinking.

Case: planning and execution of the seminar *Accounting: The Operative Cycles* (15 hours, 30 students).

Teaching objectives: fostering collaboration among students and encouraging the practice of what was learnt in theory.

When planning the seminar, the teacher expressed the will to improve students' understanding of some themes learnt as theory, connected to the relations existing among the phases of the operative cycles in accounting.

In this case, we let the students experiment directly with an integrated accounting management software used in firms.

The students were not left alone in this work. During the practice, the teacher created some teams which would play as if they were the different offices of a firm dealing with the activities connected, for example, to the management of purchase orders.

A double objective was reached this way: students collaborated to achieve some results, as required by the exercise, and they were given the opportunity, too, to put the theory of accounting into practice.

Case: planning and execution of the seminar *Managing complexity through group behaviour and team work* (15 hours, 20 students).

Teaching objectives: teaching to gather and analyze information, and stimulating active participation and collaboration among students.

In collaboration with the teacher, we have developed a multimedia hypertext, made up of some animated presentations, a film with an interview about a real case, an interactive text for self-assessment, and a forum at the students' disposal.

The approach was that of problem-solving, starting from the information given in the presentation, which the students had to analyze and elaborate in order to derive their own opinions. In this process, they were also helped by the film with the case, which provided some insights that were more connected to the real world.

In this seminar, rather than for discussion within the classroom, the forum was used by the students during the week between the lessons, to exchange the material they had to gather as homework and to prepare the interventions of the different groups for the following lesson.

Case: planning and execution of the course *Foundations of computer science* (40 hours, 40 students).

Teaching objectives: encouraging the practice of what has been learnt in theory, but also fostering a *learning-by-doing* approach.

The teacher developed a hypertextual structure, an e-book—to be used as a basic teaching tool for the academic course—which includes several interactive objects, presenting a problem and requiring to analyze it, or allowing the learner to freely experiment around the topic under presentation. Being developed to support and incentive learning in all the phases of the learning process, it is applied as a multifunctional teaching instrument:

- in the context of the classroom, the e-book is used by the teacher as a replacement for the course slides and a complement for the blackboard, but also by the students as a tool

enabling them to experiment the contents that the teacher introduces with the emphasis a problem solving approach; in this way the classroom begins an interactive laboratory, in which any student can actively participate in the lesson;

- in the context of the off-line learning, the e-book allows the students a self-paced access to the course contents, enabling them to test their acquisition of the contents and possibly to compare their solutions with other students.

Therefore the same object is adopted within and outside the classroom, by the teacher and by the learners, for both the presentation and the experimentation of the contents.

Case: planning and execution of the seminar *Cases of business information systems* (15 hours, 30 students).

Teaching objectives: teaching to analyze information, fostering a learning-by-doing approach, and stimulating active participation and collaboration among students.

All 15 hours of the seminar were managed as a virtual community, with a basic difference compared to the traditional use of this tool. In fact, while the VC is usually employed by the participants at a distance, our choice was that of experimenting with it within the classroom. This generated from the interest of proving that distance learning and e-learning are not synonyms, and that the use of ICT can be very effective also in class. In this particular case, we managed discussions among students where, potentially, everybody was talking to everybody. Such a scenario is clearly not feasible in a traditional classroom, due to the voice and content interferences that would inevitably be produced among the participants.

In the course of the seminar, the students were introduced to the problem of business information flow analysis through two case studies, presented in electronic format (an e-book) and structured, thanks to a VC, so that the students, organized in groups formed by two people, could experiment with a problem-solving approach. The constant interactivity deriving from this, together with the quality of the proposed cases with respect to both realism and complexity, made it possible to build a situation where the students progressively played the part intended for them by the VC, simulating the role of analysts/business consultants, and experimenting with new communicative situations.

THE EVALUATION

In order to gain a more scientific value, the considerations expressed up to now need to find a counterpart in their concrete application. This is what we tried to determine in the phase of evaluation of the experience, through the opinions of those who were actors in it: students and teachers.

The Questionnaire for the Students

At the end of each seminar, the students were asked to fill in a questionnaire, whose prelimi-

nary section consists of an inquiry on the level of familiarity the student has with ICT, in order to evaluate how her/his previous experience may influence the comprehensive response and the possible difficulties.

The questionnaire includes then three questions, asking each respondent to give a rating on a 1-to-7 Likert scale, where the value 1 corresponds to the worst judgment and 7 to the best. For each question, students were required to answer by choosing a single value.

The results are shown in Figures 1-3, for about 2000 questionnaires: the histograms show the cumulative number of answers for each Likert category.

The final part of the questionnaire contains some open questions, where the respondents are asked to give their opinions about:

- the quality of the materials and tools used (animations, simulations, self-evaluation tests, ...), with reference to the teaching objectives of each course;
- the change in didactics caused by the new modalities of e-learning;
- the effectiveness of the new way of learning compared to traditional didactics.

Figure 1. How much has e-learning made it possible for you to put the themes of the seminars into practice?

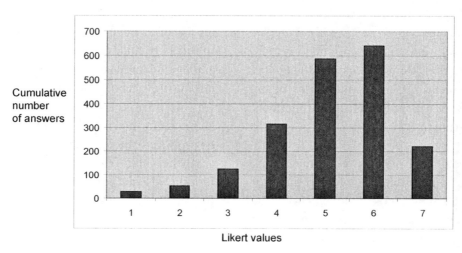

Figure 2. How much could you actively and collaboratively participate in the seminar?

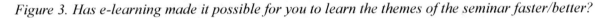

Figure 3. Has e-learning made it possible for you to learn the themes of the seminar faster/better?

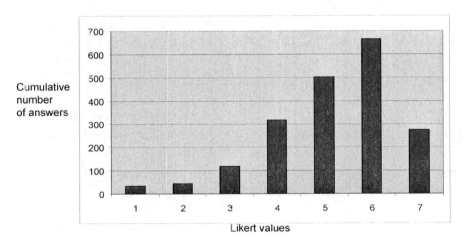

Trying to summarize some constant ideas which emerge from the analysis of the answers, "the use of the computer for learning" is the characteristic which is mostly considered as valuable and of special interest, no matter what the contents are. In particular, the use of tools such as simulators, self evaluation tests, forums, software like Visio, and the elements of multimediality, hypertextuality, interactivity, made possible by technologies, are an appreciated enrichment in relation to traditional lessons and materials: "e-learning as innovative way of teaching", so that the difference in the comparison with faculty courses is considered an advantage itself.

Interactivity, in particular, is often highlighted from different points of view: the computer is "a tool for interaction", it makes it possible to "actively take part in the lesson", "you feel more involved", and in this approach "real and concrete cases are considered, "you put into practice what you have learnt in theory". Students as well as teachers, thus, consider the *learning-by-doing* approach as totally positive in this context.

The students confirm the hypothesis that "the application of theoretical notions to concrete cases, in order to understand the problems you meet in real life, helps to learn better and quicker".

From the point of view of contents, the seminars are often seen as an occasion to get deeper into themes which were already presented during the lessons, and the possibility to find more material online is very positively valued.

Considering the management of the class and the relations in the course of the lessons, many express a high appreciation of the possibility of working in groups and of the better interaction with the teachers, which are reflected also in a "higher involvement and attention", and in the "flexibility the students have in the process of learning" (from the answers the students gave to the questionnaire).

The Questionnaire for the Teachers

Since the number of teachers involved was quite limited, we found it a more effective solution for our purposes to focus on the qualitative evaluation of the experience, and submitted some questions to them, about which we "left them carte blanche", so that they might freely express their opinions and provide us with useful suggestions and hints for a further development of our project.

The questions were the following:

- In what kind of teaching scenarios (e-learning seminar included) have you used the e-learning material? Can you briefly describe your experience?

- Do you think that the e-learning material can be properly reused in future teaching situations?
- If so, can you briefly describe the scenarios which may likely occur? (for example: students individually using the material as an integration to their study for a course; teacher using the material during the lessons of a course; ...)
- In the case of reusability, do you think that your e-learning material is already in a proper state for this reuse, from the point of view of both the contents and the technological structure?

Considering the teaching objectives we have pointed out, the evaluation of the experience by the teachers was very positive, both for the effectiveness of a "non typically academic" lesson, and for the better involvement of students: in this direction, the usefulness of the "practical", *learning-by-doing* approach was underlined (the students "... were able to experience the trial-and-error approach in a concrete an playful way..."; "... the forum enabled the students to work in a team and to better interact with the teacher..."; "... The use of multimedia and interactive tools made it possible to carry out a seminar where the students were part of a pseudo-real context in which they were requested to be active actors, and to support the theoretical lesson with a case...").

Some teachers have expressed a particular appreciation for the self-evaluation exercises, pointing out also how the time for learning is reduced ("... The material at students' disposal has considerably reduced the time necessary for learning, helping in particular the sessions with exercises in groups...").

Also concerning the possible reusability of the e-learning material, the opinion was generally positive: "... teachers can reuse the material, for sure (we have!)..."; "... The e-learning structure gives the material a good characteristic of reusability in future teaching situations…" (from the answers the teachers gave to the questionnaire).

Some respondents underlined the necessity to adapt the material when it is used in a new context or in a context different from that where it was created, but on the whole the general hope is to use it in other editions of the e-learning seminars and also for the traditional courses. In this perspective, several teachers expressed the wish to improve the quality of their material and to extend and complete it.

CONCLUSION

The main lesson we have perhaps learned in these three years of experience is a very general one: more than other teaching tasks, e-learning activities must be carefully designed, to consistently mix the three dimensions of the option space about which the teacher is called to decide: the teaching objectives of the activity under design, the ways the involved subjects will interact, the technological tools that will be exploited.

Our interest in the quality and effectiveness of the teaching process, and not only in a higher efficiency, has led us to look for an integrated approach, which aims at blending and enriching the process of traditional learning, rather than substituting it, as instead typical in the distance learning. The framework presented here offers some guidelines to the teacher who adopts the same approach, supporting her / his design work for a new e-learning project. The appropriateness of this framework seems to be confirmed by the comments of teachers and students, which testify a satisfying achievement of the objectives underlying their experiences held in the context of the eLearning@LIUC project. Such a project has allowed us to gather some evidence to the crucial hypothesis: that through e-learning it is in fact possible to change the way of learning, in order to learn better.

REFERENCES

Aster (2001, May). *An educational framework for reflecting on the use of electronic resources for small-group teaching.* Retrieved from http://ctipsy.york.ac.uk/aster/resources/framework/ASTER_Educational_Framework.htm

Ausubel, D. (1968). *Educational psychology: A cognitive view.* New York: Reinehart and Winston.

Badii, A., & Truman, S. (2001). Cognitive factors in interface design: An e-learning environment for memory performance and retention optimisation. In D. Remenyi, & A. Brown (Eds.), *Eighth European Conference on Information Technology Evaluation* (pp. 479-490). United Kingdom: Oriel College.

Bloom, B. (1956). *Taxonomy of educational objectives. Handbook 1: Cognitive domain.* New York: Davis McKJay Lo Inc.

Calvani, A., & Varisco, B. (1995). *Costruire/decostruire significati. Ipertesti, micromondi e orizzonti formativi.* Padova: CLUEP (in Italian).

Calvani, A., & Rotta, M. (1999). *Comunicazione e apprendimento in Internet. Didattica costruttivistica in rete.* Centro Studi Erickson (in Italian).

Cenarle, M., & Biolghini, D. (2000). *Net learning—Imparare insieme attraverso la rete.* Milano: Etas (in Italian).

Colazzo, L., & Molinari, A. (1996). Using hypertext projection to increase teaching effectiveness. *Journal of Educational Multimedia and Hypermedia, 5*(1), 23-48.

D'Angelo, A. (2003). Analisi economica di un sistema di eLearning. *Form@re, Newsletter per la formazione in rete.* Retrieved from http://formare.erickson.it/archivio/aprile_03/editoriale.html (in Italian).

Dede, C. (1990). The evolution of distance learning: Technology-mediated interactive learning. *Journal of Research on Computing in Education, 22*(3).

Gokhale, A. (1995). Collaborative learning enhances critical thinking in digital library and archives. *Journal of Technology Education, 7*(1). Retrieved from http://scholar.lib.vt.edu/ejournals/JTE/jte-v7n1/gokhale.jte-v7n1.html

Gross, D. (1993). *Tools for teaching.* San Francisco: Jossey-Bass.

Guttormsen Schär, S., & Krueger, H. (2000). Using new learning technologies with multimedia. *IEEE Multimedia, 7*(3), 40-51. Retrieved from http://computer.org/multimedia/mu2000/u3040abs.htm

Huba, M., & Freed, J. (1999). *Learner-centered assessment on college campuses: Shifting the focus from teaching to learning.* Allyn & Bacon.

Keeton, M., Sheckley, B., & Krejci-Griggs, J. (2002). Effectiveness and efficiency in higher education for adults. *Council on adult and experiential learning.* Chicago: Kendall-Hunt.

Jacobson, M. (1994). Issues in hypertext and hypermedia research: Toward a framework for linking theory-to-design. *Journal of Educational Multimedia and Hypermedia, 3*(2), 141-154.

Johnson, D., Johnson, R., & Smith, K. (1998). *Active learning: Cooperation in the college classroom* (2nd ed.). Edina, MN: Interaction Book Company.

Laeng, M. (1996), La multimedialità da ieri a domani. *Rivista dell'istruzione, 6,* 905 (in Italian).

Maragliano, R. (1998). *Nuovo manuale di didattica multimediale.* Bari: Laterza (in Italian).

Mari, L., Mari, C., Moro, J., Ravarini, A., Tagliavini, M., & Buonanno, G. (2002). Multifunctional eBook: A tool to innovate learning situations. In *ECEL 2002, The European Conference on eLearning,* London.

Marold, K., Larsen, G., & Moreno, A. (2000). Web-based learning: Is it working? A Comparison of student performance and achievement in Web-based courses and their in-classroom counterparts. In M. Khosrow-Pour (Ed.), *Challenges of information technology management in the 21st century* (pp. 350-353). Hershey, PA: Idea Group Publishing.

Moore, M. (1993). Three types of interaction. In K. Harry, M. Hohn, & D. Keegan (Eds.), *Distance education: New perspectives.* London: Routledge.

Palloff, R., & Pratt, K. (1999). *Building learning communities in cyberspace: Effective strategies for the online classroom.* Jossey-Bass.

Piccoli, G., Ahmad, R., & Ives, B. (2001, December). Web-based virtual learning environments: A research framework and a preliminary assessment of effectiveness in basic IT skills training. *MIS Quarterly, 25*(4), 401-426.

Saunders, P., & Werner, K. (2000). *Finding the right blend for effective learning.* Western Michigan University, Center for Teaching and Learning. Retrieved from http://www.wmich.edu/teachlearn/new/blended.htm

Schank, R. (2000). *Engines for education. Learning by doing.* Retrieved from http://www.engines4ed.org/hyperbook/nodes/NODE-120-pg.html

Smith, G., Ferguson, D., & Caris, M. (2001). Teaching college courses online vs face-to-face. *T.H.E. Online Journal,* Retrieved from http://www.thejournal.com/magazine/vault/A3407.cfm

Syed, M. (2001). Diminishing the distance in distance education. *IEEE Multimedia, 8*(3), 18-20. Retrieved from http://computer.org/multimedia/mu2001/u3018abs.htm

Tinzmann, M., Jones, B., Fennimore, T., Bakker, J., Fine, C., & Pierce, J. (1990). *What is the collaborative classroom?* Oak Brook, IL: NCREL. Retrieved from http://www.ncrel.org/sdrs/areas/rpl_esys/collab.htm

Trentin, G. (1998). *Insegnare e apprendere in rete*. Bologna: Zanichelli (in Italian).

Trentin, G. (2000). Lo spettro dei possibili usi delle reti nella formazione continua e a distanza. In *Lettera Asfor, Offerta Formativa Technology Based: Linee di Sviluppo e Criteri di Qualità* (pp. 1-6) (in Italian). Milano.

Tsichristzis, D. (1999). Reengineering the university. *Communications of the ACM, 42*(6), 93-100.

Weigel, V. (2000). E-learning and the tradeoff between richness and reach in higher education change. *The Magazine of Higher Learning*. Retrieved from http://www.heldref.org/html/body_chg.html

Weimer, M. (2002). *Learner-centered teaching: Five key changes to practice*. Jossey-Bass.

Wilson, B. (1996). *Constructivistic learning environments*. Englewood Cliffs, NJ: Educational Technology Publications.

Chapter 8.18
Virtual Schools

Glenn Russell
Monash University, Australia

INTRODUCTION: THE EMERGENCE OF THE VIRTUAL SCHOOL

Until recent times, schools have been characterised by the physical presence of teachers and students together. Usually, a building is used for instruction, and teaching materials such as books or blackboards are often in evidence. In the 20th century, alternatives to what may be called "bricks-and-mortar" schools emerged. These were forms of distance education, where children could learn without attending classes on a regular basis. The technologies used included mail, for correspondence schools, and the 20th century technologies of radio and television.

Virtual schools can be seen as a variant of distance education. Russell (2004) argued that they emerged in the closing years of the 20th century and can be understood as a form of schooling that uses online computers to provide some or all of a student's education. Typically, spatial and temporal distancing is employed, and this results in students being able to use their computers at convenient times in their homes or elsewhere, rather than being subject to meeting at an agreed upon time in a school building.

The concept of a virtual school is agreed upon only in broad terms, as there are a number of variants. Some virtual schools insist on an agreed upon minimum of face-to-face contact, while others are so organized that a student might never set foot in a classroom. It is possible for a virtual school to have no physical presence for students to visit, and an office building in one state or country can be used to deliver virtual school services to interstate or international students.

One way of categorizing virtual schools is by imagining where they might be placed on a scale of face-to-face contact between students and teachers. At the conservative end of this scale, there would be conventional schools, where students use online computers in classrooms or labs for some of their lessons. A trained teacher in the same subject area might be available to help students, or other teachers, volunteers, or parents could supervise them.

Toward the middle of such a scale would be mixed-mode examples, where some subjects are offered in virtual mode, but students are asked to visit the school on a regular basis to monitor their progress or to participate in other face-to-face subjects, such as sport, drama, or art. At the other end of the scale are virtual schools where the student and teacher never meet, and there is no requirement for the student to enter a school building for the duration of the course. One example of such a virtual school is Florida High School, where, as the Florida High School Evaluation (2002) noted, there is no Florida High School building, and students and teachers can be anywhere in the world.

FACTORS PROMOTING THE RISE OF VIRTUAL SCHOOLS

The principal factors that account for the growth of virtual schools include globalisation, technological change, availability of information technology (IT), economic rationalism, the model provided by higher education, perceptions about traditional schools, and the vested interests of those involved in them.

The first of these factors, globalisation, refers to a process in which traditional geographic boundaries are bypassed by international businesses that use IT for globally oriented companies. It is now possible for curriculum to be delivered remotely from across state and national borders. Educational administrators can purchase online units of work for their school, and parents in developed countries can sometimes choose between a traditional school and its virtual counterpart.

As IT continues to develop, there is a correspondingly increased capacity to deliver relevant curricula online. As broadband connections become more common, students will be less likely to encounter prolonged delays while Web pages load or other information is downloaded. Advances in computers and software design have led to developments such as full-motion video clips, animations, desktop videoconferencing, and online music. Collectively, what is referred to as the Internet is already very different from the simple slow-loading Web pages of the early 1990s.

Economic rationalism also drives the spread of virtual schools, because the application of economic rationalism is associated with productivity. For education, as Rutherford (1993) suggested, the collective or government provision of goods and services is a disincentive to private provision, and deregulation and commercialisation should be encouraged. Consistent with this understanding is the idea that schools, as we know them, are inefficient and should be radically changed. Perelman (1992) argued that schools are remnants of an earlier industrial age that ought to be replaced with technology.

The ways in which higher education has adopted online teaching provide an example of how online education can be accepted as an alternative. The online courses provided by universities in recent years have proliferated (Russell & Russell, 2001). As increasing numbers of parents complete an online tertiary course, there is a corresponding growth in the conceptual understanding that virtual schooling may also be a viable alternative.

Those convinced that existing schools are unsatisfactory can see virtual schools as one alternative. Criticism of schools for not adequately meeting student needs, for providing inadequate skills required for employment, or not preparing students for examinations and entrance tests, are continuing themes that can be identified in a number of educational systems. Discussions related to school reform can include funding, resourcing, teacher supply, curriculum change, and pedagogy, but they can also include more radical alternatives, such as virtual schooling.

PROBLEMS OF VIRTUAL SCHOOLS AND THEIR SOLUTIONS

Virtual schools face a number of challenges related to the way that teaching and learning are implemented in online environments. While similar problems can also be identified in conventional schools, the different natures of virtual schools serve to highlight these concerns. These problems include authenticity, interactivity, socialzsation, experiential learning, responsibility and account-ability, teacher training, certification, class sizes, accreditation, student suitability, and equity.

The first of these problems, authenticity, relates to the verification of the student as the person who has completed the corresponding assignments and tests from a virtual school. Virtual schools may assign students a secure password to use over the Internet, but this procedure would not preclude students from giving their passwords to a parent or tutor who completed the work on their behalf. A possible solution that may have to be considered is to independently test students to confirm that they have the understanding, knowledge, and skills suggested by their submitted work.

Interactivity describes the relationship between the learner and the educational environment. For virtual school students, there is an interactive relationship involving the multimedia, the online materials used, and the teacher. Students would typically access materials on the World Wide Web, respond to them, and send completed work electronically to their teachers. The preferred way for students to become involved in online learn-ing is to have an active engagement involving a response. If a student is directed to a static Web page containing a teacher's lecture notes, learn-ing may be less effective, unless other teaching methods are used to supplement it. The solution to this problem will be found in both the increased capability of students' online computers to operate in a rich multimedia environment, and the rec-ognition by course designers that virtual schools should take advantage of advances in learning theory and technological capability.

Socialization continues to be a problem with virtual schools, because there is an expectation in conventional schooling that students will learn how to work cooperatively with others and will internalize the norms and values necessary for living in a civilized community. Moll (1998) is concerned with disruption to the tradition of public education as the primary vehicle for the transfer-ence of national narratives and humanistic and democratic values. Clearly, socialization will still occur if students use online learning supplemented by some contact with teachers and opportunities for organized sports. However, students' ability to relate to others in society is likely to change. Despite this concern, a type of virtual school that routinely insists on organized face-to-face learn-ing and social situations, with peers, teachers, and other adults, will reduce the problems that otherwise are likely to arise

A related concern to that of socialization is the belief that Web culture is inherently isolating, and that by encouraging students to pursue their education with a virtual school, an existing trend toward loss of community may be exacerbated. Kraut et al. (1998) originally suggested that Internet use could be associated with declines in participants' communication with family members in the household, declines in the size of their social circles, and increases in depression and loneliness. However, more recent research (Kraut, Kiesler, Boneva, Cummings, Helgeson, & Crawford, 2002) found that negative effects had largely dissipated.

There are some teaching activities in conven-tional schools referred to as experiential. These usually involve some form of hands-on activity or physical interaction with others. Typically, a teacher will provide a demonstration, explana-tion, or modeling of what is to be learned, and activities that follow provide opportunity to correct errors. While virtual schools commonly offer subjects such as mathematics and social studies, the study of physical education, drama, art, and the laboratory component of science is more problematic. Sometimes the problem does not arise, because students will enroll only for subjects that they missed or that they need for credit toward a qualification.

A common solution to these problems is for the virtual school to provide online or print-based teaching materials, as with other subjects in the range to be offered. Students complete the activities and send evidence of the completed work to the school. The Open School (2002) in British Columbia, Canada, offers art in both elementary and secondary school levels. At the Fraser Valley Distance Education Centre (2002), students are invited to participate in a science fair by sending in digital pictures and a digital video clip of their project to the supervising teacher.

Changing notions of responsibility, accountability, and student discipline are also likely to arise in virtual school environments (Russell, 2002). In a traditional school, teachers accept responsibility for the students in their charge, including the prevention of physical injury, and accountability for using appropriate teaching techniques. When there is a spatial and temporal distance between teacher and student, teachers are unable to exercise some of their accustomed responsibilities. While there is still a requirement to act ethically, and to ensure that appropriate teaching materials and methods are used, much of the responsibility shifts to parents, students, and to the suppliers of the online materials.

Teacher training is also emerging as an area of concern. Virtual teachers will find that some new skills are required, while others are less important. Class management skills in a face-to-face environment will differ from their online equivalents, as will many of the teaching practices. Salmon (2002) identified a number of skills that will be required by online teachers in the future. It is clear that there will be an ongoing need to use technological skills and to apply these skills to an appropriate educational context. However, it is unlikely that many teachers' colleges and other providers of trained teachers have modified their courses to reflect these changes, as mainstream teacher education is still focused on conventional school education. There are, nevertheless, some hopeful signs. The California Virtual School Report (2002) provided evidence of the use of online modules for teachers at Durham Virtual High School, in Canada, and a 15-week teacher-training program in Fairfax County School District.

Parents would normally expect that the virtual teacher working with their child would be a competent online teacher and be certified or registered with the corresponding school system. Where a student is working from home, and the principal contact with the teacher is by e-mail, the anonymity of the communication mode could conceivably cover the use of unqualified teachers. The necessity for demonstrating that a high-quality educational experience is being supplied is, however, likely to reduce this possibility. Florida Virtual High School uses only certified classroom teachers (Schnitz & Young, 2002, p. 4). As the online environment becomes more competitive, it is likely that virtual schools will provide evidence of their teachers' certifications.

With conventional schools, the issue of class sizes is a perennial problem. The diversity of virtual schools means that it is not easy to determine corresponding workloads. The evaluation of Virtual High School (VHS; Kozma et al., 2000) revealed that some of the teachers involved in the case study had to complete their VHS work at home in addition to their normal teaching load during the day. When teachers are asked to take responsibility for large groups of students, the time available for individual attention is likely to be reduced, and the quality of the educational service provided may be less satisfactory. There are indications that some virtual schools have recognised this problem. Louisiana Virtual School (2002), for example, is limited to 20 students per course.

Accreditation of courses across geographic regions will also become an increasing problem. Palloff and Pratt (2001) noted concerns with the quality of online high school programs as early as 2001. Varying standards can mean that a course in one area is not recognized in another. Students will increasingly be able to choose programs across state and even national borders and complete their schoolwork by sitting at home with their computers.

An important item relating to the quality of a student's educational experience in a virtual school is the recognition that not all students are suited to online learning. Already, some virtual schools try to determine whether the prospective student is suited to online learning by using questionnaires. Typically, these questionnaires ask students about their independent learning skills, motivation, time management abilities, and comfort with technology.

If virtual schools are perceived to be advantageous for those enrolled in them, there are also concerns as to when the access to them is seen as inequitable. Bikson and Paris (1999) found that there were "highly significant differences in household computer access based on income" (p. 9), in the United States. It is reasonable to assume that households with children will have less access to computers to use in a virtual school if they are part of a disadvantaged group. Unless there is careful planning, the use of technology-mediated education is likely, in the short term, to further entrench those inequalities that exist in society.

FUTURE TRENDS IN VIRTUAL SCHOOLS

Two broad trends can be identified in the growth of virtual schools. These are the continued expansion in the number of virtual schools, and the trend from virtual high schools to virtual K–12 schools. Research by Clark (2001, p. 3) indicated that more virtual schools began their operations in the United States during the period 2000–2001 (43%) than in the previous 4 years combined. Fifty-one percent of virtual schools surveyed offered junior high and middle school courses as well as high school courses, and about one in four schools offered courses across the whole K–12 spectrum (Clark, 2001, p. 4). In Canada, there is also evidence of

growing demand for virtual schools. The 2-year cumulative growth rate for Alberta virtual schools was 125% (SAEE, 2002).

Collectively, the implication of these trends is that there will be increased attention devoted to those problems that arise from virtual schooling across the K–12 range. When virtual schools made their first appearance, it would have been possible for some educators to dismiss them because they were experimental, or ignore their existence because they catered only to a niche market of high school students. In some cases, this suggestion may still be valid, but support for virtual schooling is increasing, rather decreasing, and the nature of what is offered is becoming more comprehensive.

CONCLUSION

Virtual schools continue the tradition whereby students learn at a distance from their teachers. The availability of online courses through the Internet has simultaneously reduced the emphasis given to older forms of distance education, while it increased the opportunities for students to explore alternatives to traditional school education. It is likely that there will be an increase in the number of virtual schools, and that they will continue to attract students. The expected increase in the number and type of virtual schools is likely to provide both exciting possibilities and daunting challenges.

REFERENCES

A national survey of virtual education practice and policy with recommendations for the state of California. Available online: VHS_Report_lowres.pdf

Bikson, T. K., & Paris, C. W. A. (1999). *Citizens, computers and connectivity: A review of trends.*

Available from http://www.rand.org/publications/MR/MR1109/mr1109.pdf

California Virtual School Report. (2000). *The California Virtual High School report.*

Clark, T. (2001). *Virtual schools: Trends and issues—A study of virtual schools in the United States.* Available from http://www.WestEd.org/online_pubs/virtual schools.pdf

Florida High School Evaluation. (2002). *The Florida High School Evaluation: 1999–2000 year-end report for the Orange County School Board.* Tallahassee, FL: Center for the Study of Teaching and Learning, Florida State University. Available from http://www.flvs.net/_about_us/pdf_au/fhseval_99-00.pdf

Fraser Valley Distance Education Centre. (2002). Available from http://www.fvrcs.gov.bc.ca/

Kozma, R., Zucker, A., Espinoza, C., McGee, R., Yarnell, L., Zalles, D., & Lewis, A.(2000). *The online course experience: Evaluation of the Virtual High School's third year of implementation, 1999–2000.* Available from http://www.sri.com/policy/ctl/html/vhs.html

Kraut, R., Kiesler, S., Boneva, B., Cummings, J., Helgeson, V., & Crawford, C. (2002). Internet paradox revisited. *Journal of Social Issues, 58*(1), 49–74.

Kraut, R., Patterson, M., Lundmark, V., Kiesler, S., Mukopadhyay, T., & Scherlis, W.(1998). Internet paradox: A social technology that reduces social involvement and psychological well-being? *American Psychologist, 53*(9), 1017–1031.

Louisiana Virtual School. (2002). Available from http://www.icet.doc.state.la.us/distance

Moll, M. (1998). No more teachers, no more schools: Information technology and the "de-schooled" society. *Technology in Society, 20,* 357–369.

Open School. (2002). *Open School in British Columbia, Canada.* Available from http://open-school.bc.ca

Palloff, R. M., & Pratt, K. (2001). *Lessons from the cyberspace classroom: The realities of online teaching.* San Francisco: Jossey-Bass.

Perelman, L.(1992). *School's out: Hyperlearning, the new technology and the end of education.* New York: William Morrow and Company.

Russell, G. (2002). *Responsibility for school education in an online globalised world.* Focus paper presented to Technology Colleges Trust Vision 2020 Online Conference (United Kingdom).

Russell, G. (2004). Virtual schools: A critical view. In C. Cavanaugh (Ed.), *Development and management of virtual schools: Issues and trends* (pp. 1–25). Hershey, PA: Information Science Publishing.

Russell, G., & Russell, N. (2001). Virtualisation and the late age of schools. *Melbourne Studies in Education, 42*(1), 25–44.

Rutherford, T. (1993). Democracy, markets and Australian schools. In C. James, C. Jones, & A. Norton (Eds.), *A defence of economic rationalism* (pp. 151–159). St Leonards: Allen and Unwin.

SAEE. (2002). *Executive summary of e-learning: Studying Canada's virtual secondary schools.* Available from http://www.saee.bc.ca/vschool-sum.html

Salmon, G. (2000). *E-moderating: The key to teaching and learning online.* London: Kogan Page.

Schnitz, J., & Young, J. E. (2002). *Models of virtual schooling.* Available from http://www.can.ibm.com/k12/pdf/Virtualschool.pdf

KEY TERMS

Bricks-and-Mortar Schools: These are traditional schools, where students attend at a physical school building.

Distance Education: A generic term referring to education where teachers and students are geographically separate. Modes employed include print and nonprint technologies.

Experiential Learning: Learning based on direct and unmediated instruction or on physical interaction with people and materials.

Globalization: The bypassing of traditional geographic borders using information technology to enable global orientation of business and remote curriculum delivery.

Interactivity: The relationship between the learner and the educational environment.

Socialization: The process by which students internalize the norms and values necessary for living in a civilized community.

Virtual School: A form of schooling that uses online computers for part or all of a student's education.

This work was previously published in the Encyclopedia of Information Science and Technology, Vol. 5, edited by M. Khosrow-Pour, pp. 3002-3006, copyright 2005 by Idea Group Reference (an imprint of IGI Global).

Chapter 8.19
Future Directions of
Course Management Systems

David Mills
ANGEL™ Learning, Inc., USA

ABSTRACT

Course management systems will unquestionably become one of the most critical enterprise systems in higher education. This is because these systems are more closely aligned with the core mission of teaching and learning than any others. Although these systems have already undergone extraordinary transformation in just a few short years, we are at only the very beginning of the evolutionary process. It is critical that CMS vendors look to the students, educators, and administrators that interact with these systems to identify what new tools and features they need. Consequently, the next stage of innovation in course management systems should therefore focus more on features specifically related to promoting better and more efficient processes for teaching and learning online. More flexible administration options should make these systems easier to maintain.

Emerging standards will continue to simplify communications and data exchange with other systems. Finally, the infusion of sound principles of instructional design and learning theory into the tools themselves promises to transform today's course management systems into tomorrow's expert systems for teaching and learning.

INTRODUCTION

Course management systems (CMS) have exploded onto the scene of higher education. In the late 1990s, the course management system was a fragile, loosely coupled set of Internet-based communication tools organized to support teaching a course. Each course had its own user accounts that, in most cases, were entered manually. The systems consisted of general communication and collaboration tools that were not in any way specifically tailored to online teaching and learning. Even with these limitations, however, course management systems have become one of the fastest growing enterprise systems ever.

Today's CMS evolution has been driven by user demand. Now, students and teachers can use the same user name and password to access all their courses from a single site. In most cases, the user name is even the same for all systems. Today's CMS are also expected to automatically integrate with other enterprise systems to synchronize course catalog and enrollment data. These advances have helped to catapult CMS adoption not only for distance education courses, but also for blended learning and to complement traditional classroom courses. In its current state, the CMS space has only tapped the surface of what these systems can offer. Now that the foundational issues of accounts and roster management are in place, future evolutions of CMS will be much more interesting. They will certainly become more integrated with other institutional systems. The benefits of automated synchronization of information are significant and tangible. They will also be engineered to be more specifically dedicated to the process of teaching and learning. But, for CMS technology to continue to grow, it must find ways to save time, save money, or offer tangible benefits for teaching and learning not otherwise attainable.

To answer the question of how next-generation CMS should or will be extended to add even more value, one must consider the stakeholders of these systems and clearly identify their needs. As the market and the community it serves evolve, active listening will become an increasingly important tool in designing these systems. Only by identifying the needs of each group of stakeholders will CMS technology be propelled forward in the necessary and appropriate direction. The primary stakeholders are students taking online courses or using the CMS as a resource for their traditional classes, faculty who use the CMS to both develop and teach their courses, and the information technology (IT) administrators who

support these systems. To that end, the personas that follow attempt to provide a snapshot of each group of stakeholders, what motivates them, and some of the things they want and need from a course management system.

THE STUDENT

John is an 18-year-old freshman at Sunny College. The year is 2004, and John has grown up with technology. He has been using the Internet for both fun and school work since he was 10. Hardly a day goes by that John does not chat with his online buddies. He also enjoys online gaming where he is quite comfortable using the 17-button game pad control and simultaneously chatting with his virtual teammates about the virtual world they are conquering. When he is not gaming, John is either chatting online or using the Internet to research a class project. When John does not know a word, his first thought is to go online. Using a traditional dictionary does not even cross his mind. He is used to a world that provides him what he wants when he wants it. From on-demand movies to online access to his new bank account, the virtual world is an integral part of his life. John's parents are amazed at his ability to watch a television show, instant message (IM) with friends, and do his homework simultaneously.

John was amazed at the process he had to endure when he enrolled at Sunny College. He actually had to walk to three different buildings collecting and delivering paperwork. He could not believe what a waste of time it was. It took almost five hours to enroll in four traditional and one online course. Thankfully, John is told that from this point forward he will be able to enroll online. On the first day of classes, John learned that two of his traditional classes would make heavy use of the CMS for quizzes and homework submissions. Another of his classes would be using a stand-alone Web site not hosted in the CMS.

It took nearly a week after classes started before John was given his individual sign-on user name and password. When he was finally signed on, only three of his courses showed up. At first, he thought something must have gone wrong with his registration. He then realized that only his classes with an online presence were listed.

As a result of the delay in getting his logon information, John was behind in his online course as well as the two courses that were using the CMS for quizzing and homework submissions. His instructors were accustomed to this, so they did not penalize him. However, it still made life difficult in the second week trying to complete two weeks worth of assignments while still getting acclimated to his classes.

All four of John's courses that used the Internet required that he check the course site every day for important announcements and assignments. This quickly became a tedious task because John had to actually log in to the CMS, enter each course, exit the CMS, and then go to the Web site for the course that was using a separate site. The entire process only took about ten minutes, but John quickly grew tired of repeating the steps every day. Moreover, after three weeks of doing this, only once was there an important announcement. As a result, John became more lax and only checked three times a week. This cost him dearly when one of his instructors posted a pop quiz that was only available for one day. While frustrating, it was clearly stated that students were supposed to check the site every day, so John had no recourse.

Navigating the course sites was also a problem. The course using a stand-alone Web site had a completely different navigation system than the courses using the CMS. Even the three courses using the CMS had completely different layouts. As a result, when John needed to find lecture notes or take a quiz, it took him much longer than it should have.

At one point in the semester, John missed a day of class due to illness. It was easy enough to get the lecture notes and assignments in his courses that had a course site. However, the course that had no site was a problem. He sent an e-mail to the instructor but had not yet received a response, and he did not know a soul in the class. While he had the lecture notes from his other courses, there were still a couple of things he did not understand. For his courses on the CMS, he simply sent an e-mail to his fellow students with his questions. However, in the course with a stand-alone Web site, there was no way to communicate with his fellow students.

In general, John was pleased that most of his courses had at least some online presence. However, he felt there was room for improvement. He would have liked to be able to contact his classmates in all of his courses. He was also frustrated by the expectations some of his teachers placed on his use of the course site. It seemed like an unfair burden having to spend ten minutes every day checking for new items just in case.

Analysis of the Student Experience

We can derive from John and his experiences several important details. Today's young adults have grown up in a world where they are immersed in technology every day. They are keenly aware of the productivity technology can provide and harsh critics when technology does not effectively solve a problem. These learners will likewise have very high expectations of what a course management system can and should offer them.

System Interoperability

John's initial registration experience points to a need for better integration among and between all of a school's enterprise systems. It also points to the need for standardized formats that allow data to be electronically exchanged between institutions easily, securely, and efficiently. Imagine the time, money, and frustration that would be saved if student records could be electronically transferred from a high school directly to a college or university. Agreeing to a set of schemas for data and communication protocols would make such a solution possible.

While high schools and higher education institutions will undoubtedly have different requirements and subsequently adopt different standards, this does not mean such transparency is impossible. For example, The School Interoperability Framework (SIF)[1] is a popular K-12 framework. Higher education has been drawn more to the specifications set forth by IMS.[2] However, agreeing to a simple mapping between these two standards will allow appropriate data to be electronically transferred. These standards are still in the early stages, and adoption is not yet widespread. Some of these initiatives will undoubtedly die off or become merged with others. Once it is clear which standards will prevail and more schools begin to adopt them, tremendous potential for interoperability will begin to be realized.

Single Point of Access

Many of John's problems stemmed from the fact that not all courses used the same system and, as a result, it was difficult to keep up with his courses. This also meant that communication tools that could have proven useful were not always available. These difficulties illustrate the benefits of having a single place students can go to access all their courses. This may be a function of the course management system or could be handled by another portal application adopted at the institution. In either case, all courses should be accessible from a single access point. Kentucky Virtual University (KYVU)[3] provides such a portal for their students. KYVU instructors choose from a number of course management systems. However, when the students log on, they see all of their courses listed under the "My Courses" section of their profiles. Selecting a course takes

them directly to the course on the appropriate CMS without requiring additional authentication.

Learner Productivity

John's experiences also show that next-generation course management systems should focus on minimizing the effort it takes for students to keep up-to-date with their courses. The system should deliver the appropriate information, at the appropriate time, using an appropriate channel instead of requiring students to constantly log on to check. Much of this simply has to do with the way data are organized and presented within the system. For example, in many of today's course management systems, you must not only log on to the system, but actually enter each specific course to see calendar items, announcements, and course e-mails, and to check for new discussion forum postings or assignments. Learner productivity could be greatly enhanced if learners had easy access to a unified, top-level summary of these items outside their individual courses.

PDA Synchronization

This simple collation of information goes a long way toward making it easier for students to keep up with their courses. However, the portal page is only one medium for presenting such information. Personal digital assistants (PDAs) have become a valuable tool for both work and school. This same information could be synchronized with a learner's PDA to provide instant access to the information. With the explosion of wireless technologies, students could even access the items directly from their PDAs when necessity dictates. Automated e-mails or SMS (short message service) messages could also be sent to provide critical updates when appropriate. Of course, with so many communication channels, what is delivered, where, and when should ultimately be configurable by each individual learner.

THE INSTRUCTOR

Anne is a professor of biology who teaches 100-level and 200-level courses. Anne has been using the school's current course management system to post her syllabi and class notes. She taught an online course in 2002 but was disappointed. Creating the course was no small task. She had not anticipated the amount of time this would take and spent many late nights early in the semester putting her course together. She had expected to be able to easily import content from the publisher of the text she was using, but the publisher did not have content available for her CMS. Instead, she had to cut and paste, retype, and upload most of the files individually. Anne was sure she would find ample content on the Web to supplement the text. She was quickly disappointed when search after search resulted in useless materials. No doubt, there was appropriate content on the Web somewhere, but she could not find it. Anne asked a fellow biology professor who taught the same course online the previous semester if she could use some of his content. He was more than happy to oblige, but much of his content was created in the CMS system, and he could not figure out how to share it with her.

This was Anne's first attempt at creating an online course. She received no training about effective pedagogical approaches for online teaching. As a result, she fell into the same pitfall as countless others before her — ineffective and inappropriate use of the tools available. She added discussion boards, chat rooms, and more without really knowing how these tools should be used or what purpose they served.

Once Anne started actually teaching the course, she was overwhelmed by the amount of work it entailed. She had not realized how much time and effort it would be to keep up with the postings in all the discussion forums she had created. On top of that, the constant stream of mail was almost more than she could bear. Students seemed to expect her to respond immediately to every post. While Anne had dedicated a large portion of the overall grade to participation in forums, chat, and mail, she quickly realized that the reports the system provided allowed her to do little more than count the number of items each student had submitted. A qualitative review, though necessary, was not feasible given the time it would take.

Anne found having assignments submitted by e-mail was a welcome change, but the electronic drop boxes she had set up in her course had their own problems. It took far too many clicks to view and grade assignments. There was also no easy way to provide students with inline feedback to their submissions. Students also kept submitting work late, requiring her to go back and check drop boxes for assignments that were long since due.

Anne felt like she was letting her students down when she could not intervene quickly enough to get them back on track. Twice as many students dropped her online class as her traditional class, and teaching it consumed significantly more of her time. The lack of personal interaction made it difficult to identify students at risk. The fact that the reports available in the CMS were not well suited to the task of assessing students exacerbated the problem. Anne also grew tired of writing the same e-mail ten times for ten different students, and grading work submitted online by her students was a laborious task that required far too many clicks. In addition, some of her students had significant problems learning the CMS. Anne often felt more like a technical support person than she did a teacher.

At the end of the semester, Anne was disappointed to learn that her final grades could not be submitted through the CMS. Though all of her students' final grades were already in the online grade book, Anne was forced to transcribe them manually onto the official final grade submission form to provide to the registrar.

The next semester Anne opted not to teach any online courses. It was just too much work, and she did not feel she could adequately engage her students. Anne's experience online did, however, illustrate how effective the Internet can be for low-stakes assessments such as weekly quizzes. It was also a little easier to manage assignments submitted online instead of on paper. As a result, Anne chose to continue using electronic methods for homework submissions and low-stakes assessment even in her traditional courses. But when Anne completed an online section of one of her courses and wanted to re-use some of the items from that section in her next semester's traditional courses, she was disappointed to learn that the CMS had no easy way for her to re-use bits and pieces of the course. Her only option was to download the entire course as an archive on her computer and then import the entire course into each of her new sections. This was a painfully long process because the course had a number of high-resolution images of cells that were huge. The export file was nearly 200 megabytes compressed. It took her hours to transfer the materials into each new section and then delete two-thirds of what she had just imported.

Analysis of the Instructor Experience

Anne is representative of many of today's instructors. She is motivated to provide a rich learning experience for her students but feels hamstrung by the technology. As with the student experience, many instructor issues could be resolved simply by providing the appropriate data in the appropriate context. This task-oriented approach dictates that the system conform to the user and the task he or she is trying to accomplish.

Data Filtering and Presentation

Consider, for example, the data that is collected from an online quiz. Most of today's course management systems simply provide a listing of the submissions to the quiz and their respective grades. With such a view, it is extremely difficult to answer questions such as how many people scored less than 70, or who did worse on this quiz than they did on the last one? These data are far more useful when they are filtered and transformed to deal with the particular task at hand.

When instructors are ready to grade manually graded items such as essay questions or drop box submissions, all they really want to see are those items that have not yet been graded. Furthermore, if all the responses to a particular essay or drop box were grouped together, the task of grading would be that much easier. At some other point in time, an instructor will undoubtedly want to assess an individual learner. In this case, it would be beneficial to have all of the learner's quiz scores listed graphically, ordered by completion date, and compared to the class average as illustrated in Figure 3. With such a presentation, the instructor would immediately understand how the student has been performing. On the day following an exam, the instructor will undoubtedly want to see which students had trouble with the exam and are subsequently at risk. In this case, it would be helpful if the instructor could easily see a list of those students who did not perform at or above a specified threshold as depicted in Figure 4. Once identified, it is likely the instructor will want to contact these students, assign them some remedial content, or take another appropriate action.

All of the above cases deal with the same quiz data. However, by filtering the data and changing how they are presented, they become far more useful with far less effort. The ANGEL™ system contains a number of tools designed around this framework. The "Learner Profile" provides an easy way to assess individual learners, while the he "WhoDunIt" agent allows instructors to query for people who have or have not completed item X or scored at least Y on item Z and send an appropriate message. The "What's New" agent identifies submissions that need to be graded. Lastly, the "Actions and Triggers" framework allows the instructor to automate the delivery of messages, release of content, and more in response to user performance and interaction with the system.

While these tools are a great start, much more can be done. With the volumes of data CMS can handle, the potential exists for truly intelligent systems. Agents can be developed that understand the relationships between data elements, can derive information automatically, and make it available to the appropriate people at the appropriate time. When considering such solutions, it is important to balance the automation of the system with the autonomy of the instructor. Consider, for example a learner assessment agent. Instead of requiring the instructor to review the progress of all students in the course, such an agent could highlight those students who fall outside an acceptable range of performance on one or more items. But, what is an acceptable range of performance and who defines it? A reasonable guess would be to inform the instructor when a student scores below a 70 percent on an item. However, what if this is a physics class and the mean score was a 40? Or, what if the instructor wants to intervene when a student scores anything less than an 80? These questions illustrate that, while it may be acceptable to make some default assumptions, those parameters should ultimately be configurable by the end user to meet his or her specific needs.

Features Specific to Teaching and Learning

Another issue in using CMS is illustrated by Anne's experience with the course drop box. For the most part, the tools available in today's course management systems were not originally designed for teaching and learning. As a result, many features that would be useful in a teaching and learning context are either underdeveloped or missing altogether. Consider, for example, threaded discussion tools. The ability to identify the type of message being posted, such as an assertion or rebuttal, is missing from most of these systems, though the benefit in a teaching and learning context is evident. Likewise, the ability

to model different types of discussions such as "the hot seat" or "fish bowl" is either difficult or impossible. The current drop box functionalities in many of today's systems also have some deficiencies because they were modeled after the shared file spaces often used for such purposes before the advent of the CMS, rather than modeling the solution based on the actual work flow of homework submission and grading. As these systems mature, these tools will naturally become more specialized to the task of teaching and learning. Additionally, new tools will surely be introduced that facilitate other teaching and learning tasks.

Domain-Specific Tool Sets

It should also be recognized that some departments, such as mathematics and the languages, will need access to highly specialized tools. The next generation of course management systems will need to do a much better job at providing appropriate tool sets based on the type of course, its department, and other such criteria. Consider the following example. The math department has three advanced quizzing tools they want accessible in the CMS. The language department has two tools specifically for creating language exercises. The biology department has a virtual lab tool they want to integrate. None of these tools serves any practical purposes outside the context of their particular domains. If all of these tools are made available to all courses, the interface will be so crowded and confusing that faculty will quickly become frustrated. The other choice is not to integrate these tools at all, which would dramatically reduce the usefulness of the system for faculty in these departments. If the system can provide these tools only to those courses with which the department is associated, adoption could occur within these departments without affecting the usability of the system for others.

ANGELÔ has some support for this capability through its support for custom extensions, custom objects, and environment variables. Using these capabilities, the system can be customized so different options appear under the "Add Content" menu for a particular course. While the current capabilities make such customizations possible, more can be done to make these customizations practical. Imagine a system in which you could simply select the tools you wanted to use for a course from a categorized menu. In addition to domain-specific tools, tools that promote specific learning theories could be made available. Custom systems developed at the institutions could be seamlessly integrated. You could even set up your default preferences, so any course could be customized to your preferences with a single click. This may sound farfetched, but it is actually just around the corner. This level of customization is critical if enterprise-level course management systems are going to be adopted by departments with very specific needs.

System Interoperability

The inability of the CMS to export grades to the registrar system illustrates another area where the next-generation course management systems couldn't offer improvement. An institution should be able to easily and securely pass such data between its enterprise systems using standardized communication protocols and data structures. The IMS Enterprise specification currently offers support for representing much of the required data for such operations. New standards are also emerging for communication protocols to be used for the actual passing of data between systems. Web-based simple object access protocol (SOAP) services are at the core of most of these emerging standards.

Content Re-Use

Re-use of content is yet another key area where future versions of course management systems can make great strides. Instructors managing multiple sections of a course are currently subjected to the task of manually replicating content across these courses. Semester transitions are another problem area in this respect. As richer content is developed, institutions increasingly want to share these resources across larger segments of the university such as departments, schools, or campuses. One example currently available in ANGELÔ is the ability to create question banks in a departmental resource library. These questions are then available for use on quizzes in any course offered by the department. In this way, the same questions can be made available to multiple sections of the same course or even altogether different courses. Projects such as Merlot (*www. merlot.org*) are currently developing global resource repositories of valuable content that could be used by these systems. Future systems must find ways to address these needs.

Course Management System Integration

Some vendors have opted to integrate resource sharing functionality in the CMS. While this solution may work for simple resources such as graphics, documents, and Web pages, it may prove problematic when dealing with more advanced resources such as quizzes and communication tools. Moreover, a solution based on a content management system will most likely never be able to support global repositories.

ANGEL™ currently supports a flexible resource library system that allows advanced resources such as quizzes, discussion boards, and even entire modules to be shared across course sections, departments, schools, campuses, or the entire system. Experience with this feature has helped to highlight where special consideration needs to be given when sharing such resources. For example, should library administrators be able to view and manage all responses to a quiz? What access to managing results should instructors have in a course that links to a quiz? Where should results be stored? When should results be purged and by whom? Unfortunately, the answer to all of these questions is that it depends. The details of responses to a course evaluation should most likely not be available to the instructors in the individual courses. Alternately, the results and submissions to a quiz should be managed by the individual course instructors, perhaps exclusively. The next generation of systems will need to be able to support such nuances to maximize the utility of such resources.

Library System Integration

Another way content re-use can be improved is by providing better integration with library systems. Some universities, such as Penn State, have been able to achieve tight integration between their course management system and their library system. In fact, in Penn State's case, librarians even use the CMS to author collections of resources for courses, departments, and schools. However, there are currently no accepted standards for generalizing such integration. Simply adopting a global authentication system such as Shibboleth® or Pubcookie would allow a much higher level of integration than is currently possible. While some course management systems, such as ANGELÔ, can be configured to use such a global authentication model, the value only comes when other systems support it as well.

Global Resource Sharing

In addition to better support for internal sharing of resources systems, better support for sharing resources globally is an inevitable requirement. This is really only now becoming possible, thanks to the emergence of standards that will allow interoperability of disparate systems. IMS has developed specifications for digital repositories and Web services that offer promise in this area. Shibboleth[4] is a promising candidate for a global authentication and authorization solution. Other standards are emerging to help with intellectual

property rights issues. As these and other standards become more widely adopted, the vision of secure global repositories will become a reality. Thanks to still other standards such as SCORM,[5] QTI,[6] and the IMS Learning Design[7] specification, it should be possible to share even the richest content in platform-neutral formats.

Digital Rights Management

Better support for digital rights management will be essential in promoting the re-use of content across boundaries of courses, departments, schools, and institutions. It is essential that the appropriate copyright and fair use information stay with an asset as the asset is used in these different contexts. Not only should systems preserve this information, they should be able to analyze it and act upon it accordingly when requests are received to view, reference, or copy the asset.

Infusing Instructional Design Principles into Content Creation Tool Sets

Just as the tools delivered by course management systems need to be designed more specifically for teaching and learning, so too should the tools and interfaces for creating the content. One example of this specification raised at the 2003 National Learning Infrastructure Initiative (NLII) focus session on next-generation course management systems is infusing instructional design principles and learning theory into the content creation tools themselves. This would undoubtedly result in the creation of better content by a majority of faculty. Just imagine — an instructor clicks the link to add a discussion forum to his or her course and is prompted with background information on effective use of forums and a series of questions about what he or she wants to accomplish. After answering these questions, a new discussion forum is created with appropriate settings. The instructor has learned about how to effectively use forums and has had the settings adjusted for his or her particular objective with a minimum of effort.

Now imagine the instructor needs to create a new forum for each of the twenty teams in the course. That useful information and series of questions will become very old in short order. Moreover, what learning theory should be used? Should this be up to the vendor? Should it be institutionally defined or be left up to the instructor? These questions illustrate that any such integration must be accompanied by a degree of flexibility. Instructors should have easy access to the assistance when they want it, but should not be burdened by it when they do not. Likewise, the environment should be flexible enough to support a variety of learning theories and allow wizards to be customized or extended as they see fit. An instructor could have the option of specifying his or her preferred learning theory and subsequently be directed to wizards that are consistent with that theory.

Recent releases of ANGEL◊ provide the hooks for adding some such custom extensions. The "Add Content" page of the lessons section can be extended to include template and wizards sections. Moreover, which items are displayed under this section can be customized based on the course being accessed or the individual accessing the tool.

THE IT ADMINISTRATOR

Frank has been a system administrator for 10 years. The university where Frank works has about 8,000 students and 500 faculty members. Frank started as a server administrator in the School of Business, where he maintained the departmental servers. The business school was an early adopter of course management software, and Frank handled all the details of maintaining the system. When the university decided to implement an enterprise

course management system in 1999, Frank was transferred to IT services with a promotion and put in charge of the new enterprise initiative. While pleased with the promotion and honored to have been selected, Frank sometimes wonders if it was worth it. Frank never imagined there could be so many issues involved in maintaining a single system. Before the first class was added to the system, committee meetings were held for weeks trying to resolve issues related to the Family Educational Rights and Privacy Act, single sign on, account management, course catalog synchronization, and enrollment synchronization.

The registrar took a very strict interpretation of the act, commonly known as FERPA, and decided that students should not see other students unless they explicitly give their consent. Unfortunately, this capability was not possible with the CMS. As a result, the school had to disable all features that allow students to see fellow classmates.

Due to limitations of both the registrar system and the CMS, the process for synchronizing both course catalog and enrollment information is extremely complicated. Every night the system is synchronized with the latest course catalog and enrollment information from the student information system. Special handling is required for cross-listed courses, nonstandard schedule courses, and lecture/lab combination courses. A batch job extracts the necessary information from the student information system and writes it to text files. The files are then transferred to another machine where a scheduled script processes the files and imports the data into the CMS. The process is not pretty, but it works, usually.

The number of restores he has had to do to recover data accidentally deleted by instructors has frustrated Frank. Usually the problem is that an instructor accidentally removed a student from his or her course, which subsequently deletes all data associated with that student. Sometimes, it is an important discussion forum or quiz that has been deleted along with the associated responses.

In either case, the recovery process takes far too long. Frank must restore an earlier version of the huge database to another system, log on to the course, export it, and import it back into the production system. If the deletion just occurred, the course is restored over the current production course. If the deletion happened some time ago, the restoration is done to an alternate course, and the instructor must manually transfer the appropriate information.

In addition to doing database restores, Frank spends a lot of time running custom reports for the provosts and deans of various departments. The system is very limited in its ability to retrieve the specific data he wants. As a result, Frank spends hours massaging and filtering the data in Excel to get it into an acceptable format. Frank knows he could extract the information he needs directly from the database, but the licensing prohibits this.

Many of the management and administration tasks that Frank does on the CMS could actually be done by other staff members. The problem is that the CMS does not allow Frank to specify that a user should have some administrative rights and not others. As a result, it is just too risky to give administrative rights to these additional staff members.

The start of a new semester is always frantic. There are always around 100 new accounts that, for whatever reason, are not correctly synchronized with the CMS. Frank also gets a number of support tickets from faculty and students about users not showing up in this section or that. Occasionally, this is a hiccup in the system. More often, users do not show up because they are not officially enrolled, they have not paid their tuition, or the instructor has simply not activated the course for student access. Another recurring theme is that the wrong instructor is listed as the faculty of record for a course. When this happens, the system enrolls the wrong instructor as an editor, and Frank is responsible for manually overriding to correct this situation.

The university policy is to archive courses that are more than two semesters old. This responsibility also falls on Frank. The system has some tools for archiving and deleting courses, but they could stand some improvement. The task is made much more complicated by the nonstandard schedule and open enrollment courses that are hosted on the system. Before Frank runs the scripts to delete old courses, he must manually review the list and remove these courses. Invariably, he misses one or two and ends up having to do a system restore to the courses.

Analysis of IT Administrator Experience

Frank's experience illustrates that there is room for improvement on the administration side of these systems as well. The FERPA issues make visible that each institution will have its own policies, procedures, guidelines, and requirements. Any enterprise system should adapt and conform to these requirements rather than require the institution to conform to the system. If the system is not flexible enough, institutions will choose not to use it, in whole or in part.

Support for Standards for Data Interchange and Communication

The issues surrounding the synchronization of the CMS with the registrar and other enterprise systems indicate the need for better support of standards for data interchange and communication. Several such standards exist, but many products have yet to support them. The IMS Enterprise specification is probably the most widely supported standard for exchange of these types of data. However, IMS has only recently begun addressing which communication channel systems should be used to send the data. As standards organizations extend these specifications and standards are more widely adopted, the data synchronization process will be significantly simplified.

Roster Synchronization

The reference to cross-listed courses highlights an important detail when dealing with this data. The data stored in the student information system and other registrar systems may not be in an optimal format for the content management system. Sometimes a course is cross-listed because it is offered under two or more departments. Other times the course is cross-listed because it is offered on both a standard track and an honors level. Combined lectures and labs are sometimes listed as a single section and other times are listed separately. These variances exist to support distinctions required by degree audit systems and the like. They are problematic when synchronizing the data with a CMS. If the cross-listing of a course is simply a matter of it being listed under two departments, it should most likely appear as a single course section in the CMS. However, if the course is listed as a standard track and an honors track, an instructor may choose to have these sections represented distinctly within the CMS. Likewise, a lecture/lab combination may need to be treated as a single section in the CMS or could potentially need to be treated as two separate spaces in the system.

CMS vendors must begin to address these tricky issues to minimize the customization required when implementing synchronization solutions. By dealing with these exceptions as part of the CMS rather than the synchronization process, the synchronization process is simplified to the point where standardized solutions are possible. The ANGELÔ Merged Roster Manager shown in Figure 5 is an example of this sychronization. This tool provides instructors the ability to treat multiple courses in the student information system as a single section in the CMS or create a new section in the CMS based on the enrollment information of multiple sections in the SIS data. For example, if 13 students are enrolled in a course under a standard track, and three are enrolled under the honors track, the instructor

can configure the CMS to have all the students appear in a single course space. Moreover, as students are added or dropped in the standard or honors sections, the roster in the merged course is automatically updated. Alternately, if multiple lab sections share a common lecture, the instructor can create a common lecture section that is associated with one or more lab sections so a user enrolled in any lab section is automatically enrolled in the common lecture section.

The roster synchronization issue is just one example of a situation in which course management systems need to be more flexible and tolerant of exceptions than other enterprise systems. Another example is allowing nonstudents to access a course. In a traditional classroom setting, when instructors want to use course assistants or invite guest lecturers to speak to the class, they do so at their own discretion. A course management system should be capable of affording this same luxury to instructors in the online world. Furthermore, systems should be able to support this level of flexibility regardless of whether the students in the course were added manually or through synchronization with another system. For the most part, the course management system is a consumer of integration data rather than a publisher. In the few exceptions to this rule, such as reporting of final grades back to the registrar system, the process should be structured to ensure that only valid entries are made back to the other system.

Data Recovery

Recovery of data is another area in which course management systems need improvement. Most enterprise systems restrict access to changing or deleting the data in the system to a very tightly controlled, highly trained group of users. How many people have rights to add or delete courses in the registrar system? How many people have rights to officially enroll or unenroll users in a course section? How many people have rights to delete a user's e-mail account? More than likely, the answer to each of these questions is very few. Contrarily, course management systems have hundreds of faculty only marginally trained on the system who have the ultimate power to delete content and remove users from a course. With some course management systems, the removal of a user from a course even deletes all of the submissions and data associated with that user. Course management systems need to consider this difference. Systems could flag items as deleted rather than physically deleting them and only purge the data when instructed by a system administrator. Alternately, systems could automatically back up data when appropriate, so data administrators or even course editors could restore the data with far less effort than is currently required.

Distribution of Administrative Responsibilities

Many of today's course management systems offer little or no flexibility with respect to the administrative features of the system. To allow institutions to effectively distribute responsibilities without exposing themselves to unnecessary risks, these systems need to allow much more granular control over the assignment of administrative features. For example, it should be possible to grant users the right to create and manage courses without also granting them administration rights. Likewise, these systems should support administrators at the campus, school, and department levels that have access to administrative tools for only the appropriate subset of users and courses.

Open Systems

Frank's issues with custom reports are directly attributable to the closed and proprietary nature of many of today's course management systems. In the future, course management systems should strive to be as open as possible. There will always be unforeseen requirements or requirements that

apply to such a small set of users they are not practical to implement. If vendors design course management systems in a manner that allows low-level access to the data they contain, institutions are empowered to address these issues on their own. By providing deeper access into the system, institutions should be able to develop custom-tailored solutions that meet their specific needs. In addition to low-level access, higher-level interfaces should be provided that allow customizations to be written in a manner that requires little or no modification when transitioning from one version of the software to the next.

CONCLUSION

Course management systems have rapidly evolved from fragile experimental systems to indispensable enterprise applications. Most of the enhancements to course management systems to date have been concerned with meeting the foundational requirements of an enterprise system. Now that these underpinnings are in place, course management systems can begin to focus on features specifically related to promoting better and more efficient processes for teaching and learning online. It is therefore critical that CMS vendors look to the stakeholders in these systems to identify what new tools and features are needed. Having considered the personas of John the student, Anne the educator, and Frank the administrator, it is clear much more can be done to accommodate these users.

Course management systems will unquestionably become one of the most critical enterprise systems in higher education. This is because these systems are more closely aligned with the core mission of teaching and learning than any others. Although these systems have already undergone extraordinary transformation in just a few short years, we are at only the very beginning of the evolutionary process. These systems will become more sophisticated as they are continually refined to meet the unique needs of teaching and learning. Infusion of sound principles of instructional design and learning theory into the tools themselves will help transform today's course management systems into tomorrow's expert systems for teaching and learning. Emerging standards will continue to simplify communications and data exchange with other systems. More flexible administration options should make these systems easier to maintain.

ENDNOTES

[1] http://www.sifinfo.org
[2] http://imsproject.org
[3] http://www.kyvu.org/home.htm
[4] http://shibboleth.internet2.edu
[5] http://www.adlnet.org/
[6] http://www.imsglobal.org/question/index.cfm
[7] http://www.imsglobal.org/learningdesign/index.cfm

Chapter 8.20
Next Generation:
Speculations in New Technologies

Bryan Alexander
Middlebury College, USA

ABSTRACT

Next-generation course management systems (CMS) are likely to take advantage of today's applications' structural and pedagogical limitations, supporting student and inter-collegiate collaboration. They should also be influenced by developments in social software and pre-existing information-sharing projects. CMS will reach out to the larger world to integrate with global informatics initiatives.

INTRODUCTION

The Internet will reveal the true hierarchy of good, because what is at stake is the essence of language: freedom. This hierarchy is complex: hyper-textual, interwoven, alive, mobile, teeming and spinning like a biosphere. (Pierre Levy[1])

Had Levy invented CMS, perhaps he would have imagined the now-available next-generation systems differently:

Catherine visits her course spaces after morning coffee, two hours before her first class of the day. Opening the main Web browser on her tablet, she scans her portal for today's information. Content feeds from sociology and French show activity, including an argument about Habermas (again), notes from the verb study group's leader, and three responses to her blog writing: one comment and two trackbacks to other blogs. But they aren't urgent this morning, so she marks them for later perusal. The morning class reading reappears, a chapter from Mary Shelley's last novel, with further comments and annotations attached, largely from her classmates. As she considers these collegial intertexts, a flurry of instant messages, or IMs, erupt alongside the reading. She greets several (two friends, a high school student with a question), then updates her away message to insist that she's "really busy studying." Switching over to her research project's feed, Catherine finds from video imagery and data streams that the Icelandic volcano has cooled slightly, and that her Swedish and Malaysian colleagues consider this well within their models. On a creative impulse, she grabs a screen capture of the caldera, adds it to her current video autobiography, and sets the editor to "rendering." She also copies the image to her course space profile, thinking it a dramatic yet economical way of representing herself and her geological interests. Maybe another volcanophile will inquire about joining her team—they really want a librarian this month.

This reminds her of her reading for the impending class. Catherine reflexively searches for commentary on The Last Man *(1826), adding to her personal wiki notes, then posting annotations with links to sources: a London professor's semantic analysis (both data set and commentary) and a Toyko high school class's discussion with some interesting reactions. Her trackbacks to their Web sites might trigger follow-up IMs, e-mails, or posts to her blog. Thinking about how this will look to future readers within the learning object her professor assures her the class discourse will become, she revises her prose to a more scholarly pitch, then races out the door, tablet under her arm.*

During the discussion, Catherine builds out loud on a point she made in course space. While conversation moves on, she takes notes but copies class notes on her earlier topic into a spin-off space, add links from her professor and fellow students, pastes in a copy of her earlier wiki and annotation comments, links to a social search query for two keyword combinations, then saves the new entity in an encrypted folder for later development. It might be the materials for a paper. Just before the end of class, Catherine notices that her video editor has finished rendering her updated film, so she uploads the entire clip to her blog, checks its permissions ("use freely, with acknowledgment") and awaits comments from friends around the world.[2]

As we've seen throughout this book, the current generation of CMS has grown rapidly, succeeding in being adopted across higher education with impressive speed. In order to apprehend the CMS landscape, it is important to consider the dynamics of that success in order to examine each application's formal features. Extensive adoption stems from several factors, each of which strikes at the heart of campus informatics. To begin with, compared with many other digital tools used for various purposes in higher education, CMS are relatively easy to support. Although stories about slow customer response are widespread, the software does not require external training (compare with Oracle™ geographic information system tools, or Director™), massive installation processes (compare Banner™), or complex interactions with rapidly changing software and hardware environments (such as digital video). Additionally, CMS have been embraced not by early adopters, but also by the broad technological middle of campuses, generally, extending the reach of computer-mediated teaching and learning, a major instructional technology goal.

These CMS have also taken the lead for a major external reason, in that they serve as shields from the copyright struggles currently raging throughout the United States and intervening in many levels of campus life. The Digital Millennium Copyright Act (DMCA), which came into effect in 2000, poses a restriction on fair use (itself enshrined in the 1976 Copyright Act) in its blanket prohibitions of unauthorized access through anticircumvention technologies.[3] At the same time, the Motion Picture Association of America (MPAA) and, especially, the Recording Industry Association of America (RIAA) have fostered a climate of copyright wariness, and sometimes fear, with their subpoenas and take-down campaigns aimed at colleges. Yet the leading CMS offer a rare bright spot in this gloomy landscape by taking advantage of the TEACH Act (2002). This law allows a fair use defense for educational use of digital materials, so long as such usage occurs within a closed classroom environment. That is, materials can be copied under fair use intent if they are accessible only to a class's students and instructor, and only for the duration of that class (a semester or equivalent), and if outsiders (the rest of the Web and world) are blocked by technological means, such as a decently strong login and password system.[4] The leading CMS do precisely that, providing a shielded, single-class space, requiring minimal effort on the part of instructors. In a sense, these CMS articulate a specific copyright stance, using TEACH to protect accessed content by supporting a "walled garden."

Such informatics structures necessarily embody pedagogical principles, which also tie into the success of CMS. Unlike the classic decentralizing, antiauthoritarian pedagogy of computer-mediated teaching, where the sage on the stage gives way to the guide on the side, Blackboard™ retains that sage's position. CMS make it very easy for an instructor to upload documents, such as a syllabus or readings; put another way, CMS enable the hierarchical transmission of information, from expert to learner. The reverse flow is also present in the form of drop boxes. The broadcast model of learning is what these are about, rather than any collaborative or nonauthoritarian one. Students are not encouraged to be content creators or participants in the construction of learning materials, based on the shape of the interface; their participation is limited to dropping off work in the instructor's box, posting to discussions, and checking the grade book. Peer learning and collaboration approaches are available through discussion tools (themselves traditional forms, dating back to the Daedalus software in education, or UseNet and the first list servers in the Internet), but these remain relatively underused. As Glenda Morgan put it in her 2003 ECAR study:

Faculty look to course management systems to help them communicate easily with students, to give students access to class documents, and for the convenience and transparency of the online gradebook... Fifty-nine percent of faculty surveyed reported that their communication with students increased as a result of using the CMS. This communication is broadcast in nature, from the faculty member to the student.

Morgan's study also notes that student demand for CMS rarely played a role in faculty decisions to adopt it. We should note that, at the same time, those students have already adopted other tools for communicating with each other, and that those are not broadcast, but peer-to-peer, such as instant messaging and file-sharing applications. Students are already living a collaborative digital environment but not through leading CMS.

Collaboration is also underplayed between classes. As the TEACH-compliant password protection restricts the online environment to a single class, it also produces a speed bump for sharing between classes. Multiple sections of the same large class can collaborate, much as students share the physical space of a large lecture hall. But communication between different classes on the same campus is difficult, and even more problematic between campuses, especially as different CMS are in play. It is difficult for an intercampus class, or a virtual program, to work through these tools. Blackboard™'s founder aims for a global reach for education (Pittinsky, 2003) but within the segmentations of classes.

It is important to note the pedagogical implications of these thickened digital barriers between classes. While they don't prevent students from browsing the Web (Carmean & Haefner, 2002), they nevertheless present an interface speed bump to reaching the full, open Web of Tim Berners-Lee's vision. That creator of the Web sought easy and direct connections between users and the world of documents, where technological mediation would be enabling rather than shaping and restrictive. The screen of WebCT™, its presentation of options, and the emphasis on a separate space lead the user to focus on materials available within that framework. The larger implication of this barrier is not that it is an obstacle to move from CMS to the Web, but that it is very hard to reach from the Web into a CMS. As a default, materials within this form of virtual class are not accessible to the outside world. Although it might be construed as a radical shift or regression, this is not a pedagogical innovation, as it recalls the medieval origins of the university of creating a safe, secluded zone for study, followed by the residential campus' sense of focused learning.

Leading CMS also appear something other than radical when we consider concurrent cyber-cultural developments. The rise of visually rich, interactive games and simulations, especially in the massively multiplayer online sense, has led to the creation of large virtual worlds, some with global reach (e.g., EverQuest™, Star Wars Galaxies™). Collaborative social software has blossomed into a multilevel movement, including social networking tools (Friendster™, Orkut™, LinkedIN™, Flickr™), publication technologies (blogs, wikis), and Web applications and standards (trackback, RSS). Peer-to-peer (p2p)-oriented technologies and practices have threatened business models and altered the norms of collaboration, from Napster to BitTorrent™. Large data sets continue to grow in size and applications, while searchability and metadata remain problematic. Learning objects persist as a focus for digital material production, usage, and sharing. This complex, dynamic mix of technologies and practices has shaped the generation entering colleges and using CMS, while offering many different approaches to the administrative, pedagogical, and communicative problems CMS attempt to solve. We can draw on cyberculture to get a sense of possible futures for courseware.

At the largest level, consider the notion of cyberinfrastructure. In 2003, the National Science Foundation's (NFS) Advisory Committee for Cyberinfrastructure released *Revolutionizing Science and Engineering through Cyberinfrastructure*, which describes the developing complex system linking scientists and their students as they taught and researched:[5]

Like the physical infrastructure of roads, bridges, power grids, telephone lines, and water systems that support modern society, "cyberinfrastructure" refers to the distributed computer, information and communication technologies combined with the personnel and integrating components that provide a long-term platform to empower the modern scientific research endeavor.

Rather than articulating a series of classroom and lab spaces organized conservatively, this report instead relies on an advanced sense of information-based networks, along the lines of Pierre Levy's global intelligences, or Berners-Lee's vision of a world of researchers sharing their documents. These networks share a variety of heterogeneous objects across multiple platforms, from digital video to simulation runs, handheld devices to distributed computing. The concept of cyberinfrastructure enables the organization of this vast array of objects into a dynamic field for teaching and learning. The widespread interest in this paradigm has led to a follow-up study for other disciplines, the Mellon Foundation-funded Commission on Cyberinfrastructure for the Humanities & Social Sciences, which is in progress at the time of this writing.[6]

This approach offers several levels of interaction with CMS. First, there is the sense that CMS might participate in such a structure. The problem becomes importing fluid content into these class spaces and being able to expose and publish CMS content to the wider infrastructure. Second, given the collaborative nature of the NSF's model, partly driven by the collaborative tendencies of the sciences, the challenge becomes building CMS capable of communication across class barriers. Third, the emphasis on collaboration also raises the problem of interoperability between applications and systems — that is, how does a FirstClass™ class talk with a Blackboard™ one? Fourth, one wonders about the information habits learned in a now traditional CMS, which focus on content presented to the learner, rather than materials sought out or constructed. Can such a learner dive into a planetary cyberinfrastructure, or would he or she suffer a literacy gap requiring acculturation and training (Lynch, 2003)?

A similarly grand or architectonic approach is that of information ecology. This springs largely out of the knowledge management field, and is predicated on developing strategies for sharing information and knowledge effectively within a group. Information can be explicit, as in documents, as well as implicit, in the sense of knowledge rarely voiced or shared. The ecological element involves considering a community's information as a dynamic, interrelated, holistic system, where multiple agents play multilevel roles, producing, sharing, and consuming information.[7] Not so open and wide-ranging as the NSF's cyberinfrastructure, the information ecology model nevertheless shares its heterogeneity. Davenport and Prusak (1997) argue that this model includes politics, behavior and culture, staffing, materials, practical processes, and information architecture. In this context, traditional CMS are small, closed systems, each constituting its own ecology, largely withdrawn from the information ecosystem of a campus, discipline, or, more broadly, academe. How will future course management tools approach the larger levels of information ecologies? Will students' skills in information fluency, honed within the micro-environments of classes, apply when seeking implicit knowledge in larger environments?

A more recent[8] approach to organizing social information at the macro level is the social software movement, which encompasses technologies, practices, and the general sense of improving our ability to collaborate by using digital networks.[9] The roots of this approach include social network analysis (SNA), which analyzes and maps out connections between people in social, informational contexts. The most popular application of SNA is the "Kevin Bacon game," where players seek to build connections between Hollywood people

and the hard-working actor. Networks display uncanny similarities across a variety of venues, from virus propagation to news story coverage to the operations of terrorist organizations, which helps us understand human interaction more precisely and proactively (Barabasi, 2002; Watts, 2003). For example, we now have a working body of knowledge about mapping informal social connections that cut across the grain of formal organizational structures. We understand the spread of stories across a population better, knowing to look for a certain proportion of connecting "hubs" rather than "spokes." Social network applications, structured in part by SNA insights, have emerged and grown dramatically in the past year and seek to connect people for politics (Meetup, used most notably with the Dean and Clark presidential campaigns), friendship and dating (Friendster™, Orkut™), and business (LinkedIN™) (Boyd, 2003). Other technologies have affiliated with social software, including collaborative writing tools such as blogs and wikis.[10]

Social software has already been repurposed for campus needs. On the social level, TheFaceBook has appeared as a Friendster™ for college campuses.[11] In terms of courseware, a growing number of classes have used wikis and blogs as virtual classrooms. Such tools enable nearly all pedagogical functions found in CMS: document presentation, discussion, and communication. They add many other affordances, depending on the implementation. For example, wikis allow users to edit others' documents directly, which is valued by some writing classes (Rick, Guzdial, et al., 2002).[12] Blogs can support discussion postings from beyond the classroom, allowing students to upload their multimedia work while encouraging hyperlinking to the full Web, building a more porous class boundary.[13] Blogs as CMS offer a different culture than that supported by Blackboard™ or WebCT™, one that is more collaborative and less hierarchical (Long, 2002). In one sense, then, social software tools are alternatives or competitors to leading CMS, supporting different pedagogies and support models. In a different, more prospective sense, upcoming CMS could build in social software functionality. One could imagine a campus-wide discovery tool, *a la* LinkedIN™, where members of the campus community could present their interests and connect with the similarly minded. Faculty could develop an alternative method for soliciting class interest beyond the registrar. Over time, a blog-based CMS would create archives of previous classes, which could serve as learning objects for subsequent iterations of that class or related ones, distributing learning over time as well as space.

Remaining at this higher level of analysis, and considering the preceding set of contemporary movements in digital informatics, it is clear that boundary issues are critical for the development of CMS. While the leading examples of the form are very conservative in forming barriers around the classic classroom, other aspects of cyberculture are transforming what academic boundaries have meant. Classroom, community sector (library, IT, faculty, student, administration, and so on), and academic discipline all connect in different ways under the cyberinfrastructure paradigm, when considered part of the same information ecology, or when students join instructors in blogging. Separate campuses and educational sectors (state school, research I, small liberal arts, community college, for-profit) begin to see cross-fertilizations by these practices. A community college student, a research I librarian, and a faculty member at a large state school can end up reading and posting to the same blog, learning together. Similarly diverse populations already make use of Massachusetts Institute of Technology's OpenCourseWare (Diamond, 2003). If we speculate on what forms such collaborations can take, some form of CMS could emerge to encompass and support them. Anxieties around border-crossing[14] might be alleviated by a powerful, flexible, and accessible application. Indeed, one possible reason for the rapid adoption of Blackboard™ and WebCT™ might be their refusal to challenge these borders, rendering them even more palatable.

We have some precedents for organizing hybrid groups around information-centric needs. The Internet has contained discussion forums since the 1970s, with the advent of Usenet and list servers. Their organizational innovations are deeply underrated, given the long-term, sustained successes of both, and especially Usenet, in bringing together very heterogeneous populations from many nations for the purpose of conversation. More recently, as noted above, social networking tools continue in this vein, using innovative approaches to connect people for business, friendship, and romance. Perhaps it is time for a Learnster™ or LinkedEdu™ CMS, where teachers and would-be learners could present themselves and discover each other, and then find ready-at-hand the collaborative tools for extending learning.

Gilles Deleuze and Felix Guattari (1987) offer a powerful metaphor for describing the conceptual shift suggested above by opposing the games of chess and go. In the western classic, pieces possess carefully demarcated roles, ranked in specific hierarchies. Their spatial positioning shapes the board and its early movements quite reliably. In contrast, the Japanese game's pieces are entirely equal at start, all with precisely the same capabilities. It is only when played onto a similarly unmarked board that new patterns and connections become apparent, emerging from patterns and formations iterated through play. The entire board can be revised and transfigured at a stroke, in contrast with chess' unfolding, gradual nature. Our current CMS, like chess, neatly map out specific hierarchical positions and roles, separated carefully by class and campus. New CMS might operate like go, connecting across the board, driving new formations and organizations.

At the same time, it is important to distinguish between connection and collaboration, discovery and follow-up operations. While the Internet is quite good at announcing projects, successful projects are less common, as a glance at Source-Forge will indicate. We do know of a series of collaborative projects that have succeeded to various levels. The Wikipedia, for example, is now a large, collaboratively edited encyclopedia, where any user can edit any of the more than 300,000 entries.[15] Despite its openness to vandalism, the site's community and content have persisted. In the gaming world, alternate reality games (ARGs) have grown in recent years, as teams of players distributed around the world combine to solve complex puzzles pieced together into narratives. Players conduct Web research, translate texts in foreign languages, apply steganography, analyze video, decrypt codes, and build multiple, searchable archives for their work, along with social practices for welcoming new players. The Beast, a promotion for the film *A.I.*, is the most recognized of these games.[16]

Beyond textual collaborations, distributed pedagogy has gradually emerged, despite the various setbacks for distance learning. For example, the Associated Colleges of the South's Sunoikisis project, a "virtual classics" department, supports teaching, research, and study abroad for nearly a dozen liberal arts colleges. Faculty lecture to other schools' classes, students compare notes with classmates in different states, and teams across two time zones learn archaeology before coalescing at a dig in Turkey.[17] Proving that new forms of CMS can grow to meet new pedagogical needs to support inter-campus classes, the ACS built its own open-source CMS, the Course Delivery System, which readily integrates classes on two physically separate campuses. In terms of research, the China Filtering Project distributed data-gathering around the world to assess China's network strategy.[18]

Taken together, these information-centered collaborations demonstrate ecological responses to information needs, combining multiple digital tools around a social, purposeful nexus. Each enables multiple positions along a continuum of participation and textual production, with participants reading, writing, or both to various levels. Mechanisms appear to explain the process, from archive presentations to staged participation levels. In a sense, all of this is what an institution of higher learning has done for decades, from the admissions process through college archives—and now CMS. From another perspective, the invention and fluidity of online information collaborative describes potential forms for new CMS.

A deeper challenge to current CMS is to address self-organizing learning. How does a campus link its preferred courseware to a group that emerges to study a topic on its own, such as a reading group, a film club seeking to expand its knowledge, or a religious exploration? More importantly, how does a CMS grapple with self-organization within pre-existing institutions? Small group work within a class, for instance, can be a dramatically effective practice for participants, yet that success remains locked behind a WebCT™ password, hidden from the Web, and possibly inaccessible after time passes, depending on the campus. Beyond that class, CMS could grow to support connections between students in different classes sharing similar interests, as noted above; a greater challenge is to link over time and between classes. Such connections are hard to make in a face-to-face, nondigital environment, and become even more difficult as a campus size increases. Yet social software and collaborative information practices surely offer guides to opening pathways between two students—one in 1999, the other in 2004—interested in the later Byzantine Empire in the Balkans. Web-presented archives, searchable postings, and collaborative filtering could add to our abilities to interconnect members of our communities. Portals can, in theory, support such functions at the cost of some intensive dataveillance, which gives rise to privacy and process concerns.

Further, a full course format might not be the best mode for such connected students. One response to this may be found in Middlebury College's Segue CMS, which allows users to create content pages in various formats, including class, blog, Web site, and research project.[19] Users can spin off from one format to another, launching a blog from a class, a research project from a Web site, and so on. While Blackboard™ and WebCT™ speak in terms of modularity, perhaps the next step for CMS includes a deeper modular function, where CMS pieces can be disaggregated and reassembled by users.

Along these lines, then, may be found an alternative approach to the copyright problems. As we noted above, traditional CMS afford a solution to digital fair use, at least in the United States, by using the class restriction to create a TEACH Act shield. This has been one reason for the widespread adoption of these CMS. But an alternative response to intellectual property struggles has emerged over the past several years, in favor of shared content. There is a growing openness to alternative intellectual property systems, such as the GNU code license and the burgeoning Creative Commons.[20] Despite limitations due to the extension of copyright terms in 1998, the public domain has been increasingly celebrated.[21] Shouldn't the next generation of CMS be able to partake of the Creative Commons, such as by allowing the publication of a class under a CC license? Moreover, could a new CMS support sharing copyright-free content across campuses, including public domain materials, as well as works created for sharing? WebCT™'s Vista™ might offer a seed for this, with its ability to share content between the same class on networked campuses. Students who want to gain exposure for their creative works, for example, could make them available through a CMS network. Expanded collaboration tools could then facilitate feedback, along with the creation of derivative and follow-up works.

So many options for functional growth in CMS suggest changes in their architecture, or the creation of radically different ones. Perhaps a CMS could follow in the path of the peer-to-peer collaboration tool Groove and be based on shared document spaces.[22] Instead of broadcasting content to students, a class would consist of materials shared between students and instructors, passed back and forth, modified and grown. This would free up modularity in that exported documents could be shifted to other p2p spaces, allowing students and instructors to repurpose the materials or archive them for personal use. Encrypting content should reiterate the TEACH copyright shield, creating a safe space for learning. Focusing on p2p relationships should drive the development of a more collaborative tool set within such a CMS, including by now familiar options such as image sharing, profile searching, instant messaging, co-browsing, and co-authoring pages. Truly extensible design would allow these tools to be disaggregated at will and repurposed for new functions, depending on the learner.

Turning from the class to the world suggests a different architecture based on exposure to the Web. While CMS have boomed, multiple projects have developed and connected for improving the searchability of educational content. The Open Archives Initiative, for example, has been working with repository projects to expose content to spiders, while discussing metadata standards to improve discovery.[23] IMS has been developing and promulgating standards for sharing information.[24] CMS could play a role in giving feedback to these movements, as their work often ends up aimed at class content and experience. Moreover, CMS should have much to gain by working to enhance the incorporation of, and access to, such materials within the class environment. Additionally, these large projects could serve as venue to win greater exposure for class content, produced by any members of the college community. Collaborative informatics projects constitute a complex network already in process, and collaborating with them could help connect students with the larger networks of the world.

That world has advanced in many ways since CMS first made their way into our classrooms and campuses. Paying attention to those developments and their ambitions is a powerful way to imagine the next generation of courseware, especially as it addresses common issues and dynamics. The risks are large, especially as the world's defensive tendencies manifest in a growing drive for walled gardens, between fears of copyright, terrorism, and compromised privacy. Yet the chance to expand teaching and learning, to deepen our communities of knowledge, and to play a role in the growth of global networks for collaboration demonstrates the powerful situation and potential of CMS. Given their present success, we should imagine their next iterations boldly.

REFERENCES

Barbasi, A.L. *(2002). Linked.* New York: Perseus.

Boyd, S. *(2003).* Are you ready for social software? *Darwinmag,* March. Available at http://www. darwinmag.com/read/050103/social.html

Carmean, C. & Haefner, J. (2002). Mind over matter: Transforming course management systems into effective learning environments. *Educause Review,* November/December. Available at http://www.educause.edu/ir/library/pdf/erm0261.pdf

Davenport, T. H. & Prusak, L. (1997). *Information ecology: Mastering the information and knowledge environment.* New York: Oxford University Press.

Deleuze, G. & Guattari, F. (1987). *A thousand plateaus: Capitalism and schizophrenia,* B. Massumi. (Trans.). Minneapolis: University of Minnesota Press. Originally published as *Mille plateaux* (Paris: Éditions de minuit, 1980).

Diamond, D. (2003). MIT everywhere. *Wired,* September. Available at http://www.wired.com/wired/archive/11.09/mit_pr.html

Levy, P. (2000). *Collective intelligence: Mankind's emerging world in cyberspace.* R. Bononno (Trans.). New York: Perseus.

Long, P. (2002). Blogs: A disruptive technology coming of age? *Syllabus,* October. Available at http://www.syllabus.com/article.asp?id=6774

Lynch, C. (2003). Life after graduation day: Beyond the academy's digital walls. *Educause Review,* September/October, 12-13. Available at http://www.educause.edu/ir/library/pdf/erm0356.pdf

Morgan, G. (2003). Faculty use of course management systems. *ECAR Key Findings.* Available at http://www.educause.edu/ir/library/pdf/ecar_so/ers/ers0302/ekf0302.pdf

Nardi, B. & O'Day, V. L. (1999). *Information ecologies: Using technology with heart* (2nd ed.). Cambridge: MIT.

National Science Foundation. (2003). *Revolutionizing science and engineering through cyberinfrastructure.* Available at http://www.community technology .org/nsf_ci_report/

Pittinsky, M. S. (2003). *The wired tower.* New York: Financial Times.

Rick, J., Guzdial, M., Carroll, K., Holloway-Attaway, L. & Walker, B. (2002). Collaborative learning at low cost: CoWeb use in English composition. Paper published in the *Proceedings of CSCL,* Boulder, CO. Available at http://newmedia.colorado.edu/cscl/93.pdf

Watts, D. J. (2003). *Six degrees: The science of a connected age.* New York: Norton.

ENDNOTES

http://webnetmuseum.org/html/en/reflexion/reflexion_levy_seo ul_en.htm

2   A fine introduction to Trackback is at http://www.movabletype.org/trackback/beginners/. The Creative Commons Web site, http://creative commons.org/, offers a good survey of alternative copyright permissions. The full text of The Last Man is available at http://www.rc.umd.edu/editions/mws/lastman/.

3   http://www.educause.edu/issues/issue.asp?Issue=DMCA

4   The TEACH Act toolkit is probably the Web's best resource for this law. Find it at http://www.lib.ncsu.edu/scc/legislative/teachkit/.

5   http://www.communitytechnology.org/nsf_ci_report/

6   http://www.acls.org/cyberinfrastructure/cyber.htm

7   http://en.wikipedia.org/wiki/Information_ecology has a good, hyperlinked discussion. Cf also Nardi and O'Day (1999).

8   Arguably, social software is a conservative movement, in that it recapitulates older visions of computer-mediated collaboration. JFC Licklider is perhaps the leading historical figure for this, from his visionary published work ("Man-Computer Symbiosis," 1960, and "The Computer as a Communications Device," 1968) to his crucial role in funding the initial Internet research from the Advanced Research Projects Agency (ARPA) during the 1960s.

9   The Many-to-Many blog is a fine source of information and news on social software. Find it at http://www.corante.com/many/.

10   In a valuable and influential article, Clay Shirky argues that most Internet technologies can be repurposed for social software needs, including e-mail. Read the article at http://www.shirky.com/writings/group_politics.html.

11   http://thefacebook.com/

12   For example, Denham Gray's human-computer interaction class (http://www.voght.com/cgi-bin/pywiki?HciSummer) or Georgia Tech's campus-wide wiki, CoWeb (http://c2.com/w2/bridges/CoWeb).

13   For example, Barbara Ganley's 2003 Irish literature class (http://wl.middlebury.edu/irishf03/). http://www.weblogg-ed.com/ is an excellent blog for keeping up with this approach.

14   See Gloria Anzaldua's work, most notably Borderlands/La Frontera (San Francisco, CA: Spinsters/Aunt Lute, 1987).

15   http://en.wikipedia.org/wiki/Main_Page

16   See http://www.nytimes.com/2001/05/03/technology/03GAME.html? searchpv=site01.http://unfiction.com is a fine resource for information and collaboration.

17   http://www.sunoikisis.org/ and http://www.nitle.org/tr_lm_segue_cds.php

18   http://cyber.law.harvard.edu/filtering/china/test/

19   http://segue.middlebury.edu/index.php?&action=site&site=segue and www.nitle.org/tr_lm_segue_cds.php

20   http://www.creativecommons.org/

21   http://www.law.duke.edu/cspd/ is one leading academic focus of public domain study.

22   http://www.groove.net/home/

23   http://www.openarchives.org/

24   http://www.imsglobal.org/

This work was previously published in Course Management Systems for Learning: Beyond Accidental Pedagogy, edited by P. McGee, C. Carmean, and A. Jafari, pp. 359-373, copyright 2005 by Information Science Publishing (an imprint of IGI Global).

Chapter 8.21
The Emerging Use of E-Learning Environments in K-12 Education:
Implications for School Decision Makers

Christopher O'Mahoney
Saint Ignatius' College, Australia

ABSTRACT

Virtual learning environments (VLEs) and managed learning environments (MLEs) are emerging as popular and useful tools in a variety of educational contexts. Since the late 1990s a number of 'off-the-shelf' solutions have been produced. These have generally been targeted at the tertiary education sector. In the early years of the new millennium, we have seen increased interest in VLEs/MLEs in the primary and secondary education sectors. In this chapter, a brief overview of e-learning in the secondary and tertiary education sectors over the period from 1994 to 2004 is provided, leading to the more recent emergence of VLEs and MLEs. Three models of e-learning are explored. Examples of solutions from around the world are considered in light of these definitions. Through the case of one school's journey towards an e-learning strategy, we look at the decisions and dilemmas facing schools and school authorities in developing their own VLE/MLE solutions.

INTRODUCTION

The history of adoption of technological innovations in schools is characterised by a mixture of enthusiasm and apprehension. The adoption of information and communications technology (ICT) in schools is no exception. Governments, educational authorities, individual schools, and educationalists have recognised the tremendous potential of ICT to transform teaching and learning. At the same time, there has been a collective intake of breath as social, financial, industrial, political, pedagogical, and logistical implications have emerged (Cuban, 2000). Increasingly, ICT literacy is a requirement in the K-12 education sector, for both staff and students.

In educational ICT, change is the one constant. For the most part, educational institutions have been on the receiving end of ICT innovation, responding to change rather than driving change. As a result, the journey towards literacy with ICT innovations in schools more often follows ad-hoc diffusion models, rather than as an outcome of specific decision-making strategies. Thus, investments by schools in products/solutions such as school administration systems, e-mail systems, local area networks, laptop programmes, intranets, virtual private networks (VPNs), and the like, can be isolated decisions rather than forming elements of some wider e-learning strategy (Jones, 2003).

Over the past 10 years, many schools have worked hard to begin integrating these disparate solutions and streamline their ICT management. With the increasing ubiquity of the World Wide Web and browser-based educational resources, the integration of various e-learning components became a possibility. In the late 1990s, early versions of integrated learning management systems emerged, predominantly targeted at the tertiary education sector. Now, in the first decade of the new millennium, a variety of solutions are being developed with the primary and secondary sector in mind (BECTA, 2001a, 2001b, 2001c).

This chapter reviews the emergence of e-learning technology components over the period from 1994 to 2004 and their implementation in the K-12 education sector, with particular reference to attempts to integrate these various components into a broader e-learning strategy. By analysing literature concerning models of e-learning in schools, it is shown that many schools, although mapping closely to these models, do so more by coincidence than design. It is suggested that school

e-learning strategies evolve to accommodate specific ICT components and capabilities as they emerge. The challenge for schools, as always, is to have the agility to respond appropriately to these innovations, while at the same time exercising wisdom and discernment in their implementation (Dowling, 2003).

LITERATURE

The concepts of computer-based training (CBT) and computer-assisted learning (CAL) have been in circulation since the 1980s, initially in industry. The reality of distance teaching and distance learning has been with us much longer. It has only been since the ubiquity of the World Wide Web in the early 1990s that we have seen a convergence of these domains. A number of factors have assisted this convergence. Increasingly sophisticated Web browsers; increasingly sophisticated Web scripting languages; increasing bandwidth; improved data compression techniques; reducing costs; increased access to powerful personal computing devices; and increased levels of user knowledge and understanding are some of these factors.

A selective chronology of events related to e-learning innovations is shown in Table 1.

Definitions

A variety of definitions of e-learning exist, but most have a common theme. The key distinction between definitions of e-learning and previous definitions of CAL, CBT, and the like are a focus on Web-enabled technologies. For instance:

E-learning is online training that is delivered in a synchronous (real-time, instructor-led) or asynchronous (self-paced) format. (Jones, 2003)

E-learning is:

.... an innovative approach for delivering electroni-

Table 1. A brief chronology of e-learning innovations, 1990-2003

Year	Event
1990	• Tim Berners-Lee proposes his idea for a World Wide Web
mid-1994	• Mosaic communications founded by Mark Andreesen and Jim Clark • Schoolsnet Australia launched - early ISP.
October 1994	• Early version of Netscape launched
1995	• Netscape v 1.1 launched, quickly followed by v 1.2
Late 1995	• Windows 95 launched, which included the first version of Internet Explorer
1996	• Early version of WebCT launched - Vancouver. • Internet Explorer 3 vs Netscape Navigator 3
1997	• Blackboard Inc launched - Washington D.C. • digitalbrain plc incorporated – London • Impaq launched as an ISP – Australia • Internet Explorer 4 vs Netscape Communicator
1998	• AOL buys Netscape
1999	• Microsoft launches Internet Explorer 5
2000	• Netscape 6 vs Internet Explorer 5.5
2001	• Microsoft releases Internet Explorer 6.0 • Granada launches its "Learnwise" portal – UK • Schoolsnet Australia re-badged as 'myinternet'
2002	• Netscape 7 released
2003	• Thousands of working with E-learning products

cally mediated, well-designed, learner-centred and interactive learning environments to anyone, anyplace, anytime by utilising the Internet and digital technologies in concert with instructional design principles. (Morrison & Khan, 2003)

More functional definitions describe the typical components found in an e-learning system. Barnes and Greer (2002) suggest that Web-based learning environments include "Web browsing and authoring, file transfer, e-mail, chat, discussion groups (communities) and shared whiteboards." A more detailed description is provided by BECTA (2001a) as follows:

Although there is some confusion about the definition of Virtual Learning Environments, they are generally a combination of some or all of the following features:

- *communication tools such as e-mail, bulletin boards and chat rooms*
- *collaboration tools such as online forums, intranets, electronic diaries and calendars*
- *tools to create online content and courses*
- *online assessment and marking*
- *integration with school management information systems*
- *controlled access to curriculum resources*
- *student access to content and communications beyond the school.*

As well as multiple definitions, multiple labels exist for describing e-learning systems. Such terms as *learning management system* (LMS), *course management system* (CMS), *virtual learning environment* (VLE), *managed learning environment* (MLE), and *portal* are often used interchangeably. In the UK, the Joint Information Systems Steering Committee (JISC) makes a clear distinction between a virtual learning environment and a managed learning environment, thus:

*While recognising that the world at large will continue to use terminology in different and often ambiguous ways, the term **Virtual Learning Environment (VLE)** is used to refer to the 'online' interactions of various kinds which take place between learners and tutors. The JISC MLE Steering Group has said that VLE refers to the components in which learners and tutors participate in 'online' interactions of various kinds, including online learning. The JISC MLE Steering Group has said that the term **Managed Learning Environment (MLE)** is used to include the whole range of information systems and processes of a college (including its VLE if it has one) that contribute directly, or indirectly, to learning and the management of that learning.* (JISC, 2001)

A useful model displaying the components encapsulated in the JISC definition is shown in Figure 1.

The JISC model thus separates VLE and MLE components as seen in Example 1.

Whereas such platforms as WebCT and Blackboard have dominated the e-learning market in the tertiary sector (Beshears, 2000), solutions in the primary and secondary education sectors are not so homogeneous. As seen in Table 1, a number of vendors supplying e-learning components and complete solutions have emerged since the late 1990s. Impaq and myinternet (in Australia) and digitalbrain and Learnwise (in the UK) are just some of the vendors in this domain providing off-the-shelf solutions for schools. Other vendors such as Microsoft (Sharepoint and ClassServer) and Novell (Extend) offer development platforms for highly customisable e-learning solutions. Concerns have been raised by some commentators, however, that some schools are attempting to implement e-learning solutions without sufficient cognizance of issues for learners, administrators, support staff, and the institution itself (Barnes & Greer, 2002; Kilmurray, 2003; Morrison & Khan, 2003).

In the early years of the new millennium, educational authorities in developed countries and

Figure 1. The JISC managed learning environment model (JISC, 2001)

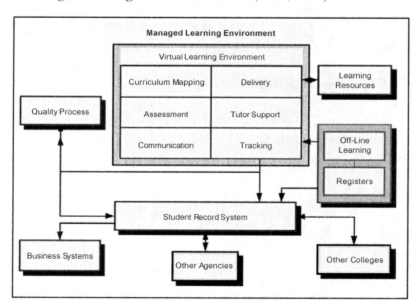

Example 1. Components of VLE and MLE

VLE Components
Curriculum mapping
Assessment
Communication
Delivery
Tutor support
Tracking

MLE Components
Learning Resources
Student Record System
Business Systems
Off-line learning
Registers
Quality process

third-party vendors simultaneously recognised that if e-learning environments were to succeed, then appropriate content needed to be developed. Furthermore, to ensure interoperability and re-usability of this content, standards needed to be developed. Initiatives such as Curriculum Online (UK) and The Learning Federation (Australia) are indicative of government-sponsored approaches to the development of e-learning content, now commonly known as learning objects. Commercial developers of learning objects include XSIQ, SchoolKit Enactz, and Granada LearnWise. In terms of e-learning standards, we are now seeing the deployment of such standards as SIF (the Student Interoperability Framework) and SCORM (the Scalable Content Object Reference Model). At the time of writing this chapter, the boundary between developers of e-learning platforms and the developers of e-learning objects is blurred.

Components of an E-Learning Strategy

Badrul Khan has been an active commentator in the e-learning domain since 1997 (Khan, 1997). He notes: "A successful e-learning system involved a systematic process of planning, design, development, evaluation and implementation to create an online environment where learning is actively fostered and supported" (Morrison & Khan, 2003).

Khan has proposed a model that identifies eight dimensions organisations need to address in developing an e-learning strategy (see Figure 2). These dimensions are as follows:

- Institutional
- Pedagogical
- Technological
- Interface Design
- Management
- Ethical
- Resource Support
- Evaluation

Figure 2. Khan's eight-point model of e-learning

Figure 3. Hitch and MacBrayne's (2003) model for effectively supporting e-learning

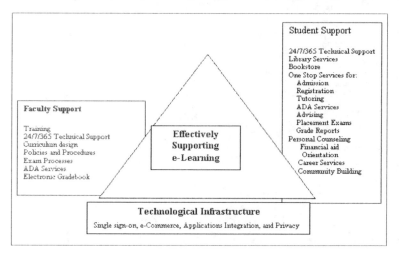

Each dimension is broken down into issues related to specific aspects of a successful e-learning environment. Khan's model is not specific to schools or universities, but rather is applicable to any organisation engaged in e-learning for its constituent community, and thus is put forward as a "global e-learning framework" (Morrison & Khan, 2003).

Requirements for E-Learning Support

Another model proposed by Hitch and MacBrayne (2003) concentrates on the support required for a successful e-learning strategy. Their model has three key components: (a) Faculty Support and (b) Student Support, "glued" together by (c) Technological Infrastructure (see Figure 3). They note:

Despite the rising use of information technology in instruction, both in the traditional classroom and at a distance, there remains a substantial gap in providing off-campus students with an array of academic and support services equivalent to

the on-campus services...Institutions that provide e-learning, whether they offer totally online or hybrid (i.e., blending face-to-face with online instruction) courses, must provide concurrent e-student support mechanisms. (Hitch & Mac-Brayne, 2003)

Potential Benefits of E-Learning Environments

The implementation of an e-learning environment in a school or school authority offers potential benefits to a variety of stakeholders. These can be summarised as follows:

General Benefits

- Can make it easier for staff and students to use ICT within an integrated environment.
- Brings together a variety of features in one piece of software with a consistent look and feel, which is consequently easier to learn and manage.
- Offers a different communication dimension through e-mail, discussion groups, and chat

rooms, in addition to face-to-face classroom interaction.

- Can improve the learning environment and standard of discussion, if the communications are managed effectively by the teacher.

Benefits for Students

- Offers the flexibility of "anytime, anywhere" access.
- Encourages gains in student ICT skills in general, and in journalistic writing, understanding, and presentation skills.
- Encourages development of higher levels of deep and strategic learning styles.
- Encourages the discovery of successful approaches to learning through trial and error in discussion, and through expressing ideas in a written but public way.

Benefits for Teachers

- Supports teacher confidence, and enhances practice and collaboration.
- Fosters self-study by teachers willing to make the commitment to the technology and to sharing personal views and experiences.
- Increases teacher participation in online seminars, which can lead to increased performance in group work.

Benefits for Parents

- Provides a communication gateway between home and school.
- Allows parents to monitor their children's progress.
- Provides access to online content that can help parents to support homework studies out of school hours.

To summarise this section, it can be seen that the literature as of 2004 describes an e-learning milieu that is not fully mature. In this context, attempts by schools to devise e-learning strategies and implement them are necessarily subject to the vagaries of an evolving marketplace. The following section describes the journey of one school on the path towards a full e-learning solution.

ONE SCHOOL'S JOURNEY TOWARDS AN E-LEARNING STRATEGY

Background

The case study school is an independent day and boarding college for boys in Sydney, Australia. Although initially hesitant to embrace ICT innovations in the early 1990s, the school's management realised in 1994 that a number of push and pull factors were at work which required a whole-school strategy for ICT. After engaging external consultants, the school tabled its first ICT Strategic Plan in late 1995. This plan made provision for an extensive rollout of fibre-optic and category5 cabling throughout the school site, an ongoing programme of investment in end-user hardware and software, a review of curriculum outcomes to incorporate ICT elements, and provision for staff training and support.

It is interesting to note the extent to which ICT has become embedded in the school's culture by considering the comparisons between 1994 and 2004 (see Table 2).

In addition to a high level of ICT provision within the school, members of the school community (staff, students, and parents) also exhibited high levels of access to ICT outside school. In 2003, 95% of staff reported access to the Internet and e-mail from home. Students and parents reported high levels of access to computers

Table 2. Evolution of ICT in case study school

Dimension	1994	2004
Students	1100	1550
Academic Staff	120	170
Support Staff	50	80
Student Computers	60	450
Student-Computer ratio	1:19	1:3.5
Staff Computers	15	120
(Academic) Staff-Computer ratio	1:8	1:1.5
% Computers with internet access	5%	100%
% students with email accounts	0%	100%
% staff with email accounts	0%	100%
Servers	2	35
ICT Support staff	1	10
Annual ICT spend (as % of total spend)	1%	8%

(97%), and high levels of access to the Internet and e-mail in the home (89%). These metrics are consistent with similar statistics from other developed countries around the world (Research Machines, 2000; DfES, 2001; National Statistics, 2002; Ofsted, 2002).

ICT Management Model

There has been an emerging perception throughout the school that ICT is an enabler, providing increased efficiencies and effectiveness in administration, and adding value to teaching and learning (Mumtaz, 2000; Kennewell, Parkinson, & Tanner, 2000; Passey, 2002). Work by O'Mahony (2000) has noted that in educational institutions, ICT can become either a bridge or a chasm. ICT can be a bridge insofar as it has the potential to:

- directly support pedagogical efforts in the teaching and learning context, in terms of delivering educational and applications software to the classroom;
- directly support the back-office functions of the school; and

- indirectly minimise the impact of necessary administrative functions that teachers, students, and parents are required to perform.

ICT can be a chasm in the sense that it:

- so frequently falls short of the expectations and overblown promises of suppliers, developers, and purchasers; and
- is often perceived as a weapon for administrative control, rather than a tool for educational empowerment.

Arising from the school's efforts to embed ICT into its operations is a growing recognition of key factors that enable ICT to flourish, and thus for the school to be more effective in its overall development (Kirkman, 2000). These factors are presented here as a six-point model for achieving confident use with ICT, as seen in Figure 4. Within the school, this model exists in a specific organisational context, whereby ICT has increasingly become a fulcrum for change in the organisation's culture.

Figure 4. Core components of the school's ICT management

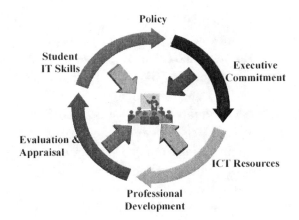

In summary, the six-point-model comprises:

1. Policy
2. Executive Commitment
3. ICT Resources
4. Professional Development
5. Evaluation/Appraisal
6. Student Learning

Details of the 6-point model are outlined in the following sections.

Curriculum ICT Policy (Strategic)

The school, through its ICT Strategic Plan, makes a clear statement of intent and direction concerning the use of information and communication technologies in curriculum areas. This statement is visible in school documentation at the senior management level, and is internalised throughout the curriculum (Kennewell et al., 2000). "In...successful schools, senior management do more than provide support for the IT Coordinator's policy; rather the IT policy is viewed as emanating from the senior management" (ACCAC, 1999). The curriculum ICT policy seeks to articulate well with the school's business and strategic development plans.

Department Commitment (Tactical)

At the department level, ICT policies exist which articulate with the wider ICT strategy, and provide necessary detail and context for the respective curriculum area. These policies express the department's commitment to ICT professional development, and specify expectations of ICT use in the classroom, both in terms of minimum hours and ICT-based tasks (Newton, 2003; Lambert & Nolan, 2003).

ICT Resources

A prerequisite to success with school ICT is the provision of sufficient, reliable, and up-to-date resources (ACCAC, 1999). These resources include network infrastructure, workstation and peripheral hardware, software, and human resources. Table 2 indicates the school's investment in resources over a 10-year period. Strong project management methodologies have been applied to ensure that the school gains good value for money, recognising that inferior products do more damage than good. Rigorous criteria are used in selecting hardware and software applications, such as ease of integration and ease of use (Stevenson, 1997).

Teacher Professional Development

Hiring external trainers has often been the only option for schools, but increasingly, schools are considering the appointment of dedicated training staff within the overall ICT function (Watson, 2001), as is the case with the subject school (O'Mahony, 2002). As well as having ICT resources and policies regarding the use of ICT in teaching, learning, and administration, the school has implemented a robust and measurable professional development programme (Donnelly, 2000; Russell, Finger, & Russell, 2000). This programme has six main components: (1) initial ICT orientation, (2) formal timetabled ICT training (one period per fortnight), (3) ICT surgeries, (4) a regular 'ICT tips & tricks' newsletter, (5) evening master classes (once per term), and (6) provision for external ICT courses.

Staff Appraisal and Review

The appraisal and review process gives crucial feedback for all aspects of the model. To drive home the message concerning the school's commitment to ICT, effective classroom use of ICT is a performance indicator for staff. The reviewer can flag the reviewee's ICT training needs, which is communicated to the training function/coordinator, who organises/delivers the required training. Once completed, confirmation of training is passed back along the chain. As well as providing feedback on staff ability, the appraisal and review process offers the opportunity to flag any issues concerning ICT resourcing or access. These issues, too, are forwarded to the relevant person/function. Collectively, these items will assist in the formation of subsequent ICT strategies.

Student Learning

The ultimate aim of this model, and in particular the Staff ICT professional development programme, is the improvement of student learning. Thus, complementary to a Staff ICT skills programme is a cross-curricular student ICT skills programme. Transcending the use of ICT in specific subjects, this programme provides broad-based exposure to generic ICT skills, including keyboard familiarity, word processing, spreadsheets, presentation graphics, Internet searching, critical analysis of Web-based data, and "appropriate ICT use" (NGfL, 2002a, 2002b; Hruskocy, Cennamo, Ertmer, & Johnson, 2000).

The Evolving E-Learning Strategy

As mentioned in the first section of this chapter, the development of an e-learning strategy at the subject school has been an evolutionary process, which has been modified over time as new e-learning technologies have emerged. Between 1994 and 1996, the school's main focus was on baseline ICT connectivity. That is, fundamental infrastructure decisions were a priority, establishing and maintaining a reliable local area network and associated intra-mural ICT services. Although early Web browsers became available in 1995, the school was not at that stage in a position to leverage these new technologies.

The school's first triennial ICT Strategic Plan (1997-1999) began to focus on value-added ICT functions, including early stages of 'extra-mural' connectivity. The school was fortunate in gaining access to the Internet through a broadband cable modem, and strategies were agreed for implementing e-mail for staff, a rudimentary intranet, and a school Web site. At the same time, the school continued its investment in end-user hardware and software, maintaining a focus on traditional classroom-based teaching and learning.

A second triennial ICT Strategic Plan covered the 2000-2002 period. This plan identified three core challenges:

- responding proactively to the pace of technological change,
- providing opportunities for staff ICT training, and
- providing extra-mural services for school staff and students.

In this plan, there was a clear recognition by management that the school needed to be continually renewing and reviewing its ICT efforts, replacing an earlier belief that one-off capital investment would be sufficient. Parallel to this was an understanding that staff needed further

Table 3. Chronology of integration of e-learning components in case study school as of 2002

Components (from BECTA 2001a)	Installed for staff	Installed for students	Integrated
Communication tools such as email, bulletin boards and chat rooms	1996	2002	2002
Collaboration tools such as online forums, intranets, electronic diaries and calendars	1997	2001	2001
Tools to create online content and courses	2001	N	N
Online assessment and marking	2001	N	N
Integration with school management information systems (target for 2005)	N	N	N
Controlled access to curriculum resources	1996	1996	2000
Student access to content and communications beyond the school.	2001	2001	2001

encouragement to build ICT into their teaching and administration. The third core challenge recognised the value to be gained from implementing remote access technologies, and the school actively sought out potential solutions for both staff and students.

In 2001, the school entered into an agreement with an e-learning solution provider (Impaq Australia) for the provision of VPN remote access and portal services. This initiative was the beginning of the school's attempts to integrate previously disparate ICT components into a more cohesive single e-learning solution (Weatherly & McDonald, 2003). Although not without challenges and frustrations (see Boydell & Kane, 2002), by the end of 2002 the school had implemented e-learning components according to Table 3.

The school is now in the middle year of its third triennial ICT Strategic Plan (2003-2005). Key strategies identified in this plan include:

- Identity management and database integration
- Network security
- Continuing commitment to staff training and support
- Continuing commitment to remote access services

Although the school continued to integrate e-learning components during 2003, a significant setback occurred late in the year when the e-learning solution provider ceased trading. This meant the potential loss of some critical remote services and, by extension, a potential loss of confidence in the school's e-learning efforts. By early 2004, however, the school had engaged alternative providers. It is indicative of the rapid development of e-learning solutions that this 'second time around' implementation was much smoother than the first.

One of the projects being addressed by the school involves integrating a variety of administrative databases (such as student administration, library, business operations, and human resources) with the main network directory service (network authentication). This project has multiple phases and constraints, but has the potential to deliver increased efficiency and effectiveness across the whole school. It will also add value to the school's e-learning efforts by enabling integration of other e-learning components to the school's management information systems through the implementation of SIF standards (as mentioned earlier; see Table 3).

The school has encouraged staff to reflect on changes to their teaching practice as a result of

Table 4. Staff reflections on e-learning environments

Stakeholder	Reflections
Benefits of utilizing an e-learning environment for me as a teacher:	• Allows me to be organised, which encourages my students to be organised. • Reduces my photocopying / paper waste. • Students can't 'lose' handouts, thus improving use of my time. • Material is 'always' available for revision purposes. • Easy to update on-the-fly before, during and after class.
Benefits of utilizing an e-learning environment for my faculty:	• Allows the sharing of resources. • Common files can be loaded into one area for all classes to access. • Excellent medium for communicating outcomes, programs, etc between faculty members. • Reduces photocopying / paper waste
Benefits of utilizing an e-learning environment for my students:	• They have an area where they know they can find all the information needed for a specific subject. • They can see important due dates quickly and easily • Assignment / handout information can never be 'lost'. • Material can be collected / viewed even if they miss a class. • Test content / exam revision / class work can be accessed anytime, anywhere. • Forgot your homework? - Look it up on the portal.
Benefits of utilizing an e-learning environment for the wider community:	• Parents can access information about the course, such as programs, assessment schedules, etc. • Parents can look at assignment details and homework tasks. • Allows parents to see what is happening in their child's class. • Wider community can access photos, current events, school newsletter etc.

utilising e-learning elements. One example of this reflective process is shown in Table 4.

ISSUES AND IMPLICATIONS

A number of considerations can be perceived for an institution planning the implementation of an e-learning strategy. Articulating with the work of Khan (Morrison & Khan, 2003) as discussed above, Maguire (2003) has identified an analysis grid for online teaching and learning that compares three main categories across five main stakeholders. This grid assists institutions to analyse issues and implications, as in Table 5. For each institution, the grid would be populated differently, according to local needs and contexts. This example table is populated with a typical response drawn from the case study school.

In terms of infrastructure, one of the fundamental questions facing schools that are planning the implementation of an e-learning environment is where to locate host servers and services. Some providers of e-learning solutions offer hosting services as part of their solution. In this instance, client school(s) are only obliged to provide Internet access to the provider's wide area network. This remote hosting model can offer cost-effective advantages to schools, as well as enabling multiple schools to share common resources. An alternative to this is the local hosting model, whereby individual schools buy a licence to operate the e-learning solution, and provide all hardware and network infrastructure themselves. This model gives the client school more control over the development and deployment of the solution, and can deliver cost savings under certain upload/download conditions.

Another major issue facing school decision makers is the degree of customisation required in an e-learning environment. As discussed earlier, suppliers of e-learning solutions fall into two main categories—those that provide an out-of-the-box solution requiring no customisation by the school

Table 5. Considerations for implementing an e-learning strategy

	Pedagogical	**Technological**	**People Factors**
Students	How will this help me learn?	How do we provide 1-to-1 access to e-learning resources?	What is the most appropriate 'class size'?
Teachers	How will this help me teach?	How can I ensure the engagement of the students, and reliability of the technology?	How do we build staff skill and enthusiasm for these initiatives?
School Managers	How will this improve our outcomes?	How do we measure the cost benefits?	What 'critical mass' of people is required before this will work?
Teacher Educators	What pedagogical models are best suited to this environment?	What skills do we need to include in our training programmes?	How do we ensure our people have the agility to respond positively to change?
Institutions	How does this fit our core business?	Is this sustainable?	How do we retain our unique 'brand' of education?

Table 6. E-learning implementation questions for schools

Can your school sustain an e-learning environment financially, technically and administratively?	Will introducing an e-learning environment add value to the teaching and learning process, over and above current systems?
Should your school look for intermediate solutions first, such as providing the content only?	Is it worth considering working in a consortium with other schools, to share costs and resources?
Penetration of Internet / e-mail access among target community	Extent of existing intranet/extranet services
IT Competence of staff	Available bandwidth for upload and download
Cost / Total Cost of Ownership	Reliability
Training requirements	Robustness
Service Level Agreements	User Interface
Currency of content	Upgrade path
Maintenance	Security

(the buy option), and those that provide an open platform requiring heavy customisation by the school (the 'build' option). Both the build and the buy options have their positive and negative dimensions, depending on a wide variety of institutional factors. For instance, there has been a trend over the past few years for state-run schools to favour suppliers offering a buy solution, whereas independent schools have tended to favour platforms offering a build solution.

A third option gaining prominence in 2003 and 2004 takes the best of both options. Some institutions find that 'blending' an off-the-shelf framework with D-I-Y content offers a happy compromise. A number of vendors are happy to offer conversion/migration of existing intranet content, however significant issues need to be addressed to enable the "blend" approach to be successful. Other implementation questions facing schools are summarised in Table 6.

Schools considering the introduction of an e-learning solution need to reflect deeply, consult widely, and plan carefully using rigorous project methodologies in order to ensure a successful and effective implementation.

CONCLUSION

This chapter has looked closely at the emergence of e-learning environments in the K-12 education sector. A review of the literature has traced the development of separate e-learning components over the period from 1994 to 2004, and the parallel development of models used to describe, prescribe, and predict these phenomena.

An effective e-learning strategy brings together components that have evolved separately, and integrates them in a way that successfully supports learning. E-learning environments, when implemented well, can greatly assist staff and students to achieve ICT literacy. By integrating components into one environment, with a seamless and attractive interface, students and staff can become engaged, enthusiastic, and competent with a variety of information and communication technologies.

Reflecting on the case described in the third section of this chapter, it is possible to identify elements in common with models proposed by other commentators. Although individual components had been implemented separately as technologies emerged, the process of integration since 2002 demonstrates strong parallels with the e-learning components described by BECTA (2001a). Furthermore, the three-dimension support model suggested by Hitch and MacBrayne

(2003) has clear synergies with the case school's emerging strategy (see Figure 3). That is, there is a close mapping between the school's policy and practice, and the model's three dimensions of faculty support, student support, and technological infrastructure.

Similarly, a close relationship can be perceived between the school's six-point Strategic ICT model and Khan's eight-point model of e-learning development. Table 7 demonstrates a concordance between the two, showing some overlap where some items of the school's plan incorporate more than one of Khan's dimensions.

Despite the parallels between theoretical models and the lived experience described in the case study, it is suggested that this is more the result of coincidence than design. School managers devise their e-learning strategies using the best information at hand at the time, but by necessity these strategies must be modified to accommodate a rapidly changing ICT landscape. In addition, the implementation of e-learning strategies requires significant changes to teaching practice (McLoughlin, 2000; Kilmurray, 2003; Weatherly & McDonald, 2003). A recent newspaper article notes:

With issues such as teaching effective new media literacy and critical analysis, achieving depth and breadth in students' e-research and ensuring fair

Table 7. E-learning model concordance

Khan's 8-point model	The case study school's 6-point plan
Institutional	ICT Policy (Strategic)
Ethical	ICT Policy (Strategic)
Management	Department Commitment (Tactical)
Technological	ICT Resources
Interface Design	ICT Resources
Resource Support	ICT Resources / Professional Development
Pedagogical	Student learning
Evaluation	Appraisal and Review

Figure 8. Relevant Web sites

Organisation	Web site
The British Educational Communications and Technology Agency - BECTA	www.becta.org.uk
Blackboard (USA)	www.blackboard.com
Curriculum Online (UK)	www.curriculumonline.gov.uk
The Department for Education and Skills (UK) - DFES	www.dfes.gov.uk
Digitalbrain (UK)	www.digitalbrain.com
Granada LearnWise (UK)	www.learnwise.com
The Joint Information Systems Committee (UK) - JISC	www.jisc.ac.uk
Microsoft ClassServer (USA)	www.microsoft.com/education/ClassServer.aspx
Microsoft Sharepoint (USA)	www.microsoft.com/sharepoint
Myinternet (Australia)	www.myinternet.com.au
SchoolKit	www.schoolkit.com
StudentNet (Australia)	http://portals.studentnet.edu.au/studentnet/
The Learning Federation (Australia)	www.thelearningfederation.edu.au
WebCT (Canada)	www.webct.com
XSIQ (Australia)	www.xsiq.com.au

assessment and equity, it is apparent there is not only a gap between the generations in e-learning, there is also a gap between teachers and administrators. (Friedlander, 2004)

As some commentators have noted: "E-learning is by no means as simple to develop as it appears" (Bechervaise & Chomley, 2003). By the same token, the implementation of e-learning strategies can be an epiphany event for some educational institutions (Paoletti, 2003; McKay & Merrill, 2003). The roles of teachers and educational leaders are especially critical to the success of these innovations (Yee, 2000; Schiller, 2002; Webb & Downes, 2003).

Like many other ICT innovations, e-learning can be a mixed blessing. As these innovations mature and the requirements for achieving true ICT literacy evolve, school strategies will continue to be reactive rather than proactive, responding with agility and discernment to this volatile domain. One thing is certain: "The human act of teaching is more than the sum of its parts" (Dowling 2003).

REFERENCES

ACCAC. (1999). *Whole school approaches to developing ICT capability.* Cardiff: ACCAC.

Barnes, A., & Greer, R. (2002, July). Factors affecting successful R-12 learning communities in Web-based environments. *Proceedings of the Australian Computers in Education Conference (ACEC2002).*

Bechervaise, N. E., & Chomley, P. M. M. (2003, November). E-lusive learning: Innovation, forced change and reflexivity. *Proceedings of the E-Learning Conference on Design and Development.* Melbourne: RMIT.

BECTA (British Educational Communications and Technology Agency). (2001a). *A review of the research literature on the use of managed learning environments and virtual learning environments in education, and a consideration of the implications for schools in the United Kingdom.* Retrieved from http://www.becta.org.uk/page_documents/research/VLE_report.pdf

BECTA. (2001b). *Primary schools of the future—achieving today. A report to the DfEE by BECTA.* Coventry: British Educational Communications and Technology Agency.

BECTA. (2001c). *The secondary school of the future. A preliminary report to the DfEE by BECTA.* Coventry: British Educational Communications and Technology Agency.

Beshears, F. M. (2000). *Web-based learning management systems.* Retrieved from http://ist-socrates.berkeley.edu/~fmb/articles/web_based_lms.html

Boydell, S., & Kane, J. (2002, July). All aboard!—Creating a ubiquitous intranet. *Proceedings of the Australian Computers in Education Conference (ACEC2002).*

Cuban, L. (2000). *Oversold and underused: Computers in the classroom.* Cambridge, MA: Harvard University Press.

DfES. (2001). *ICT access and use: Report on the benchmark survey.* DfES Research Report No 252. London: Department for Education and Skills.

Donnelly, J. (2000). *Information management strategy for schools and local education authorities—Report on training needs.* Retrieved from http://dfes.gov.uk/ims/JDReportfinal.rtf

Dowling, C. (2003). The role of the human teacher in learning environments of the future. *Proceedings of the IFIP Working Groups 3.1 and 3.3 Working Conference: ICT and the Teacher of the Future,* Melbourne.

Friedlander, J. (2004). Cool to be wired for school. *Sydney Morning Herald,* (April 16).

Hitch, L. P., & MacBrayne, P. (2003). *A model for effectively supporting e-learning.* Retrieved from http://ts.mivu.org/default.asp?show=article&id=1016

Hruskocy, C., Cennamo, K. S., Ertmer, P. A., & Johnson, T. (2000). Creating a community of technology users: Students become technology experts for teachers and peers. *Journal of Technology and Teacher Education, 8,* 69-84.

JISC. (2001). *MLEs and VLEs explained.* Retrieved from http://www.jisc.ac.uk/index.cfm?name=mle_briefings_1

Jones, A. J. (2003). ICT and future teachers: Are we preparing for e-learning? *Proceedings of the IFIP Working Groups 3.1 and 3.3 Working Conference: ICT and the Teacher of the Future,* Melbourne.

Kennewell, S., Parkinson, J., & Tanner, H. (2000). *Developing the ICT-capable school.* London: Routledge Falmer.

Khan, B. H. (1997). *Web-based instruction.* Englewood Cliffs, NJ: Educational Technology Publications.

Kirkman, C. (2000). A model for the effective management of information and communications technology development in schools derived from six contrasting case studies. *Journal of IT for Teacher Education, 9*(1).

Kilmurray, J. (2003). *E-learning: It's more than automation.* Retrieved from http://ts.mivu.org/default.asp?show=article&id=1014

Lambert, M. J., & Nolan, C. J. P. (2003). Managing learning environments in schools: Developing ICT capable teachers. In I. Selwood, A. Fung, & C. O'Mahony (Eds.), *Management of education in the Information Age—The role of ICT.* London: Kluwer for IFIP.

Maguire, M. (2003, October). Questions for Web-based teaching and learning. *Proceedings of the Australian Catholic University School of Education Seminar Series* (unpublished).

McKay, E., & Merrill, M. D. (2003, November). Cognitive skill and Web-based educational systems. *Proceedings of the E-Learning Conference on Design and Development.* Melbourne: RMIT.

McLoughlin, C. (2000). Creating partnerships for generative learning and systematic change: Redefining academic roles and relationships in support of learning. *International Journal for Academic Development, 5*(2).

Morrison, J. L., & Khan, B. H. (2003). *The global e-learning framework: An interview with Badrul Khan.* Retrieved from http://ts.mivu.org/default. asp?show=article &id=1019

Mumtaz, S. (2000). Factors affecting teachers' use of information and communications technology: A review of the literature. *Journal of Information Technology for Teacher Education, 9*(3).

National Statistics. (2002). Retrieved from www. dfes.gov.uk/statistics/db/sbu/b0360/sb07-2002. pdf

Newton, L. (2003). Management and the use of ICT in subject teaching—Integration for learning. In I. Selwood, A. Fung, & C. O'Mahony (Eds.), *Management of education in the Information Age—the role of ICT.* London: Kluwer for IFIP.

NGfL. (2002a). *Impact2: The impact of information and communication technologies on pupil learning and attainment* (ICT in School Research and Evaluation Series—No 7). Annesley: DfES.

NGfL. (2002b). *Impact2: Learning at home and school: Case studies* (ICT in School Research and Evaluation Series—No 8). Annesley: DfES.

Ofsted. (2002). *ICT in schools, effect of government initiatives.* Retrieved from *www.ofsted.gov. uk/public/docs01/ictreport.pdf*

O'Mahony, C. D. (2000). *The evolution and evaluation of information systems in NSW secondary schools in the 1990s: The impact of values on information systems.* Unpublished PhD Thesis, Macquarie University, Australia.

O'Mahony, C. D. (2002). Managing ICT access and training for educators: A case study. *Proceedings of the Information Technology for Educational Management Conference* (ITEM2002), Helsinki.

Paoletti, J. B. (2003). *Wanted: Course revision without pain.* Retrieved from http://ts.mivu.org/ default.asp?show=article&id=1034

Passey, D. (2002). *ICT and school management: A review of selected literature.* Unpublished Research Report, Department of Educational Research, Lancaster University, UK.

Research Machines PLC. (2000). *The RM G7 (8) Report 2000 comparing ICT provision in schools.* Abingdon: RMplc.

Russell, G., Finger, G., & Russell, N. (2000). Information technology skills of Australian teachers: Implications for teacher education. *Journal of IT for Teacher Education, 9*(2).

Schiller, J. (2002). Interventions by school leaders in effective implementation of information and communications technology: Perceptions of Australian principals. *Journal of Information Technology for Teacher Education, 11*(3).

Stevenson, R. (1997). Information and communications technology in UK schools: An independent inquiry. *The Stevenson Report.*

Watson, G. (2001). Models of information technology teacher professional development that engage with teachers' hearts and minds. *Journal of IT for Teacher Education, 10*(1-2).

Weatherly, G., & McDonald, R. (2003). *Where technology and course development meet.* Retrieved from http://ts.mivu.org/default.asp?show=article&id=951

Webb, I., & Downes, T. (2003). Raising the standards: ICT and the teacher of the future. *Proceedings of the IFIP Working Groups 3.1 and 3.3 Working Conference: ICT and the Teacher of the Future,* Melbourne.

Yee, D.L. (2000). Images of school principals' information and communications technology leadership. *Journal of IT for Teacher Education, 9*(3).

Chapter 8.22
Academic, Economic, and Technological Trends Affecting Distance Education

Nathan K. Lindsay
University of Michigan, USA

Peter B. Williams
Brigham Young University, USA

Scott L. Howell
Brigham Young University, USA

A FOUNDATION FOR INFORMED PLANNING

A number of prominent distance learning journals have established the need for administrators to be informed and prepared with strategic plans equal to foreseeable challenges. This article provides decision makers with 32 trends that affect distance learning and thus enable them to plan accordingly. The trends are organized into categories as they pertain to academics (including students and faculty), the economy, technology, and distance learning.

Recently, Beaudoin (2003) urged institutional leaders "to be informed and enlightened enough to ask fundamental questions that could well influence their institution's future viability" (p. 1). Decision makers often rely on long-term demographic and economic projections, based on current trends and foreseeable influences, in their strategic planning (Reeve & Perlich, 2002). While identifying trends does not offer solutions to distance learning challenges, decision makers will benefit by carefully considering each trend as it affects their institution and goals.

METHODOLOGY

The trends presented in this article were identified during an integrative literature review, conducted to summarize the current state and future directions of distance education. Resources were selected based on their currency and relevance to distance education, information technology, and impact on the larger, higher education community. As themes emerged, the citations were then ordered in sub categories and specific trends, and condensed for publication.

ACADEMIC TRENDS

Knowledge and Information are Growing Exponentially

One cannot dispute that there is a proliferation of new information: "In the past, information doubled every 10 years; now it doubles every four years" (Aslanian, 2001, p. 5; see also Finkelstein, 1996). This growth in information will certainly continue to dramatically impact higher education and learning in general.

The Institutional Landscape of Higher Education is Changing: Traditional Campuses are Declining, For-Profit Institutions are Growing, and Public and Private Institutions are Merging

Changes in institutional landscape may magnify competition among educational providers and allow new models and leaders to emerge. Currently, only 4-5% of all higher education students are enrolled with for-profit providers, but 33% of all online students are enrolled with these same providers (Gallagher, 2003). Dunn (2000) projected that by 2025, "half of today's existing independent colleges will be closed, merged, or significantly altered in mission," and that "the distinctions between and among public and private, for-profit and nonprofit institutions of higher education will largely disappear" (p. 37).

There is a Shift in Organizational Structure Toward Decentralization

Much of a distance education program's success or failure can be attributed to how it is organized. Hickman (2003) has observed a movement "from a highly centralized core of administrators, coordinators, [and] marketing and support staffs to a more 'institutionalized' approach in which continuing education personnel were assigned to academic units within a university" (p. 6).

Instruction is Becoming More Learner-Centered, Non-Linear, and Self-Directed

Instructional approaches are becoming more learner-centered, "recursive and non-linear, engaging, self-directed, and meaningful from the learner's perspective" (McCombs, 2000, p. 1). Whereas most instructors previously followed a "transmission" or lecture-style approach to teaching, more instructional diversity is occurring among teachers who are trying a larger variety of approaches (Eckert, 2003).

There is a Growing Emphasis on Academic Accountability

In a recent poll by the North Central Association of Colleges and Schools, university presidents, administrators, and faculty members rated increasing demands for accountability (80%) and expanding use of distance education (78%) as the highest impact trends on future NCA (i.e., regulatory) activities (de Alva, 2000). Distance educators must plan to accommodate this emphasis on accountability to maintain accreditation and meet consumer demands.

Academic Emphasis is Shifting from Course-Completion to Competency

Related to accountability trends, there is a slight shift from "theoretical" and "seat-based time" to "outcomes-based" or "employer-based" competency. In many cases, "certification is becoming more preferable than a degree" (Gallagher, 2003). Diplomas are less meaningful to employers; knowledge, performance, and skills are what count to them (Callahan, 2003). With an emphasis on competency, course content will be dictated more "by what learners need, [than] by what has been traditionally done" (de Alva, 2000, p. 38).

Education is Becoming More Seamless Between High School, College, and Further Studies

As universities shift toward competency and institutions cater more closely to learners' specific needs, distinctions between educational levels will dissolve. "Incentives will be given to students and institutions to move students through at a faster rate [and] the home school movement will lead to a home-college movement" (Dunn, 2000, p. 37). As leaders in the effort to cater to learners' needs, distance education programs may be a dominant influence in this trend.

Higher Education Outsourcing and Partnerships are Increasing

Universities are traditionally independent, free-standing, and competitive (Hawkins, 2003). In contrast, distance learning institutions have been more cooperative and accommodating with partner institutions. Interestingly, Rubin (2003) has noted that "traditional universities are becoming more like distance learning universities and not the opposite" (p. 59). With this shift, more institutions are creating partnerships with other colleges, universities, and companies to share technology and to produce and deliver courses (Dunn, 2000; Carnevale, 2000a; Cheney, 2002; Primary Research Group, 2004).

Some Advocate Standardizing Content in Learning Objects

Frydenberg (2002, para 38) noted that "the central issue in courseware development at the moment is the potential for developing reusable learning objects, tagging them in a systemic way, storing them in well-designed databases, and retrieving and recombining them with other objects to create customized learning experiences for specific needs." Such customized learning, allowing for "true" individualized learning, is the future and strength of educational technology (Saba, 2003).

STUDENT/ENROLLMENT TRENDS

The Current Higher Education Infrastructure Cannot Accommodate the Growing College-Aged Population and Enrollments, Making More Distance Education Programs Necessary

Callahan (2003) noted at a recent UCEA conference that the largest high school class in US history will occur in 2009. In corroboration of this projection, a survey conducted by the National Center for Education Statistics predicted that college enrollment will grow 16% over the next 10 years (Jones, 2003). With this growth in population and enrollments, and the need for more lifelong learning, many institutions acknowledge that within the decade there will be more students than their facilities can accommodate (Oblinger, Barone, & Hawkins, 2001).

Students are Shopping for Courses that Meet their Schedules and Circumstances

More learners are requiring flexibility in program structure to accommodate their other responsibilities, such as full-time jobs or family needs (Penn State Strategic Plan, 1998). With these constraints, students are enrolling for courses that best accommodate their schedules and learning styles, and then transferring the credit to the university where they will earn their degrees (Carnevale, 2000a; Johnstone, Ewell, & Paulson, 2002; Paulson, 2002).

Higher Education Learner Profiles, Including Online, Information-Age, and Adult Learners, are Changing

Online students are "generally older, have completed more college credit hours and more degree programs, and have a higher all-college GPA than their traditional counterparts" (Diaz, 2002, pp. 1-2). Information-age learners prefer doing to knowing, trial-and-error to logic, and typing to handwriting. Adult learners need to know the rationale for what they are learning and are motivated by professional advancement. However, they tend to feel insecure about their ability to succeed in distance learning, find instruction that matches their learning style, and have sufficient instructor contact, support services, and technology training (Diaz, 2002; Dortch, 2003; Dubois, 1996).

The Percentage of Adult, Female, and Minority Learners is Increasing

While the number of 18- to 24-year-old students increased only 41% between 1970 and 2000, the number of adult students increased 170% (Aslanian, 2001). More women than men now enroll in college (57% of students are women), a trend supported by the fact that growing numbers of women are entering the workforce (UCEA, 2002). If enrollment follows population projections, higher education can expect the increase in minorities to continue—for example, the Hispanic population in the US is expected to increase 63% by 2020, reaching 55 million people (UCEA, 2002).

Completion and Retention Rates Concern Administrators and Faculty Members

A Chronicle of Higher Education article in 2000 reported that "no national statistics exist yet about how many students complete distance programs or courses, but anecdotal evidence and studies by individual institutions suggest that course-completion and program-retention rates are generally lower in distance education courses than in their face-to-face counterparts" (Brady, 2001, p. 352). However, many concerns are unwarranted, and institutional results are mixed. Brigham (2003), in a benchmark survey of four-year institutions' distance education programs, found that 66% of the distance learning institutions have an 80% or better completion rate for their distance education courses; 87% have 70% or better completion.

FACULTY TRENDS

Traditional Faculty Roles are Shifting or "Unbundling"

Paulson (2002) noted that "rather than incorporating the responsibility for all technology- and competency-based functions into a single concept of 'faculty member', universities are disaggregating faculty instructional activities and [assigning] them to distinct professionals" (p. 124). Doing this involves a "deliberate division of labor among the faculty, creating new kinds of instructional staff, or deploying nontenure-track instructional staff (such as adjunct faculty, graduate teaching assistants, or undergraduate assistants) in new ways" (p. 126).

The Need for Faculty Development, Support, and Training is Growing

In Green's (2002) survey on computing and information technology in US higher education, chief academic and information technology officials rated "helping faculty integrate technology into their instruction" the single most important IT issue confronting their campuses over the next two or three years (p. 7). An EDUCAUSE survey supported the issue's importance: "faculty development, support, and training" was rated the fifth overall strategic concern, as well as the fifth IT issue most likely to become even more significant in the next year (Crawford, Rudy, & the EDUCAUSE Current Issues Committee, 2003).

Faculty Tenure is Being Challenged, Allowing for More Non-Traditional Faculty Roles in Distance Education

Faculty tenure status is coming under fire as new state, private, and for-profit distance learning universities are created. The results of de Alva's 2000 survey support this trend: governors rated "maintaining traditional faculty roles and tenure" as the least desirable characteristic of a 21st century university (p. 34). Currently, contributions to distance education seldom move faculty members toward tenure, a problem in the present system that needs to be rectified by administrators and faculty.

Some Faculty Members are Resisting Technological Course Delivery

As long as distance education contributions are not considered in tenure and promotion decisions, and as long as professors have their own, traditional ways of delivering courses, many faculty members will hesitate to participate in online courses (Oravec, 2003). Concerning this reluctance, Dunn (2000) predicted that many faculty members will revolt against technological course delivery and the emerging expectations their institutions will have of faculty members.

Faculty Members who Participate in Distance Education Courses Develop Better Attitudes Toward Distance Education and Technology

Despite some resistance, the results of a study by McGraw-Hill showed a strong increase in overall faculty support for technology in education, with only 22% viewing it as important in 1999 and 57% in 2003 (Chick et al., 2002). Another 2002 study showed that "most teachers (85%) were not philosophically opposed to distance education" (Lindner, Murphy, Dooley, & Jones, 2002, p. 5). Further, teaching at a distance improves perceptions of distance education: "Faculty members who had not taught distance education courses perceived the level of support as lower than those who had" (Lindner et al., 2002, p. 5).

Instructors of Distance Courses Can Feel Isolated

Despite growing support among faculty members for distance learning, there are acknowledged drawbacks. "Design teams and instructors must anticipate isolation that can be felt by instructors who are separated from their students. This isolation may affect instructor satisfaction, motivation, and potential long-term involvement in distance learning" (Childers & Berner, 2000, p. 64).

Faculty Members Demand Reduced Workload and Increased Compensation for Distance Courses

An NEA survey reported that faculty members' top concern about distance education was that they will do more work for the same amount of pay, which apparently is a merited concern. The NEA (2000) found that most faculty members do spend more time on their distance courses than they do on traditional courses, and 84% of them do not get a reduced workload. Similarly, 63% of distance faculty members receive no extra compensation for their distance courses.

ECONOMIC TRENDS

There are Competing Interests and Limited Resources for Higher Education and Higher Education Initiatives, Such as Distance Education

The Washington-based Center on Budget and Policy Priorities recently calculated the combined deficits of the nation's 50 state governments to total $85 billion within the next year, "the highest number since the Great Depression" (White, 2003, p. 54). These scarce resources will prompt all universities to seek additional external sources of funding. To worsen the problem, university costs and enrollments are growing (UCEA, 2001).

Funding Challenges are the Top IT Concern for Many

A study from the Colorado Department of Education reported that "the cost per student of a high-quality online learning program is the same as or greater than the per-student cost of physical school [i.e., traditional] education" (Branigan, 2003, p. 1). EDUCAUSE reported similar results: "IT Funding

Challenges has become the number-one IT-related issue in terms of its strategic importance to the institution, its potential to become even more significant, and its capture of IT leaders' time" (Crawford et al., 2003, p. 12).

Lifelong Learning is Becoming a Competitive Necessity

Some have estimated that people change careers, on average, every 10 years (Cetron, 2003). Undoubtedly, "the changing nature of the workforce in the Information Age...[will require] a continuous cycle of retraining and retooling" (Dasher-Alston & Patton, 1998, p. 12). In such circumstances, "the opportunity for training is becoming one of the most desirable benefits any job can offer," and employers are coming to "view employee training as a good investment" (Cetron, 2003, pp. 6, 22).

TECHNOLOGY TRENDS

Technological Devices are Becoming More Versatile and Ubiquitous

One obvious trend affecting distance education is the advancement of technology. Infrastructures are expanding, computers are doubling in speed while decreasing in cost, and high-speed network connections are continuing to increase. Computer, fax, picture phone, duplication, and other modalities are merging and becoming available at ever cheaper prices (Cetron, 2003). IT functionalities not imagined 10 years ago are being realized.

There is a Huge Growth in Internet Usage

Not only is technology becoming more ubiquitous, it is being used more competently by more people from all nationalities, age groups, and socioeconomic levels (Murray, 2003). As Cetron (2003) reports, the number of current Internet users is approximately 500 million worldwide (1/12 of the population) and will almost double by 2005. A primary reason for the expansion is a growing percentage of users outside the US.

Technological Fluency is Becoming a Graduation Requirement

Since the networked world is dominating the economy, increasing the power of the individual, and changing business models, no one can afford to be without computer competence (Oblinger & Kidwell, 2000). Accordingly, universities are beginning to list the fluent use of technology as an outcome skill, encouraging students to take online courses, and even requiring students to take at least one online course before they graduate (Young, 2002).

DISTANCE LEARNING TRENDS

More Courses, Degrees, and Universities are Becoming Available through Distance Education Programs

The literature is replete with evidence of the growing demand for distance education, and organizations from within and outside higher education are adapting to accommodate such growth. The annual market for distance learning is currently $4.5 billion, and it is "expected to grow to $11 billion by 2005" (Kariya, 2003, p. 49). Some analysts predict that demand for distributed education will grow from "five percent of all higher education institutions in 1998 to 15 percent by 2002" (Oblinger & Kidwell, 2000, p. 32).

The Internet is Becoming Dominant among other Distance Education Media

Distance education has always existed in one form or another. However, accompanying the growth in Internet usage, "today's distance education focus has dramatically shifted toward network-based technologies (in general) and Internet-based delivery (more specifically)" (Kinley, 2001, p. 7). Not only is online learning more common now, but it is increasing 40% annually (Gallagher, 2002).

The Distinction Between Distance and Local Education is Disappearing

As universities digitally enhance more courses, the distinction between distance and local education is becoming blurred (Primary Research Group, 2004; Dunn, 2000). In fact, most online students live in the local vicinity of the institution offering their course (Carr, 2000). Traditional in-state, out-of-state, and international student distinctions are being eliminated, as are the course delivery formats distinctions, and the corresponding fee structures for the respective groups are breaking down (Carnevale, 2000b, 2000c).

The Need for Effective Course Management Systems and Web Services is Growing

Web services is "a relatively new term used to describe new software standards that allow for integration of different applications as well as the secure exchange of data over the Internet" (Crawford et al., 2003, p. 24). Web services ranked number six on the EDUCAUSE list of IT issues becoming more significant in 2003-2004, and instructional/course management systems were ranked number nine on the same list (Crawford et al., 2003).

There is an Increasing Need for Strategies that Better Utilize the Capabilities of Technology

Technological advancements have caused distance educators to ask how "new technologies such as wireless, mobile laptop computing, personal digital assistants (PDAs), videoconferencing, videostreaming, virtual reality, and gaming environments enhance distributed learning" (Crawford et al., 2003, p. 24). Distance learning research should focus on delivery strategies that improve instructional effectiveness and help solve capacity constraints, economic concerns, and higher education consumer needs.

CONCLUSION

In response to trends outlined in this article, distance learning has the potential to respond to student needs and overcome funding challenges that traditional institutions cannot. Although higher education institutions are changing to favor distance education, the complexities of major transformations will require time and patience. As Bates (2000) suggests, perhaps "the biggest challenge [in distance education] is the lack of vision and the failure to use technology strategically" (p. 7). Institutions will strengthen their distance learning strategic plans by identifying and understanding distance education trends for student enrollments, faculty support, and larger academic, technological, and economic issues.

ACKNOWLEDGMENTS

An earlier version of this manuscript originally appeared in the Online Journal of Distance Learning Administration, Volume VI, Number III, Fall 2003.

REFERENCES

Aslanian, C.B. (2001). *Adult students today.* New York: The College Board.

Bates, T. (2000). *Distance education in dual mode higher education institutions: Chall enges and changes.* Retrieved May 8, 2004, from http://bates. cstudies .ubc.ca/papers/challengesa ndchanges. html

Beaudoin, M.F. (2003). Distance education leadership for the new century. *Online Journal of Distance Learning Administration, 6*(2). Retrieved May 8, 2004, from http://www.westga. edu/%7Edistance/ojdla/summer62/beaudoin62. html

Brady, L. (2001). Fault lines in the terrain of distance education. *Computers and Composition, 18*, 347-358.

Branigan, C. (2003). Forum addresses virtual schooling myths. *eSchool News*, (June 2).

Brigham, D. (2003). *Benchmark information survey.* Unpublished Presentation, Excelsior University.

Callahan, P.M. (2003, March 28-30). *Proceedings of the UCEA 88th Annual Conference*, Chicago, IL.

Carnevale, D. (2000a). *Accrediting bodies consider new standards for distance education programs.* The Chronicle of Higher Education (September 8).

Carnevale, D. (2000b). Southern educators seek to cut tuition rates for online courses. *The Chronicle of Higher Education*, (March 31).

Carnevale, D. (2000c). 2 models for collaboration in distance education. *The Chronicle of Higher Education*, (May 19).

Carr, S. (2000). Many professors are optimistic on distance learning, survey finds. *The Chronicle of Higher Education*, (July 7).

Cetron, M.J. & Daview, O. (2003). *50 trends shaping the future*. Special Report, World Future Society.

Cheney, D.W. (2002, November). *The application and implications of information technology in postsecondary distance education: An initial bibliography*. Special Report, National Science Foundation.

Chick, S., Day, R., Hook, R., Owston, R., Warkentin, J., Cooper, P.M., Hahn, J., & Saundercook, J. (2002). *Technology and student success in higher education: A research study on faculty perceptions of technology and student success*. Toronto: McGraw-Hill Ryerson Limited.

Childers, J.L. & Berner, R.T. (2000). General education issues, distance education practices: Building community and classroom interaction through the integration of curriculum, instructional design, and technology. *The Journal of General Education, 49*(1), 53-65.

Crawford, G., Rudy, J.A., & the EDUCAUSE Current Issues Committee. (2003, November). *Fourth annual EDUCAUSE survey identifies current IT issues* (pp. 12-26).

Dasher-Alston, R.M. & Patton, G.W. (1998). Evaluation criteria for distance learning. *Planning for Higher Education*, 11-17.

de Alva, J.K. (2000). *Remaking the academy*. EDUCAUSE, (March/April), 32-40.

Diaz, D.P. (2002, May/June). *Online drop rates revisited*. The Technology Source. Retrieved May 8, 2004, from http://ts.mivu.org/default.asp?show =article&id=981

Dortch, K.D. (2003, April 13-15). How to get learners to learn. *Distance Education and Training Council: Report on the DETC 77th Annual Conference*.

Dunn, S. (2000). The virtualizing of education. *The Futurist, 34*(2), 34-38.

Dubois, J.R. (1996). Going the distance: A national distance learning initiative. *Adult Learning, 8*(1), 19-21.

Eckert, E. (2003). Review—New directions for adult and continuing education: Contemporary viewpoints on teaching adults effectively. *Adult Basic Education, 13*(1), 62-64.

Finkelstein, M.J., Frances, C., Jewett, F.I., & Scholz, B.W. (Eds.). (2000). *Dollars, distance, and online education: The new economics of college teaching and learning*. Phoenix, AZ: The American Council on Education and the Oryx Press.

Frydenberg, J. (2002). Quality standards in eLearning: A matrix of analysis. *International Review of Research in Open and Distance Learning,* (October). Athabasca University. Retrieved February 28, 2005 from http://www.irrodl.org/content/v3.2/fry denberg.html

Gallagher, S. (2002). *Report—Distance learning at the tipping point: Critical success factors to growing fully online distance learning programs*. Boston: Eduventures.

Gallagher, R. (2003, March 28-30). *The next 20 years: How is online distance learning likely to evolve?* Proceedings of the UCEA 88th Annual Conference, Chicago, IL.

Green, K.C. (2002). *Campus Computing 2002: The 13th national survey of computing and information technology in American higher education*. Encino, CA: Campus Computing.

Hawkins, B. (2003, March 28-30). Distributed learning: Promises and pitfalls. *Proceedings of the UCEA 88th Annual Conference*, Chicago, IL.

Hickman, C.J. (2003, March 29). *Results of survey regarding distance education offerings*. University Continuing Education Association (UCEA) Distance Learning Community of Practice, Research Committee Report.

Johnstone, S.M., Ewell, P., & Paulson, K. (2002). *Student learning as academic currency.* ACE Center for Policy Analysis. Retrieved June 1, 2003, from http://www.acenet.edu/bookstore/pdf/distributed-learning/distributed-learning-04.pdf

Jones, R. (2003). A recommendation for managing the predicted growth in college enrollment at a time of adverse economic conditions. *Online Journal of Distance Learning Administration, 6*(1). Retrieved May 8, 2004, from http://www.westga.edu/%7Edistance/ojdla/spring61/jones61.htm

Kariya, S. (2003). Online education expands and evolves. *IEEE Spectrum, 40*(5), 49-51.

Kinley, E.R. (2001). I*mplementing distance education, the impact of institutional characteristics: A view from the department chair's chair.* Unpublished Doctoral Dissertation, University of Nebraska-Lincoln, USA.

Lindner, J.R., Murphy, T.H., Dooley, K.E., & Jones, E.T. (2002). The faculty mind and how to read it. *Distance Education Report, 6*(14), 5.

McCombs, B.L. (2000). A*ssessing the role of educational technology in the teaching and learning process: A learner-centered perspective.* The Secretary's Conference on Educational Technology. Retrieved June 1, 2003, from http://www.ed.gov/Technology/techconf/2000/mccombs_paper.html

Murray, C. (2003). Study reveals shifts in digital divide for students. *eSchool News,* (May), 36-37.

NEA (National Education Association). (2000, June). *A survey of traditional and distance learning higher education members.* Retrieved June 1, 2003, from http://www.nea.org/he/abouthe/dlstudy.pdf

Oblinger, D. & Kidwell, J. (2000). Distance learning: Are we being realistic? *EDUCAUSE,* (May/June), 31-39.

Oblinger, D., Barone, C.A., & Hawkins, B.L. (2001). *Distributed education and its challenges: An overview.* American Council on Education (ACE). Retrieved June 1, 2003, from http://www.acenet.edu/bookstore/pdf/distributed-learning/distributed-learning-01.pdf

Oravec, J. (2003). Some influences of online distance learning on U.S. higher education. *Journal of Further and Higher Education, 27*(1), 89-104.

Paulson, K. (2002). Reconfiguring faculty roles for virtual settings. *The Journal of Higher Education, 73*(1), 123-140.

Penn State Strategic Plan. (1998). *An emerging set of guiding principles and practices for the design and development of distance education.* Retrieved May 8, 2004, from http://www.outreach.psu.edu/de/ide/

Primary Research Group, Inc. (2004). *The survey of distance and cyberlearning programs in higher education, 2004.*

Reeve, R.T. & Perlich, P.S. (2002). Utah economic and business review. *Bureau of Economic and Business Research (BERB), 62*(9-10), 1-15.

Rubin, E. (2003). Speaking personally—with Eugene Robin. *The American Journal of Distance Education, 17*(1), 59-69.

Saba, F. (2003). *Report on visiting Brigham Young University.* Unpublished Report, Brigham Young University, USA.

UCEA (University Continuing Education Association). (2001). *Survey summary: Results from the 2001 UCEA management survey.*

UCEA (University Continuing Education Association). (2002). *Lifelong learning trends: A profile of continuing higher education* (7th edition).

White, L. (2003). Deconstructing the public-private dichotomy in higher education. *Change, 35*(3), 48-54.

Young, J.R. (2002). 'Hybrid' teaching seeks to end the divide between traditional and online instruction. *Chronicle of Higher Education,* (March 22). Retrieved May 8, 2004, from http://chronicle.com/free/v48/i28/28a03301.htm

KEY TERMS

Academic Accountability: The emphasis from society, government, and academia that education should lead to beneficial outcomes and learning that can be measured.

Competency: The recent focus on competency that comes from employers stands in contrast to previous ways of acknowledging learning, such as seat-based time or diplomas. To an increasing degree, graduates are being judged by what they can do, not by what they know.

Decentralization: Represents the move away from a tightly grouped core of administrators and personnel that facilitate distance education, to a system that is more integrated into the different units of an institution.

Learner-Centered: Education that focuses on students and their learning, rather than on teachers and their methods. There has been a significant paradigm shift toward learner-centered education in the last decade.

Learning Objects: Available information (usually on the Web) that is reusable and applicable to many different learning contexts.

Lifelong Learning: Learning that extends beyond formal instruction and beyond the classroom. Distance education is facilitating the education of countless individuals in later stages of their lives.

Outsourcing: The growing practice in distance education of using external organizations to perform functions necessary to postsecondary institutions or programs.

Seamless Education: Seamless education refers to learning where boundaries between educational levels dissolve. For example, the transition between high school and college is becoming less distinct.

Technological Fluency: In addition to traditional literacy, technological literacy is increasingly becoming a necessity in higher education and in society. With the abundance of available information, information literacy is also growing in importance.

Unbundling of Faculty Roles: Entails the division of traditional faculty tasks. No longer are all faculty designing their instruction, implementing it, and then conducting the assessment of learning. More and more, different people or technological devices are performing these and other functions.

This work was previously published in Encyclopedia of Distance Learning, Vol. 1, edited by C. Howard, J. Boettcher, L. Justice, K. Schenk, P.L. Rogers, and G.A. Berg, pp. 7-15, copyright 2005 by Idea Group Reference (an imprint of IGI Global).

Chapter 8.23
Social Change Research and the Gender Gap in Computer Science

Jane Margolis
University of California Los Angeles, USA

Allan Fisher
iCarnegie, Inc., USA

INTRODUCTION

At Carnegie Mellon University, home to one of the top computer science departments in the country, only 7% of the students in the entering computer science class in 1995 were women. By the fall of 2000, that proportion had risen to 42%. While the percentage of women entering has declined slightly, likely reflecting the bursting of the Internet bubble, Carnegie Mellon's female computer science enrollment remains at about 30%, far higher than the average among research departments of computer science. Today, in 2005, the Carnegie Mellon School of Computer Science, with its increased number of female students, is a changed place. What sparked this development?

The story of the research that served as a catalyst for these increased numbers can be found in our book Unlocking the Clubhouse: Women in Computing (Margolis & Fisher, 2002). In this book we lay out the blueprints—the walls, doors, and windows—of the "boys' clubhouse" of computing education. We describe some specific institutional changes, enacted both by us and by others at Carnegie Mellon, which resulted in increasing the recruitment and retention of women students. These changes range from rethinking admissions criteria; contextualizing computer science ("computing with a purpose"); paying attention to students' experiences and the department culture; accommodating a wide range of previous computing experience; recognizing that women and students who do not fit the prevailing norm are disproportionately affected by problems like poor teaching, unapproachable faculty, or hostile peers; providing students with a broader picture of what it means to be in computer science, other than the hacker stereotype; outreach to high schools; and the formation of a vibrant women's organization.

In this article we offer reflections about some of the critical factors that contributed to our research becoming an instrument for social change. We provide some "lessons learned" for other institutions that are thinking about addressing the gender gap in their computer science departments. While the Carnegie Mellon developments began with a body of research, we do not believe that extensive research is necessary for all institutions. However, it is important to understand the local situation well enough to customize a general set of strategies. While people rightfully want to learn from successful initiatives, and not "reinvent the wheel," the constitution of each department—its history, the culture, the demographics, the leadership, the pressure points, what is known and not known about the experiences for women students—will differ from institution to institution. In most cases, but not all, initiatives can be modeled after existing programs by understanding the commonalties and differences between the situations. To achieve this, some straightforward data gathering, as opposed to in-depth research, is usually called for.

Here we present brief summaries of "lessons learned" from our research on the gender gap in computing. We believe that what we learned applies to planning an intervention as well as to conducting research.

ADDRESSING THE PROBLEM

- **Understand Your System and Know Your Numbers:** While lessons from other settings and other "diversity projects" can be instructive, the critical question is how this all applies to your own institution. The management truism that "you can't improve what you do not measure" applies here. Where is the bottleneck in your department?

Is it in admissions? Is it in retention? When are people being lost? How many women students are in the department? How many women faculty? How does this compare to other technical departments in your institution? What are the retention rates of women in computer science? What have the trends been? What is the culture of your department? How do women experience the department? And, where are the relevant points of intervention within your department? Local information is also critical to community engagement. While information about the gender gap from other places can be imported, especially when you have a "convinced audience," there is nothing like shining the light on your own backyard, and providing evidence from your own students, to make an institutional community take notice. The Carnegie Mellon research was based on some 300 Carnegie Mellon student interviews, over a 4-year period of time with a core sample group of 50 male and 50 female computer science students. We also conducted observations of computer science classes, and held interviews with computer science administrators and faculty. We lived in the department (Allan as Associate Dean of Undergraduate Education and Jane as Visiting Research Scientist) and were familiar with it from the inside.

- **Leverage Interdisciplinary Expertise:** Our research was conducted by an interdisciplinary team. Jane is a qualitative researcher with a background in Education and Women's Studies. Allan is a computer scientist and at the time of our research was Associate Dean of Undergraduate Education in the Carnegie Mellon School of Computer Science. While we originally referred to our research partnership as an "insider-outsider" collaboration (with Allan as the insider and Jane as the outsider), we quickly realized that since we were studying the dynamics of the gender gap in computer science, both perspectives were at the core of the problem. Neither perspective was on the margins. Each of us had a key to the puzzle that the other lacked. This both equalized our collaboration and opened up the range of issues that could be investigated. Our collaboration explored the traditionally unspoken issues that impact women's experiences such as confidence, sense of belonging, and different male and female motivations for studying computer science.

- **Listen to Students Holistically:** It was students' experiences (and our eagerness to hear from women's perspectives) that led us to the trouble spots in the department. But, to learn about those experiences we had to construct an interview guide that allowed us to learn more than the "party line," and more than what was "safe" to talk about. We also needed to construct a process that allowed interviewees, speaking in confidence to third parties, to talk about topics not commonly discussed in computer science culture. Open-ended questions that encourage interviewees to describe and shape their own accounts of their experiences (such as "Can you tell me the story of you and computers?" or "Can you tell me about your decision to major in computer science?"), rather than choosing amongst pre-selected generic answers, were critical to this process.

- **Take the Long View vs. a Single Snapshot:** We conducted multiple interviews with our sample of students, following some students over a four-year period of time. This longitudinal approach allowed us to take more than just a single snapshot of students' experiences. The multiple interviews allowed us to observe the evolution of students' relationship to computing. At any point along

the way had we drawn premature conclusions about the student, we would have an incomplete and misleading story. We learned that we cannot think in terms of a static set of influences on students' experiences but rather must understand students' stories in terms of a web of influences and a sequence of turning points, at each of which a different set of factors may be critical. These webs of influence were only apparent over time. Long term funding from the Alfred P. Sloan Foundation made this possible, and senior faculty member Lenore Blum and Carol Frieze have been able to use other sources of support to investigate ongoing cultural changes within the department since. We emphasize again, though, that large-scale research is not required for effective interventions—but tracking the numbers is.

- **Understand the Dynamics: One of our interview questions was:** "Can you describe your fellow computer science students?" Male and female students gave similar descriptions of their colleagues: myopically obsessed, living and breathing computers 24/7, emerging occasionally from behind the computer with a "monitor tan." Students' responses to this "geek mythology" were interesting. Despite the fact that both male and female students had similar descriptions of their fellow computer science students, about two thirds of the women and one third of the men explicitly dissociated themselves from the stereotype: "But that's not me." Yet the widely held perception of computer science students as being interested in nothing but computing became a set of expectations against which students judged themselves. Listening to the students tell of their experiences, we heard how each student's self-evaluation becomes a critical part of his or her sense of belonging in computer science. We heard how the obsessed computer whiz kids became the reference group—a frame of reference for each student's self-assessment. As a result, some students felt a good fit between their preferences and this model of what it is like to be a computer science student and others did not. Women fell disproportionately into this latter category.

We then saw how this sense of being outside of the norm makes women students especially vulnerable to other injuries such as poor teaching, inhospitable learning environments, and unhelpful instructors. When compounded by feeling outside the norm, seemingly small and sometimes unintended slights often are magnified. All of these chip away at a student's confidence. This, in turn, often leads to a loss of interest in the discipline. We saw once-enthusiastic students, mostly female, in a descending spiral of eroding interest and confidence, driven by negative comparisons to peers and by a variety of environmental insults.

One key observation on these influences is something that we have come to view almost as the First Law of Educational Diversity: in a situation with in-groups and out-groups, "everything bad happens worse" for the members of the out-groups. Because of doubts about fit, comparisons with members of the in-groups, and the feedback between confidence and interest, bumps in the road—poor teaching, lack of advising, weed-out experiences, and so forth—disproportionately create disaffection and attrition among the out-groups. Note that a corollary of this observation is that many effective interventions in favor of diversity are good for all students.

SHAPING A RESPONSE

- **Make a Leadership Commitment:** Because the initial impetus for this project came from Allan, who was an "insider" with authority over the undergraduate program, there was an unusual level of legitimacy associated with this research. It has been all too common for gender investigations to be marginalized and not taken seriously. Because of Allan's position in the department and the fact that the project was part of the department (with a department office and a title granted to Jane as Visiting Research Scientist), a legitimacy was bestowed on the project that we believe helped facilitate information gathering and cooperation from different members of the community. This commitment was continued after our departure from the scene by Lenore Blum, who has had extensive experience in gender projects, and Peter Lee, Allan's successor as Associate Dean, with support from University leaders as well.

- **Focus on the Bottlenecks:** At Carnegie Mellon, we were "losing women" in two main ways: at admission time, where all three of application, acceptance, and matriculation rates were lower for women than for men; and in the early years of the curriculum, where negative experiences and a sense of "lack of fit" created disproportionate attrition among women. In other settings we have seen, introductory courses, processes for choosing one's major, and "weed-out" courses have posed bottlenecks. We believe it is critical to monitor such bottlenecks over time, and to focus interventions there.

- **Attend to the Basics:** A few powerful heuristics apply to almost all settings. Programs should provide mentoring and community, multiple pathways into the curriculum for students with differing levels of experience, a high-quality and positive learning environment, and should develop a culture that supports and celebrates multiple approaches to the study of computer science.

- **Paint a Broad Picture of Computer Science:** Most CS faculty think of computer science as a dynamic, multi-disciplinary field that combines aspects of mathematics, engineering and science and has application in nearly every field of human endeavor. However, many prospective students, including some of the most enthusiastic, inherit from high school and society a narrow notion of computer science as focused on computers and on coding. Addressing this ongoing legacy is a key challenge for the computing community. Further, the introductory sequences of traditional curricula often reinforce narrow images of the field, by focusing primarily on equipping students with the programming tools they will use in later, more diverse courses. Carnegie Mellon's response to this issue has included the addition of an "immigration course" introducing new students to the breadth of the field. Other institutions have developed introductory courses that use integrative projects, that focus on principles over programming, or that link computer science to applications, to help to broaden students' vision.

- **Catalyze and Support Women's Community:** Upon her arrival at Carnegie Mellon in 1999, Lenore Blum led the creation of Women@SCS, an organized group of women in computer science. A professional group like this plays several important roles. Perhaps foremost, it provides an environment for women to experience being female computer scientists together with others, without feeling the need to "learn to speak 'boy'" (as eloquently phrased by Anita Borg) in order to be in the field. In this vein, it provides a venue both for professional development experiences and for mutual support. The most successful instances of such groups seem to combine substantial student leadership with ongoing faculty support.

Beyond its direct impact on its membership, a women's group increases the visibility and influence of women in the larger community. At Carnegie Mellon, the women's group has developed representation on standing committees, has organized events for the entire community, and has developed a variety of recruiting and outreach activities—even assisting in the creation of women's groups at other institutions.

SUSTAINING PROGRESS

- **Leverage Critical Mass and "Virtuous Cycles":** At Carnegie Mellon, changes in admission policy (removing previous computing experience as a preference factor, and emphasizing leadership potential in addition to numeric predictors) were an important factor leading to the increased enrollment of women. With more women classmates, female students no longer felt as isolated. And, as these talented and able women students had more of a presence in the department, faculty and administrators began to recognize how the increasing numbers of women in the program made the program even stronger and enhanced its competitive advantage. This, in turn, helped to make the environment ever more appealing to women students.

- **Watch the Student Experience Like a Hawk:** In most academic settings, especially in large institutions, key interactions with students are factored across multiple organizations: admissions, academics, student affairs, housing, career counseling, and so forth. Each of these areas presents opportunities to foster or weaken a student's affiliation with a discipline. We believe it is critical to work cross-functionally both to provide students with positive experiences and to head off the oversights that can miss such opportunities or, worse yet, drive students away. While Carnegie Mellon's reputation and recruiting power played a key role in the rapid increase in the involvement of women in computing, we believe that the university's culture of working across organizational boundaries was also an essential factor; at various times, we were able to work closely with admissions staff, other colleges, the student affairs office, and others to address specific issues.

- **Adapt:** Especially in a field like computing, with its rapid technological change and dynamic business cycle, change is a constant. The students whom we first studied were among the first generation to grow up with personal computers as a pervasive presence in the home; ten years later, a new generation has grown up with the Internet and all it implies. Ten years ago we saw the first inklings of the Internet boom, and now we have been through boom, bust, and consolidation. As we write this, perhaps the key human resource challenge to the discipline of computing in the developed countries is the public perception that "all the computing jobs are going offshore" to the developing world.

Just as the external environment changes, communities change. Lenore Blum and Carol Frieze have observed a shift in the Carnegie Mellon computer science student culture, in which both men and women are likely to take a broad and connected view of the field, and in which the traditional gender stereotypes of computing are largely defused. In light of internal and external changes, it is necessary to adapt dynamically to new sources of challenge and advantage.

CONCLUSION

Our research challenges the assumptions we often heard (and still hear): that women are less suited than men to do computer science, or that the subject is "just boring" for women and girls. Instead, our research shows the weighty institutional influences that steal women's interests in computer science away from them. It is critical to recognize the "First Law" effect, that women and other students who do not fit the prevailing norm are disproportionately affected by problems within the "computer science pipeline."

The goal should not be to fit women into computer science as it is usually conceived and taught. Instead, as we suggest in Unlocking the Clubhouse, "a cultural and curricular revolution is required to change [the culture of] computer science so that the valuable contributions and perspectives of women are respected within the discipline." Ultimately, this revolution serves not only the interests of the women involved, but those of the discipline itself.

REFERENCES

Blum, L., & Frieze, C. (2005). The evolving culture of computing: Similarity is the difference. *Frontiers: A Journal of Women Studies, 26*(1), 110-125. Retrieved January 14, 2006, from http://muse.jhu.edu/journals/frontiers/v026/26.1blum.html

CRA Taulbee Survey. Retrieved January 14, 2006, from http://www.cra.org/statistics/

Dweck, C. S. (1999). *Self-theories: Their role in motivation, personality and development.* Philadelphia: Psychology Press.

Margolis, J., & Fisher, A. (2002). *Unlocking the clubhouse: Women in computing.* Cambridge: MIT Press.

Rosser, S. (1990). *Female friendly science: Applying women's studies methods and theories to attract students.* New York: Pergamon Press.

Seymour, E., & Hewitt, N. (1997). *Talking about leaving: Why undergraduates leave the sciences.* Boulder, CO: Westview Press.

Steele, C. (1997). A threat in the air: How stereotypes shape intellectual identity and performance. *American Psychologist, 52*(6), 613-629.

Chapter 8.24
simSchool and the Conceptual Assessment Framework

David Gibson
David Gibson, CurveShift.com, USA

ABSTRACT

simSchool is a game-based simulation developed with funding from the Preparing Tomorrow's Teachers to Use Technology (PT3, 2003) program of the United States Department of Education. The simulation provides users with a training environment for developing skills such as lesson planning, differentiating instruction, classroom management, special education, and adapting teaching to multiple cognitive abilities. This chapter uses simSchool as an example to present and discuss an application of the Conceptual Assessment Framework (CAF) of Almond, Steinberg, and Mislevy (2002) as a general model for building assessments of what users learn through games and simulations. The CAF organizes the theories of teaching as well as the inferential frameworks in simSchool that are used to provide feedback to players about their levels of knowledge and abilities as teachers. The framework is generally relevant and useful for planning how to assess gains made by users while playing games or using simulations.

INTRODUCTION

An assessment is a machine for reasoning about what students know, can do, or have accomplished, based on a handful of things they say, do, or make in particular settings. (Mislevy, Steinberg, & Almond, 2003, p. 4)

Assessment is a broad concept. It encompasses small decisions such as whether to have dinner out or eat in (e.g., when we might assess our refrigerator and pocketbook) as well as larger decisions such as whether to become a rock star or an accountant (e.g., when we might assess our lifetime chance of success given our skills). Its essence is that we size up a situation by gathering data, apply some criteria to make inferences that are meaningful to us, and then decide what to do next. When assessing what someone has learned from playing a game or simulation, the same steps are taken, increasingly with automated help from networked computers.

Confusion and debate is often created, however, when the relatively straightforward process of making inferences and decisions expands to include technical issues and the politics of formal educational assessment. Questions arise about audience (who is giving and taking this assessment?), purpose (how will the results be used?), and ownership (who has the control here?), as well as about the fairness, reliability, and validity of the methods. SimSchool has its own answers to these questions. Your situation will most likely be different. This chapter cannot hope to discuss everything about assessment, but will endeavor to provide you with a framework of ideas that you can use in your setting. It will try to do this by calling attention from a detailed level of explanation of how simSchool is thinking about its challenges, to general statements that are valid for most assessments.

The audience for simSchool's assessment has two important constituents: future teachers and the professors guiding them into the profession, which in a general setting might be called users and their supervisors. There are many other possible audiences for assessment, but we do not deal with them in this chapter.

The purpose of the simSchool assessment focuses on making inferences about what the user knows and can do as a teacher. There are other purposes of assessing games and simulations. Program assessment focuses on determining if an investment in a program is paying off. Formative assessments are used to make improvements. Opinion surveys are used to find out how people feel. And there are others. We do not address these alternative purposes.

Concerning ownership, in simSchool the supervisor and user both have access to the assessment information, but the supervisor owns the data. Users see the results, hopefully analyze them, and base their future learning and action on them. But, they cannot withhold their data from the primary owner, the supervisor, who is interested in determining the extent of learning. There are many other ways to make the decision about ownership of assessment information (e.g. an institution, the public, the user), and as with the numerous options for audience and purpose, we understandably cannot deal with them in this chapter.

The plan of the chapter is to present the main concepts of an assessment framework and show in a general sense how simSchool uses the framework to organize its assessment capabilities. Undertaking both of these tasks, the chapter illustrates how games and simulations in general can assess what a user learns.

Conceptual Assessment Framework

Recent work stemming from adaptive testing (Almond, Steinberg, & Mislevy, 2002) expresses and shares a core of ideas with other research on assessment, which holds that every assessment of student learning involves three fundamental components: "a model of how students represent knowledge and develop competence in the subject domain, tasks or situations that allow one to observe students' performance, and an interpretation method for drawing inferences from the performance evidence thus obtained." (Pellegrino, Chudowsky, & Glaser, 2001, p. 2). Mislevy et al.

Figure 1. Initial components of the conceptual assessment framework

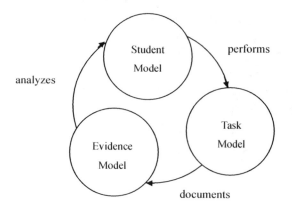

(2003) refer to these as three components of a Conceptual Assessment Framework (CAF): the student model, task model, and evidence model (Figure 1).

Student Model

All simulations and games simplify or exaggerate aspects of the real world. The student model simplifies an ideal user into a handful of variables that are central to the assessment. It represents the ideal configuration or goal state of the variables used in the assessment, for example, what the ideal user would do if they knew what they needed to and could do what they needed to in order to produce evidence of what the game was teaching. The ideal state of the variables is compared to the actual input from the user's interactions to determine rightness, closeness to expert performance, on track, on target and so forth.

The student model can also be thought of as the "ideal user" model. As a real user performs tasks, responds to a prompt, or explores a networked, hyperlinked space of resources, the current state of variables of his or her moves is compared with the ideal user model and used to construct a scoring record that is analyzed using the evidence model. The user performs tasks by interacting with the task model, which in turn documents that performance for the evidence model's analysis (Figure 1).

A simple example of the student model can be illustrated by the "objective test," which is admittedly over simplified. In an objective test, the right answers are what an ideal user would answer. A user who knows and can do everything on the test will get all answers correct. The correct answers are the "student model" and are compared to the actual responses of a user to determine a metric distance (e.g., getting 40% correct is further away from the ideal than 90% correct). Since some of the potential answers can be "close but wrong" because they are constructed to distract the user, an analysis can distinguish between the user's guesses, weaknesses in concepts and skills, and right answers. The concept of the student model can be extended to include fuzzy, uncertain, socially determined, and evolving variable configurations, to help deal with situations and kinds of knowledge that are not amenable to objective testing, for example, "soft skills" such as leadership, cultural responsiveness, and forming evidence-based opinions.

In developing simSchool, we created two "student models." This was needed because the concepts we are trying to teach through the game concern how to become better at assessing student learning in a classroom. A good classroom teacher needs to build an accurate mental model—their own private student model—of each of their students. This adds a second layer of complexity (possibly confusion) to the assessment, but we will try to explain. Inside the rule-based artificial intelligence engine of each simStudent, the variables of importance to the teacher's assessment of that student (e.g., emotional and academic variables) have a beginning and current state. The game player can assess how much learning has taken place in the simStudents by comparing the beginning with middle and ending states.

This is similar to what teachers need to do in the real world in order to assess whether a student has learned something from one's teaching. simSchool also makes available variables that are much harder, uncertain, or impossible to track and display in the real world—for example, the emotional stability, extraversion, and intellectual openness of the simStudent. To assess what the user knows and is able to do as a result of playing the game, one of the inferences we want to be able to make is how well the player understands the students and knows what to do to adjust things so that all of them can learn. The simSchool "student model" is thus a picture of how the user develops and uses their own "student model" for each of the SimStudents. This double entendre

Table 1. Example task model characteristics

Prompt or Challenge	How does the game or sim set a context? What is the story line or set-up?
Content	What does the learner need to already know vs. acquire during play in order to succeed?
Validity	What assumptions about the real world are embedded in the game engine's world model?
Performance opportunity	Is relevant performance evidence produced?
Affordance	What does the object, tool, situation allow the user to do that is valid for and maps with the analysis of the evidence?
Rich Picture	Are several kinds and pieces of evidence elicited?
Trustworthiness	Is the user's interaction evidence reliable for making inferences about what they know and can do?
Ideal vs. Real variables	What are the ideal game states for performances that exhibit the knowledge and skills of interest, and how does each user's record compare and change over time?

may not be needed in many assessment plans of games and sims, but is important in assessing many soft skills.

Task Model

The task model is at first glance simply the prompt or challenge given to the user. In general, games and simulations present task situations through a world or project-wide model that contextualizes the game or simulation. The variables associated with the states of the simulated world, objects, and interactive actors collectively represent the potential states of interaction with the user.

But there is another deeper way to look at the task model too. Each situation or task prompt embeds content in the form of theories and assumptions provided by the creators of the game. For example, if there is a ball to pick up, then does it bounce, and if so, does it follow earth gravity or some other rule? In addition to the content assumptions, each task model has other characteristics that impact assessment (Table 1).

The job of the task model is to represent the problem space and store and update variables that document how the interactions of the user have impacted the model and changed over time. The

variables sent from the task model to the evidence model are more grist for the analysis mill in the evidence model.

Inside the game engine of simSchool, there are currently two forms of tasks faced by each simStudent. There are classroom assignments from and verbal interactions with the user. We plan to add neighbor interactions with other simStudents in the future. The ideal states for maximum learning are tasks that have settings for each variable that are slightly above the current settings of the SimStudent, which can be further enhanced by conversations with a similar profile. When the player understands and applies these facts to improve his or her game score, it reinforces several actions that are central to the main lessons we want the player to get out of the experience—choosing tasks carefully matched to student needs and understanding what kind of supportive interactions students need.

The central challenge for the user, as in all games and sims, is to "figure out how the game engine works," and by doing so, internalize the task model characteristics. Thus, the nature and quality of the task model design determines much of what the assessment can say that the user learned through the game or sim.

Evidence Model

The third stop in the CAF sequence is the evidence model, which contains the inference rules that relate the task and student model variables into an analysis of what the user knows and can do. A simplistic view of the evidence model is that it compares the student model to the current scoring record's storage of variables that were changed by the user's interaction with the task model. The differences are then used to create an output to control new settings and interactions (e.g. new prompts or items in an online testing system, new body positions, attitudes, and language in the simStudents) as well as to characterize the metric distance between what the user has done and what the task called for.

The evidence model specifies analytic conditions and outputs (e.g. a graphic depiction, delivery of messages such as "You've won!" or a narrative about the results of game or sim play) that are called for by clusters of behavioral sequences, timings and game states that are represented in the variables of the student and task models.

In simSchool, a teacher can choose a task that helps some students but causes other students to fail. A key concept of teaching is to understand what needs to change in the task so that the failing students can learn. The evidence model compares what the player actually does with the combination of tasks that would have caused the best results for all simStudents. The analysis engine then prepares scores, narratives, and graphics to help point out what happened, what the failing students

needed to have happen, and how it impacted the overall results of the game. More details about this process are described later.

The three elements of the assessment framework—student, task, and evidence models—operate at two levels at least. Level 1 deals with the user experience and the operational level of the models. In the case of a computer-based simulation like simSchool, this amounts to things such as the code, databases, and graphic user interface. Level 2 deals with the evidence-based reasoning that allows us to make inferences from the artifacts of the user's interactions at Level 1. Level 2 reasoning, which we turn to next, allows us to infer what users know and can do based on how they work with and perform in a game or simulation.

Evidence-Based Reasoning

Assessment results are supported by an evidence-based generalization of the scientific method that includes dealing with issues such as observation, inference, and verification. Artifacts made by the user, when observed and classified, become evidence that relate "the particular things students say or do, to what they know or can do as more broadly conceived; that is, in terms that have meanings beyond the specifics of the immediate observations" (Mislevy et al., 2003, p. 1.). Note the Level 1 aspect of the artifact itself a physical object (even if virtual!) and the Level 2 aspect of the "evidence value" of the artifact after it has been classified for use within an assessment context.

Figure 2. Rules of evidence

Rules of Evidence

Evidence rules classify observables into patterns that can then be used to make inferences or claims about the learner (Figure 2). In both classification and inference there may be fuzzy as well as operationally well-defined rules of evidence present and contributing to the classification or inference. When we say "rules" we mean a set of "If...then..." statements that connect patterns or sets of observables into meaningful relationships that allow us to make claims.

Reasoning from evidence in order to support inferences about a learner is grounded in claims about the nature of knowledge for a particular domain, situations that evoke evidence of that knowledge, and how we can connect these to the claims we want to make about learners for the purpose of a particular assessment. This structure of reasoning mirrors the three Conceptual Assessment Framework components, except that the nature of knowledge replaces the student model. In simSchool, among the claims we make about teaching are that one's knowledge of content and how to teach plus one's attitudes or predispositions toward learners influences how one performs as a teacher. We view the simSchool game as an example of someone practicing the skills of teaching and potentially evoking evidence of this knowledge.

All games and simulation assessments that focus on what users have learned have to claim that the evidence created by a user interacting with the application provided realistic evidence of performance in some other context. In other words, it would not be very helpful for an assessment to only be able to say that the learner had learned to "play the game." We want to be able to say that by playing the game, we infer that this user knows and is able to do something in the real world. To create appropriate evidence in simSchool, we place the learner in a classroom of simStudents that evokes or invites realistic behaviors and responses. We then record and classify the responses and artifacts according to our model of teaching. We evaluate the learner's responses against our model of teaching and make judgments about what they know and can do.

In what follows, we attempt to make clear each of these aspects of the relationship of simSchool to the CAF, and by extension how you can use the framework to think about your game or simulation assessment challenges.

simStudents, Users, and the CAF of simSchool

There are two paths we want to follow. In one, the CAF framework is used to analyze the internal alignment of simSchool teaching with its impact on simStudents, and shed light on how the sim models the realities of teaching and classrooms. On the second path, the CAF is the framework by which we can draw inferences about the user's knowledge and abilities as a teacher, and evaluate the validity of our inferences.

To hopefully keep things clear, we will use the term "simStudent" to represent a database profile of a student and "Teacher" or "User" to represent the player who is using simSchool to practice teaching. We want to deal with both analyses, even though there is a chance for some confusion, for two reasons. By focusing on the "simStudents," we can present how the sim embeds theories of instruction, classroom management, learning theory, and psychology aligned with assessment results and other evidence of student learning. By focusing on the Teacher or User, we explain how simSchool can be used as a platform for professional development.

Generalizing to all educational games and sims that want to assess the extent of learning, there will be an internal model of the knowledge and skills intended to be transferred or learned, an interface that leverages important intervention points in the model while representing realistic actions and knowledge needed in an analogous real-world situation, and ways of comparing beginning, middle, and ending states. These data are then used to build a picture of what level of knowledge and skill the user demonstrated. At this point we should be able to answer the question "What did the user demonstrate that they know and can do during this round of play?" There are of course many questions to answer beyond this simplified schema of an assessment. Did the user know these things before playing, or did they develop the knowledge during play?

Over how many games did the user's knowledge change and by how much? Were there patterns of change in knowledge such as different speeds of learning under different circumstances? And so on. These kinds of questions require a record of the user that is stored and analyzed over time, but otherwise utilize the same framework we have been developing.

We turn now to the specifics of simSchool as a platform designed to teach teachers, and

Table 2. Conceptual Assessment Framework for the SIMSTUDENT MODEL

Component Name	Definition	simSchool Features
Student Model (Defines the SimStudent as a personality and learner, how the simStudent "thinks," "learns," and "feels")	Specifies the dependencies and statistical properties of relationships among variables that lead to claims about the knowledge, skills, and abilities of the learner. A scoring record holds the values of those variables at a point in time.	simSchool uses dynamic variables to represent and store simStudent behavior and performance in two broad areas: emotional and academic matters. The state of the simStudent includes representations of each factor as a continuum from –1 to 1, where 0 represents "on grade level" or "the norm" for the factor, 1 is well above, and –1 is well below. The emotional factors are taken from the OCEAN model of personality, also known as the Big Five in personality theory. Academic factors are taken from assessed domains of a subject area (e.g. in mathematics: problem solving, computation, and communication might be used. Or if a finer grain is needed, then specific skills within arithmetic might be used).
Task Model (Defines the tasks for the SimStudents determining how they react to the player's choice of tasks and verbal interactions)	Specifies variables used to describe key features of tasks (e.g. content, difficulty), the presentation format (e.g. directions, stimulus, prompts), and the work or response product (e.g. answers, work samples)	simSchool organizes the variables of the task model to map 1:1 with the student model variables (e.g. the specific emotional and academic factors required by a task) and provides an administrative interface to facilitate a variety of settings. Tasks are further organized by Depth of Knowledge levels (Webb, 2002). The task model settings act as "point attractors," causing the student model variables to change over time in the direction of the task model's variables.
Evidence Model (Defines inferences about the simStudents that we want the future teacher to be able to make)	Specifies how to identify and evaluate features of the work or response product, and how to update the scoring record.	Pattern matching routines and relational algorithms are used to compute metric distances between the initial arrays of the student and the arrays of the task and verbal interventions chosen by the teacher. The simsStudents then display by body position and verbal statements how they are doing. If the task is in the sweet spot for learning (what educators call the Zone of Proximal Development) then they stay on task and improve over time. In general there are three ways to get off task; tasks that are completed but then not replaced by other tasks cause "down time;" tasks that are too high to be completed in a reasonable amount of time cause "frustration;" and those that are too low to create a positive learning challenge cause "boredom."

attempt to continue to draw out generalizable lessons about assessment that might be useful in any educational game or simulation

Level 1: Game Play

At Level 1—the Game Play level—we will explain the internal model of the simStudents and the simulated classroom as a model of instructional planning, teaching, and classroom realities. We will call this the SIMSTUDENT MODEL. It shows how simSchool embeds theories of aligned instruction and assessment, psychology of inter-personal interactions, and cognitive growth. In general terms, this might be thought of as the game world model, the game engine.

We will then focus on a model that we will call the TEACHER MODEL in which the three components of the CAF are intended to build the foundation for making inferences about what the user knows and can do as a teacher. In general terms, this might be thought of as the user model, what we expect users to do while playing the game.

simSTUDENT MODEL

Table 2 lists the three CAF components (student, task, and evidence models), their definition, and how simSchool basically and broadly embodies the concepts. The definition column is taken from Mislevy et al. (2003). Following the table, we discuss more of the details of each of the framework's models.

Additional Notes About the simStudent Model

The major factors of the SimStudent personality are independent of one another but can also be clustered or related by rules at different aggregation points. For example agreeableness for working with others and emotional stability operate

independently so that a task that requires some emotional risk but does not require social inter-action can be differentiated from a similar task in a group context. But when a teacher speaks harshly and in a voice loud enough for the whole class to hear, both factors are impacted as though they are linked.

The representational schema for simStudents is highly flexible, which allows the selection of students for any conceivable school context (e.g., mixtures of race, gender, and performance profiles) and allows a wide range of classroom behaviors. This allows specific teaching strategies and adaptations to be matched for timely, respon-sive teaching adjustments in special settings such as special education, ESL, low literacy, poverty, within a single class session or across several ses-sions. Alternative dimensions can be added to the student profile, as needed for particular simula-tions, for example, learning style, self-perception, and subject-area specifics. There is no limit on the number of dimensions of personality, which in the base game is set at five emotional and one generic academic dimension.

The simsStudents attempt to close the gap be-tween their internal settings and the requirement of the tasks and verbal interventions and if given enough time, will come to rest on the requirements. This approximates student achievement and is made more realistic by the fact that not all of the simStudents can reach those goals in the allotted time of a classroom. Some students reach the task goal too quickly and then get bored; others will take too long to see any improvement and will get frustrated. The challenge for any teacher is getting the balance and mix right so that students remained challenged appropriately for most of the class and do not spend too much unproductive time sitting around.

The "game challenge" in simSchool is that even when many of these variables can be brought up to the surface for monitoring, they are difficult to control. The ideal task for one student is not going to work for all students, and as time marches on in the classroom, everyone can potentially get bored or frustrated. As expertise develops, more of the variables can be hidden, as many of them are in real life, which raises the challenge level.

TEACHER MODEL

Table 3 presents Level 1 issues for the users, for example, teachers who might use simSchool in teacher development programs. The CAF component names and definitions are the same as above, but the analysis of features, rather than outlining how the artificial simStudents learn, points to an evidence-based chain of reasoning needed to make inferences and claims about what the user knows and can do as a teacher.

Level 2: Inferences about the User

A chain of reasoning from Level 1 to Level 2 allows us to make inferences about teaching knowledge. The inferences are based on a claim that a simulated set of typical teaching tasks (making instructional decisions, making adjustments and adaptations during instruction, talking to students) elicits user actions and related artifacts that stand as evidence of knowledge of teaching. The kinds of knowledge that evidence from simSchool game play potentially refers to include:

Knowledge of Students

- Reading and using student records to make instructional decisions
- Pre-planning assessment and instruction to meet individual and group needs
- Observing in-classroom behavior and making inferences about adaptations needed in instruction and assessments

Pre-Planning Instruction

- Knowing what subject one is prepared to teach
- Knowing how many and what kinds of tasks are suited and fit with a subject
- Estimating the number of class sessions needed to teach a particular set of tasks

Making and Using Tasks

- Designing appropriate tasks
- Sequencing tasks for best effect

Making and Using Assessments

- Aligning assessment items to assess a given objective
- Estimating the number of and what kinds of assessment items/measures are suited and fit for a particular set of objectives
- Understanding the data produced by administration of a pre-assessment

Re-Planning Instruction

- Prior to instruction, choosing whole-class instructional strategies based on (aligned with) pre-assessment results
- Prior to instruction, choosing individual strategies based on (aligned with) student records and individual pre-assessment results.

Classroom Decision-Making

- Interpreting in-class performance (on task versus off task behaviors) as academic versus emotional issues
- "Reading" students via participation clues and language
- Speaking to students in effective and appropriate ways

Table 3. Conceptual Assessment Framework for the TEACHER MODEL

Component Name	Definition	simSchool Features
Student Model (Models the user of the simulation)	Specifies the dependencies and statistical properties of relationships among variables that lead to claims about the knowledge, skills, and abilities of the learner. A scoring record holds the values of those variables at a point in time.	For any selection of simStudents, there are "best choices" of tasks, "best order" of tasks, "best timing" for conversations, and "best attitude" for verbal interactions. These "bests" are generally not repeatable, since the context of the game changes constantly, so the game cannot lead to a simplistic level of "learning the trick" of the game. Instead, the player has to form heuristics and strategies that pay-off most of the time. As the user makes choices of task and talk.
Task Model (Defines the tasks for the user of the simulation)	Specifies variables used to describe key features of tasks (e.g. content, difficulty), the presentation format (e.g. directions, stimulus, prompts), and the work or response product (e.g. answers, work samples).	The player has many options for action, but at the heart of the assessment, there are just two fundamental tasks: matching tasks to students and speaking to students at the right time and in the right ways to help them. Both of these tasks have complex subtask levels, evidence of which is tracked from user movements and stored in a complex time-based scoring record. For example, if the user never "reads" the student profile, we expect to see longer times and inconsistent effects from verbal interactions and more mismatches of classroom tasks.
Evidence Model (Defines inferences about the user of the simulation)	Specifies how to identify and evaluate features of the work or response product, and how to update the scoring record.	Several metrics are used to form a full analysis. For example, total number of simStudents academically gaining and emotionally happy during the lesson, clusters of simStudents and collections of best tasks for each group, specific moment-by-moment timeline graphs that illustrate the impact of player moves. These metrics form sub-analyses that are collected over time and compared with earlier versions, to make inferences about growth of the user.

- Grouping students for differentiated instruction
- Adjusting instructional strategies based on in-class performance
- Individualizing tasks
- Focusing talk and discussion on improved student performance

Making and Using a Post-Assessment

- Designing appropriate and aligned test items to assess a given "unit of study" (objectives plus the instructional strategies and adaptations that have occurred during a number of class sessions)
- Estimating the number of and what kinds of assessment items/measures are suited and fit for the unit of study
- Understanding the data produced by administration of a post-assessment

Reflections on Teaching

- Making mental notes (and possibly written records such as grade book notations) about the evolution of a unit of study—the interaction of one's plans with the realities of teaching
- Abstracting and articulating lessons learned from the whole experience

In general, to make a Level 2 claim, there is a mapping of artifacts produced by the evidence model, which aggregates artifacts from the comparison of the student and task models. For example, to make the claim that a user has improved in his or her ability (or now "knows how") to, for example "interpret in-class performance as academic versus emotional issues," we can lookup the current level of the user on several sub-analyses and categorize those levels on a

Table 4. Example of mapping Level 2 inferences from Level 1 evidence

Level 2 Claims	Level 1 Artifacts	Evidence Model
User knows how to interpret in-class performance as academic versus emotional issues. And so forth [Other claims are proposed to provide a complete picture of the "assessable" knowledge and skills central to the game or sim.]	When a simStudent slumps in the desk: 0—user ignores the student 1—user speaks to the student about their behavior…making an assertion 2—…making an observation 3—…asking a question 4—user speaks to the student about their academic performance…making an assertion 5—…making an observation 6—…asking a question 7—user changes the task, selecting one with more challenge 8—…one with equal challenge 9—…one with less challenge And so forth [Similar lists of artifact options are created for other conditions, such as "When the user selects a task," "When an individual student is grouped with other students," etc.]	Example rules of inference related to the claim include: If the user shows evidence of 0 for most students most of the time, he or she is not exhibiting how to diagnose in-class performance. If the user shows evidence of 1 or 4 and the student does not improve, the user is making assumptions without relating to or understanding the student. If the user shows evidence of 1 or 4 and the student does improve, the user is successful in interpreting in-class performance. If the user shows more evidence of either 1 or 4 with better success as defined above, then if 1, the user interprets behavioral issues or if 4, interprets academic issues more successfully. And so forth [Other Evidence Model sub-analyses are created as rule systems and are used to classify the Level 1 artifacts in clusters that support the Level 2 claims.]

continuum of development (called a rubric in educational assessment). Table 4 shows a partial example of the mapping process for one claim important to simSchool.

The same body of evidence can consistently support multiple claims by clustering parts of the Level 1 evidence into new configurations. For example, evidence of asking a lot of questions of students might show more openness than making a lot of assertions, if everything else is going well. But if the questions are not working to improve student performance, then it is not a good strategy for this particular group of students, and might be a sign of weakness and uncertainty. This points out the ambivalent role of evidence at Level 1. Data is just data. At Level 2, data becomes knowledge with the addition of context that is provided by change over time and point of view applied to various clusters of the

data through the "if…then" rules that map the evidence to the claims.

A general rule of assessment is to consider several sources of data when making a claim or assessment decision. The corroborating evidence helps establish the validity of the finding. For knowledge and skills that are complex—such as becoming a skilled teacher—the assessment will have many high level claims, supported by a higher number of evidence model rules, that in turn utilize an even higher number of artifacts and artifact clusters. In traditional testing and measurement theory, this structure is described by "item response theory." For assessment in on-line games and sims, item response theory can be expanded via the CAF to include approaches such as neural net analysis, complex systems theory, semantic Web mechanisms, evidence-based inference, and performance assessment methods.

Some of these approaches are explored in other chapters in this book. A full exposition is beyond the scope of this chapter.

CONCLUSION

Assessing learning in educational games and simulations requires a formalization of familiar everyday reasoning. The assessment of what users know and can do based on artifacts they create involves three basic phases: sizing up the situation based on gathered data, applying some criteria to make inferences that are meaningful to the desired claims and inferences, and then deciding what to do next. The Conceptual Assessment Framework, a broad and flexible way of thinking about assessment possibilities, prompts us to make clear the ideal user model, relationships among the user's actions and states during game play, the affordances of the task model, and the potential inferences we can make from comparing these data sources.

REFERENCES

Almond, R. G., Steinberg, L. S., & Mislevy, R. J. (2002). Enhancing the design and delivery of assessment systems: A four process architecture. *The Journal of Technology, Learning, and Assessment, 1*(5). Retrieved from http://www.jtla.org

Mislevy, R. J., Steinberg, L. S., & Almond, R. G. (2003). On the structure of educational assessments. *Measurement: Interdisciplinary Research and Perspectives, 1*(1), 3-62.

Webb, N. (2002). *Alignment: Depth of knowledge level definitions*. Retrieved from http://facstaff.wcer.wisc.edu/normw/state%alignment%20page%20one.htm

This work was previously published in Games and Simulations in Online Learning: Research and Development Frameworks, edited by D. Gibson, pp. 308-322, copyright 2007 by Information Science Publishing (an imprint of IGI Global).

Index

A

Q